Springer Series on

ATOMIC, OPTICAL, AND PLASMA PHYSICS 70

Springer Series on

ATOMIC, OPTICAL, AND PLASMA PHYSICS

The Springer Series on Atomic, Optical, and Plasma Physics covers in a comprehensive manner theory and experiment in the entire field of atoms and molecules and their interaction with electromagnetic radiation. Books in the series provide a rich source of new ideas and techniques with wide applications in fields such as chemistry, materials science, astrophysics, surface science, plasma technology, advanced optics, aeronomy, and engineering. Laser physics is a particular connecting theme that has provided much of the continuing impetus for new developments in the field. The purpose of the series is to cover the gap between standard undergraduate textbooks and the research literature with emphasis on the fundamental ideas, methods, techniques, and results in the field.

For further volumes:
http://www.springer.com/series/411

J.T. Mendonça • Hugo Terças

Physics of Ultra-Cold Matter

Atomic Clouds, Bose Einstein Condensates and Rydberg Plasmas

Springer

J.T. Mendonça
Instituto Superior Tecnico
Av. Rovisco Pais 1
Lisbon, Portugal

Hugo Terças
Université Blaise Pascal
Aubière Cedex, France

ISSN 1615-5653
ISBN 978-1-4614-5412-0 ISBN 978-1-4614-5413-7 (eBook)
DOI 10.1007/978-1-4614-5413-7
Springer New York Heidelberg Dordrecht London

Library of Congress Control Number: 2012947410

Printed on acid-free paper

Springer is part of Springer Science+Business Media (www.springer.com)

Preface

This book aims to present an unified view of three different areas of ultra cold matter, stressing its differences and similarities. We have also tried to reconcile our two complementary and sometimes conflicting attitudes, related with the research motivation and the pedagogical purpose. Most of the book is written with a pedagogical intent, where priority is given to analytical derivations of the main physical concepts and results, over detailed information about the published research literature. Every chapter contains, nevertheless a short discussion on selected experiments and theoretical work. A large number of books have been published on Bose Einstein condensates, but we have approached this subject from a different and somewhat unconventional perspective. Our aim is to suggest a new approach to the problems involved with the production of ultra-cold atoms, Rydberg plasmas and Bose condensates.

We were surprised, during the preparation of this work, with the strong similarities that can be found in the collective processes that can take place in both classical and quantum gases, in neutral gas atoms was well as in plasmas. We have therefore explored the problems of collective modes in condensed and non-condensed cold matter. By covering such different fields and such large areas of knowledge, we could not be exhaustive and complete. We have nevertheless given, for each subject, and whenever possible, references to recent review articles where the reader can find a more detailed account of the literature.

We would like to thank our co-authors in this area, and the stimulating discussions we had with many researchers, with both theoretical an experimental backgrounds. We specially want thank Robin Kaiser who gave us guidance in the field of atom cooling and confinement, Robert Bingham and Charles Wang for their fresh and innovative approach to quantum coherence, and Padma Shukla for his inspiration in the approach of novel aspects of ultra-cold and strongly coupled plasmas. We also would like to thank Thomas Pohl for very helpful discussions on Rydberg atoms, and Jorge Loureiro for his precious contributions in the exploration of new aspects of collective processes. We also thank Gert Brodin and Mattias Marklund for their support on quantum vacuum problems, and Antonio Serbeto and Gordon Robb for their occasional collaboration on condensates. Research being a collective process itself, our work results from the contributions of several and sometimes anonymous people. We thank them all.

Lisbon, Portugal J.T. Mendonça
Aubiere Cedex, France Hugo Terças

Numa incerta hora fria
perguntei ao fantasma
que força nos prendia

Perguntas, **Carlos Drummond de Andrade**

Contents

1 Introduction ... 1
1.1 Three Phases of Ultra-cold Matter ... 1
1.2 Historical Perspective ... 3
1.3 Book Overview ... 4
References ... 6

Part I Atomic Clouds

2 Laser Cooling ... 9
2.1 Atom in the Laser Field ... 9
2.2 Laser Cooling Force ... 14
2.3 Doppler Limit ... 17
2.4 Magnetic Traps ... 18
2.4.1 Multipolar Field Configuration ... 19
2.4.2 Helmholtz Configuration ... 20
2.4.3 Ioffe Configuration ... 20
2.4.4 Anti-Helmholtz Configuration ... 21
2.5 Sisyphus Cooling ... 23
2.6 Evaporative Cooling ... 25
2.7 Sympathetic Cooling ... 29
References ... 33

3 Wave Kinetic Approach ... 35
3.1 Wigner-Moyal Procedure ... 35
3.1.1 Quasi-distributions ... 36
3.1.2 Weyl Transformation ... 39
3.1.3 Wave Kinetic Equation ... 41
3.1.4 The Quasi-classical Limit ... 43
3.2 Center-of-Mass Equation ... 44
3.3 Wave Kinetic Description of the Laser-Atom Interaction ... 47
3.4 Two-Level Atom ... 48

3.5 Links with Dynamics and Statistics ... 50
3.5.1 Quasi-classical Limit ... 50
3.5.2 Momentum Diffusion and the Doppler Limit ... 51
3.6 Lambda Configuration ... 53
3.7 Two Coupled Radiative Transitions ... 55
3.8 Influence of a Blue-Detuned Pump ... 58
References ... 60

4 Atomic Clouds ... 63
4.1 Atom-Atom Collisions ... 64
4.2 Feshbach Resonances ... 69
4.3 Collective Forces ... 74
4.4 Equilibrium Profiles ... 79
4.4.1 Qualitative Discussion ... 80
4.4.2 Quantitative Model ... 82
4.5 Coulomb Expansion ... 84
References ... 88

5 Waves and Oscillations in Clouds ... 89
5.1 Hybrid Sound ... 89
5.1.1 Fluid Description ... 89
5.1.2 Kinetic Approach ... 91
5.2 Tonks-Dattner Modes ... 96
5.3 Large Scale Oscillations ... 99
5.3.1 The Centre-of-Mass Oscillation ... 99
5.3.2 Normal Modes ... 100
5.4 Nonlinear Mode Coupling ... 103
5.5 Quasi-linear Diffusion ... 107
5.6 Phaser, the Phonon Laser ... 110
References ... 114

6 Photons in the Ultra-cold Gas ... 115
6.1 Optical Properties ... 116
6.2 Modulational Instability ... 118
6.3 Photon Bubbles ... 120
6.4 Roton Instability ... 125
6.5 Density Fluctuations ... 131
6.6 Collective Laser Scattering ... 135
References ... 138

Part II The Physics of Bose-Einstein Condensates

7 Bose Einstein Condensates ... 143
7.1 Uniform Gas ... 144
7.2 Trapped Gas ... 145
7.3 Atom Correlations ... 149

7.4 Mean Field Approximation 153
7.5 Thomas-Fermi Approximation 155
7.6 Fluid and Kinetic Formulations 158
7.6.1 Quantum Fluid Equations 158
7.6.2 Wave Kinetic Equation 160
References 161

8 Elementary Excitations in BECs 163
8.1 Sound Waves 163
8.2 Global Oscillations 165
8.3 Kinetic Processes 168
8.4 Landau Damping 169
8.5 Dynamical Instabilities 171
8.6 Wakefields in Bose-Einstein Condensates 174
References 180

9 Solitons 181
9.1 Effective One-Dimensional Gross-Pitaevskii Equation 182
9.2 One-Dimensional Dark and Grey Solitons 184
9.2.1 Energy of the Soliton 187
9.3 The Inverse Scattering Transform 188
9.4 Interaction Between Two Dark Solitons 189
9.5 Bright Solitons 193
9.6 Dark Solitons in Harmonic Traps 194
9.7 The Soliton Gas 197
9.8 Solitons in Two Dimensions 200
References 202

10 Quantum Field Theory of BECs 205
10.1 Bogoliubov Theory 205
10.2 BEC Depletion 209
10.3 Phonon Pair Creation 211
10.3.1 Time Refraction 211
10.3.2 Dynamical Casimir Effect 215
10.4 Acoustic Black Holes 217
10.4.1 Hawking Radiation 218
10.4.2 Effective Metric in a Condensate 219
10.4.3 Acoustic Hawking (Unruh) Radiation 220
References 222

11 Superfluidity 225
11.1 Phonon Kinetics 225
11.2 Phonon Fluid Equations 228
11.3 Slow Perturbations in the Superfluid 230
11.4 Superfluid Currents 232
11.5 Phonon Landau Damping 234

11.6 Roton Excitation ... 235
11.6.1 Wave Kinetic Equation with Dipolar Interactions ... 236
11.6.2 Dispersion Relation ... 237
11.6.3 Roton Instability ... 238
References ... 240

12 Rotating BECs ... 241
12.1 Quantum Vortices ... 241
12.2 Vortex Nucleation ... 244
12.3 Tkachenko Modes ... 245
12.4 Rossby Waves ... 246
12.5 Rossby-Tkatchenko Modes ... 252
12.6 Coupling with Photon OAM States ... 254
References ... 256

13 Quantum Coherence ... 257
13.1 Atom Interferometry ... 257
13.2 Time Interferometry ... 259
13.3 Decoherence Processes ... 262
13.4 Gravitational Decoherence ... 265
13.5 Josephson Tunneling of a Condensate ... 269
References ... 274

Part III The Physics of Ultracold Plasmas

14 Ultra-cold Plasmas ... 277
14.1 Different Plasma Regimes ... 278
14.2 Basic Plasma Properties ... 280
14.3 Ionization Processes ... 283
14.4 Single Particle Motion ... 285
14.5 Adiabatic Invariants ... 289
14.6 Plasma Equations ... 293
14.6.1 Klimontovitch Equation ... 294
14.6.2 Vlasov Equation ... 296
14.6.3 Kinetic Equations with Collisions ... 298
14.7 Fluid Equations ... 299
References ... 302

15 Physics of Rydberg Plasmas ... 305
15.1 Plasma Expansion in the Collisional Regime ... 306
15.1.1 Free Diffusion ... 306
15.1.2 Ambipolar Diffusion Regime ... 308
15.1.3 Recombination in Volume ... 309
15.2 Collisionless Plasma Expansion ... 310
15.3 Strongly Coupled Ions ... 313
15.3.1 Ion-Neutral Coupling ... 313

15.3.2 Ion-Ion Coupling 316
15.3.3 Phase Transitions 317
15.4 Disorder Induced Heating 321
15.5 Quasi-equilibrium States 323
15.6 Rydberg Atoms 326
15.6.1 Basic Properties 326
15.6.2 Rydberg Blockade 329
15.7 Three-Body Recombination 334
References 336

16 Waves in Rydberg Plasmas 339
16.1 Isotropic Plasmas 340
16.2 Polaritons and Slow Light 343
16.3 Ponderomotive Force 349
16.4 Electron Drift Instability 351
16.5 Drift Waves at Plasma Gradients 353
16.6 Waves in Magnetized Cold Plasmas 355
16.6.1 General Dispersion Relation 355
16.6.2 Parallel Propagation 358
16.6.3 Perpendicular Propagation 360
16.7 Waves in Expanding Plasmas 361
16.8 Waves in Strongly Coupled Plasmas 364
References 366

17 Kinetic Theory of Waves 367
17.1 Kinetic Dispersion Relation 368
17.2 Electron Plasma Waves 371
17.3 Ion Acoustic Waves 373
17.4 Waves in Quantum Plasmas 374
17.5 Sound Waves in a Turbulent Plasma 378
17.5.1 Plasmon Kinetic Equation 378
17.5.2 Ion Oscillations 380
References 381

18 Conclusions 383

Appendix 385
A.1 Atomic Structure 385
A.2 Quantum Theory of Radiative Transitions 389

Index 395

List of Figures

Fig. 2.1 Basic absorption-emission cycle of atomic laser cooling: two complementary pictures. (a) Energy picture, with a virtual energy level $E_2 - |\mathbf{k} \cdot \mathbf{v}|$; (b) Momentum picture, with an intermediate atom velocity $v' = v - \hbar k/M$ 14

Fig. 2.2 A magnetic trap with Ioffe configuration. For experimental purposes, two compensation coils are used to control the bias field B_0 21

Fig. 2.3 Schematic representation trapping mechanism in one dimension MOT for a $J = 0 \rightarrow J' = 1$ transition in an atom 22

Fig. 2.4 (a) Schematic representation of a simple multi-level structure sensitive to the light shifts induced by the polarization gradient. (b) Optical pumping mechanism in an optical potential with position-dependent polarization....... 25

Fig. 2.5 Evolution of the phase-space density of both buffer (*dashed line*) and target (*full line*) atoms. The *red dashed line* corresponds to the phase-space density necessary to occur Bose-Einstein condensation ($\varrho_{\text{BEC}} = 2.612$). (a) Condensation of both species. (b) No condensation. The parameters are discussed in the text and correspond to the experiment performed in Ref. [32] 31

Fig. 2.6 Four different scenarios illustrating the features of sympathetic cooling, obtained for $\omega_2 = \sqrt{2}\omega_1$ 32

Fig. 3.1 Gaussian wave function $\psi(x)$, versus the corresponding Wigner function $W(x,k)$.......... 37

Fig. 3.2 Diagram of the theoretical approach followed in this chapter, with a link between the main equations for the atom centre-of-mass motion 52

Fig. 3.3 Lambda configuration: this is a three-level atom model with two allowed radiative transitions.......... 53

Fig. 3.4 Two coupled radiative transitions: this is a four-level atom model, with two radiative transitions, which are coupled by some unspecified (*C*) process, due to collisions or to radiative pumping 55
Fig. 3.5 Two coupled radiative transitions: this is a four-level atom model, with two radiative transitions, which are coupled by some unspecified (*C*) process, due to collisions or to radiative pumping 58
Fig. 3.6 Parallel distributions for blue-detuned excitation of a laser cooled atom: initial thermal distribution $G_0 = G(t = 0)$, and asymptotic equilibrium distribution $G_{eq} = G(t \to \infty)$, as a function of the normalized parallel velocity $z = Mu/\hbar k_b$ 60

Fig. 4.1 Fashbach resonance: two potential courses, one closed $V_c(r)$ and the other open $V_0(r)$. Twp atoms colliding at energy $E \to 0$ can excite an internal state $E_c \simeq E$ 69
Fig. 4.2 Effect of the long-range interaction on the density profile for different polytropic exponents. The *black dashed line* is the gaussian profile obtained for the isothermal case ($\nu_{\text{eff}} = 0$ and $\gamma = 1$) and is used here to normalize the numerical solutions. *Thick lines* are obtained for $\gamma = 1$ (*black full line*), $\gamma = 2$ (*blue dashed line*), $\gamma = 3$ (*orange dotted line*) and $\gamma = 4$ (*green dash-dotted line*). (a) $\nu_{\text{eff}} = 0$. (b) $\nu_{\text{eff}} = 0.5$. (c) $\nu_{\text{eff}} = 0.99$ 85

Fig. 5.1 Dispersion relation ω plotted against $k\lambda_D$. The *red dashed line* represents the acoustic asymptotic behavior of the hybrid waves 91
Fig. 5.2 Dispersion relation, representing the dimensionless variables, frequency ω/ω_P versus wavenumber $k\lambda_D$. The *dashed curve* was obtained for the dimensionless quantum parameter $H \equiv \hbar^2/4M^2u_s^2\lambda_D^2 = 1$. The *dashed line* was obtained for $H = 0.1$ and represents the quasi-classical regime.. 95
Fig. 5.3 Normalized modes $\omega_{\nu,l}^2/\omega_p^2$ plotted against the nodes $r_{\nu,l}/a$ of the radial solution for the density perturbation $\tilde{n}$, for $\lambda_D/a = 0.1$ and $1 < \nu < 10$. *Blue squares* ($l = 0$), *red circles* ($l = 1$), *violet stars* ($l = 2$), *yellow triangles* ($l = 3$), *green open circles* ($l = 4$). The *full line* is plotted at $l = 0$ and scales as $1/\nu^2$ 99

Fig. 5.4 Dependence of the plasma frequency ω_p on the detuning δ, plotted at $I/I_s = 0.3$ for different values of the magnetic field gradient. *Solid line*, $b' = 15$ G/cm, *dashed line*, $b' = 0.8$ G/cm, *dotted line*, $b' = 0.4$ G/cm 104

Fig. 5.5 Stability chart of Eq. (5.72) obtained numerically with a $n = 5$ Hill determinant. The *full lines* represent π-period solutions and the *dashed lines* represent the 2π-period ones. The *shadowed areas* correspond to stable (bounded) solutions. The *dots* represent the thresholds for the observation of unstable, corresponding to $\nu = 12$ 106

Fig. 5.6 Phase diagram of the nonlinear coupling of center-of-mass and collective oscillations, obtained for $I = 0.3I_s$ and for a stochastic perturbation of $\xi/2\pi = 223$ Hz. The *shadowed areas* are stable 108

Fig. 5.7 Phonon laser scheme: (a) The gas is cooled down by red-detuned laser beams with frequency ω_a and wavevector $\mathbf{k}_a$ into a velocity distribution $W_0(v)$; (b) the ultra-cold gas is pumped, creating an inversion of population in velocity space; (c) Phonons with frequency ω and wavevector $\mathbf{k}$ are coherently emitted by the atoms, which decay into a lower kinetic energy state (The width of the distributions and the distance between the peaks are generic and provide only a schematic perspective of the instability mechanism. See discussion in the text) 112

Fig. 5.8 Phonon laser scheme in the classical limit. The two upper levels are replaced by a region of negative derivative, where inverse Landau damping will occur........ 112

Fig. 5.9 Saturation of the laser instability, as described by the quasi-linear equations. Temporal evolution of the normalized population inversion in velocity space $z(\tau) \equiv \Delta G(u,\tau)/\Delta G(u,0)$ (*dashed line*), and the phonon amplitude square $y(\tau) \equiv |V_k(\tau)|^2$ (*full line*), as a function of the time variable τ, for $a = 1$ and $\bar{\nu} = 1/2$. For illustration purposes, we take $y(0) = 0.001$ 114

Fig. 6.1 Schematic representation of spherical bubbles resulting from the unstable coupling of laser light with the atom mean field........ 124

Fig. 6.2 Threshold curves for photon bubble formation in the ultra-cold gas: (*a*) Oscillating bubbles, photon characteristic scale L/L_n, versus normalized wavenumber $z = q/k_a$ with $L_n = (\omega_p/\omega_s)/\sqrt{2}k_a$; (*b*) Purely growing bubbles, L/L_n versus $z = \sqrt{\tau\nu}$, with normalization factor $L_n = (\omega_p/\omega_s)(l/\sqrt{2})$. Instability occurs below theses curves 126

Fig. 6.3 Illustration of the real (*left panel*) and imaginary (*right panel*) parts of the polariton dispersion relation in the macroscopic regime for $D_0 = 2.0\lambda_D^2\omega_p$. We can observe the emergence of a roton minimum for $\omega_d = 1.9\omega_p$ (*full black line*) and $\omega_d = 1.99\omega_p$. The rotons softens the frequency to $\Omega(q_{\rm rot}) = 0$ at the critical value $\omega_d^{(c)} = 2.0\omega_p$. Roton instability is illustrated here for $\Omega = 2.2\omega_p$. The short-wavelength oscillations corresponding to $q\lambda_D > 1$ are kinematically damped. The usual plasma dispersion relation of hybrid phonons is presented here for comparison (*dashed gray line*) 129

Fig. 6.4 Static structure factor $S(k)$ depicted for $D_0 = 2.0\lambda_D^2\omega_p$. $\omega_d = 0$ (*gray dashed line*), $\omega_d = 1.9\omega_p$ (*black full line*) and $\omega_d = 1.99\omega_p$ (*black dashed line*). A pick emerges in the static structure factor around $k_{\rm rot} = \lambda_D/\ell_d^2 = 0.5\lambda_D^{-1}$ 130

Fig. 6.5 Two-point correlation function $g(r)$ depicted for $D_0 = 2.0\lambda_D^2\omega_p$. $\omega_d = 0$ (*gray dashed line*), $\omega_d = 1.9\omega_p$ (*black full line*) and $\omega_d = 1.99\omega_p$ (*black dashed line*). The correlation function oscillates with the period a of $2\pi/q_{\rm rot} \simeq 4\pi\lambda_D^{-1}$ 131

Fig. 6.6 Dynamical structure factor $S(\alpha, \omega')$ for different values of the scattering parameter α: *blue full line* ($\alpha = 0.85$), *black dashed line* ($\alpha = 1.0$), and *red full line* ($\alpha = 5.0$). Both coherent and incoherent scattering processes are represented 138

Fig. 7.1 Bose condensation: number of particles in the zero energy state N_0, in bold, and number of atoms in positive energy states N', as a function of the temperature T, below the critical temperature T_c 146

Fig. 7.2 Normalized density profile of a condensed ($n_0(x)$) and a thermal gas ($n_{\rm th}(x)$) over a given direction x, for a spherical trap. We have used $T/\hbar\omega = 4$ 148

Fig. 7.3 Comparison between the Thomas-Fermi profile (*dashed line*) and the numerical solution (*full line*) of time-independent Gross-Piatevskii equation (7.61), obtained for $\mu = 10\hbar\omega_0$ and $g = 9.5\mu/n(0)$. The small difference at the vicinity of the Thomas-Fermi radius R is of the order of few ξ 156
Fig. 7.4 Variational Thomas-Fermi energy in (7.70) as a function of the width of the cloud. Stable (*full line*) and unstable (*dashed line*) regimes are observed as a function of the condensed number of particles N 158

Fig. 8.1 Stability diagram for the two-stream BEC. The shadowed are is unstable. The *upper curve* corresponds to $KH = 1$. The *dashed lower curve* corresponds to the case $K = (\beta^2 - 1/H^2\beta^2)^{1/2}$, for $\beta = 1.01$ 173
Fig. 8.2 Normalized wave growth rate $\tilde{\Gamma} = \text{Im}(\tilde{\Omega})/\omega_0$ for different values of the parameters β and H. *Blue lines* ($\beta = 0.5$), *black lines* ($\beta = 1.0$) and *red lines* ($\beta = 1.5$). *Full lines* ($H = 0.5$), *dashed lines* ($H = 1.0$) and *dotted lines* ($H = 1.5$). The maximum growth rate, corresponding to the most unstable mode, occurs for the lowest values of both β and H 174
Fig. 8.3 Wakefield excitation: (a) Wakefield oscillation (*bold curve*) excited by a condensate beam with a Gaussian shape $N_0(\eta) = \exp[-(\eta/\sigma)^2]$ (*thin curve*), for $\sigma = 3$; (b) Two cases of wakefield excitation, for two different widths of the Gaussian beam, $\sigma = 4$ (*bold curve*) and $\sigma = 1$ (*thin curve*).... 177
Fig. 8.4 Modulational instability of a condensate moving in a background non-condensate gas, with a velocity equal to the sound speed of the non-condensate gas, $u_0 = u_s$, as determined by equations. Equation (8.63) for a transverse Gaussian density profile.... 179

Fig. 9.1 Solitary solution (9.2) propagating from *left* to the *right* ($v > 0$) 182
Fig. 9.2 *Dark* ($v = 0$) and *grey* ($v = 0.3c_s$) solitary solutions obtained from (9.21) (*left panel*) and (9.22) (*right panel*) 186
Fig. 9.3 Collision dynamics of two identical dark solitons. The panels on the *left* correspond to the density $n_{12}(z,t) = |\psi_{1,2}(z,t)|^2$, while the respective phase portraits are located in the *right* side. From *top* to *bottom*: $v = 0.3c_s$, $v = 0.5c_s$ and $v = 0.8c_s$. The *arrows* indicate the direction of propagation before (*red*) and after (*blue*) the collision. The *black dashed line* indicates both the density and phase profiles at the collision instant $t = 0$.... 192

Fig. 9.4 Illustration of the dynamics of a dark soliton performing slow oscillations in a harmonic trap 196
Fig. 9.5 Real (*full*) and imaginary (*dashed*) parts of the dispersion relation of a dark soliton gas with $N_0 = 0.15n_0$. The *right panel* illustrates the dependence of k^* on α 199
Fig. 9.6 Snapshots of a rectilinear dark-soliton propagating in a two-dimensional BEC 201
Fig. 9.7 Dependence of the critical wavevector k_c on the soliton speed 201
Fig. 9.8 Snapshots of a ring dark-soliton propagating in a two-dimensional BEC 202

Fig. 10.1 Time transmission T, and reflection R coefficients, as a function of the temporal discontinuities of the atom density, represented by the ratio f_0/f_1 213
Fig. 10.2 Number of phonon pair created from vacuum, (*a*) for a sudden change in the atom density, as described by $n(t) - n_0 = [1 + \tanh(2t - 1)]/2$; (*b*) for a sinusoidal perturbation such that $\eta_0 = 1$ 216
Fig. 10.3 de Laval nozzle configuration 221

Fig. 11.1 Dispersion relation of Bogoliubov oscillations in a condensate, modified by the long dipole-dipole interactions, (*a*) U_d as described by a C_6 van der Waals potential; and (*b*) $U_d = 0$ for reference 239

Fig. 12.1 The approximate vortex solution (12.8) is shown in normalized radial coordinates. The size of the vortex is determined by the healing length ξ 243
Fig. 12.2 Formation of a vortex array, as shown by solving numerically the Gross-Pitaevskii equation for a rotating BEC..... 244
Fig. 12.3 Dispersion relation of the Rossby waves in a rapidly rotating BEC, for $v_R = 0.1c_s$. It is clearly shown that phase speed is generally negative. The *blue full line* corresponds to the Thomas-Fermi case, $\xi = 0$. The *black dashed* and *red full lines* respectively correspond to $\xi = 0.7r_0$ and $\xi = 1.3r_0$ 250
Fig. 12.4 Nonlinear stationary solitary wave resulting from the saturation of the triad resonant decay of Rossby waves, obtained for $\Omega = 2.4\omega_\perp$ and $\beta = 1.6$: (a) $\mu = 0.2\hbar\omega_\perp$ and (c) $\mu = -0.2\hbar\omega_\perp$. Plots (b) and (d) respectively compare the radial structures (*full lines*) of (a) and (c) with the corresponding Thomas-Fermi equilibria (*dashed lines*) discussed in the text, obtained for the same set of parameters 251

Fig. 12.5 Dispersion relation for the Rossby-Tkachenko modes in a rotating vortex, for $c_T = 0.2c_s$. (a) Homogeneous case. for comparison we reprent the Rossby dispersion curve (*dashed line*) and the Tkachenko dispersion curve (*full line*); (b) Trapped case: Eigen-frequencies obtained for the compatible wavenumber $k_0 = 5.45/T_\perp$. From *top* to *bottom*, $l = 5$, $l = 3$ and $l = 0$ 253

Fig. 13.1 Illustration of the interference signal (13.20) as function of the delay time for a time-interferometry setup with ^{8}5Rb atoms cooled down to $\sim 150\,\mu$K, and a period of $T \sim 800\,\mu$s. *Black*, *red* and *blue curves* are respectively obtained for $N = 1, 2$ and 3 echos 262

Fig. 13.2 Young's two-slit experiment for matter waves, with a path detector system 263

Fig. 13.3 Illustration of the double well axial potential, for $\mu = \lambda = 2$, (a) symmetric case, $h = 0$, and (b) asymmetric case, $h = 1$ 270

Fig. 14.1 Two-step photoionization process, using photons with energies $\hbar\omega_1$ and $\hbar\omega_2$. The second photon energy exceeds by a small amount ΔE the ionization energy 284

Fig. 14.2 Coordinate system for guiding centre motion of a charged particle in a static magnetic field $\mathbf{B}_0(\mathbf{r})$ 288

Fig. 15.1 (a) Single ion sphere, with radius a_i; (b) two-ions sphere, with radius $2^{1/3}a_i$ 318

Fig. 15.2 (a) Thomas-Fermi potential obtained in the strong trap regime $\phi \gg 1$ for two different ion gaussian profiles: $a = \lambda_D$, (*dashed line*) and $a = 0.5\lambda_D$ (*full line*). (b) Ion (*dashed lines*) and electron (*full lines*) profiles in the early stages of a Rydberg plasma: $a = \lambda_D$ (*black lines*) and $a = 0.5\lambda_D$ (*gray lines*) 326

Fig. 15.3 Energy levels of two nearby atoms: if atom 1 is excited by a two-photon transition, from the ground level $|g\rangle_1$ to the Rydberg state $|n\rangle_1$, atom 2 located at a distance R will not be excited into the same state by the same photon transitions, due to the energy shift $\Delta(R)$ 329

Fig. 16.1 Cut-off frequency for electromagnetic waves propagating in a Rydberg plasma, for $\alpha = \pm 0.5$ or $\tau_a = \pm 1/2\omega_{pe}$. Dimensionless parameters $X = \omega/\omega_{pe}$ and $Y = \omega/\omega_a$ are used 346

Fig. 16.2 Dielectric function $\epsilon'(z) = 1 + \chi'(z)$ are represented (in *bold*) as a function of the wave frequency detuning $z = \Delta/\gamma$, for $a = 1$, $g = \eta = 0.1$, and (a) $b = -1$ and (b) $b = 1$. For comparison, the curves of pure atomic dispersion corresponding to a non-ionized gas, are also shown 347
Fig. 16.3 Polariton dispersion relation in a Rydberg plasma, obtained for $\omega_a = 6\omega_p$ and a inverted-population parameter $\tau_a = 0.5\omega_p^{-1}$. (a) $\Omega = 0$, (b) $\Omega = 0.1\omega_p$ and (c) $\Omega = 0.5\omega_p$. In the above plots, the *thin dashed line* represents the usual electromagnetic wave dispersion relation $\omega^2 = \omega_p^2 + k^2c^2$ 348
Fig. 16.4 Dimensionless magnetic field b, as a function of the relative wave frequency $x = \omega/\omega_a$, for two different values of the natural bandwidth parameter $g = \gamma^2/\omega_a^2$, for $g = 0.01$ and $g = 0.001$ (*bold*) 351
Fig. 16.5 (a) Dispersion relation of ion acoustic waves in a strongly coupled plasma, and (b) in the absence of ion coupling 365

List of Tables

Table 2.1 Comparison of relevant atomic parameters for the main cooling transitions for the isotopes of elements used in laser cooling experiments. The data are taken from Ref. [13] 25

Table A.1 Ground state and first ionization energy of different atomic elements 389

Chapter 1
Introduction

1.1 Three Phases of Ultra-cold Matter

The advent of laser cooling of atoms [1] led to the discovery of ultra-cold matter, with temperatures well bellow the liquid Helium, which displays a variety of new physical phenomena. It is the aim of this book to give an overview of this recent area of science, with a discussion of its main results and an description of its theoretical concepts and methods.

Ultra-cold matter can be considered in three distinct phases. First, we have the *ultra-cold gas*, which can be created and confined in a magneto-optical trap (MOT). Such a trap consists typically of three pairs of laser beams, slightly detuned with respect to a given atomic transition frequency, and a pair of Helmholtz coils. Typical temperatures are in the micro-Kelvin range, and the gas can be maintained in nearly steady-state conditions inside the trap. This is a gas bubble of 10^8–10^{10} atoms, with a nearly spherical shape and a diameter of few millimeters. Surprisingly, this neutral atomic cloud behaves as a one-component plasma, because each atom in the cloud scatters photons from the cooling laser beams, and these secondary photons can push away the nearby atoms by radiation pressure. From such an exchange of photons between nearby atoms results a repulsive force, which can be described by an *effective electric charge* of the neutral atoms. This repulsive force competes with an attractive force due to photon absorption: the atoms in the outer part of the cloud, closer to the laser source, will experience a larger laser intensity, thus pushing them towards the centre. This compressive force is a shadow type of force, which strongly resembles the gravitational mechanism proposed by Le Sage [2] in the eighteenth century, as alternative to Newton's universal attraction. A similar kind of shadow force is also known in dusty plasmas [3], as a mechanism of attraction between charged dust particles. In usual experimental conditions, the repulsive force acting on ultra-cold atoms dominates over the shadow force, and nearby atoms tend to repel each other. From this repulsive effect it results a plasma-like behavior, similar to that of non-neutral plasmas, which lead to a variety of collective processes, some of which have already been observe experimentally.

J.T. Mendonça and H. Terças, *Physics of Ultra-Cold Matter*, Springer Series on Atomic, Optical, and Plasma Physics 70, DOI 10.1007/978-1-4614-5413-7_1,

Second, we consider the *Bose Einstein condensate* (BEC), which results from an evaporative cooling of the ultra-cold gas. This is achieved by switching-off the laser cooling beams, which therefore cancels the long-range interaction between atoms, and the latter now interact at a much shorter scale defined by the wave *scattering length* a_s. Evaporation of the warm atoms in the cloud will leave just the coolest atoms in place. The resulting temperature of the remaining atomic cloud can then reach the nano-Kelvin domain, does allowing for the occurrence of Bose Einstein condensation [4]. This state of ultra-cold matter also displays very interesting and novel properties, which have been extensively explored both in theory and experiments. It particular, BECs possess fluid-like properties which lead to similar, but not identical, collective atomic processes that those of cold atomic clouds in MOTs. The similarity and differences between the condensed and the non-condensed phases of ultra-cold matter will be deeply discussed in this book.

Finally, we have the *Rydberg plasmas*, which can be formed by photo-ionization of the ultra-cold gas [5]. They are usually called *ultra-cold neutral plasmas* in the atomic physics community, which is redundant and misleading. Redundant, because plasmas are usually in a state of charge neutrality, and only the exceptions are explicitly called *non-neutral*. Misleading because the neutral atoms in the ultra-cold (and non-ionized) gas already display a plasma-like behavior. The atomic gas can therefore be seen as an exotic version of a non-neutral plasma, as noticed before. For that reason, we have proposed to call *Rydberg plasmas* to the ionized phase of ultra-cold matter, because they contain a large number of atoms excited in high Rydberg states, as will also be shown in the present work. These ultra-cold plasmas have typical electron temperatures of a few Kelvin, and ion temperatures in the milli-Kelvin range. We notice that the traditional view of a plasma is that of a hot ionized gas, with temperatures of the order or above an electron-Volt. These Rydberg plasmas therefore extend the concept of plasma state to the domain of very low temperatures, which in some sense is counterintuitive and opens the way for new plasma phenomena. In the same way as the Bose Einstein condensates, this new plasma state is a transient state, because the ionized plasma cloud tends to expand and dissipate after a few milli-seconds. Comparatively, the condensates can live much longer.

We therefore describe three distinct phases: the steady state ultra-cold gas in MOTs, the transient Bose Einstein condensate, and the equally transient Rydberg plasma. These three phases of ultra-cold matter will be reviewed here. We will stress the similarities and fundamental differences between them, as well as the similarities (sometimes unexpected) of their respective theoretical methods. From our description of the ultra-cold matter, a global and coherent picture will hopefully emerge of a new area of Physics, somewhere at the boundary of atomic physics, condensed matter and plasma physics.

1.2 Historical Perspective

Physics was always attracted by extremes, the infinitely small and the Planck scale, the infinitely large and the size of the universe, the extremely hot matter and the big-bang. The extremely cold matter, and the absolute zero of temperature, are our present concern. We briefly review the history of the concepts and results to be discussed in the present work.

By the middle of the nineteenth century, Lord Kelvin had clarified the concept of temperature, and defined the absolute zero at $-273.15\,^{\circ}\mathrm{C}$, or 0° *Kelvin* (K), a temperature at which, according to the Nernst principle, the entropy also vanishes. At the beginning of the nineteenth century, an important step towards the absolute zero was taken when Kamerlingh Onnes liquified Helium-4, at a temperature of 4.2 K and atmospheric pressure. This discovery boosted the search for new properties of cold matter, such as superconductivity and superfluidity.

An important new ingredient was introduced by Bose in 1924 [6], when he proposed his statistical distribution, valid for an ideal gas of photons, launching the concept of *boson*, as a particle with zero or integer spin. His distribution was generalized by Einstein [7] to particles with a finite mass, who first noticed that for such particles a critical temperature T_c should exist, below which a fraction of the gas would condense into a state of zero momentum. This phenomenon is now called Bose-Einstein condensation. In a series of experiments performed by Kapitza [8] and by Allen and Misiner [9] in 1938, superfluidity of Helium-4 was achieved, for temperatures below 2.17 K. This new state of matter was soon understood [10, 11] as due to the occurrence of partial Bose-Einstein condensation. For many years, superfluid Helium-4 remained as the paradigmatic example of a BEC. However, in contrast with the Einstein paradigm, valid for an ideal gas, such a state was obtained for strongly interacting particles.

Things dramatically changed with the discovery of laser cooling techniques, which produced another spectacular boost in the search for the absolute zero. The idea of using light to change the state of motion of material particles predates the invention of the laser and goes back to the early nineteenth century. The experiments of Lebedev [12], and Nichols and Hull [13] in 1901, and later of Frisch in 1933 [14], on light pressure, are usually recognized as important precursors.

Laser cooling was first mentioned in the work of Hänsch and Schawlow [15] in 1975. Two years later, Ashkin [16] proposed a theoretical model for atom cooling using a nearly resonant laser beam. The technique for cooling and trapping a large number of atoms using magnetic traps and laser beams was developed during the 1980s of the last century, by Chu [17], Cohen-Tannoudji [18], Philips [19] and others. Temperatures in the micro-Kelvin range became possible, which paved the way for the condensation of alkaline vapors a decade later, in the line of Einstein's prediction.

Bose-Einstein condensation in vapors of alkaline atoms was first reported in 1995 [20, 21]. Using evaporative cooling of atomic gases contained in MOTs, temperatures in the nano-Kelvin range are presently being achieved, thus allowing for condensation of a large variety of atomic gases, such as Rubidium, Sodium, Lithium, Cesium, Yterbium, or Hydrogen. This new state of matter presents very interesting new properties, which were explored both theoretical and experimentally by an increasingly large scientific community, as stated in many review papers [22, 23] and books [4, 24].

Unfortunately, BECs made of alkaline vapors possess a finite lifetime and can hardly survive in steady state conditions, and attention has been also given in recent years to the non-condensed steady state gas which is produced by laser cooling alone. Collective behavior was first considered by Dalibard in 1988 [25], who called the attention for a shadow force due to laser absorption in a magneto-optical trap. Another important step was the discovery of an *effective electric charge* for the neutral atoms [26]. This leads to an electrostatic type of interactions between nearby atoms, which is revealed by the Coulomb-like explosions of the atomic cloud when the magnetic confinement is switched off [27]. More recently, the existence of *hybrid phonons* was predicted, which are the quasi-particles associated with a new type of sound waves displaying a cut-off frequency [28]. This however has not yet been confirmed experimentally. The first part of this book will be focused on the collective properties of this non-condensed ultra-cold gas, which to our knowledge have not yet been extensively reviewed in the literature.

Finally, spontaneous or photo-ionization of the ultra-cold gas, leads to the creation of ultra-cold plasmas, which will be considered in the last part of this book. This cold plasma state can be achieved by photo-ionizing a cold cloud in a MOT [29], but it can also be achieved by spontaneous evolution on a gas of neutral atoms, excited in Rydberg quantum states, into a plasma [30]. The plasma ions are strongly coupled, with their Coulomb interactions well exceeding their kinetic energy. This leads to de occurrence of new phenomena, similar in some respects to those of a liquid metal [31]. The ultra-cold plasma also displays similarities with the highly compressed plasma created by laser fusion experiments, although in a quite different range of parameters.

1.3 Book Overview

Given the historical perspective, it is not surprising to adopt the present book structure, where a couple of introductory chapters on laser cooling is followed by three different groups of chapters, devoted to the three different phases of ultra-cold matter. More specifically, Chaps. 2 and 3 will be described the physics of laser cooling and magnetic trapping. In these chapters, we also introduce the wave kinetic description, which will serve as a guideline for comparison between these three phases. Such a description is based on the Wigner distribution, first proposed in 1935 [32], and then explored in many areas of physics, including laser cooling processes [33, 34]. Although other theoretical methods are also used and discussed, the wave kinetic approach will be used as a leitmotiv throughout this book. This is an useful tool for comparison between the three phases of ultra-cold matter, in their semi-classical and quantum regimes.

From Chaps. 4 to 6 we discuss the properties of the non-condensed ultra-cold matter, with particular emphasis on the repulsive inter-atomic forces and the resulting oscillations and collective processes. We discuss the effective charge of neutral atoms, and the associated Coulomb type of potential of the neutral gas, as well as their basic oscillation modes, such as the hybrid acoustic modes. We then consider the equilibrium configurations, and the influence of plasma boundaries on the internal mode structure. Finally, we discuss resonant wave-atom effects, nonlinear mode coupling and wave scattering processes. Quasi-linear atom velocity distributions are considered, and the influence of a turbulent spectrum of acoustic oscillations on the temperature limit for laser cooling is established. Quantum kinetic regimes and their quasi-classical limits are considered.

Chapters 7–12 deal with Bose-Einstein condensation, and the main properties of the condensed phase. We first mention the basic ideas related with the condensation phenomena, in homogeneous and trapped configurations. We then focus on the mean field approximation, which can be described by the celebrated Gross-Pitaevskii equation. This well-established theoretical approach is compared with the wave kinetic description, which is exactly equivalent but can lead to a quasi-classical formulation, more refined than the usual fluid approximation. Further advantages of the Wigner-Moyal formalism will be discussed later. The elementary oscillations of the condensate, and their linear and nonlinear properties will be considered.

Chapter 13 is devoted to atomic interferometers and to quantum coherence. This implies both the single atom interferometers, as well as collective interferometers based on BECs. Time interferometry will be discussed and compared with the traditional spatial configurations. The questions associated with quantum decoherence will be analyzed, given their basic importance for the understanding of the quantum description of isolated systems, non-locality and entanglement. We also give relevance to the detection of gravitational fields and the possible manifestation of quantum space-time fluctuations at the atomic level. Such a manifestation can be seen in the light of a brownian motion model, where the brownian particle is replaced by a BEC and the microscopic fluctuations are due to the background gravitational field.

Chapters 14–17 concern the state of ultra-cold plasmas. Plasma physics is a vast and old area of science, and here we concentrate on problems specific to the ultra-cold matter. This includes plasma formation and expansion, Rydberg atoms and new dispersive wave properties. We also compare the plasmon and phonon modes in a plasma with the hybrid phonons of the neutral gas, and with the Bogoliubov oscillations of the condensates. Another interesting aspect is related with the strong coupling of the plasma ions, eventually leading to the formation of coulomb crystals.

Finally, in the concluding Chap. 18, we give an overview of the entire field and its possible future developments. It should be noticed that the condensed phase of ultra-cold matter is at present more developed than the other two, but interesting bridges and comparisons can nevertheless be established between them. The emphasis of this book will be on the physical understanding of the results, concepts and methods of ultra-cold matter, with less concern for possible applications.

References

1. H.J. Metcalf, P. van der Straten, *Laser Cooling and Trapping* (Springer, Berlin, 1999)
2. G.L. Le Sage. *Lucrèce Newtonien* (Nouveaux Mémoires de l'Académie Royale des Sciences et Belles-Lettres, Paris, France, 1782), p. 404
3. A.M. Ignatov, Plasma Phys. Rep. **22**, 585 (1996)
4. C.J. Pethick, H. Smith, *Bose-Einstein Condensation in Dilute Gases*, 2nd edn. (Cambridge University Press, Cambridge/New York, 2008)
5. T.C. Kilian, T. Pattard, T. Pohl, J.M. Rost, Ultracold neutral plasmas. Phys. Rep. **449**, 77 (2007)
6. S.N. Bose, Z. Phys. **26**, 178 (1924)
7. A. Einstein, Sitzunsber. Preuss. Akad. Wiss. Phys. Math. Kl **22**, 261 (1924); ibid. **23**, 3 (1925)
8. P. Kapitza, Nature (London) **141**, 74 (1938)
9. J.F. Allen, A.D. Misiner, Nature (London) **141**, 75 (1938)
10. F. London, Phys. Rev. **54**, 947 (1938)
11. N.N. Bogoliubov, J. Phys. (Moscow) **11**, 23 (1947)
12. P. Lebedev, Ann. Phys. (Leipzig) **6**, 433 (1901)
13. E.F. Nichols, G.F. Hull, Phys. Rev. **17**, 26 (1903)
14. O.R. Frisch, Z. Phys. **86**, 42 (1933)
15. T.W. Hänsch, A.L. Schawlow, Opt. Commun. **13**, 68 (1975)
16. A. Ashkin, Phys. Rev. Lett. **40**, 729 (1978)
17. S. Chu, The manipulation of neutral particles. Rev. Mod. Phys. **70**, 685 (1998)
18. C.N. Cohen-Tannoudji, Manipulating atoms with photons. Rev. Mod. Phys. **70**, 707 (1998)
19. W.D. Philips, Laser cooling and trapping of neutral atoms. Rev. Mod. Phys. **70**, 721 (1998)
20. K.B. Davis, M.O. Mewes, M.R. Andrews, N.J. van Druten, D.S. Durfee, D.M. Kurn, W. Ketterle, Phys. Rev. Lett. **75**, 3969 (1995)
21. M.H. Anderson, J.R. Ensher, M.R. Matthewa, C.E. Wieman, E.A. Cornell, Science **269**, 198 (1995)
22. A.J. Leggett, Bose-Eintein condensation in the alkali gases: some fundamental concepts. Rev. Mod. Phys. **73**, 307 (2001)
23. F. Dalfovo, S. Giorgini, L.P. Pitaevskii, S. Stringari, Theory of Bose-Einstein condensation in trapped gases. Rev. Mod. Phys. **71**, 463 (1999)
24. L. Pitaevskii, S. Stringari, *Bose Einstein Condensation* (Clarendon, Oxford, 2003)
25. J. Dalibard, Opt. Commun. **68**, 203 (1988)
26. T. Walker, D. Sesko, C. Wieman, Phys. Rev. Lett. **64**, 408 (1990)
27. L. Pruvost, I. Serre, H.T. Duong, J. Jortner, Phys. Rev. A **61**, 053408 (2000)
28. J.T. Mendonça, R. Kaiser, H. Terças, J. Loureiro, Phys. Rev. A **78**, 013408 (2008)
29. T.C. Killian, J.J. Kulin, S.D. Bergeson, L.A. Oroczo, C. Orzel, S.L. Rolston, Phys. Rev. Lett. **83**, 4776 (1999)
30. M.P. Robinson, B. Laburthe Tolra, M.W. Noel, T.F. Gallagher, P. Pillet, Phys. Rev. Lett. **85**, 4466 (2000)
31. S. Ichimaru, Strongly coupled plasmas: high-density classical plasmas and degenerate electron liquids. Rev. Mod. Phys. **54**, 1017 (1982)
32. E.P. Wigner, Phys. Rev. **40**, 749 (1932)
33. S. Stenholm, The semiclassical theory of laser cooling. Rev. Mod. Phys. **58**, 699 (1986)
34. J. Dalibard, C. Cohen-Tannoudji, Atomic motion in laser light: connection between semiclassical and quantum description. J. Phys. B **18**, 1661 (1985)

Part I
Atomic Clouds

Chapter 2
Laser Cooling

In this introductory chapter, we first consider the interaction of a two-level atom with a monochromatic laser field, by reviewing the emission and absorption radiation processes, as described by the semi-classical optical Bloch equations. We then discuss the basic principles of laser cooling, using both the momentum and the energy pictures. A simple expression for the laser cooling force is derived and the concept of Doppler temperature limit, characterizing the laser cooling process, is introduced. This is followed by a discussion of magnetic traps, with particular emphasis on the Helmholtz and Ioffe configurations, and a description of the magneto-optical trap.

With these basic ingredients in mind, we can then discuss the sub-Doppler cooling processes, which can lead the atomic gas to temperatures well below the Doppler temperature limit, and in particular, to Bose Einstein condensation. The first of these sub-Doppler processes is the celebrated Sisyphus cooling, which makes use of counter-propagating circularly polarized laser beams. The second is the evaporative cooling, which is known for long time to work at macroscopic level, but is also extremely efficient at cooling a confined atomic cloud, at the cost of a severe reduction of the number of confined atoms. The chapter finishes with a discussion of the sympathetic cooling, which can be used to cool down atomic and molecular gases for which the principles of laser cooling cannot directly be applied.

2.1 Atom in the Laser Field

In order to understand the main properties of the atom-laser interaction, we first consider the semi-classical approach, where we use a quantum description for a two-level atom but describe radiation as a classical field. The full quantum description of both atom and radiation is given in the Appendix. For a detailed account see, for instance [1–3]. This semi-classical description allows us to understand the emission

J.T. Mendonça and H. Terças, *Physics of Ultra-Cold Matter*, Springer Series on Atomic, Optical, and Plasma Physics 70, DOI 10.1007/978-1-4614-5413-7_2,

and absorption mechanisms of light by atoms. However, spontaneous emission can only be understood with the full quantum model, and its contribution is introduced phenomenologically here.

The semi-classical Hamiltonian operator relevant to this problem has two distinct parts

$$H = H_a + H_{\text{int}}, \tag{2.1}$$

where H_a is the unperturbed Hamiltonian of the atom, with the eigenvalue equation $H_a|j\rangle = E_j|j\rangle$. Here E_j are the energy values of the atomic quantum levels, with the corresponding eigenstates $|j\rangle$. We consider a simple two-level atom model, where $j = 1, 2$. In Eq. (2.1), we have also used the interaction Hamiltonian H_{int}, which describes the coupling between the atom and the laser field, as specified below. In the absence of the laser field, i.e. $H_{\text{int}} = 0$, the atomic state vectors have solutions of the form $|j, t\rangle = \exp(-iE_j t/\hbar)|j\rangle$. When $H_{\text{int}} \neq 0$, these atomic states become coupled by the radiation field, and the atomic state vector will contain a superposition of the two states, as described by

$$|\psi(\mathbf{r}, t)\rangle = \sum_{j=1,2} C_j(t) \exp(-iE_j t/\hbar)|j\rangle, \tag{2.2}$$

where the coefficients $C_j(t)$ satisfy the normalization condition $\langle\psi(\mathbf{r}, t)|\psi(\mathbf{r}, t)\rangle = |C_1(t)|^2 + |C_2(t)|^2 = 1$. The quantities $|C_j(t)|^2$ give the probability for the atom to be in the state j. In order to determine these probabilities, we consider the time dependent Schrödinger equation

$$i\hbar \frac{\partial}{\partial t}|\psi\rangle = (H_a + H_{\text{int}})|\psi\rangle. \tag{2.3}$$

Using the superposition of states (2.2), we can derive from here the following two evolution equations

$$i\hbar \frac{\partial C_1}{\partial t} = C_1 H_{11} + C_2 H_{12} e^{-i\omega_0 t}, \quad i\hbar \frac{\partial C_2}{\partial t} = C_1 H_{21} e^{i\omega_0 t} + C_2 H_{22} \tag{2.4}$$

where we have introduced the transition frequency $\omega_0 = (E_2 - E_1)/\hbar$, and used the matrix elements of the interaction Hamiltonian operator, $H_{ij} = \langle i|H_{\text{int}}|j\rangle$.

Before studying the solutions of Eq. (2.4), we need to discuss the explicit form of these matrix elements. Let us consider the generic case where the two quantum levels $|1\rangle$ and $|2\rangle$ are associated with the electron eigenstates in the atom. It is known that, in the Coulomb gauge, the interaction Hamiltonian is given by

$$H_{\text{int}} = \frac{e}{m}\mathbf{p} \cdot \mathbf{A}(\mathbf{r}, t) + \frac{e^2}{2m} A^2(\mathbf{r}, t). \tag{2.5}$$

where $\mathbf{A}(\mathbf{r}, t)$ is the vector potential associated with the radiation field. This expression can be simplified in the following way. First, for a low intensity radiation field, the quadratic term A^2 can be neglected. Second, we can usually assume that the wavelength of the electromagnetic radiation is much larger than the dimensions of the atom. This is the so-called *dipole approximation*, which allows us to develop the vector potential $\mathbf{A}(\mathbf{r}, t)$ around the atom position $\mathbf{r}_0 = 0$, as follows: $\mathbf{A}(\mathbf{r}, t) = \mathbf{A}(t) \exp(i\mathbf{k} \cdot \mathbf{r}) \simeq \mathbf{A}(t)$. The interaction Hamiltonian is then reduced to $H_{\text{int}} = (e/m)\mathbf{p} \cdot \mathbf{A}(t)$. We can also write it as

$$H_{\text{int}} = -(p_{12}|1\rangle\langle 2| + p_{21}|2\rangle\langle 1|)E(t), \tag{2.6}$$

where $\mathcal{E}(t) = -\partial \mathbf{A}/\partial t$, is the laser electric field, and we have introduced the electric dipole moment of the atomic transition, as $p_{12} = p_{21}^* = -e\langle 1|(\mathbf{r} \cdot \mathbf{e})|2\rangle$, where $\mathbf{e} = \mathcal{E}/|\mathcal{E}|$ is the unit polarization vector. In the particular case of linear polarization along the x-axis, we can simply write $(\mathbf{r} \cdot \mathbf{e}) = x$.

It can easily be realized that the off-diagonal terms of the interaction Hamiltonian are $H_{12} = H_{21}^* = p_{12}E(t)$. We should also notice that the diagonal terms of this Hamiltonian are equal to zero, because they contain factors of the form $\langle 1|2\rangle = 0$. Replacing this in the evolution equations (2.4), and using an electric field of the form $\mathcal{E}(t) = \mathcal{E}_0 \cos(\omega t)$, where $\mathcal{E}$ is the amplitude an ω the laser frequency, we get

$$\frac{\partial C_1}{\partial t} = -i\Omega_R e^{-i\omega_0 t} \cos(\omega t) C_2, \quad \frac{\partial C_2}{\partial t} = -i\Omega_R^* e^{i\omega_0 t} \cos(\omega t) C_1, \tag{2.7}$$

where the new quantity $\Omega_R = p_{12}E_0/\hbar$ is called the *Rabi frequency*. The solutions of these coupled equations are well known. First, we consider the case of an atom which is initially in the lower energy level $|1\rangle$, and is submitted to a low intensity laser field. This means that coupling with the excited state $|2\rangle$ ia weak, and we can take $C_1(t) \simeq 1$, leading to the simple solution

$$C_2(t) = \frac{1}{2}\Omega_R^* \left[\frac{1 - e^{i(\omega_0+\omega)t}}{(\omega_0 + \omega)} + \frac{1 - e^{i(\omega_0-\omega)t}}{(\omega_0 - \omega)} \right]. \tag{2.8}$$

The transition probability from the lower to the upper state is $|C_2(t)|^2$. For long interaction times, the second term in this equation becomes dominant, and we can approximately write

$$|C_2(t)|^2 \simeq \frac{|\Omega_R|^2}{4} \left| \frac{1 - e^{i(\omega_0-\omega)t}}{(\omega_0 - \omega)} \right|^2 = |\Omega_R|^2 \frac{\sin^2[(\omega_0 - \omega)t/2]}{(\omega_0 - \omega)^2}. \tag{2.9}$$

The neglection of the non-resonant term is known as the *rotating wave approximation*. The transition probability attains a maximum at resonance, when $\omega = \omega_0$, with a value that evolves proportional to the square of time $|C_2(t)|^2 = \frac{1}{4}|\Omega_R|^2 t^2$. A more general solution can be obtained in the rotating wave approximation, where the coupled equations (2.7) reduce to

$$\frac{\partial C_1}{\partial t} = -\frac{i}{2}\Omega_R e^{-i\Delta t} C_2, \quad \frac{\partial C_2}{\partial t} = -\frac{i}{2}\Omega_R^* e^{i\Delta t} C_1, \tag{2.10}$$

where $\Delta = \omega_0 - \omega$ is the frequency detuning. Solving this we get

$$|C_2(t)|^2 = \frac{|\Omega_R|^2}{\Omega^2} \sin^2\left(\frac{t}{2}\Omega\right). \tag{2.11}$$

This shows that the occupation of the energy levels oscillates with a frequency $\Omega = \sqrt{|\Omega_R|^2 + \Delta^2}$. In the resonant case of $\omega = \omega_0$, or $\Delta = 0$, we are reduced to $\Omega = |\Omega_R|$. This shows that the Rabi frequency $|\Omega_R|$ is the natural frequency of oscillation of the energy level occupation probability, at resonance. In this case, the above expression reduces to $|C_2(t)|^2_{\Delta=0} = \sin^2(|\Omega_R|t/2)$. This means that the occupation probability attains its maximum value after multiples of the time interval $\Delta t = \pi/2|\Omega_R|$. Noting that $|C_1(t)|^2 + |C_2(t)|^2 = 1$, we conclude that the occupation probability oscillated between the two states of the atomic transition, and that, on the average, its occupation is $1/2$ for both states. These qualitative features are confirmed by a full quantum description (see the Appendix).

Another important aspect of the atom laser interaction is the existence of an energy shift of the lower energy level. This can be determined in the following way. The use of Eq. (2.8) on the first of Eq. (2.7) leads to the following result

$$\left\langle \frac{\partial C_1}{\partial t} \right\rangle = -\frac{i}{4}|\Omega_R|^2 \left[\frac{1}{(\omega_0 + \omega)} + \frac{1}{(\omega_0 - \omega)}\right], \tag{2.12}$$

where the simbol $\langle \cdot \rangle$ represents here a time average taken over one cycle of the radiation field. For non-resonant transitions $\omega \neq \omega_0$, we can identify this quantity with the average energy shift of the ground energy level, $\Delta E = i\hbar \langle dC_1/dt \rangle$, which leads to

$$\Delta E = \hbar |\Omega_R|^2 \frac{\omega_0}{(\omega_0^2 - \omega^2)}. \tag{2.13}$$

If we take into account all the possible non-resonant radiative transitions, from several energy levels $|j\rangle$ to the same ground state, we can generalize it to

$$\Delta E = \mathcal{E}_0^2 \sum_j \frac{|p_{j1}|^2 \omega_{j1}}{(\omega_j^2 - \omega^2)}, \tag{2.14}$$

where $\mathcal{E}_0$ is the amplitude of the laser electric field, $\omega_{j1} = (E_j - E_1)/\hbar$ the different atomic transition frequencies and p_{j1} the corresponding electric dipole moments. We should notice that, in the zero frequency limit, $\omega \to 0$, this reduces to the well known Stark shift due to a static electric field. For this reason, the energy shifts described by Eqs. (2.13) or (2.14) are usually called the *ac Stark effect*.

It is sometimes useful to describe the Rabi oscillations in terms of the *density matrix operator*, which can be defined as $\rho = |\psi\rangle\langle\psi|$. For a two-level atom model, such that $|\psi| = C_1(t)|1\rangle + C_2(t)|2\rangle$, the diagonal matrix elements are

$$\rho_{11} = \langle 1|\rho|1\rangle = |C_1(t)|^2, \quad \rho_{22} = \langle 2|\rho|2\rangle = |C_2(t)|^2 \tag{2.15}$$

and the off-diagonal terms

$$\rho_{12} = \langle 1|\rho|2\rangle = C_1(t)C_2^*(t), \quad \rho_{21} = \rho_{12}^*. \tag{2.16}$$

We can see that the diagonal terms are real quantities and their sum equal to one, $\rho_{11} + \rho_{22} = 1$. We can also see that, for a collection of identical N atoms per unit volume, there will be a population of atoms in the lower state $N_1(t)$, and a population in the upper state $N_2(t)$, such $\rho_{11} = N_1(t)/N$, and $\rho_{22} = N_2(t)/N$. The temporal evolution of the density matrix elements can be derived from the above equations for the coefficients $C_j(t)$ and, in the rotating wave approximation, we get

$$\frac{d\rho_{11}}{dt} = -\frac{d\rho_{22}}{dt} = \frac{i}{2}\Omega_R^* e^{-i\Delta t}\rho_{12} - \frac{i}{2}\Omega_R e^{i\Delta t}\rho_{21} \tag{2.17}$$

and

$$\frac{d\rho_{12}}{dt} = \frac{d\rho_{21}^*}{dt} = i\Omega_R e^{i\Delta t}(\rho_{11} - \rho_{22}). \tag{2.18}$$

These are known as the *optical Bloch equations*. The original Bloch equations, which are formally identical, were derived to describe the spin states in a magnetic field. The interest of this new formulation in terms of the density matrix elements is that it can easily be extended to a mixed state, such that $\rho = \sum_j p_j|\psi_j\rangle\langle\psi_j|$, where p_j are the probabilities for the quantum system to be in a state $|\psi_j\rangle$. For initial conditions $\rho_{11}(0) = 1$ and $\rho_{22}(0) = \rho_{12}(0) = 0$, we get the solutions (2.11) for $\rho_{22}(t)$, and

$$\rho_{12}(t) = \frac{|\Omega_R|^2}{\Omega^2}\sin\left(\frac{\Omega t}{2}\right)\left[i\Omega\cos\left(\frac{\Omega t}{2}\right) - \Delta\sin\left(\frac{\Omega t}{2}\right)\right]e^{-i\Delta t}. \tag{2.19}$$

Although spontaneous emission is not described in the frame of the present semi-classical approach to the atom-laser interaction, we can include it by adding a phenomenological term into the above evolution equations. Defining Γ as the spontaneous decay rate of the upper energy level, the above Eqs. (2.17) and (2.18) become

$$\frac{d\rho_{11}}{dt} = -\frac{d\rho_{22}}{dt} = \Gamma\rho_{11} + \frac{i}{2}|\Omega_R|\,(\tilde{\rho}_{12} - \tilde{\rho}_{21}) \tag{2.20}$$

and

$$\frac{d\tilde{\rho}_{12}}{dt} = \frac{d\tilde{\rho}_{21}^*}{dt} = i|\Omega_R|(\rho_{11} - \rho_{22}) - (\Gamma/2 + i\Delta)\tilde{\rho}_{12}, \tag{2.21}$$

where we have used $\tilde{\rho}_{12} = \rho_{12}\exp(-i\Delta t - i\varphi)$, where φ is the phase of the Rabi frequency, $\Omega_R = |\Omega_R|\exp(i\varphi)$. Steady state solutions of these equations will be an important ingredient of the laser cooling precess, as discussed next.

2.2 Laser Cooling Force

The process of atom cooling by laser radiation can be very easily understood with the help of two complementary pictures: the energy and the momentum pictures, which are illustrated in Fig. 2.1. The energy picture can be described in the following way. Assume an atom with two internal energy states, with energies E_1 and E_2, interacting with a slightly red-tuned radiation field with frequency $\omega < \omega_0 \equiv (E_2 - E_1)/\hbar$, and wave vector $\mathbf{k}$. Resonant radiative transition from the lower to the upper energy level, $|1\rangle \rightarrow |2\rangle$, can occur by photon absorption if the atom moves with velocity $\mathbf{v}$, such that the Doppler shifted photon frequency becomes equal the transition frequency, $(\omega - \mathbf{k} \cdot \mathbf{v}) = \omega_0$.

If then the upper energy level spontaneously decays, by photon emission in a direction perpendicular to the atom velocity, the radiation field will gain an energy of $\hbar|\mathbf{k} \cdot \mathbf{v}|$, and due to energy conservation, the atom will loose the same amount of kinetic energy, given that it returns to the initial internal energy state $|1\rangle$. If the fluorescence lifetime of the upper state $|2\rangle$ is short, such a cycle of "photon absorption followed by radiative decay" can repeat several times in a second, the kinetic energy of the atom will reduce on the average and the atom will slow down, leading to a significant gas cooling.

Let us now consider the more commonly used momentum picture, which describes the same absorption-emission cycle from a different perspective. Due to momentum conservation, when an atom emits one photon with frequency ω and wavevector $\mathbf{k}$, it acquires a momentum $-\hbar\mathbf{k}$, which is called the *momentum recoil*. Inversely, when an atom absorbs a photon, it will absorb the photon momentum $\hbar\mathbf{k} \equiv \hbar k\mathbf{n}$, where $\mathbf{n}$ is the unit vector along the direction of $\mathbf{k}$. If such an absorption

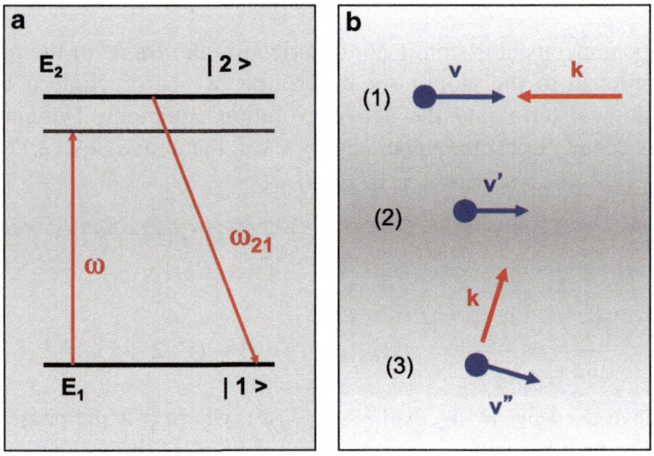

Fig. 2.1 Basic absorption-emission cycle of atomic laser cooling: two complementary pictures. (**a**) Energy picture, with a virtual energy level $E_2 - |\mathbf{k} \cdot \mathbf{v}|$; (**b**) Momentum picture, with an intermediate atom velocity $v' = v - \hbar k/M$

is followed by spontaneous emission at the same frequency $\omega = k/c$, the emitted photon will propagate in an arbitrary direction defined by $\mathbf{n}' = \mathbf{k}'/k$. As a result, in the emission-absorption cycle the atom translational momentum is not conserved, and it will receive a net momentum given by $\Delta\mathbf{p} = \hbar(\mathbf{k} - \mathbf{k}') = \hbar\omega(\mathbf{n} - \mathbf{n}')/c$. If such a cycle is repeated many times by the same atom, the momentum loss by spontaneous emission will be averaged to zero and a net momentum gain will be obtained. We can write the total momentum balance after r successive cycles of absorption-emission processes, as

$$\Delta\mathbf{P} = r\left(\hbar\mathbf{k} - \hbar\sum_{j=1}^{r}\mathbf{k}'_j\right). \tag{2.22}$$

Taking the average for a large number $r \gg 1$, the last term averages to zero, and we are reduced to $\langle\Delta\mathbf{P}\rangle \simeq r\hbar\mathbf{k}$. If r is the fluorescence rate, we can define the *ponderomotive force* acting on the atom (the momentum variation per unit time) as

$$\mathbf{F}_p = r\hbar\mathbf{k}. \tag{2.23}$$

It is obvious that this fluorescence rate is determined by the product of the spontaneous decay rate Γ with the upper energy level population of the atom, or $r = \Gamma\rho_{22}$. Let us go back to the evolution equations for the density matrix elements, describing the radiation coupling between the two energy levels $|1\rangle$ and $|2\rangle$, as given by (2.20) and (2.21). In steady state, $d/dt = 0$, these equations yield

$$\tilde{\rho}_{22} = -\frac{i}{2\Gamma}\Omega_R(\tilde{\rho}_{21} - \tilde{\rho}_{12}), \quad \tilde{\rho}_{12} = -i\frac{\Omega_R}{2}\frac{2\tilde{\rho}_{22} + 1}{(i\Delta - \Gamma/2)} = \tilde{\rho}_{21}^*, \tag{2.24}$$

with $\tilde{\rho}_{ij}$ standing for the steady-state values of the matrix elements ρ_{ij}. Using this result in Eq. (2.23), we can then write the ponderomotive force of the incident laser beam on the atom as

$$\mathbf{F}_p = \hbar\mathbf{k}\Gamma\tilde{\rho}_{22} = \hbar\mathbf{k}\Gamma\frac{\Omega_R^2}{2\Omega_R^2 + 4\Delta^2 + \Gamma^2}. \tag{2.25}$$

This result is valid for an atom at rest in the laboratory frame, or if the atom is moving, in the atom frame. We can now explicitly introduce the motion of the atom in the expression of the ponderomotive force. As noticed before, due to the Doppler effect, the laser photons are perceived as having a shifted frequency $\omega' = \omega - \mathbf{k}\cdot\mathbf{v}$. This introduces an additional detuning with respect to the transition frequency ω_0. The resulting expression for the force acting on the moving atom will then be

$$\mathbf{F}_p = \hbar\mathbf{k}\Gamma\frac{\Omega_R^2}{4(\Delta + \mathbf{k}\cdot\mathbf{v})^2 + \Gamma^2 + 2\Omega_R^2}. \tag{2.26}$$

For low intensities such that the radiative transition is not saturated, $\Omega_R^2 \ll \Gamma^2/2$, we can neglect the Rabi frequency in the denominator of the force. Using the following approximation

$$\frac{1}{4(\Delta + \mathbf{k}\cdot\mathbf{v})^2 + \Gamma^2} \simeq \frac{1}{(4\Delta^2 + \Gamma^2) + 8(\mathbf{k}\cdot\mathbf{v})\Delta} \simeq \frac{1}{(4\Delta^2 + \Gamma^2)}\left[1 - \frac{8\Delta(\mathbf{k}\cdot\mathbf{v})}{(4\Delta^2 + \Gamma^2)}\right], \tag{2.27}$$

the force acting on the atom can be approximately determined by

$$\mathbf{F}_p = \mathbf{F}_0 - \frac{M\beta\mathbf{k}}{2k^2}\mathbf{k}\cdot\mathbf{v}, \tag{2.28}$$

where M is the atomic mass, and

$$\mathbf{F}_0 = \hbar\mathbf{k}\Gamma\frac{\Omega_R^2}{(4\Delta^2 + \Gamma^2)}, \quad \beta = \hbar\Gamma k^2\frac{2^4\Omega_R^2\Delta}{M(\Delta^2 + \Gamma^2/4)}. \tag{2.29}$$

Let us consider the particularly interesting case when the atom moves in the direction of the laser propagation, $\mathbf{k} \parallel \mathbf{v}$. We are then reduced to

$$\mathbf{F}_p = \mathbf{F}_0 \mp \frac{1}{2}\beta M\mathbf{v}. \tag{2.30}$$

We can then see that two different terms occur in the ponderomotive force $\mathbf{F}_p$. The first term, $\mathbf{F}_0$, is a constant radiation pressure term, which is independent of the atom velocity. In contrast, the second term can be seen as a friction term which depends on the atom velocity. It is positive for an atom moving in the opposite direction with respect to the laser beam (the atom slows down), and negative for an atom moving along the beam (the atom is pushed forward). If now the atom is located in a region with two opposite laser beams, and moves along the laser beam axis, it will experiment a total force in the $\mathbf{k}$ direction as given by

$$\mathbf{F}_p = \left(\mathbf{F}_0 - \frac{\beta}{2}M\mathbf{v}\right) - \left(\mathbf{F}_0 + \frac{\beta}{2}M\mathbf{v}\right) = -\beta M\mathbf{v}. \tag{2.31}$$

This shows that, with two counter-propagating beams, the pressure term is zero and the friction force is always negative, leading to a deceleration. The atom will therefore slow down along that direction. If the same occurs for every atom in the gas, the gas will then cool down. This is the laser cooling effect, which leads to extremely low temperatures in the micro-Kelvin domain, as first observed in the experiments with optical *molasses* by Chu and collaborators [4]. The term 'molasse' was coined because of the resemblance of the resulting viscous photon fluid to real molasses.

2.3 Doppler Limit

Let us now discuss the theoretical limits of this cooling process. These limits can be discussed by using a statistical approach to the spontaneous emission. For that purpose, we study the evolution of the atomic momentum distribution of the cold gas, $f(p,t)$, in a one-dimensional description. This evolution is determined by a Fokker-Planck equation of the form

$$\frac{\partial}{\partial t} f(p,t) = \frac{\partial}{\partial p}\left[A(p) + D(p)\frac{\partial}{\partial p}\right] f(p,t), \tag{2.32}$$

where $A(p)$ and $D(p)$ are the friction and diffusion coefficients. A steady state solution of this equation is given by

$$f(p) = f_0 \exp\left[-\int_0^p \frac{A(p')}{D(p')} dp'\right], \tag{2.33}$$

We know that the friction coefficient is related to the averaged particle momentum by $d\langle p\rangle/dt = -A(p)$. Comparing with the results obtained in the previous section, we can write the following relation

$$A(p) = \beta M v = \beta p = 8\hbar k^2 \Gamma \Delta \frac{\Omega_R^2}{M(4\Delta^2 + \Gamma^2))^2}. \tag{2.34}$$

Noting that, in typical experimental situations we have $\Omega_R \sim \Delta \sim \Gamma$, we can, for the sake of magnitude estimates, use a much simpler expression, $A(p) \sim \hbar k^2 p/M$. Turning now to the diffusion coefficient $D(p)$, we know that by definition, it is given by

$$D(p) = \frac{1}{2}\frac{\langle \Delta p\rangle^2}{\Delta t} \sim \frac{1}{4}\hbar^2 k^2 \Gamma \tag{2.35}$$

The integration of the steady state solution (2.33) then leads to

$$\int_0^p \frac{A(p')}{D(p')} dp' \simeq \int_0^p \frac{\hbar k^2 p'}{M} \frac{4}{\hbar^2 k^2 \Gamma} dp' = \frac{2}{M\Gamma} p^2 \tag{2.36}$$

We can then reduce the steady state distribution $f(p)$ to a Maxwell distribution such that

$$f(p) = f_0 \exp\left(-\frac{p^2}{2MT_D}\right) \tag{2.37}$$

with the effective temperature (in energy units)

$$T_D \sim \frac{\hbar}{2}\Gamma, \tag{2.38}$$

known as the *Doppler temperature* limit. We can see that this temperature is limited by the natural life-time of the atomic transition. This result has a simple physical meaning. We note that the spontaneous emission implies a time uncertainty of $\Delta t \sim 1/\Gamma$. Therefore, according to the uncertainty principle, the energy uncertainty is of order $\Delta E \sim \hbar\Gamma$ which nearly coincides with (2.38). For a detailed discussion of the Doppler limit see [5].

2.4 Magnetic Traps

Atoms can be confined by static magnetic fields, by oscillating fields and by laser light. Confinement of neutral atoms in static magnetic traps is a well established technique, first reported by Migdall et al. [6]. In recent years, this technique has evolved into the fabrication of atomic chips at the micro-meter scale, using several magnetic trap configurations [7]. It is well known that magnetic fields produce an energy shift in the internal atomic levels. For inhomogeneous magnetic fields, this leads to the occurrence of local energy minima where the atoms can be confined. For an atom with a permanent magnetic moment μ, at rest in a static magnetic field $\mathbf{B}$, the energy shift can be defined as

$$\Delta E(\mathbf{r}) = -\mu \cdot \mathbf{B}(\mathbf{r}) = g_J \mu_B m_J |\mathbf{B}(\mathbf{r})| \tag{2.39}$$

where m_J is the projection of the total angular momentum $\mathbf{J}$, g_J is the corresponding Landé factor, and $\mu_B = e\hbar/2m_e$ is the *Bohr magneton*. A well known general theorem prevents the occurrence of static magnetic field maxima in free space, in the absence of currents. This means that only local minima can be created in the experiments. For atomic states such that $g_J m_J > 0$, this interaction allows to create energy minima around which the atoms can be confined. If the atom is moving with velocity $\mathbf{v}$, it will perceive a time varying magnetic field, which can eventually induce transitions between these magnetic states, thus allowing a trapped atom to become untrapped. According to Eq. (2.39), the typical energy difference between magnetic states is $\hbar\omega_L$, where $\omega_L = |\mu B|/\hbar$ is the *Larmor frequency*. The time scale for field variation in the atom rest frame is determined by $dB/dt = \mathbf{v} \cdot \nabla B$. If this time scale remains much larger than $1/\omega_L$ the magnetic dipole of the atom will have time to adapt to the slowly varying field, and the atom will stay in the same magnetic state. Therefore, the existence of such an *adiabatic regime* will be defined by the inequality

$$\frac{1}{B}\frac{dB}{dt} \ll \omega_L. \tag{2.40}$$

If, at some given position this inequality is not verified, transitions to other magnetic states can take place, which means that magnetic confinement can fail. This is the case, in particular for regions where the magnetic field amplitude, and the Larmor frequency, tend to zero. This is the reason why the magnetic minimum of a magnetic trap cannot be zero. A non-zero minimum has to exist in other to fulfill the adiabatic condition at the centre of the trap.

2.4.1 Multipolar Field Configuration

Simple magnetic configurations with a magnetic field minimum can be constructed, using the basic elements of conducting currents, straight wires and coils. Let us first consider the *multipolar field* configuration produced by a number n of linear and parallel conductors, with the same current I flowing all of them in the same direction. The conductors are assumed over an imaginary cylindrical surface with radius a, and parallel to the axis of the cylinder, and can be described by the current density $\mathbf{J}(\mathbf{r}) = \sum_{j=1}^{n} I \mathbf{e}_z \delta(\mathbf{r} - \mathbf{r}_j)$, where $\mathbf{r}_j$ is a polar vector with components $x_j = a\cos(2\pi j/n)$ and $y_j = a\sin(2\pi j/n)$. The corresponding vector potential $\mathbf{A}$ can be written as

$$\mathbf{A} = -\frac{\mu_0 I}{4\pi}\mathbf{e}_z \ln R\,, \quad R^2 = \Pi_{j=1}^{n} r_j^2 = \left(a^{2n} + r^{2n} - 2r^n a^n \cos n\theta\right) \tag{2.41}$$

In the inner region, close to the cylindrical axis where $r \ll a$, the resulting magnetic field $\mathbf{B} = \nabla \times \mathbf{A}$ can be written in polar coordinates as

$$B_r \simeq -n\frac{\mu_0 I}{2\pi a}\left(\frac{r}{a}\right)^{n-1} \sin(n\theta)\,, \quad B_\theta \simeq -n\frac{\mu_0 I}{2\pi a}\left(\frac{r}{a}\right)^{n-1} \cos(n\theta) \tag{2.42}$$

From here we conclude that the magnetic field tends to zero when we approach the cylindrical axis, with a field strength varying with distance as

$$B = \sqrt{B_r^2 + B_\theta^2} \simeq n\frac{\mu_0 I}{2\pi a}\left(\frac{r}{a}\right)^{n-1}. \tag{2.43}$$

This result shows that B increases in the radial direction. In this sense it can be considered a *minimum B* configuration. It is also possible to consider a similar multipolar field produced by $2n$ parallel conductors, where alternatively the current reverses its direction. This means that, we will have a current I along the z-axis, for straight wires located on the same imaginary cylindrical surface at $r = a$ and $\theta = 2\pi j/n$, for $j = 1, 2, \ldots n$, and a current $-I$ for the remaining n wires located at $r = a$ and $\theta = (2j + 1)\pi/n$, for $j = 1, 2, \ldots n$. The resulting vector potential is now determined by

$$\mathbf{A} = -\frac{\mu_0 I}{4\pi}\mathbf{e}_z \ln\left(\frac{R_-}{R_+}\right)\,, \quad R_\pm^2 = \Pi_{j=1}^{n} r_j^2 = \left(a^{2n} + r^{2n} \pm 2r^n a^n \cos n\theta\right) \tag{2.44}$$

In the region close to the cylindrical axis, $r \ll a$, the magnetic field is approximately given by twice the previous value in Eq. (2.43), which results from the fact that we have twice as much wires. In this sense, there is no significant change for the field near the axis, if we revert the current alternatively in half of them. We still get a minimum field configuration. However, the same is not true for $r \sim a$ where significant field changes will occur.

2.4.2 Helmholtz Configuration

Another simple magnetic configuration is produced by two parallel coils, with radius a, located at a distance $2l$ from each other, at $z = \pm l$. We consider the case where the currents in the two coils are equal to I and flow in the same direction. The resulting vector potential is purely poloidal, and determined by the expression

$$A_\theta \simeq \frac{\mu_0 I a^2}{(a^2 + l^2)^{3/2}} r \left[1 + \frac{15}{2} \frac{l^2 z^2}{(a^2 + l^2)^2} \right]. \tag{2.45}$$

This field configuration now exhibits a minimum at $z = 0$, which in contrast with the multipolar configuration is not zero. A closer analysis shows that this minimum only exists for $a < 2l$. In the opposite case, the minimum is replaced by a maximum at $z = 0$. The corresponding magnetic field close to the magnetic axis $r \ll a$ is purely axial, $B_r \sim 0$ and $B_\theta \sim 0$, with a minimum given by

$$|B_z|_{\min} \simeq \frac{\mu_0 I a^2}{(a^2 + l^2)^{3/2}}. \tag{2.46}$$

On the other hand, the field will be maximum near the coils, at $z = \pm l$. Quite often, we characterize the magnetic mirror by the ratio $R = |B_z|_{\max}/|B_z|_{\min}$. This configuration is known as a simple *magnetic mirror*. It can easily be recognized that the two configurations discussed above are in some sense complementary, because near the origin, the field increases radially in the multipolar case, and axially in the magnetic mirror.

2.4.3 Ioffe Configuration

A superposition of the two basic configurations was first proposed by Ioffe for plasma confinement [8], and latter applied to neutral atoms [9]. This is called a *Ioffe trap* and it is able to created a local minimum at the origin. The resulting field can be described by

$$B(z, r) = B_0 + \frac{r^2}{2} \left(\frac{B_1^2}{B_0} - B_2 \right) + B_2 z^2, \tag{2.47}$$

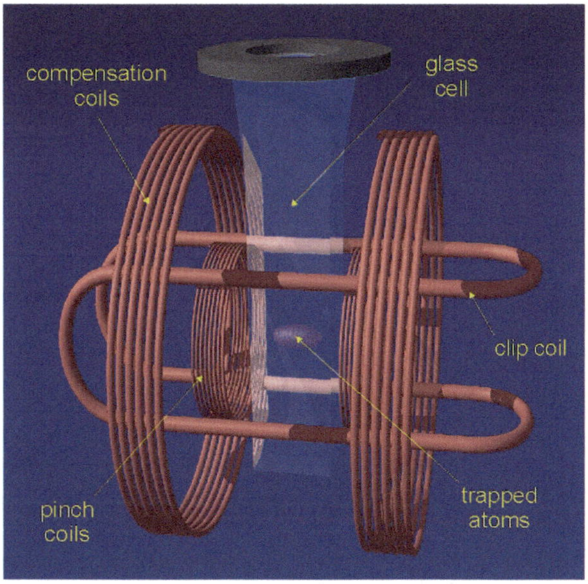

Fig. 2.2 A magnetic trap with Ioffe configuration. For experimental purposes, two compensation coils are used to control the bias field B_0

where B_0, B_1 and B_2 are positive constants. The field is minimum on axis ($z = 0$) if $B_1^2 > B_0 B_2$. The quantity B_0 is usually referred as the *bias field*, and gives the magnetic field minimum at the centre of the trap. Atoms with temperature T can be trapped if $\mu_J B_0 \gg T$. The trapping potential seen by an atom of mass M can then be written as

$$V(\mathbf{r}) = \frac{1}{2} M \omega_z^2 z^2 + \frac{1}{2} \omega_r^2 r^2, \tag{2.48}$$

where ω_z and ω_r are the axial and radial oscillating frequencies, as determined by

$$\omega_z = \sqrt{|\mu| \left(\frac{B_1^2}{B_0} - B_2 \right) / M}\,, \quad \omega_r = \sqrt{|\mu| B_2 / M}. \tag{2.49}$$

The Ioffe trap configuration is topologically equivalent to a *baseball* coil, as illustrated in Fig. 2.2, where the pinch and clip coils create the basic magnetic configuration.

2.4.4 Anti-Helmholtz Configuration

One of the simplest ways to produce a magnetic minimum is to revert the currents in the simple magnetic mirror. When the currents are equal in the coils, but flow in opposite directions, the vector potential (2.45) is replaced by

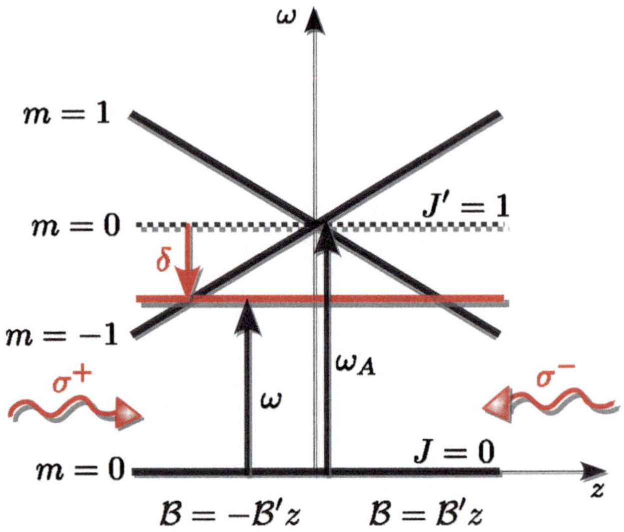

Fig. 2.3 Schematic representation trapping mechanism in one dimension MOT for a $J = 0 \rightarrow J' = 1$ transition in an atom

$$A_\theta \simeq -A_0 rz\,, \quad A_0 = 3\left(\mu_0 I a^2\right)\frac{l}{(a^2+l^2)^{5/2}}. \tag{2.50}$$

This is again valid for the region $r \ll a$. The corresponding magnetic field has components $B_\theta = 0$, $B_r = A_0 r$ and $B_z = 2A_0 z$. And the total field strength in the vicinity of the origin is determined by

$$B(\mathbf{r}) = A_0\sqrt{r^2+4z^2}. \tag{2.51}$$

This field satisfies a minimum criterion, and increases in all directions, but with a slope which is different for the axial and the radial directions. In the case where $a = l$ this is usually called the *anti-Helmholtz configuration*. An important difference with respect to the Ioffe trap is that the magnetic field minimum here is equal to zero, which increases the trap losses near the centre.

Let us finally consider a magneto-optical trap (MOT) which results from the simultaneous use of laser cooling beams and a magnetic trapping, or from an atomic physics perspective, a combination of Zeeman and Doppler shifts. The magnetic configuration is basically made of two anti-Helmholtz coils as discussed above.

In order to describe the principle of a magneto-optical trap, we consider an atom ground state, with total angular momentum $J = 0$, and an excited state with angular quantum number $J = 1$, with three sub-levels $m_J = 0, \pm 1$. Two counter-propagating laser beams are red-detuned with respect to the atomic transition, with left and right circular polarizations $\sigma^\pm$. The inhomogeneous magnetic field created by the two anti-Helmholtz coils is zero at the centre of the trap and increases linearly, as shown in Fig. 2.3. The laser beam with polarization σ^+ (propagating along the negative z-direction in this figure) can couple the ground state to the $m = +1$ excited state, due to the additional Doppler shift associated with atoms moving outwards in the positive region $z > 0$. A similar process occurs with the atoms moving outwards in the region $z < 0$, with transitions allowed to the state $m = -1$ by the laser beam with polarization σ^-. The net result is cooling and radiative pressure pushing the atoms towards the centre of the trap, $z = 0$. Minimal temperature is attained for a detuning of the order of $\Delta = (\Gamma/2)\sqrt{1 + I/I_{\text{sat}}}$, where Γ is the atomic transition line width, I is the laser intensity and I_{sat} the saturation intensity. In typical experimental conditions, we can choose $I/I_{\text{sat}} \sim 10 - 100$, which corresponds to $\Delta \sim (1 - 10)\Gamma$. This is obviously much larger than the requirement for attaining the Doppler limit temperature, and indeed temperatures well below this limit are obtained. This is due to the existence of additional cooling processes that can take place in a MOT, as those described next.

The magneto-optical trap was first experimentally used by Raab et al. [10] and now became a common technique for the production and confinement of an ultra-cold gas. In such a device, the laser and magnetic confinement forces considered so far are complemented by collective forces, resulting from long range atomic interactions, which will be discussed later in Chap. 4.

2.5 Sisyphus Cooling

The limitations of the Doppler cooling theory, based on the two-level atom configuration, became rather evident as experiments performed with optical molasses soon revealed that the temperatures were actually much lower than the limit given in Sect. 2.3. This happens because two important ingredients were ignored within the simple Doppler cooling theory: first, the intensity of the radiation field produced by two counter-propagating lasers is inhomogeneous; second, real atoms are not two-level systems; in fact, they have Zeeman sub-levels in the ground state.[1] The physics behind cooling in this configuration is, however, rather involved [11], for which we draw some of the most important lines here.

Let us consider an atom interacting with two counter-propagating laser beams, along the z axis, possessing the same frequency ω_0 but in orthogonal polarization states, such that

$$\mathbf{E}(\mathbf{r}, t) = E_0 \left(\mathbf{e}_1 e^{i\mathbf{k}_0 \cdot \mathbf{r}} + \mathbf{e}_2 e^{-i\mathbf{k}_0 \cdot \mathbf{r}} \right) e^{-i\omega_0 t} + \text{c.c.}, \tag{2.52}$$

with $(\mathbf{e}_1 \cdot \mathbf{e}_2) = 0$, and $(\mathbf{e}_1 \cdot \mathbf{k}_0) = (\mathbf{e}_2 \cdot \mathbf{k}_0) = 0$. For instance, we can set $\mathbf{k}_0 = k_0 \mathbf{e}_z$ and $\mathbf{e}_1 = \mathbf{e}_x$ and $\mathbf{e}_2 = \mathbf{e}_y$. This is equivalent to write

$$\mathbf{E}(\mathbf{r}, t) = \sqrt{2} E_0 \mathbf{e}(\mathbf{r}) e^{i\mathbf{k}_0 \cdot \mathbf{r} - i\omega_0 t} + \text{c.c.}, \tag{2.53}$$

[1] Actually, in the absence of magnetic fields, alkali atoms have degenerate ground states.

where $\mathbf{e}(\mathbf{r})$ is a position dependent unit polarization vector, as defined by

$$\mathbf{e}(\mathbf{r}) = \frac{1}{\sqrt{2}}\left(\mathbf{e}_1 + \mathbf{e}_2 e^{-2i\mathbf{k}_0\cdot\mathbf{r}}\right). \tag{2.54}$$

This shows that the field polarization changes along the z-direction with a period of $\pi/k_0 = \lambda_0/2$, half the laser wavelength, alternating between linear and circular polarization. Such an alternation can be shown more explicitly by introducing the right and left circular polarization vectors, $\mathbf{e}_\pm = (\mathbf{e}_1 \pm i\mathbf{e}_2)/\sqrt{2}$, in terms of which the above expression becomes

$$\mathbf{e}(\mathbf{r}) = \frac{1}{2}\mathbf{e}_+\left[1 - \sin(2\mathbf{k}_0\cdot\mathbf{r})\right] + \frac{1}{2}\mathbf{e}_-\left[1 + \sin(2\mathbf{k}_0\cdot\mathbf{r})\right] + \frac{1}{\sqrt{2}}\mathbf{e}_2\cos(2\mathbf{k}_0\cdot\mathbf{r}). \tag{2.55}$$

This shows that the intensity of the right and left circular polarization modes vary in space as

$$I_\pm \propto \frac{1}{2}\left[1 \mp \sin(2\mathbf{k}_0\cdot\mathbf{r})\right]. \tag{2.56}$$

For a ground state with two degenerate magnetic quantum numbers $m_g = \pm 1/2$, this can lead to an energy shift of the ground states $g = \pm 1/2$, due to the existence of these $\sigma^\pm$ polarization modes, of the form

$$\Delta E^\pm = -U_0\left[1 \mp \sin(2\mathbf{k}_0\cdot\mathbf{r})\right], \quad \text{with} \quad U_0 = \frac{\hbar\Omega_R^2\Delta}{\Gamma^2 + 4\Delta^2}. \tag{2.57}$$

An atom moving to the right ($\mathbf{k}_0$ positive direction) in the state $m_g = +1/2$, will be excited to the upper level by the σ^+ polarization mode, near the upper part of the oscillating potential hill (see Fig. 2.4). It will then be optically pumped to the other ground sub-level at the bottom of the well, by emitting a blue shifted photon. The kinetic energy loss will therefore be of the order of U_0. This process will then repeat itself, with more probability for an atom to climb the potential hill than to go down.

The lowest temperature limit is now given by the recoil energy associated with the process of photon emission by a atom. This recoil energy is $E_{\text{rec}} = (\hbar k)^2/2M$, where $\hbar k \simeq \hbar\omega_0/c$ is the photon momentum, and M the mass of the atom. The resulting *recoil temperature* limit, T_{rec} is given in energy units by

$$T_{\text{rec}} \equiv E_{\text{rec}} = \frac{\hbar^2\omega_0^2}{2Mc^2}. \tag{2.58}$$

This value lies well below the Doppler limit, as observed in the experiments [12]. Additional cooling, leading to temperatures even lower than T_{rec} can be achieved with the evaporative cooling technique, which consists of extracting atoms in high kinetic energy from the trap, as discussed next. In Table 2.1, we present some of its values, together with other parameters, for the most important atomic isotopes used in experiments.

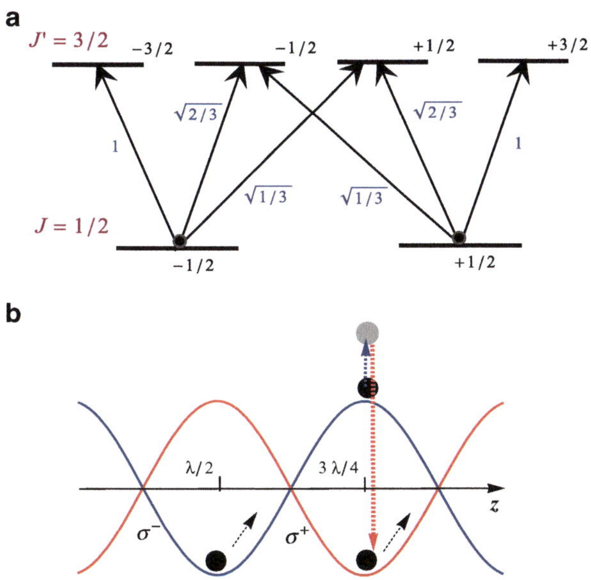

Fig. 2.4 (**a**) Schematic representation of a simple multi-level structure sensitive to the light shifts induced by the polarization gradient. (**b**) Optical pumping mechanism in an optical potential with position-dependent polarization

Table 2.1 Comparison of relevant atomic parameters for the main cooling transitions for the isotopes of elements used in laser cooling experiments. The data are taken from Ref. [13]

Isotope	Cooling transition	$\Gamma/2\pi$ (MHz)	I_s (mW/cm^2)	T_D (μK)	$T_{\rm rec}$ (nK)
^{39}K	$4^2S_{1/2} \rightarrow 4^2P_{3/2}$	6.2	1.81	148	830
^{87}Rb	$5^2S_{1/2} \rightarrow 5^2P_{3/2}$	5.9	1.67	145	370
^{40}Ca	$4^1S_0 \rightarrow 4^1P_1$	34.2	60	832	2,670
^{88}Sr	$4^1S_0 \rightarrow 4^1P_1$	31.8	42.7	767	1,020

2.6 Evaporative Cooling

Evaporative cooling is perceived in everyday life. It results from the fact that evaporation carries away the particles of a given body or system with the largest kinetic energies. It was first proposed to cool trapped Hydrogen atoms by Hess in 1985 [14], and later used to cool alkali atoms [15]. An early authoritative review was given in [16]. In order to produce evaporation, we need to corrode the borders of the magnetic trap in order to produce a shallow potential. This can be made by using radio-frequency fields. Given the magnetic field inhomogeneity, this field will connect upper and lower Zeeman ground levels, producing a magnetic hole that allows the atoms in the outer regions of the trap to escape. Evaporative cooling became a basic technique to attain the critical temperature and to achieve Bose Einstein condensation [17].

Let us briefly review the concepts of this cooling process, by assuming that an atomic gas is trapped inside a given potential well $U(\mathbf{r})$. The temperature drops due to loss of atoms, as characterized by the evaporation parameter α, defined as [16]

$$\alpha = \frac{d(\ln T)}{d(\ln N)}, \tag{2.59}$$

where T is the temperature and N the number of atoms. If evaporation is processed at a constant rate, then the final temperature at some later instant t can be simply determined by

$$T(t) = T(0)\left[\frac{N(t)}{N(0)}\right]^{\alpha}, \tag{2.60}$$

where $N(0)$ and $T(0)$ are the initial values. We can assume a potential well with a generic shape $U(\mathbf{r}) \propto r^{-\delta/3}$, where $\delta = 3/2$ for a three dimensional harmonic potential. Evaporation is controlled by limiting the depth U_0 of the confining potential. The kinetic energy of the escaping atoms is larger and of the order of U_0. In what concerns the trapped atoms, we can assume that they are approximately described by a Boltzmann distribution, with an average energy of $U_{\text{th}} = (\delta + 3/2)T$. Under these assumptions, it is then possible to establish a simple relation between the evaporation parameter and the potential depth, as

$$\alpha = \frac{U_0 - U_{\text{th}}}{U_{\text{th}}} = \frac{\eta}{(\delta + 3/2)} - 1, \tag{2.61}$$

where we have used $\eta = U_0/T$. We can see that, for $\eta = 6$ we have $\alpha = 1$ for a harmonic trap. A drop of temperature of 10^{-3} would then be achieved for a fraction of 10^{-3} atoms remaining inside the trap after evaporation. This would bridge the gap between the temperature range provided by laser cooling and the critical temperature required for Bose Einstein condensation, as demonstrated experimentally for alkaline atoms and for atomic Hydrogen.

Let us now consider the temporal evolution of the trapped atoms. The velocity of atoms of critical kinetic energy equal to $U_0 \equiv \eta T$, is equal to $\sqrt{2U_0/m} = \sqrt{\pi\eta}\bar{v}/2$, where $\bar{v}$ is the average velocity of the trapped atoms. For large values of η, the fraction of atoms with energies larger than the threshold U_0 is approximately given by $2e^{-\eta}\sqrt{\eta/\pi}$. From here we conclude that the rate of evaporation provided by collisions is

$$\frac{dN}{dt} = -\frac{N}{\tau_v}, \quad \tau_v = \frac{\sqrt{2}e^{\eta}}{\eta}\tau_{\text{col}}, \tag{2.62}$$

where τ_v is the evaporation time. We have also defined the elastic collision time $\tau_{\text{col}} = 1/(\sqrt{2}n_0\sigma\bar{v})$, where n_0 is the maximum density of atoms, σ is the cross section for elastic collisions, and $\sqrt{2}\bar{v}$ is the average relative velocity between two colliding atoms. This expression for the atom evaporation rate is confirmed by a more detailed analysis.

Because evaporative cooling is essentially a classical process, many classical models have been proposed [15, 18–23]. Such a classical description is valid if the temperature of the gas is much larger than the quantum level spacing of the atoms in the trap. In addition, the de Broglie wavelength has to be larger than the mean interatomic distance, or $n\lambda_T^3 \gg 1$, where n is the atom density and $\lambda_T = \hbar\sqrt{2\pi/MT}$ is the de Broglie wavelength of a thermal atom. Following [23], let us assume a classical distribution function $f(\mathbf{r}, \mathbf{p})$, normalized to the total number of atoms in the trap, as

$$\int d\mathbf{r} \int \frac{d\mathbf{p}}{(2\pi\hbar)^3} f(\mathbf{r}, \mathbf{p}) = N, \tag{2.63}$$

where $\mathbf{p}$ is the atom momentum. The evolution of this distribution is described by the Boltzmann equation

$$\left(\frac{\partial}{\partial t} + \mathbf{v} \cdot \nabla + \mathbf{F} \cdot \frac{\partial}{\partial \mathbf{p}}\right) f = C(\mathbf{r}, \mathbf{p}), \tag{2.64}$$

where $\mathbf{v} = \mathbf{p}/M$, and $\mathbf{F} = -\nabla U(\mathbf{r})$ is the classical force. For s-wave collisions, which are those relevant for the trapped atoms, and are independent of energy, the collision integral can be written in the form

$$C(\mathbf{r}, \mathbf{p}) = \sigma \int \frac{d\Omega}{2\pi} \int \frac{d\mathbf{p}}{(2\pi\hbar)^3} (f_1 f_2 - f_1' f_2') \Delta v, \tag{2.65}$$

where $\Delta \mathbf{v} = (\mathbf{p}_1' - \mathbf{p}_2')/2M$ is the relative velocity of two atoms before collision, σ is the atom collision cross-section, Ω specifies the direction of scattering, and the usual notation for the Boltzmann collision integral is used [24]. This kinetic description can be simplified by assuming that in the evaporation process, the distribution function only depends on the particle energy, E. We can then use the energy distribution function $f(E)$, such that

$$f(\mathbf{r}, \mathbf{p}) = \int f(E)\delta(W(\mathbf{r}) - E)dE, \tag{2.66}$$

where $W(\mathbf{r}) = U(\mathbf{r}) + p^2/2M$. Multiplying the kinetic equation (2.64) by the delta function appearing in this relation, and integrating over $\mathbf{r}$ and $\mathbf{p}$, we are reduced to the equation

$$\rho(E)\frac{\partial f(E)}{\partial t} = C(E), \tag{2.67}$$

where we have defined

$$\rho(E) = \int d\mathbf{r} \int \frac{d\mathbf{p}}{(2\pi\hbar)^3} \delta(W(\mathbf{r}) - E). \tag{2.68}$$

This quantity can be interpreted as the energy density of the atomic states. The new Boltzmann collision integral is now

$$C(E) = \frac{M\sigma}{\pi^2\hbar^3} \int dE_1 \int dE_2 \int dE_1' \delta(E_1 + E_2 - E_1' - E)\rho(E_{min})$$
$$\left[f(E_1)f(E_2) - f(E_1')f(E)\right], \tag{2.69}$$

where $E_{min} = \min\{E_1, E_2, E_1'\}$. Assuming that the atoms with energy above a critical value U_0 evaporate, we have $f(E > U_0) = 0$. In that case, it can be shown that the kinetic equation (2.64) satisfies a truncated Boltzmann distribution of the form

$$f(E) = (n_0\lambda_T^3)e^{-E/T}H(U_0 - E), \tag{2.70}$$

where $H(x)$ is the Heaviside step-function. This solution can now be replaced in (2.66) yielding

$$f(\mathbf{r}, \mathbf{p}) = (n_0\lambda_T^3)e^{-W(\mathbf{r})/T}H[U_0 - W(\mathbf{r})]. \tag{2.71}$$

It is obvious that T cannot be, strictly speaking, the temperature of the system, because this is an open system which is not in thermal equilibrium, allowing for the energetic particles to scape. However, we can still call it a 'temperature' in the broad sense. Integration over momentum then leads to the density profile

$$n(\mathbf{r}) = n_0 e^{-W(\mathbf{r})/T}\left[\mathrm{Erf}\left(\sqrt{\zeta}\right) - 2\sqrt{\frac{\zeta}{\pi}}e^{-\zeta}\right], \tag{2.72}$$

where $\zeta = U_0 - U(\mathbf{r})/T$. Notice that this is different from the free density profile and, in particular, the number of atoms at the centre of the trap is smaller than in the absence of evaporation, $n(0) < n_0$.

Let us now consider the rate of change on the number of atoms remaining inside the trap. Using the truncated distribution (2.71) to integrate over the trapped energies $E < U_0$, we obtain from the kinetic equation (2.67)

$$\frac{dN}{dt} = -\int_{U_0}^{\infty} \rho(E)\frac{\partial f(E)}{\partial t}dE = -\frac{M\sigma}{\pi^2\hbar^3}\int dE_1 \int dE_2 \int dE_1' \, \rho(E_1')f(E_1)f(E_2), \tag{2.73}$$

where the energies E_1, E_2 and E_1' are upper bounded by U_0. Noting that in this range of integration the distribution functions are simple exponentials, we can easily get

$$\frac{dN}{dt} = -n_0^2\sigma\bar{v}e^{-\eta}V_{eff}, \tag{2.74}$$

where $\bar{v} = \sqrt{8T/\pi M}$, and V_{eff} is an effective volume determined by

$$V_{eff} = \frac{\lambda_T^3}{T}\int_0^{U_0} \left[(U_0 - E - T)e^{-E/T} + Te^{-\eta}\right]dE, \tag{2.75}$$

with $\eta = U_0/T$. This is a more refined version of Eq. (2.62). We can also compute the change of the mean kinetic energy per atom, reading

$$\frac{dW}{dt} = -\int_{U_0}^{\infty} dE\rho(E)E\frac{\partial f(E)}{\partial t}. \tag{2.76}$$

Using a similar procedure as in (2.73), we then get

$$\frac{dW}{dt} = \frac{dN}{dt}\left[U_0 + \left(1 - \frac{V'_{\text{eff}}}{V_{\text{eff}}}\right)T\right], \tag{2.77}$$

where a new effective volume appears, as

$$V'_{\text{eff}} = \frac{\lambda_T^3}{T}\int_0^{U_0}\left[Te^{-E/T} - (U_0 - E - T)e^{-\eta}\right] \tag{2.78}$$

Finally, the temperature evolution can be determined by using the equation of state, $W = CT + \mu N$, where $C = (\partial W/\partial T)_N$ is the heat capacity, and $\mu = (\partial W/\partial N)_T$ is the chemical potential. This then leads to

$$\frac{dT}{dt} = \frac{1}{C}\left[U_0 + \left(1 - \frac{V'_{\text{eff}}}{V_{\text{eff}}}\right)T - \mu\right]\frac{dN}{dt}. \tag{2.79}$$

This represents a refinement of the scaling law (2.60).

The above description can be reformulated using a quantum description. Quantum theories of evaporative cooling have been published over the years [25–29]. In this case, the Boltzmann distribution is replaced by a Bose distribution, and a similar procedure can lead to a truncated function of the form

$$f(\mathbf{r},\mathbf{p}) = \frac{H(A(\mathbf{r}) - E)}{Z^{-1}\exp(E/T) - 1}, \tag{2.80}$$

with

$$A(\mathbf{r}) = U_0 - U(\mathbf{r}) \quad \text{and} \quad Z(\mathbf{r}) = \exp\{[\mu - U(\mathbf{r})]/T\}, \tag{2.81}$$

where the effective potential can also be taken into account [29].

2.7 Sympathetic Cooling

The evaporative cooling technique can be applied to species that can be brought down to low enough temperatures (typically, at the micro-Kelvin region) with laser cooling. There are others, however, for which direct laser cooling is not possible or simply ineffective (as it is the case of molecules, whose electronic structure is too complicated). It is, however, possible to make use of inter-particle interactions to induce cooling to other particles which cannot be directly laser cooled. This is the physical principle underlying the *sympathetic cooling* technique. For example, atomic ions that can be directly laser cooled are used to cool nearby ions or atoms, by way of their mutual Coulomb interaction. This includes most molecular ion species, especially large organic molecules [30]. The cooling of neutral atoms through sympathetic cooling was first demonstrated by Myatt et al. in 1997 [31]. Here, a technique with electric and magnetic fields was used, where atoms with spin in one direction were more weakly confined than those with spin in the opposite direction. The weakly confined atoms with a high kinetic energy were allowed to escape faster, lowering the total kinetic energy, resulting in a cooling of the strongly confined atoms.

A theoretical understanding of sympathetic cooling is quite simple to obtain after the discussion presented in the previous section. Let us consider a gas of N_2 particles with mass M_2 as the target specie, i.e., the one to be sympathetic cooled, and a 'buffer' gas (which can be laser cooled) of $N_1 = N_1(t)$ particles of mass M_1. The total kinetic energy in an harmonic trap ($\delta = 3/2$) is then given by $U_{\text{th}} = 3(N_1 + N_2)T$. Consider now that the buffer gas is set to evaporate with an energy cutoff $U_0 = \eta T$. The corresponding variation in the energy is given by

$$\frac{dU_{\text{th}}}{dN_1} = (\eta + \kappa)T, \tag{2.82}$$

where $0 \leq \kappa \leq 1$ is a parameter depending on the cutoff (the latter definition can be found in reference [23] and depends on the experimental details of the evaporation procedure). After rethermalization, the energy shift is determined by $E + dE = (N_1 + dN_1 + N_2)(T + dT)$. This corresponds to a temperature variation of the form

$$\frac{dT}{T} = \tilde{\alpha}\frac{dN_1}{N_1 + N_2}, \tag{2.83}$$

where $\tilde{\alpha} = (\kappa + \eta)/3 - 1$ is the new evaporation parameter. Assuming a constant cutoff η, we have

$$T = T_{\text{min}}\left(\frac{N_1 + N_2}{N_2}\right)^{\tilde{\alpha}}, \quad \text{with} \quad T_{\text{min}} = T(0)\left(\frac{N_2}{N_1(0)}\right)^{\tilde{\alpha}}. \tag{2.84}$$

As it immediately follows from Eq. (2.84), the minimum temperature T_{min} is reached when all the buffer atoms are evaporated ($N_1 = 0$). Because we are mainly interested in bringing the target specie down to quantum degeneracy, it is relevant to consider the phase space density $\varrho_j = 2.17 N_j \hbar\omega_j / T$ [32], where $j = 1, 2$ and ω_j are the trapping frequencies. Using the latter result, we can explicitly compute the phase-space of both the buffer and target gas

$$\varrho_j = 2.17 N_j \left(\frac{\hbar\omega_2}{T}\right)^3 \left[\frac{N_1(0)}{N_1 + N_2}\right]^{3\tilde{\alpha}}. \tag{2.85}$$

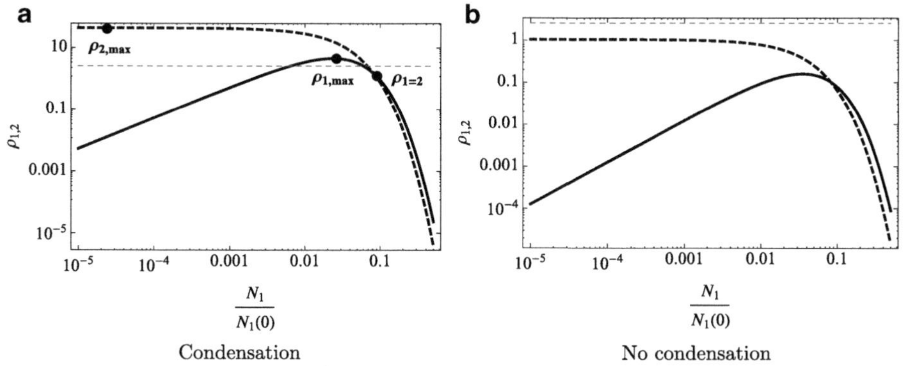

Fig. 2.5 Evolution of the phase-space density of both buffer (*dashed line*) and target (*full line*) atoms. The *red dashed line* corresponds to the phase-space density necessary to occur Bose-Einstein condensation ($\varrho_{\text{BEC}} = 2.612$). (**a**) Condensation of both species. (**b**) No condensation. The parameters are discussed in the text and correspond to the experiment performed in Ref. [32]

After the evaporation of all the buffer atoms, we observe that the phase space density for the target atoms reaches its maximum value

$$\varrho_2^{\max} = \frac{2.17}{N_2^{(3\tilde{\alpha}-1)}}\left[N_1(0)^{\tilde{\alpha}}\frac{\hbar\omega_2}{T(0)}\right]^3, \tag{2.86}$$

where we have assumed that no evaporation of the target atoms occur and therefore N_2 is constant. Also, the buffer phase-space density increases first to a maximum value

$$\varrho_1^{\max} = \varrho_2^{\max}\left(\frac{\omega_1}{\omega_2}\right)^3 \frac{(3\tilde{\alpha}-1)^{(3\tilde{\alpha}-1)}}{3\tilde{\alpha}^{(3\tilde{\alpha})}} \tag{2.87}$$

and then decreases to zero. Another important value corresponds to the point where the phase-space density of the two species are equal

$$\varrho_{1=2} = \varrho_2^{\max}\left[1+\left(\frac{\omega_1}{\omega_2}\right)^3\right]^{-3\tilde{\alpha}}. \tag{2.88}$$

Notice that the relative positions between $\varrho_1^{\max}$, $\varrho_2^{\max}$ and $\varrho_{1=2}$ can be modified by changing the different evaporation parameters. In Fig. 2.5, the curves $\varrho_{1,2}$ in Eq. (2.85) are depicted for the evaporation of $N_1(0) = 10^8$ atoms to sympathetically cool down $N_2 = 6.5 \times 10^7$ target atoms. For definiteness, we use the experimental parameters of Ref. [32], with $\omega_2/\omega_1 = 2$, an initial temperature of $T(0) = 200\,\mu\text{K}$ and $3\tilde{\alpha} - 1 > (\omega_1/\omega_2)^3$ to assure that both the species can be condensed (i.e., $\varrho_{1,2} > \varrho_{\text{BEC}} = 2.612$). The physical meaning of this critical value will be explained in future chapters. If the mutual condensation condition $3\tilde{\alpha} - 1 > (\omega_1/\omega_2)^3$ is

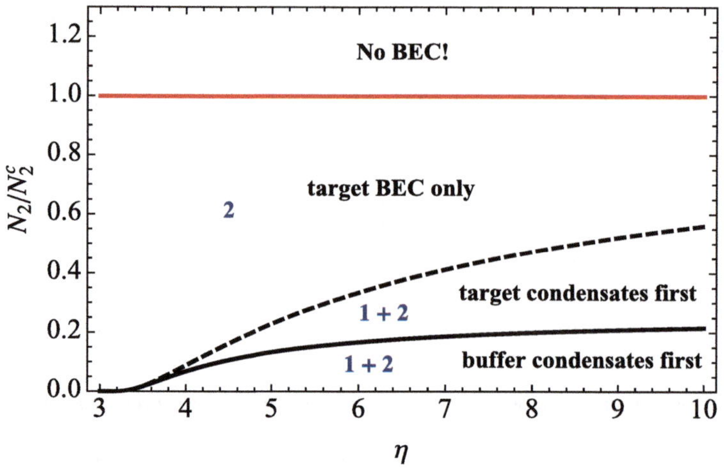

Fig. 2.6 Four different scenarios illustrating the features of sympathetic cooling, obtained for $\omega_2 = \sqrt{2}\omega_1$

satisfied, the buffer condenses first if $\varrho_{1,2} = 2.612$, otherwise the target condenses first. These previous conditions can be established as a function of the number of target atoms N_2. This allows us to define three critical numbers, N_2^a, N_2^b and N_2^c, for which $\varrho_{1=2}$, $\varrho_1^{\max}$ and $\varrho_2^{\max}$ simultaneously equal 2.612. In Fig. 2.6, we represent the critical curves N_2^a/N_2^c and N_2^b/N_2^c as a function of the cutoff η.

Another important property to be analyzed is the rate at which the cooling process occurs. So far, we have assumed that both the buffer and target atoms possess the same temperature (this is valid if the evaporation occurs slowly enough so the system can continuously thermalize). Let us now define the temperature average and difference between the two species, respectively given by $\Delta T = T_1 - T_2$ and $\bar{T} = (T_1 + T_2)/2$. Assuming that both specie obey a Maxwell-Boltzmann distribution, it is possible to compute the energy exchange $\Delta U_{\text{th}} = -2\Gamma\Delta T$, where Γ is the number of interspecies collisions per unit time, which can be approximately given by[2]

$$\Gamma \simeq \frac{N_1 N_2}{\pi^2 R_x R_y R_z} \sigma_{12} \bar{v}_{12}, \tag{2.89}$$

where $\bar{v}_{12} = \sqrt{k_B(T_1 + T_2)/\mathcal{M}}$ and $R_j = \sqrt{(k_B/\mathcal{M})(T_1/\omega_{1,j}^2 + T_2/\omega_{2,j}^2)}$, with $j = x, y, z$. Here, we have introduced the reduced mass

[2]For experimental reasons, Γ should be multiplied by a factor $e^{-\Delta^2/R_z^2}$, where $\Delta = g/\omega_{1,z}^2 - g/\omega_{2.z}^2$ represents the gravitational sag between the two clouds and $R_z = \sqrt{(k_B/\mathcal{M})(T_1/\omega_{1,z}^2 + T_2/\omega_{2,z}^2)}$ is the vertical size of the clouds. Near condensation, however, the effect of the sag is negligible and Eq. (2.89) holds. See Ref. [32] and references therein for further details.

$$\mathcal{M} = \frac{8(M_1 M_2)^2}{2(M_1 + M_2)^3}. \tag{2.90}$$

From this definition, we can easily write (for equal trapping frequencies, $\omega = \omega_1 = \omega_2$)

$$\frac{1}{\tau_{12}} = \frac{1}{\Delta T}\frac{d\Delta T}{dt} = \frac{N_1 + N_2}{3T}\frac{\omega^3 \sigma_{12} \mathcal{M}}{2\pi^2}. \tag{2.91}$$

It is possible to verify that this result is similar to that obtained in for a single specie, as discussed in the previous section. Noticing that we are here considering a trapped gas, Eq. (2.62) should be modified by replacing n_0 by $(N_1 + N_2)/\lambda_{\text{ho}}$, which would simply yield [33] $\tau_{col}^{-1} = \gamma/3$, where $\gamma = N\omega^3 \sigma M/2\pi^2 T$ is the average collision rate in a harmonic trap. There are species for which the collision frequencies σ and σ_{12} interspecies are very similar, as it is the case of ^{87}Rb [34]. This implies that

$$\frac{1}{\tau_{12}} = \frac{1}{\tau_{\text{col},1}} + \frac{1}{\tau_{\text{col},2}}, \tag{2.92}$$

which remarkably states that thermalization between two different species is faster than thermalization of a single specie. And this is the key feature of sympathetic cooling: if the evaporation ramp is applied to cooling of the buffer alone, interspecies thermalization will remain efficient.

We stress that although we have considered the case of bosonic species for definiteness, this fact does not limit the present discussion and the same theory can be applied to the case of fermions. The only question that should be addressed in the latter is the absence of condensation.

References

1. M.O. Scully, S. Zubairy, *Quantum Optics* (Cambridge University Press, Cambridge/New York, 1997)
2. R. Loudon, *The Quantum Theory of Radiation* (Noth-Holland, Amsterdam, 1958)
3. C. Cohen-Tannoudji, J. Dupont-Roc, G. Grynberg, *Photons and Atoms: Introduction to Quantum Electrodynamics* (Wiley, New York, 1987)
4. S. Chu, L. Hollberg, J.E. Bjorkholm, A. Cable, A. Ashkin, Phys. Rev. Lett. **55**, 48 (1985)
5. S. Stenholm, The semiclassical theory of laser cooling. Rev. Mod. Phys. **58**, 699 (1986)
6. A.L. Migdall, J.V. Prodan, W.D. Philips, T.H. Bergeman, H.J. Metcalf, Phys. Rev. Lett., **54**, 2596 (1985)
7. J. Fortágh, C. Zimmermann, Magnetic microtraps for ultracold atoms. Rev. Mod. Phys. **79**, 235 (2007)
8. Yu.V. Gott, M.S. Yoffe, V.G. Tel'kovskii, Nucl. Fusion Suppl. Part 3, 1045 (1962)
9. D.E. Pritchard, Phys. Rev. Lett. **51**, 1336 (1983)
10. E.L. Raab, M. Printiss, A. Cable, S. Chu, D. Pritchard, Phys. Rev. Lett. **59**, 2631 (1987)
11. J. Dalibard, C. Cohen-Tannoudji, J. Opt. Soc. Am. B **6**, 2023 (1989)
12. C. Salomon, J. Dalibard, W. Phillips, A. Clairon, S. Guellati, Europhys. Lett. **12**, 683 (1990)

13. NIST Database, http://physics.nist.gov/PhysRefData/ASD/lines$_-$form.html (2009)
14. H.F. Hess, Phys. Rev. B **34**, 3476 (1986)
15. K.B. Davis, M.-O. Mewes, W. Ketterle, Appl. Phys. B **60**, 155 (1995)
16. W. Ketterle, N.J. van Druten, Adv. At. Mol. Opt. Phys. **37**, 181 (1996)
17. K.B. Davis et al., Phys. Rev. Lett. **75**, 3969 (1995)
18. J.M. Doyle, J.C. Sandberg, I.A. Yu, C.L. Cesar, D. Kleppner, T.J. Greytak, Phys. B **194**, 13 (1994)
19. K.M. O'Hara, M.E. Gehm, S.R. Granate, J.E. Thomas, Phys. Rev. A **64**, 051403(R) (2001)
20. P.J. Tol, W. Hogerworst, W. Vassen, Phys. Rev. A **70**, 013404 (2004)
21. D. Comparat, A. Fioretti, G. Stern, E. Dimova, B. Laburthe Tolra, P. Pillet, Phys. Rev. A **73**, 043410 (2006)
22. K. Berg-Sorensen, Phys. Rev. A **55**, 1281 (1997)
23. O.J. Luiten, M.W. Reynolds, J.T.M. Walraven, Phys. Rev. A **53**, 381 (1996)
24. R. Balescu, *Equilibrium and Non-equilibrium Statistical Mechanics* (Wiley, New York, 1975)
25. J. Tempere, F. Brosens, J.T. Devreese, Solid State Commun. **102**, 691 (1997)
26. M. Holland, J. Williams, J. Cooper, Phys. Rev. A **55**, 3670 (1997)
27. H. Wu, E. Arimondo, C.J. Cooper, Phys. Rev. A **55**, 3670 (1997)
28. D. Jaksch, C.W. Gardiner, P. Zoller, Phys. Rev. A **56**, 575 (1997)
29. M. Yamashita, M. Koashi, N. Imoto, Phys. Rev. A **59**, 2243 (1999)
30. S. Schiller, C. Lämmerzahl, Phys. Rev. A **68**, 053406 (2003)
31. C.J. Myatt, E.A. Burt, R.W. Ghrist, E.A. Cornell, C.E. Wieman, Phys. Rev. Lett. **78**, 586 (1997)
32. C. Delannoy, S.G. Murdoch, V. Boyer, V. Josse, P. Bouyer, A. Aspect, Phys. Rev. A **63**, 051602(R) (2001)
33. M. Arndt et al., Phys. Rev. Lett. **79**, 625 (1997); H. Wu, C.J. Foot, J. Phys. B **29**, L321 (1996); C.R. Monroe et al., Phys. Rev. Lett. **70**, 414 (1993); D.W. Snoke, J.P. Wolfe, Phys. Rev. B **39**, 4030 (1989)
34. P.S. Julienne et al., Phys. Rev. Lett. **78**, 1880 (1997)

Chapter 3
Wave Kinetic Approach

Here we give an introductory account of the wave kinetic theory. Its basic ingredients are the Wigner function and its evolution equation. Historically it started in 1932, when Wigner proposed his function as a way to represent the quantum state of a particle in its classical phase space [1]. Later, in 1949, Moyal was able to derive an exact evolution equation for the Wigner function, starting from the Shrödinger equation [2]. In the classical limit, this evolution equation tends to the classical single particle Liouville equation. With these two pieces of knowledge, we are able today to build-up a consistent description of quantum particles in self consistent mean-fields, which are very useful to describe many different processes in quantum gases, namely, elementary excitations, collective processes and resonant interactions, as shown through many different examples in this book. This wave kinetic description has been abundantly used in the literature, and in particular for laser cooling, as discussed in the reviews [3, 4].

In the present chapter we first describe the basic aspects of the Wigner-Moyal procedure, including the concept of Weyl transformation. For a detailed review see [5]. We then illustrate this procedure by studying the interaction of a singe atom with radiation. The cases of two and three level atoms, and their relevance to laser cooling, will be considered. Attention will also be given to the quasi-classical limit, and to the links between quantum physics and the dynamical and statistical aspects of classical physics.

3.1 Wigner-Moyal Procedure

Let us consider the phase space representation of a physical system. We know that the state of a classical dynamical system can be represented by a point in phase space and, if we are dealing with an ensemble of similar systems, by a probability distribution in the same space. In contrast, a quantum system can never be represented by a single point, even when we are not considering a statistical

J.T. Mendonça and H. Terças, *Physics of Ultra-Cold Matter*, Springer Series on Atomic, Optical, and Plasma Physics 70, DOI 10.1007/978-1-4614-5413-7_3,

ensemble, because of the well known uncertainty relations. It would then be useful to find a way to represent the quantum system in the classical phase space, in order to understand the specificity of the quantum states, and the properties of their classical limits. In what follows, we motivate the definition of an appropriate distribution function able to reproduce the quantum features of a general physical system in the phase-space, by defining the Wigner function and the corresponding kinetic (or transport) equation.

3.1.1 *Quasi-distributions*

We known that the evolution of a quantum system is determined by the Schrodinger equation

$$i\hbar\frac{\partial}{\partial t}\psi = H\psi, \quad H = -\frac{\hbar^2}{2m}\nabla^2 + V(\mathbf{r}, t) \tag{3.1}$$

where $\psi \equiv \psi(\mathbf{r}, t)$ is the wave function, and H is the Hamiltonian operator for a particle of mass m submitted to a potential $V(\mathbf{r}, t)$. We also know that the probability for finding the particle at a given position $\mathbf{r}$, at a given time t, is determined by the module square of the wave function $P(\mathbf{r}, t) = |\psi(\mathbf{r}, t)|^2$. Similarly, the probability for finding the particle with a velocity $\mathbf{v}$, or a momentum $\mathbf{p} = m\mathbf{v}$ at time t, is given by the module square of the Fourier transform of this wave function $P(\mathbf{p}, t) = |\psi(\mathbf{p}, t)|^2$ where $\psi(\mathbf{p}, t)$ is defined by

$$\psi(\mathbf{p}, t) = \int \psi(\mathbf{r}, t) \exp(-i\mathbf{p} \cdot \mathbf{r}/\hbar)\, d\mathbf{r}. \tag{3.2}$$

Normalization of the wave function guarantees the conservation of the total probability, in both the coordinate and the momentum representations of the quantum states

$$\int P(\mathbf{r}, t) d\mathbf{r} = \int P(\mathbf{p}, t)\, d\mathbf{p} = 1. \tag{3.3}$$

The problem with such definitions of the quantum probability is that they cannot be used to represent the quantum state in the classical phase space of the system $(\mathbf{r}, \mathbf{p})$, because they only depend on the coordinate $\mathbf{r}$, or alternative on the momentum $\mathbf{p}$, but not on both variables simultaneously. A possible solution to this problem is the introduction of a function of both variables, as defined by

$$W(\mathbf{r}, \mathbf{p}, t) = \int \psi^*(\mathbf{r} + \mathbf{s}/2, t) \cdot \psi(\mathbf{r} - \mathbf{s}/2, t) \exp(-i\mathbf{p} \cdot \mathbf{s}/\hbar)\, d\mathbf{s}. \tag{3.4}$$

This is called the *Wigner function*. It is interesting to notice that this function contains the information given by both probabilities $P(\mathbf{r}, t)$ and $P(\mathbf{p}, t)$. By integration,

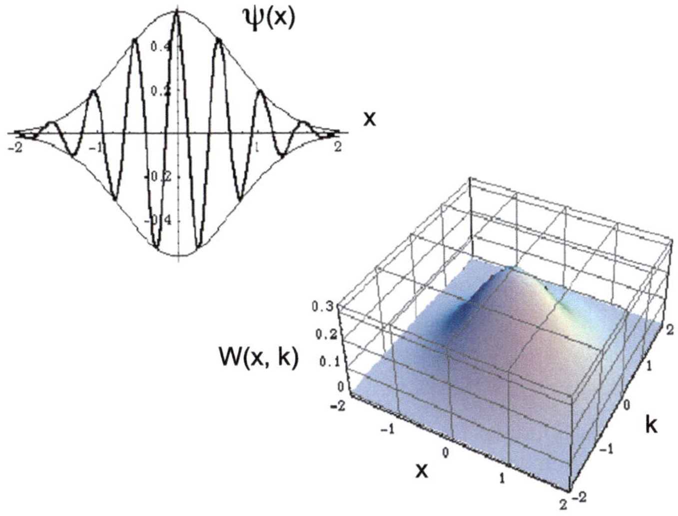

Fig. 3.1 Gaussian wave function $\psi(x)$, versus the corresponding Wigner function $W(x, k)$

we get the probability in the coordinate representation and a similar expression for the probability in the momentum representation, as given by

$$P(\mathbf{r}, t) = \int W(\mathbf{r}, \mathbf{p}, t)\, d\mathbf{p}\,, \quad P(\mathbf{p}, t) = \int W(\mathbf{r}, \mathbf{p}, t)\, d\mathbf{r}. \tag{3.5}$$

Apart from that, it contains all the information associated with the quantum correlations. In particular, if we define an auto-correlation function of the wave function as $K(\mathbf{r}_1, \mathbf{r}_2, t) = \psi(\mathbf{r}_1, t) \cdot \psi^*(\mathbf{r}_2, t)$, it can be shown that the function W is nothing but the Fourier transformation of this auto-correlation, as determined by

$$W(\mathbf{r}, \mathbf{p}, t) = \int K(\mathbf{r}, \mathbf{s}, t) \exp(-i\mathbf{p} \cdot \mathbf{s}/\hbar)\, d\mathbf{s}, \tag{3.6}$$

where $\mathbf{r} = (\mathbf{r}_1 + \mathbf{r}_2)/2$ and $\mathbf{s} = \mathbf{r}_1 - \mathbf{r}_2$. This and other interesting properties of the Wigner function will be demonstrated below. This function can then be seen as a kind of probability of a quantum system in both the coordinate and the momentum representations, and can therefore be represented in the classical phase space. It then establishes a connection between the classical and quantum formulations of dynamical and statistical physics.

However, this quantity cannot be strictly considered as a probability in the usual sense, because it is not positive definite and it can eventually take negative values. This occurs in regions of phase space where the state of the quantum system differs significantly from any classical state, and gives a signature of the irreducible nature of quantum physics. On the contrary, for quasi-classical states, its value becomes positive and tends to the usual classical probability. For this reason it is usually called a *quasi-probability* (Fig. 3.1).

In order to examine the classical representation of a quantum system in a more formal way, let us consider a pure state defined by the state vector $|\psi\rangle$. We can also introduce two identity operators, defined in the space and momentum representations, such that

$$\int |\mathbf{r}\rangle\langle\mathbf{r}| \, d\mathbf{r} = 1, \quad \int |\mathbf{k}\rangle\langle\mathbf{k}| \, \frac{d\mathbf{k}}{(2\pi)^3} = 1, \tag{3.7}$$

where $\mathbf{r}$ is the position and $\mathbf{p} = \hbar\mathbf{k}$ the momentum operators. This allows us to write the normalization condition for the state vector, $\langle\psi|\psi\rangle = 1$, as

$$\langle\psi|\psi\rangle = \int \langle\psi|\mathbf{r}\rangle\langle\mathbf{r}|\psi\rangle \, d\mathbf{r} = \int \langle\psi|\mathbf{k}\rangle\langle\mathbf{k}|\psi\rangle \frac{d\mathbf{k}}{(2\pi)^3} = 1. \tag{3.8}$$

Using the wavefunction $\psi(\mathbf{r}, t) = \langle\mathbf{r}|\psi\rangle$, or its Fourier transform $\psi(\mathbf{k}, t) = \langle\mathbf{k}|\psi\rangle$, this can also be written as

$$\langle\psi|\psi\rangle = \int \psi^*(\mathbf{r}, t)\psi(\mathbf{r}, t) \, d\mathbf{r} = \int \psi^*(\mathbf{k}, t)\psi(\mathbf{k}, t) \frac{d\mathbf{k}}{(2\pi)^3} = 1. \tag{3.9}$$

We can now introduce two additional identity operators in Eq. (3.8), and obtain the expression

$$\langle\psi|\psi\rangle = \int \frac{d\mathbf{k}}{(2\pi)^3} \int d\mathbf{r}' \int d\mathbf{r}'' \langle\psi|\mathbf{r}'\rangle\langle\mathbf{r}'|\mathbf{k}\rangle\langle\mathbf{k}|\mathbf{r}''\rangle\langle\mathbf{r}''|\psi\rangle = 1. \tag{3.10}$$

Alternatively, we could also write

$$\langle\psi|\psi\rangle = \int d\mathbf{r} \int \frac{d\mathbf{k}'}{(2\pi)^3} \int \frac{d\mathbf{k}''}{(2\pi)^3} \langle\psi|\mathbf{k}'\rangle\langle\mathbf{k}'|\mathbf{r}\rangle\langle\mathbf{r}|\mathbf{k}''\rangle\langle\mathbf{k}''|\psi\rangle = 1. \tag{3.11}$$

At this point, it is useful to introduce a well known result from elementary quantum theory which states that

$$\langle\mathbf{r}|\mathbf{k}\rangle = \langle\mathbf{k}|\mathbf{r}\rangle^* = \exp(-i\mathbf{k}\cdot\mathbf{r}). \tag{3.12}$$

This means that we can rewrite Eq. (3.10) as

$$\langle\psi|\psi\rangle = \int \frac{d\mathbf{k}}{(2\pi)^3} \int d\mathbf{r}' \int d\mathbf{r}'' \langle\psi|\mathbf{r}'\rangle\langle\mathbf{r}''|\psi\rangle \exp[-i\mathbf{k}\cdot(\mathbf{r}' - \mathbf{r}'')] = 1. \tag{3.13}$$

Similarly, from Eq. (3.11), we get

$$\langle\psi|\psi\rangle = \int d\mathbf{r} \int \frac{d\mathbf{k}'}{(2\pi)^3} \int \frac{d\mathbf{k}''}{(2\pi)^3} \langle\psi|\mathbf{k}'\rangle\langle\mathbf{k}''|\psi\rangle \exp[+i\mathbf{r}\cdot(\mathbf{k}' - \mathbf{k}'')] = 1. \tag{3.14}$$

Let us now make a transformation of the coordinates in Eq. (3.13), from $(\mathbf{r}', \mathbf{r}'')$ to the new coordinates $(\mathbf{s}, \mathbf{r})$, as defined by

$$\mathbf{s} = \mathbf{r}'' - \mathbf{r}', \quad \mathbf{r} = \frac{1}{2}(\mathbf{r}' + \mathbf{r}''). \tag{3.15}$$

We can also use new momentum coordinates in Eq. (3.14),

$$\mathbf{q} = \mathbf{k}'' - \mathbf{k}', \quad \mathbf{k} = \frac{1}{2}(\mathbf{r}' + \mathbf{r}''). \tag{3.16}$$

As a result, we can transform both equations into the following expression for the normalization condition

$$\langle \psi | \psi \rangle = \int \frac{d\mathbf{k}}{(2\pi)^3} \int d\mathbf{r}\, W(\mathbf{r}, \mathbf{p}, t) = 1, \tag{3.17}$$

where $W(\mathbf{r}, \mathbf{p}, t)$ is nothing but the Wigner function, defined now in terms of the state vector of the quantum system, as

$$W(\mathbf{r}, \mathbf{p}, t) = \int \langle \mathbf{r} + \mathbf{s}/2 | \psi \rangle \langle \psi | \mathbf{r} - \mathbf{s}/2 \rangle \exp(i\mathbf{k} \cdot \mathbf{s}) d\mathbf{s} \tag{3.18}$$

or alternatively, as

$$W(\mathbf{r}, \mathbf{p}, t) = \int \langle \mathbf{k} + \mathbf{q}/2 | \psi \rangle \langle \psi | \mathbf{k} - \mathbf{q}/2 \rangle \exp(-i\mathbf{q} \cdot \mathbf{r}) \frac{d\mathbf{q}}{(2\pi)^3}. \tag{3.19}$$

3.1.2 Weyl Transformation

Let us now consider the use of the Wigner function as a probability density, allowing in particular to calculate the mean value of a given quantum operator $\hat{A}$. By definition, we have $\langle \hat{A} \rangle = \langle \psi | \hat{A} | \psi \rangle$. We can follow a procedure similar to that used above for the normalization condition. Introducing four different identity operators in the form (3.7), we can write such a mean value as

$$\langle \hat{A} \rangle = \int \frac{d\mathbf{k}'}{(2\pi)^3} \int \frac{d\mathbf{k}''}{(2\pi)^3} \int d\mathbf{r}' \int d\mathbf{r}'' \langle \psi | \mathbf{r}' \rangle \langle \mathbf{r}' | \mathbf{k}' \rangle \langle \mathbf{k}' | \hat{A} | \mathbf{k}'' \rangle \langle \mathbf{k}'' | \mathbf{r}'' \rangle \langle \mathbf{r}'' | \psi \rangle \tag{3.20}$$

or equivalently, as

$$\langle \hat{A} \rangle = \int \frac{d\mathbf{k}'}{(2\pi)^3} \int \frac{d\mathbf{k}''}{(2\pi)^3} \int d\mathbf{r}' \int d\mathbf{r}'' \langle \psi | \mathbf{k}' \rangle \langle \mathbf{k}' | \mathbf{r}' \rangle \langle \mathbf{r}' | \hat{A} | \mathbf{r}'' \rangle \langle \mathbf{r}'' | \mathbf{k}'' \rangle \langle \mathbf{k}'' | \psi \rangle. \tag{3.21}$$

Using the identity (3.12), we can also write Eq. (3.20) as

$$\langle \hat{A} \rangle = \int \frac{d\mathbf{k}'}{(2\pi)^3} \int \frac{d\mathbf{k}''}{(2\pi)^3} \int d\mathbf{r}' \int d\mathbf{r}'' \langle \psi | \mathbf{r}' \rangle \langle \mathbf{r}'' | \psi \rangle \langle \mathbf{k}' | \hat{A} | \mathbf{k}'' \rangle e^{-i\mathbf{r}'\cdot\mathbf{k}' + i\mathbf{r}''\cdot\mathbf{k}''} \tag{3.22}$$

Alternatively, we could transform Eq. (3.21) into

$$\langle \hat{A} \rangle = \int \frac{d\mathbf{k}'}{(2\pi)^3} \int \frac{d\mathbf{k}''}{(2\pi)^3} \int d\mathbf{r}' \int d\mathbf{r}'' \langle \psi | \mathbf{k}' \rangle \langle \mathbf{k}'' | \psi \rangle \langle \mathbf{r}' | \hat{A} | \mathbf{r}'' \rangle e^{i\mathbf{r}'\cdot\mathbf{k}' - i\mathbf{r}''\cdot\mathbf{k}''}. \tag{3.23}$$

On the other hand, it can be seen from the coordinate and momentum transformations (3.15) and (3.16) that the following relation is true

$$\exp(-i\mathbf{r}' \cdot \mathbf{k}' + i\mathbf{r}'' \cdot \mathbf{k}'') = \exp(i\mathbf{k} \cdot \mathbf{s} + i\mathbf{r} \cdot \mathbf{q}). \tag{3.24}$$

This allows us to write, starting from any of the Eqs. (3.22) and (3.23), an expression for the mean value in terms of the Wigner function, as

$$\langle \hat{A} \rangle = \int \frac{d\mathbf{k}}{(2\pi)^3} \int d\mathbf{r}\, a(\mathbf{r}, \mathbf{k})\, W(\mathbf{r}, \mathbf{p}, t), \tag{3.25}$$

where we have introduced a new function $a(\mathbf{r}, \mathbf{k})$, defined by the following integral over the momentum coordinate $\mathbf{q}$, as

$$a(\mathbf{r}, \mathbf{k}) = \int \langle \mathbf{k} - \mathbf{q}/2 | \hat{A} | \mathbf{k} + \mathbf{q}/2 \rangle \exp(i\mathbf{q} \cdot \mathbf{r}) \frac{d\mathbf{q}}{(2\pi)^3}, \tag{3.26}$$

or identically, by an integral over the space coordinate $\mathbf{s}$, as

$$a(\mathbf{r}, \mathbf{k}) = \int \langle \mathbf{r} - \mathbf{s}/2 | \hat{A} | \mathbf{r} + \mathbf{s}/2 \rangle \exp(-i\mathbf{k} \cdot \mathbf{s}) d\mathbf{s}. \tag{3.27}$$

This quantity $a(\mathbf{r}, \mathbf{k})$ is clearly a classical quantity, and is known as the *Weyl tranformation* of the quantum operator $\hat{A}$. It can be seen as the classical counterpart of the operator $\hat{A}$, in the same way as the Wigner function W is the classical counterpart of the state vector $|\psi\rangle$.

It should be noticed that Eq. (3.26) determines the mean value of the quantum operator $\langle \hat{A} \rangle = \langle \psi | \hat{A} | \psi \rangle$ by using a classical averaging procedure, where W plays the role of a classical probability distribution and the Weyl transformation $a(\mathbf{r}, \mathbf{k})$ replaces the corresponding quantum operator. In the particular case where $\hat{A}$ is just the identity operator, Eq. (3.26) is reduced to the normalization condition (3.17). Similarly, the Wigner function can be seen as the Weyl transformation of the identity operator.

3.1.3 Wave Kinetic Equation

It is possible to derive an evolution equation for the Wigner function $W(\mathbf{r}, \mathbf{p}, t)$. We can call it the wave kinetic equation. It is exactly equivalent to the Schrödinger equation and carries the same amount of information. Moreover, if we take the classical limit, the wave kinetic equation can be reduced to a Liouville equation, thus establishing an interesting and useful link between Quantum Mechanics and Classical Statistical Physics.

Our first step is to consider the auto-correlation function for the wave function, defined above as $K \equiv K(\mathbf{r}_1, \mathbf{r}_2, t) = \psi(\mathbf{r}_1, t) \cdot \psi^*(\mathbf{r}_2, t)$. This suggests us to consider two arbitrary positions $\mathbf{r}_j$, for $j = 1, 2$, and write two identical versions of the Schrödinger equation for these variables, as

$$i\hbar \frac{\partial}{\partial t} \psi_j = H_j \psi_j, \tag{3.28}$$

where we have used the notation $\psi_j \equiv \psi(\mathbf{r}_j, t)$, and the Hamiltonian operator

$$H_j = -\frac{\hbar^2}{2m} \frac{\partial^2}{\partial \mathbf{r}_j^2} + V(\mathbf{r}_j, t). \tag{3.29}$$

Multiplying the equation for $j = 1$ by ψ_2^* and the complex conjugate of the equation for $j = 2$ by ψ_1, we obtain after subtraction

$$i\hbar \frac{\partial}{\partial t} K = (H_1 - H_2) K, \tag{3.30}$$

where we have assumed that the potential function $V(\mathbf{r}, t)$ is real. In order to proceed further, let us introduced two new space variables $\mathbf{r}$ and $\mathbf{s}$ defined by

$$\mathbf{r} = \frac{1}{2}(\mathbf{r}_1 + \mathbf{r}_2), \quad \mathbf{s} = \mathbf{r}_1 - \mathbf{r}_2, \tag{3.31}$$

or alternatively

$$\mathbf{r}_1 = (\mathbf{r} + \mathbf{s}/2), \quad \mathbf{r}_2 = (\mathbf{r} - \mathbf{s}/2). \tag{3.32}$$

This means that we have the differential operators

$$\frac{\partial}{\partial \mathbf{r}_1} = \frac{1}{2} \frac{\partial}{\partial \mathbf{r}} + \frac{\partial}{\partial \mathbf{s}}, \quad \frac{\partial}{\partial \mathbf{r}_2} = \frac{1}{2} \frac{\partial}{\partial \mathbf{r}} - \frac{\partial}{\partial \mathbf{s}} \tag{3.33}$$

Replacing them in the evolution Eq. (3.30), we obtain

$$i\hbar \frac{\partial}{\partial t} K = \left(-\frac{\hbar^2}{2m} \frac{\partial}{\partial \mathbf{r}} \cdot \frac{\partial}{\partial \mathbf{s}} + \Delta V \right) K, \tag{3.34}$$

where we have used the auxiliary quantity

$$\Delta V \equiv (V_1 - V_2) = V(\mathbf{r} + \mathbf{s}/2, t) - V(\mathbf{r} - \mathbf{s}/2, t). \tag{3.35}$$

At this point, it should be noticed that, for a given function $V(x + \Delta x)$, we can make a Taylor expansion around x, such that

$$V(x + \Delta x) = V(x) + \sum_{n=1}^{\infty} \frac{(\Delta x)^n}{n!} \frac{\partial^n}{\partial x^n} V(x). \tag{3.36}$$

Using the series expansion of an exponential function, we can also rewrite this equation in terms of an exponential operator, as

$$V(x + \Delta x) = \exp\left(\Delta x \frac{\partial}{\partial x}\right) V(x). \tag{3.37}$$

A triple expansion in the coordinates will then lead to the following expression for the potential functions V_j

$$V_j \equiv V(\mathbf{r} \pm \mathbf{s}/2, t) = \exp\left(\pm \frac{\mathbf{s}}{2} \cdot \nabla\right) \tag{3.38}$$

where the plus and the minus signs pertain to $j = 1$ and $j = 2$ respectively. With this result in mind, we can write the quantity ΔV in the compact form

$$\Delta V = 2 \sinh\left(\frac{1}{2} \mathbf{s} \cdot \nabla\right), \tag{3.39}$$

where the hyperbolic sine operator can be developed as

$$\sinh \alpha = \sum_{l=0}^{\infty} \frac{1}{(2l + 1)!} \alpha^{2l+1}. \tag{3.40}$$

This allow us to transform the evolution equation for the auto-correlation function into

$$\left[i\hbar \frac{\partial}{\partial t} + \frac{\hbar^2}{2m} \frac{\partial}{\partial \mathbf{r}} \cdot \frac{\partial}{\partial \mathbf{s}} - 2 \sinh\left(\frac{1}{2} \mathbf{s} \cdot \nabla\right) V \right] K = 0. \tag{3.41}$$

At this point, it is useful to introduce the Fourier transformation of K, in the variable $\mathbf{s}$, as defined by

$$W \equiv W(\mathbf{r}, \mathbf{k}, t) = \int K \exp(-i\mathbf{k} \cdot \mathbf{s})\, d\mathbf{s}. \tag{3.42}$$

Inversely, the Fourier integral given by

$$K(\mathbf{r}, \mathbf{s}, t) = \int W \exp(i\mathbf{k} \cdot \mathbf{s}) \frac{d\mathbf{k}}{(2\pi)^3}. \tag{3.43}$$

Replacing this in Eq. (3.41), we obtain

$$i\hbar \left(\frac{\partial}{\partial t} + \frac{\hbar}{m}\mathbf{k} \cdot \frac{\partial}{\partial \mathbf{r}} \right) W = 2 \sum_{l=0}^{\infty} \frac{W}{(2l+1)!} \left(\frac{\mathbf{s}}{2} \cdot \nabla \right)^{2l+1} V. \tag{3.44}$$

However, from the definition of W in (3.42), we can also write

$$\frac{\partial^m W}{\partial \mathbf{k}^m} = (-i\mathbf{s})^m W. \tag{3.45}$$

Using this result and the well known series expansion of the sine function

$$\sin x = \sum_{l=0}^{\infty} (-1)^l \frac{x^{2l+1}}{(2l+1)!}, \tag{3.46}$$

we finally get the evolution equation for the Wigner function, first derived by Moyal [2], in the form

$$\left(\frac{\partial}{\partial t} + \frac{\hbar}{m}\mathbf{k} \cdot \frac{\partial}{\partial \mathbf{r}} \right) W = \frac{2}{\hbar} W (\sin \Lambda) V, \tag{3.47}$$

where Λ is a peculiar differential operator, which acts backwards on W and forward on V, as defined by the following expression

$$\Lambda = \frac{1}{2} \left(\frac{\partial}{\partial \mathbf{k}} \right)^{\leftarrow} \cdot \left(\frac{\partial}{\partial \mathbf{r}} \right)^{\rightarrow}. \tag{3.48}$$

This equation is exactly equivalent to the Schrödinger equation from where we have started. It can be called a *wave kinetic equation*, because it determines the evolution of the quasi-probability W (and in that sense it is similar to the kinetic equations of Classical Statistical Mechanics), but on the other hand it exactly describes the same quantum properties as the wave function.

3.1.4 The Quasi-classical Limit

It is now interesting to consider an approximate expression of the wave kinetic equation (3.48), valid in the quasi-classical limit. Before taking this limit, it should be reminded that the wavevector $\mathbf{k}$ defines a quantum state with momentum $\mathbf{p} = \hbar\mathbf{k}$. The corresponding de Broglie wavelength is then $\lambda = 2\pi/k = p/h$. If the potential

V evolves very slowly, changing significantly only over a scale-length much larger then λ, we can neglect the higher order derivatives and take the approximation

$$\sin \Lambda \simeq \Lambda. \tag{3.49}$$

In this case, the above kinetic equation reduces to the following expression

$$\left(\frac{\partial}{\partial t} + \frac{\mathbf{p}}{m} \cdot \nabla - \nabla V \cdot \frac{\partial}{\partial \mathbf{p}} \right) W = 0. \tag{3.50}$$

This simply corresponds to the total time derivative of the Wigner function

$$\frac{d}{dt} W(\mathbf{r}, \mathbf{p}, t) = 0, \tag{3.51}$$

where the total time derivative operator can be explicitly written as

$$\frac{d}{dt} \equiv \frac{\partial}{\partial t} + \mathbf{v} \cdot \nabla + \mathbf{F} \cdot \frac{\partial}{\partial \mathbf{p}}. \tag{3.52}$$

Here, $\mathbf{v} = \mathbf{p}/m$ is the velocity of the particle and $\mathbf{F} = -\nabla V$ is the total force acting on the particle. This new kinetic equation is formally identical to the Liouville equation, defined in the single particle phase space. This is also commonly called a *Vlasov equation*. The classical limit of the Wigner function can then be identified with the one-particle distribution function.

3.2 Center-of-Mass Equation

We can now illustrate the above Wigner-Moyal procedure with a few explicit examples. Our previous discussion of laser cooling was mainly qualitative and phenomenological. We are now in position to consider a more systematic description of the atom-laser interaction, and of the resulting cooling process. This will allow us to derive on a more solid basis some of the laser cooling properties discussed in the previous chapter. Our description will be semi-classical, but it can also be extended to a full quantum description.

The approach followed here is based on the concept of quantum quasi-probability, as proposed by Wigner. This approach is strictly not necessary for the understanding of basic single atom physics, and it can look at first sight as unnecessary. However, it reveals to be very useful and appropriate when applied to an interacting assembly of atoms, in particular to describe the elementary excitations and collective processes in condensed and non-condensed atomic clouds, as shown in later chapters. It is also very useful to explore the similarities between the different phases of ultra-cold matter, from the non-condensed phase, to Bose-Einstein condensates and to plasmas.

The starting point of our present discussion will be the wave equation for an isolated atom. The state of a moving atom can be generally represented by a state vector $|\psi\rangle = |\boldsymbol{\rho}, \mathbf{r}\rangle$, where $\mathbf{r}$ is the atom centre-of-mass position, and $\boldsymbol{\rho} \equiv (\boldsymbol{\rho}_1, \boldsymbol{\rho}_2, \ldots \boldsymbol{\rho}_Z)$, represents the set of electronic positions with respect to the centre of mass, for a generic atom with Z electrons. We start with the non-relativistic Schrödinger equation

$$i\hbar \frac{\partial}{\partial t} |\boldsymbol{\rho}, \mathbf{r}\rangle = H |\boldsymbol{\rho}, \mathbf{r}\rangle, \tag{3.53}$$

where the Hamiltonian operator contains three different terms

$$H \equiv H(\boldsymbol{\rho}, \mathbf{r}, t) = H_e(\boldsymbol{\rho}) + H_0(\mathbf{r}) + H'(\boldsymbol{\rho}, \mathbf{r}, t). \tag{3.54}$$

The first term $H_e(\boldsymbol{\rho})$ determines the internal electronic states of the atom (described in the Appendix). The second term represents the centre of mass energy operator, as defined by

$$H_0(\boldsymbol{\rho}) = \frac{P^2}{2M} \frac{\partial^2}{\partial \boldsymbol{r}^2} + V_0(\boldsymbol{r}), \tag{3.55}$$

where M is the mass of the atom, $\mathbf{P} = -i\hbar \partial / \partial \mathbf{r}$ the centre of mass momentum, and $V_0(\mathbf{r})$ is a static confining potential. Finally, the third term in (3.53) represents the interaction Hamiltonian, which couples the atom with the radiation field, and can be written (within the dipole approximation) as

$$H'(\boldsymbol{\rho}, \mathbf{r}, t) = \frac{e}{m} \mathbf{p} \cdot \mathbf{A}, \tag{3.56}$$

where e and m are the electron charge and mass, and $\mathbf{p}$ is the total electron momentum, as defined by

$$\mathbf{p} = \sum_{j=1}^{Z} \mathbf{p}_j \equiv \sum_{j=1}^{Z} \frac{\partial}{\partial \boldsymbol{\rho}_j}. \tag{3.57}$$

In the dipole approximation, the radiation field does not depend on the internal electron coordinates $\boldsymbol{\rho}_j$, and can be represented as

$$\mathbf{A} \equiv \mathbf{A}(\mathbf{r}, t) = \int \mathbf{A} \exp(i\mathbf{k} \cdot \mathbf{r} - i\omega_k t) \frac{d\mathbf{k}}{(2\pi)^3}, \tag{3.58}$$

and $\boldsymbol{\mathcal{E}} = -\partial \mathbf{A} / \partial t$ is the electric field. Here, the frequency $\omega_k \equiv \omega(\mathbf{k})$ is determined by the photon dispersion relation. In vacuum, we simply have $\omega(\mathbf{k}) = kc$. This radiation field could in principle be divided in two distinct parts, $\mathbf{A} = \bar{\mathbf{A}} + \delta\mathbf{A}$, where the regular part $\bar{\mathbf{A}}$ represents the coherent radiation field, associated for instance with the laser field, and $\delta\mathbf{A}$ represents the electromagnetic noise which can be present in the atomic confinement region. This incoherent radiation can also be used to phenomenologically describe the quantum vacuum fluctuations, associated with the spontaneous decay. For simplicity and consistency, this incoherent term will be ignored here, but its influence will be discussed when appropriate. In particular it leads to a random force acting on the atom, which will be considered later.

We now assume that the internal electronic structure of the atom is independent of its centre of mass motion. This allows us to use $|\boldsymbol{\rho}, \mathbf{r}\rangle = |\boldsymbol{\rho}\rangle|\mathbf{r}\rangle$. This is valid for very different time scales governing the radiative transitions and the atomic motion, as discussed below. We can then use the energy representation for the electronic states, and in the spirit of perturbation theory, we write

$$|\boldsymbol{\rho}\rangle = \sum_n C_n(t)|n\rangle \exp(-iE_n t/\hbar), \tag{3.59}$$

where E_n are the energy eigenvalues of the internal atomic states, $|n\rangle$ the corresponding state vectors (where n represents a set of adequate quantum numbers), and the coefficients $C_n(t)$ obey the normalization condition

$$\sum_n |C_n(t)|^2 = 1. \tag{3.60}$$

In the absence of laser cooling, $H' = 0$, these coefficients would be constant. Replacing this in the wave equation (3.53), noting that $H_e|n\rangle = E_n|n\rangle$, and using the orthogonality condition $\langle m|n\rangle = \delta_{mn}$, we can easily obtain

$$i\hbar\left(\frac{\partial C_n}{\partial t}\right)|\mathbf{r}\rangle + i\hbar C_n \frac{\partial}{\partial t}|\mathbf{r}\rangle = C_n H_0|\mathbf{r}\rangle + \sum_l H'_{nl} C_l |\mathbf{r}\rangle \exp\left[-i(E_l - E_n)t/\hbar\right], \tag{3.61}$$

with the interaction matrix elements

$$H'_{n1} = \langle n|H'|1\rangle \equiv \int \psi_n^*(\boldsymbol{\rho}) H'(\boldsymbol{\rho}, \mathbf{r}, t)\psi_l(\boldsymbol{\rho}) d\boldsymbol{\rho}, \tag{3.62}$$

where $d\boldsymbol{\rho} \equiv \Pi_j d\boldsymbol{\rho}_j$, and $\psi_n(\boldsymbol{\rho}) = \langle \boldsymbol{\rho}|n\rangle$ are the electronic wavefunctions. Introducing the Bohr frequencies $\omega_{nl} = (E_n - E_l)/\hbar$, and defining new wavefunctions

$$\Psi_n(\mathbf{r}, t) = C_n(t)|\mathbf{r}\rangle, \tag{3.63}$$

we finally obtain

$$i\hbar\frac{\partial}{\partial t}\Psi_n = H_0\Psi_n + \sum_l H'_{nl}\Psi_l \exp\left(i\omega_{nl}t\right). \tag{3.64}$$

This is the centre of mass wave equation, which is the starting point of our model. Before concluding this section, we should notice that the interaction matrix elements H'_{nl} can also be written in the form

$$H'_{nl} \equiv H'_{nl}(\mathbf{r}, t) = -i \int \omega_k A(\mathbf{k}) p_{nl} \exp(i\mathbf{k} \cdot \mathbf{r} - i\omega_k t), \tag{3.65}$$

where we have used the dipole matrix elements p_{nl} of the appropriate radiative transition.

3.3 Wave Kinetic Description of the Laser-Atom Interaction

We now show that the standard quantum mechanical description of atom radiation interaction, as summarized in the previous section, is equivalent to a wave kinetic description based on correlation functions for centre of mass states of motion. In particular, Eq. (3.64) can be transformed into a wave kinetic equation of the Wigner-Moyal type by introducing the correlation function $K(\mathbf{r}, \mathbf{s}, t) = \langle \mathbf{r} + \mathbf{s}/2, t | \mathbf{r} - \mathbf{s}/2, t \rangle$. Taking its Fourier transformation, we arrive at the Wigner function for the centre of mass motion

$$W(\mathbf{r}, \mathbf{q}, t) = \int K(\mathbf{r}, \mathbf{s}, t) \exp(-i\mathbf{q} \cdot \mathbf{s}) \, d\mathbf{s}. \tag{3.66}$$

It is also useful to define a Wigner matrix, with elements

$$W_{nl}(\mathbf{r}, \mathbf{q}, t) = \rho_{nl}(t) W(\mathbf{r}, \mathbf{q}, t), \tag{3.67}$$

where $\rho_{nl} = \langle n | \boldsymbol{\rho} | l \rangle$ are the density matrix elements for the internal atomic states. For pure states, we simply have $\rho_{nl} = C_n^*(t) C_l(t)$.

Following the Wigner-Moyal procedure described earlier in this chapter, we can now derive an evolution equation for the matrix elements W_{nl}. Starting from Eq. (3.64), and noting the definition of $H_0(\mathbf{r})$, we obtain

$$\left(\frac{\partial}{\partial t} + \mathbf{v} \cdot \frac{\partial}{\partial \mathbf{r}} \right) W_{mn} = -\frac{i}{\hbar} \int V_0(\mathbf{k}) \Delta W_{mn} e^{i\mathbf{k} \cdot \mathbf{r}} \frac{d\mathbf{k}}{(2\pi)^3} + \tag{3.68}$$
$$i \sum_l \int h_{ml}(\mathbf{k}) \Delta W_{ml} e^{i\mathbf{k} \cdot \mathbf{r} - i\Delta_{ml} t} \frac{d\mathbf{k}}{(2\pi)^3},$$

where $V_0(\mathbf{k}')$ are the Fourier components of the external potential $V_0(\mathbf{r})$, and we have used the following quantities

$$h_{ml}(\mathbf{k}) = \frac{\omega}{2\hbar} p_{nl} A(\mathbf{k}), \quad \Delta_{ml} = (\omega_{ml} - \omega). \tag{3.69}$$

In Eq. (3.67), we have also defined the quantities

$$\Delta W_{mn} = \left[W_{ml}(\mathbf{r}, \mathbf{q} - \mathbf{k}/2, t) - W_{ml}(\mathbf{r}, \mathbf{q} + \mathbf{k}/2, t) \right]. \tag{3.70}$$

We should notice that, in this equation, the atomic wavevector $\mathbf{q}$ is simply related to the centre-of-mass velocity $\mathbf{v}$ by $\mathbf{q} = M\mathbf{v}/\hbar$, where M is the mass of the atom. This equation can also be written in a more compact form as

$$\left(\frac{\partial}{\partial t} + \mathbf{v} \cdot \frac{\partial}{\partial \mathbf{r}}\right) W_{mn} = \sum_l \int V_{ml}(\mathbf{k}) \Delta W_{ml} e^{i\mathbf{k}\cdot\mathbf{r}} \frac{d\mathbf{k}}{(2\pi)^3}, \tag{3.71}$$

where we have defined

$$V_{ml}(\mathbf{k}) = -\frac{i}{\hbar} V_0(\mathbf{k}) \delta_{nl} + i h_{ml}(\mathbf{k}) e^{i\Delta_{ml} t}. \tag{3.72}$$

The first term in this expression represents the influence of the external potential $V_0(\mathbf{r})$, and the second term is due to the interaction with the radiation field $\mathbf{A}$. From here, we can derive an evolution equation for the Wigner function W, as defined by Eq. (3.66). Taking the sum of all the diagonal elements of (3.72), with $m = n$, and noting that $\sum_n \rho_{nn} = 1$, we obtain

$$\left(\frac{\partial}{\partial t} + \mathbf{v} \cdot \frac{\partial}{\partial \mathbf{r}}\right) W = -i f_0(W) + i \sum_{n,l} \rho_{nl} \int h_{nl}(\mathbf{k}) \Delta W \exp\left[i\mathbf{k}\cdot\mathbf{r} + i\Delta_{nl}(\mathbf{k})t\right] \frac{d\mathbf{k}}{(2\pi)^3}, \tag{3.73}$$

with the external force term defined in terms of the Wigner function

$$f_0(W) = \frac{1}{\hbar} \int V_0(\mathbf{k}) \Delta W e^{i\mathbf{k}\cdot\mathbf{r}} \frac{d\mathbf{k}}{(2\pi)^3}. \tag{3.74}$$

This corresponds to the wave kinetic equation for the centre-of-mass motion. It includes an exact description of the energy and momentum exchange between the atom and the radiation field. In particular it describes what is usually called the *recoil effect*, which is a specific quantum effect. It should be noticed that such recoil can occur not only due to the atom interaction with the radiation field, but also with the low frequency acoustic modes, as will be shown in the following chapters.

Equation (3.73) can now be used to study laser cooling in different atomic configurations. Notice however that this is not a closed equation for the quasi-distribution W, because we need to determine the electronic matrix elements ρ_{nl} separately. This can be done by using a kind of *Born-Oppenheimer approximation*, where the electronic states of the atom are assumed to be independent of the centre-of-mass states, as discussed next.

3.4 Two-Level Atom

We first consider the case of a two-level atom in the presence of a single laser beam. This simple model is sufficient to describe the relevant properties of laser cooling if we ignore the Zeeman split of the atomic sub-levels. Therefore only two internal eigenstates with energies E_1 and E_2 are considered, with $\omega_{21} = (E_2 - E_1)/\hbar$. The laser field can be described by the following radiation spectrum

$$A(\mathbf{k}) = \frac{1}{2} A_0 (2\pi)^3 \left[\delta(\mathbf{k} - \mathbf{k}_0) + \delta(\mathbf{k} + \mathbf{k}_0)\right], \tag{3.75}$$

The laser frequency $\omega_0 = k_0/c$ is assumed to be nearly resonant with the atomic transition between the two energy states, $\omega_0 \simeq \omega_{21}$. All the other energy levels can be ignored. Replacing (3.75) in (3.73) we can easily get

$$\left(\frac{\partial}{\partial t} + \mathbf{v} \cdot \frac{\partial}{\partial \mathbf{r}}\right) W = -i f_0(W) + g \Delta W, \tag{3.76}$$

where the coupling coefficient is given by the expression: $g = i(h\tilde{\rho}_{21} - h^* \tilde{\rho}_{12})$, with

$$h \equiv h_{21} = \frac{\omega_0}{2\hbar} p_{21} A_0 \,, \quad \tilde{\rho}_{21} = \rho_{21} \exp\left(i\mathbf{k} \cdot \mathbf{r} + i\Delta t\right) \tag{3.77}$$

and $\Delta \equiv \Delta_{21} = \omega_{21} - \omega_0$. Here we have used the rotating wave approximation, and assumed $\Delta \ll \omega_0$. We can also write

$$g = \frac{i\omega_0 A_0}{2\hbar} \left(p_{21}\tilde{\rho}_{21} - p_{12}\tilde{\rho}_{12}\right) = -\frac{\omega_0 A_0}{\hbar} \Im(p_{21}\tilde{\rho}_{21}), \tag{3.78}$$

where we have used the properties $\tilde{\rho}_{12} = \tilde{\rho}_{21}^*$ and $p_{12} = p_{21}^*$.

In order to calculate g we need to determine ρ_{21}, which can be done by going back to Eq. (3.69). On a fast time scale, much shorter than the characteristic time for atom cooling, we can neglect the centre-of-mass motion and assume that W is nearly constant. This is equivalent to the Born-Oppenheimer approximation. On such a fast time scale, we can also neglect the influence of the confining potential, and this equation reduces to

$$\frac{d}{dt} \rho_{mn} = i \sum_l \int h_{ml} \exp(i\mathbf{k} \cdot \mathbf{r} + i\Delta_{ml}(\mathbf{k})t) \frac{d\mathbf{k}}{(2\pi)^3}. \tag{3.79}$$

For simplicity we have used $d/dt = \partial/\partial t + \mathbf{v} \cdot \partial/\partial \mathbf{r}$. In the present case of a two-level atom in the presence of the field (3.75) this reduces to the well known Bloch equations (2.20) and (2.21) of the previous chapter, with $\Omega_R = \omega_0 A_0 p_{21}/\hbar = 2h$. The steady state solutions of such equations are well known, and can be written as [6]

$$\tilde{\rho}_{21} = -\frac{1}{2} \Omega_R^* \frac{[(\Delta + \mathbf{k} \cdot \mathbf{v}) - i\Gamma/2]}{(\Delta + \mathbf{k} \cdot \mathbf{v})^2 + \Gamma^2/4 + |\Omega_R|^2/2} \tag{3.80}$$

and

$$\rho_{22} = -\frac{i}{2\Gamma} \left(\Omega_R \tilde{\rho}_{21} - \Omega_R^* \tilde{\rho}_{12}\right), \tag{3.81}$$

with the complex Rabi frequency $\Omega_R = \mathcal{E}_0 p_{21}/\hbar$, and the spontaneous decay time $1/\Gamma$. This allows us to write the coupling coefficient as $g = \Gamma\rho_{22}$, or in explicit form

$$g = \frac{\Gamma|\Omega_R|^2}{4(\Delta + \mathbf{k}\cdot\mathbf{v})^2 + \Gamma^2 + 2|\Omega_R|^2}. \tag{3.82}$$

At this point, it is useful to introduce the saturation parameter s_0 as

$$s_0 = \frac{I}{I_s} = \frac{2|\Omega_R|^2}{\Gamma^2}, \quad I = \frac{\epsilon_0}{2}\omega_0^2|A_0|^2, \tag{3.83}$$

where I is the laser intensity and I_s the saturation intensity. The coupling constant can then be rewritten as

$$g = \frac{\Gamma s_0/2}{1 + s_0 + 4(\Delta + \mathbf{k}\cdot\mathbf{v})^2/\Gamma^2}. \tag{3.84}$$

Replacing this result in Eq. (3.76), we can see that the evolution of W is now determined for long times in a closed form.

3.5 Links with Dynamics and Statistics

3.5.1 Quasi-classical Limit

It is now useful to consider the wave kinetic equation in its quasi-classical limit, where the momentum carried by an emitted or absorbed photon of momentum $\hbar\mathbf{k}$ can be considered negligible with respect to the momentum of the atom $\hbar\mathbf{q}$. In this case, we can use the approximate expressions

$$W^{(\pm)} \simeq W(\mathbf{r},\mathbf{q},t) \pm \frac{\mathbf{k}}{2}\cdot\frac{\partial W}{\partial \mathbf{q}} + \frac{\partial}{\partial \mathbf{q}}\cdot\frac{\mathbf{kk}}{4}\cdot\frac{\partial}{\partial \mathbf{q}}W \pm \ldots \tag{3.85}$$

From this, we may write

$$\Delta W = -\mathbf{k}\cdot\frac{\partial W}{\partial \mathbf{q}}. \tag{3.86}$$

We can see that no second derivative term of the form $\partial^2 W/\partial q^2$ appears in the quasi-classical limit. This means that no Fokker-Planck type of equation can be generated here (as the effects of diffusion in the momentum space are not taken into account), which has important physical consequences, as discussed next. Using this result in the external force, we obtain

$$f_0(W) = -\frac{1}{\hbar}\nabla V_0(\mathbf{r})\cdot\frac{\partial}{\partial \mathbf{q}}W. \tag{3.87}$$

We can then write the quasi-classical limit of the wave kinetic equation (3.76) as

$$\left(\frac{\partial}{\partial t} + \mathbf{v} \cdot \frac{\partial}{\partial \mathbf{r}} + \frac{\mathbf{F}}{M} \cdot \frac{\partial}{\partial \mathbf{v}} \right) W = 0, \tag{3.88}$$

where the force (per unit mass) acting on the atom has two distinct terms, one associated with the trapping potential V_0, and the other one due to the laser field, as defined by

$$\mathbf{F} = -\nabla V_0(\mathbf{r}) + g\hbar \mathbf{k}. \tag{3.89}$$

The quasi-classical kinetic equation of the atom (3.88) clearly states the conservation of the Wigner quasi-distribution W in the classical phase space $(\mathbf{r}, \mathbf{v})$, in accordance with the generic discussion of Sect. 3.1.4. This is a kinetic equation of the single-particle Liouville or Vlasov type, which shows that, in this limit, W behaves like a classical distribution. Furthermore, the classical trajectory of the atom is determined by the characteristics of the kinetic equation, which are

$$\frac{d\mathbf{r}}{dt} = \mathbf{v}, \quad \frac{d\mathbf{p}}{dt} = -\nabla V_0(\mathbf{r}) + g\hbar \mathbf{k}. \tag{3.90}$$

It is clear from Eqs. (3.89) and (3.90) that we recover here the exact result for the ponderomotive and laser-cooling force term, $\mathbf{F}_L$, as established from more conventional methods [7]. We have shown that our exact wave kinetic equation retains the atom recoil effects, and that it allows us to recover in the quasi-classical limit the familiar expressions for the laser force.

3.5.2 Momentum Diffusion and the Doppler Limit

At this point, we should consider the eventual influence of the electromagnetic noise $\delta\mathbf{A}$, which was only mentioned earlier in Sect. 3.2, and neglected afterwards. This noise, which could also include the vacuum field fluctuations, would eventually lead to a random force term $\delta\mathbf{F}$, thus transforming the equation of motion (3.90) into a Langevin equation. From there, it would then be possible to recover an equivalent Fokker-Planck equation, which is not possible to derive directly from our semi-classical approach as considered above. This stochastic force is defined by the following properties

$$\langle \delta\mathbf{F} \rangle = 0\,, \quad \langle \delta\mathbf{F}(t) \cdot \delta\mathbf{F}(t') \rangle = D\delta(t - t') \tag{3.91}$$

where D is a diffusion coefficient, and $\langle\cdot\rangle$ represents average over a statistical ensemble. It is also known that the cooling force $\mathbf{F}_L$ can be written, in the range of low detuning and low atom velocities, as $\mathbf{F}_L \simeq -M\beta\mathbf{v}$, as shown in the previous chapter. We can then write the modified (Langevin) equation of motion as

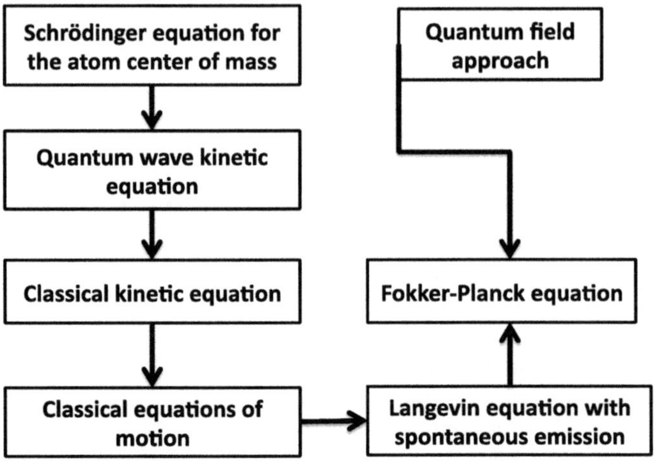

Fig. 3.2 Diagram of the theoretical approach followed in this chapter, with a link between the main equations for the atom centre-of-mass motion

$$\frac{d\mathbf{p}}{dt} = \mathbf{F}_0 - \beta \mathbf{v} + \delta \mathbf{F}. \tag{3.92}$$

It is well known that this Langevin equation is equivalent to a Fokker-Planck equation of the form

$$\frac{df}{dt} = \frac{\partial}{\partial \mathbf{v}} \cdot \left[\mathbf{A}(\mathbf{v}) + \bar{\mathbf{D}} \cdot \frac{\partial}{\partial \mathbf{v}} \right] f, \tag{3.93}$$

where $f \equiv f(\mathbf{r}, \mathbf{v}, t)$ is the classical distribution of the centre of mass velocity states, in fact the classical limit of the Wigner function W, with the friction and diffusion coefficients defined by $\mathbf{A} = -\beta \mathbf{v}$, and

$$\bar{\mathbf{D}} = \frac{1}{2} \frac{\langle \Delta v^2 \rangle}{\Delta t} \bar{\mathbf{1}} \simeq \frac{\hbar^2}{4} k^2 \Gamma \bar{\mathbf{1}}, \tag{3.94}$$

where $\bar{\mathbf{1}}$ is the unit matrix. Such a description can then be used to establish the temperature limit T_D as schematically discussed in the previous chapter.

The link between the different equations discussed in this chapter are illustrated in Fig. 3.2. Starting from the Schrödinger equation for the moving atom, and using a semi-classical coupling with the radiation field, as stated by (3.53), and (3.64) for the atom centre-of-mass states, we have derived a quantum wave kinetic equation for the Wigner function W, as stated by (3.73). This was then applied to the two-level atom model, as described (3.76). We notice that the description still remains fully quantum mechanical (as far as the atom is concerned), where the radiation field is treated as a classical external field, but both the energy and momentum exchange between the atom and the field are retained. This includes the atom recoil effects.

We have then used a quasi-classical approach, by neglecting recoil effects, which lead to the Vlasov kinetic equation (3.88). In this way, we have connected quantum mechanics to statistical mechanics, where the state of the atom centre-of-mass motion is represented in the classical single-particle phase-space $(\mathbf{r}, \mathbf{p})$. One further step is made by taking the characteristics of this kinetic equation, which are nothing but the classical equations of motion. Finally, a stochastic force term associated with spontaneous emission was introduced, leading to the Langevin equation (3.92) and to its equivalent Fokker-Planck equation (3.93). An alternative approach, more exact but also much heavier, would involve the full use of quantum theory along the whole chain of derivations, from the wave equation for the atom in a quantized radiation field, down to a Fokker-Planck equation. The connections with classical dynamics and statistics would also become much less apparent.

3.6 Lambda Configuration

Let us return to the exact quantum wave kinetic description and consider other examples of application, by generalizing the above two-level atom configuration. A straightforward generalization can be made to the so called *lambda configuration* (see Fig. 3.3). This corresponds to a three-level atom, with two allowed radiative transitions.

We assume that two different laser beams are present, with frequencies $\omega_a \simeq \omega_{31}$ and $\omega_b \simeq \omega_{32}$, and further assume that the radiative coupling between the two lower levels $|1\rangle$ and $|2\rangle$ is forbidden. In this Λ configuration, the wave kinetic equation (3.73) can be reduced to

$$\frac{d}{dt} W = -i f_0(W) + g_a \Delta W_a + g_b \Delta W_b, \tag{3.95}$$

where $d/dt = \partial/\partial t + \mathbf{v} \cdot \partial/\partial \mathbf{r}$ and

$$\Delta W_j = \left[W(\mathbf{r}, \mathbf{q} - \mathbf{k}_j/2, t) - W(\mathbf{r}, \mathbf{q} + \mathbf{k}_j/2, t) \right] \tag{3.96}$$

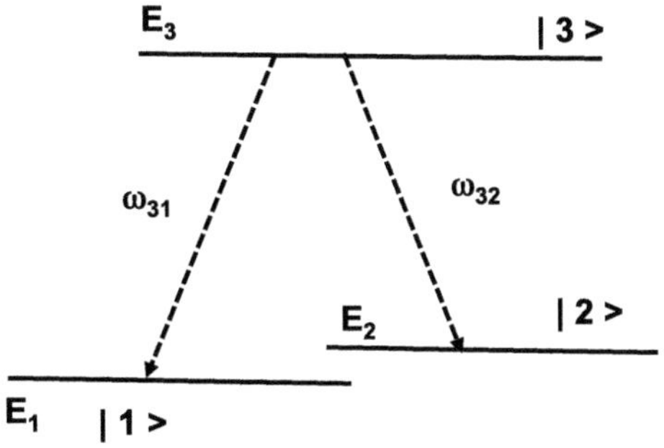

Fig. 3.3 Lambda configuration: this is a three-level atom model with two allowed radiative transitions

for $j = a, b$. This is a straightforward generalization of Eq. (3.76), where the two coupling coeffients g_a and g_b depend on different density matrix elements, according to

$$g_a = -\Im(\Omega_a \tilde{\rho}_{31}), \quad g_b = -\Im(\Omega_b \tilde{\rho}_{32}) \tag{3.97}$$

where $\Omega_a = \omega_a A_a p_{31}/\hbar$ and $\Omega_b = \omega_b A_b p_{32}/\hbar$ are the Rabi frequencies corresponding to the two radiative couplings, A_a and A_b the corresponding laser field amplitudes. We also have defined the following density matrix elements

$$\tilde{\rho}_{31} = \rho_{31} e^{i\mathbf{k}_a \cdot \mathbf{r} + i\Delta_a t}, \quad \tilde{\rho}_{32} = \rho_{32} e^{i\mathbf{k}_b \cdot \mathbf{r} + i\Delta_b t} \tag{3.98}$$

and the frequency detunings $\Delta_a = \omega_{31} - \omega_a$ and $\Delta_b = \omega_{32} - \omega_b$, can have identical or opposite signs. Now we need to find the pertinent expressions for the density matrix elements ρ_{31} and ρ_{32}. For that purpose, we go back to Eq. (3.79) and retain the appropriate terms in the sum. The resulting equations are

$$\frac{d}{dt}\rho_{33} = \frac{i}{2}\left(\Omega_a \tilde{\rho}_{31} - \Omega_a^* \tilde{\rho}_{13}\right) + \frac{i}{2}\left(\Omega_b \tilde{\rho}_{32} - \Omega_b^* \tilde{\rho}_{23}\right) - \Gamma \rho_{33} \tag{3.99}$$

and

$$\begin{aligned} \frac{d}{dt}\tilde{\rho}_{31} &= \frac{i}{2}\Omega_a^*(\rho_{33} - \rho_{11}) - [\Gamma_a + i\Delta_a]\,\tilde{\rho}_{31} \\ \frac{d}{dt}\tilde{\rho}_{32} &= \frac{i}{2}\Omega_b^*(\rho_{33} - \rho_{22}) - [\Gamma_b + i\Delta_b]\,\tilde{\rho}_{32} \end{aligned} \tag{3.100}$$

where $\Gamma_a = 2\gamma_a$ and $\Gamma_b = 2\gamma_b$ are the spontaneous decay rates into the two lower levels $|1\rangle$ and $|2\rangle$, and $\Gamma = \Gamma_a + \Gamma_b$ is the total decay rate of the upper energy level. We have also used the simplifying notation $\delta_j = (\Delta_j + \mathbf{k}_j \cdot \mathbf{v})$, for $j = a, b$. These equations generalize the usual Bloch equations for a two-level atom. Let us now consider the steady state solutions of Eq. (3.99). Putting Eqs. (3.99) and (3.97) together, we simply get $\Gamma\rho_{33} = g_a + g_b$. Moreover, from Eq. (3.101), ant noticing that for a three-level atom $\rho_{11} + \rho_{22} = 1 - \rho_{33}$, we obtain

$$\tilde{\rho}_{31} = -\frac{\Omega_a^*}{4}\frac{(\Delta_a + i\Gamma_a)}{(\Delta_a^2 + \Gamma^2/4 + |\Omega_a|^2/2}, \quad \tilde{\rho}_{32} = -\frac{\Omega_b^*}{4}\frac{(\Delta_b + i\Gamma_b)}{(\Delta_b^2 + \Gamma^2/4 + |\Omega_b|^2/2}. \tag{3.101}$$

This allows us to write the population of the upper energy level as

$$\rho_{33} = \frac{1}{\Gamma}(g_a + g_b) = \frac{1}{2}\sum_{j=a,b}\frac{|\Omega_j|^2}{4\Delta_j^2 + \Gamma^2 + 2|\Omega_j|^2} \tag{3.102}$$

from where we can easily extract the explicit expressions for g_a and g_b. It is now useful to briefly discuss the quasi-classical limit for this configuration. In this limit, the wave kinetic equation (3.95) reduces again to (3.88), where the classical force acting on the atom is now given by

$$\mathbf{F} \simeq -\nabla V_0(\mathbf{r}) + \hbar(g_a \mathbf{k}_a + g_b \mathbf{k}_b). \tag{3.103}$$

The radiation force associated with the two laser beams can be approximated by using $g_j \mathbf{k}_j \simeq -\beta_j \mathbf{v}$, where β_j are the viscous coefficients associated with the two laser beams $j = a, b$. We can see that the Λ configuration can enhance or decrease the cooling effect. Such a decrease occurs for instance when the two laser beams have the same direction of propagation, but opposite frequency detuning with respect to the two radiative transitions (i.e., one being blue- and the other red-detuned). In this case, when $|\beta_a| = |\beta_b|$ a complete inhibition of laser cooling is observed, which corresponds to the manifestation of a *dark state*.

The Λ configuration is also useful to describe the Raman cooling process, as first demonstrated by Kasevich and Chu [8]. In this case, the two laser beams are slightly red-detuned with respect with the two radiative transitions, and the two lower energy states $|1\rangle$ and $|2\rangle$ are in fact Zeeman splitted sub-levels of the ground state. The resulting recoil limit is now associated with the difference $(\mathbf{k}_a - \mathbf{k}_b)$, which is well below the recoil limit valid for the Sisyphus cooling.

3.7 Two Coupled Radiative Transitions

Of particular interest is the case of a modified Λ configuration, where we have two coupled radiative transitions operating between two distinct pairs of atom energy levels, $|1, j\rangle$ and $|2, j\rangle$, with $j = a, b$ (see Fig. 3.4). Such couple can be provided by some unspecified process, such as atomic collisions or radiative pumping. This last case is well adapted to study the Sisyphus cooling, as discussed below.

Going back to the generic wave kinetic equation (3.73), and considering the trace of the relevant density matrix, we get

$$\sum_n \rho_{nn} = \sum_{j=a,b} \rho_j = 1 \,, \quad \rho_j = \rho_{1j,1j} + \rho_{2j,2j} \,, \tag{3.104}$$

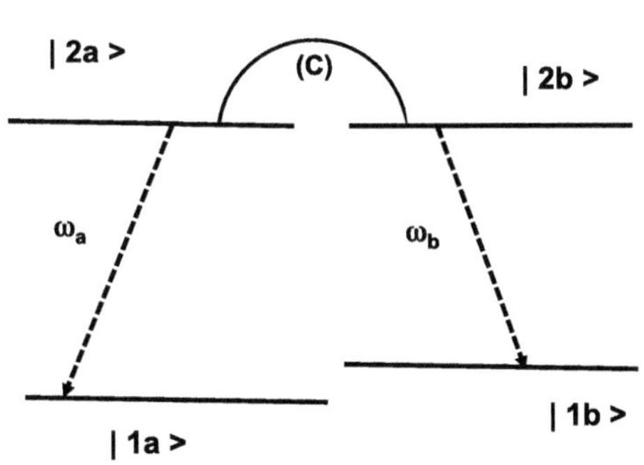

Fig. 3.4 Two coupled radiative transitions: this is a four-level atom model, with two radiative transitions, which are coupled by some unspecified (*C*) process, due to collisions or to radiative pumping

where ρ_j are the partial traces or the populations involved in the two distinct radiative processes. Accordingly, by introducing two distinct Wigner functions, $W_j = \rho_j W$, we can write two distinct kinetic equations similar to (3.76), with source terms describing the coupling process, reading

$$\left(\frac{\partial}{\partial t} + \mathbf{v} \cdot \frac{\partial}{\partial \mathbf{r}}\right) W_j = -i f_0(W_j) + g_j \Delta W_j + \left(\frac{\partial W_j}{\partial t}\right)_s , \tag{3.105}$$

where g_j and ΔW_j are defined in the same way as in the Λ configuration, and the source terms satisfy the conservation condition

$$\left(\frac{\partial W_a}{\partial t}\right)_s = -\left(\frac{\partial W_b}{\partial t}\right)_s . \tag{3.106}$$

In order to simplify the discussion, let us focus on the quasi-classical limit, where we can use the approximation

$$f_0(W_j) = \frac{1}{\hbar} \nabla V_0(\mathbf{r}) \cdot \frac{\partial}{\partial \mathbf{q}} W_j , \tag{3.107}$$

where $\hbar\mathbf{q}$ is the momentum of the atomic centre of mass. In this case, (3.105) reduces to

$$\left(\frac{\partial}{\partial t} + \mathbf{v} \cdot \frac{\partial}{\partial \mathbf{r}} + \frac{\mathbf{F}_j}{M} \cdot \frac{\partial}{\partial \mathbf{v}}\right) W_j = \left(\frac{\partial W_j}{\partial t}\right)_s \tag{3.108}$$

for $j = a, b$. The two equations written here are in fact coupled through the conservation condition (3.106). The quantities $\mathbf{F}_j$ correspond to the classical forces acting on the atom, as due to the two distinct laser beams. These equations can alternatively be established in a phenomenological way [9]. By adding these two equations, noting that $(W_a + W_b) = W$, and taking (3.106) into account, we get

$$\left(\frac{\partial}{\partial t} + \mathbf{v} \cdot \frac{\partial}{\partial \mathbf{r}}\right) W = \frac{1}{M} \left(\mathbf{F}_a \cdot \frac{\partial W_a}{\partial \mathbf{v}} + \mathbf{F}_b \cdot \frac{\partial W_b}{\partial \mathbf{v}}\right) . \tag{3.109}$$

For a number of interesting processes, such as Sisyphus cooling, we verify the symmetric force condition $\mathbf{F}_a = -\mathbf{F}_b \equiv -\mathbf{F}$. In this case, we are reduced to

$$\left(\frac{\partial}{\partial t} + \mathbf{v} \cdot \frac{\partial}{\partial \mathbf{r}}\right) W = \frac{\mathbf{F}}{M} \cdot \frac{\partial}{\partial \mathbf{v}} \delta W , \tag{3.110}$$

where we have defined the difference $\delta W = W_a - W_b$. On the other hand, by taking the difference of the two equations (3.108), and assuming a symmetric force, we get an evolution equation for the difference, as

$$\left(\frac{\partial}{\partial t} + \mathbf{v} \cdot \frac{\partial}{\partial \mathbf{r}}\right) \Delta W = \frac{\mathbf{F}}{M} \cdot \frac{\partial W}{\partial \mathbf{v}} + \left(\frac{\partial \Delta W}{\partial t}\right)_s . \tag{3.111}$$

This shows that, for a symmetric force, we can transform the coupled equations for W_a and W_b, into a new pair of coupled equations for the total distribution W and for the difference δW. Let us now defined the characteristic times for the coupling process, τ_{ab} and τ_{ba}, such that

$$\left(\frac{\partial W_a}{\partial t}\right)_s = \frac{W_b}{\tau_{ba}} - \frac{W_a}{\tau_{ab}}. \tag{3.112}$$

This shows that, in equilibrium, we should have $W_a/W_b = \tau_{ab}/\tau_{ba}$. This then leads to

$$\delta W = W \frac{\tau_{ab} - \tau_{ba}}{\tau_{ab} + \tau_{ba}}. \tag{3.113}$$

Let us apply this description to the case of Sisyphus cooling, where coupling between the two radiative cycles is due to a spatial dependence of the left and right hand polarized laser fields. In this case we can replace the above expression by a specific formula

$$\delta W = W \sin(2\mathbf{k} \cdot \mathbf{r}), \tag{3.114}$$

where k is the wavenumber associated with the two counter-propagating laser beam modes with the same frequency but orthogonal linear polarization, as discussed in the previous chapter. Such a spatially modulated population difference is imposed by optical pumping. Now, going back to Eq. (3.111), for nearly steady state conditions, and assuming that the force term on the right hand side is negligible (which is not obvious and should be verified a posteriori), we can write [9]

$$\mathbf{v} \cdot \frac{\partial}{\partial \mathbf{r}} \delta W \simeq \left(\frac{\partial \delta W}{\partial t}\right)_s \simeq \frac{\delta W}{\tau_c}, \tag{3.115}$$

where we have introduced the characteristic decay time τ_c, such that $1/\tau_c = 1/\tau_{ab} + 1/\tau_{ba}$. Inserting (3.114) on the left hand side, we get a rough estimate of the population difference ΔW as

$$\delta W \simeq 2W\tau_c(\mathbf{v} \cdot \mathbf{k}) \cos(2\mathbf{k} \cdot \mathbf{r}). \tag{3.116}$$

Replacing this estimate in Eq. (3.110), we get a closed kinetic equation for the total distribution W in the canonical form, which reads

$$\left(\frac{\partial}{\partial t} + \mathbf{v} \cdot \frac{\partial}{\partial \mathbf{r}} + \frac{\mathbf{F}_{\mathbf{eff}}}{M} \cdot \frac{\partial}{\partial \mathbf{v}}\right) W = 0, \tag{3.117}$$

where the effective force $\mathbf{F}_{\text{eff}}$ acting on the atom is given by

$$\mathbf{F}_{\text{eff}} = 2\mathbf{F}\tau_c(\mathbf{v} \cdot \mathbf{k}) \cos(2\mathbf{k} \cdot \mathbf{r}). \tag{3.118}$$

We known that in the Sisyphus configuration, **F** is the force due to the gradient of the energy shift of the lower atom levels due to the laser field. This can be written as

$$\mathbf{F} = \nabla(\Delta E) = 2U_0\mathbf{k}\cos(2\mathbf{k}\cdot\mathbf{r}), \tag{3.119}$$

where the maximum energy shift U_0 due to the ac Stark effect was calculated in the previous chapter. This shows that the effective force is modulated in space as the square of the standing wave field. By taking the average over a distance of $\pi/k = \lambda/2$, we obtain the expression

$$\langle \mathbf{F}_{\text{eff}} \rangle \simeq -\beta_{\text{eff}}\mathbf{v} \,, \quad \beta_{\text{eff}} = 2\tau_c k^2 U_0, \tag{3.120}$$

where β_{eff} is the resulting effective viscosity coefficient which describes the corresponding cooling process.

3.8 Influence of a Blue-Detuned Pump

The description of laser cooling associated with a two-level atom process can be refined by introducing the following situation: (1) we have six red-detuned laser cooling beams and, in addition, (2) we have one blue-detuned laser beam. This configuration (see Fig. 3.5) is relevant to the excitation of Doppler instabilities [10, 11], and the production of phonon laser emission, to be discussed later.

Instead of Eq. (3.75), the electromagnetic wave field spectrum is now given by

$$A(\mathbf{k}) = \frac{(2\pi)^3}{2}\sum_n A_n\left[\delta(\mathbf{k}-\mathbf{k}_n) + \delta(\mathbf{k}+\mathbf{k}_n)\right] + \frac{(2\pi)^3}{2}A_b\left[\delta(\mathbf{k}-\mathbf{k}_n b) + \delta(\mathbf{k}+\mathbf{k}_b)\right], \tag{3.121}$$

where n represents the different laser red-detuned beams. In a standard laser cooling configuration (as a magneto-optical trap), we have six laser cooling beams with equal amplitude, and we can take $A_n = A_0$, forming an orthogonal set of

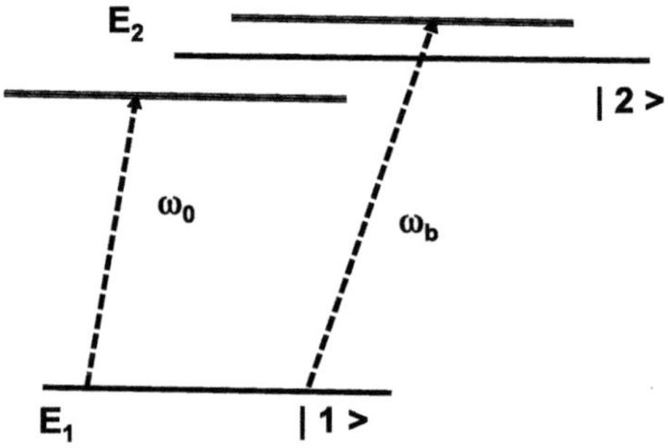

Fig. 3.5 Two coupled radiative transitions: this is a four-level atom model, with two radiative transitions, which are coupled by some unspecified (C) process, due to collisions or to radiative pumping

wavevectors $\mathbf{k}_n$. Here we assume that $\omega_0 = k_n c \leq \omega_{21}$, whereas for the blue-tuned beam we have the frequency $\omega_b = k_b c \geq \omega_{21}$. The resulting wave kinetic equation is

$$\frac{d}{dt} W = -i f_0(W) + \sum_n g_n \Delta W_n + g_b \Delta W_b, \tag{3.122}$$

where $f_0(W)$ and g_n and g_b keep the above definitions (with obvious changes). It should be noted, in particular, that the relevant frequency shifts are now given by $\Delta_n \equiv \Delta_0 = (\omega_{21} - \omega_0) < 0$ and $\Delta_b = (\omega_{21} - \omega_b) > 0$.

In the absence of the blue-detuned beam, $g_b = 0$ and $A_b = 0$, the atom can attain some steady state equilibrium W_0. In the presence of the blue-detuned pump, such that $|g_b| \ll |g_n|$, we can use a perturbative approach by assuming a solution of the form $W = W_0 + \tilde{W}$, such that

$$\frac{d}{dt} \tilde{W} = \sum_n g_n \Delta \tilde{W}_n + g_b \Delta W_{0b}, \tag{3.123}$$

where we have assumed that the perturbed distribution $\tilde{W}$ is of order $|g_b/g_n| \ll 1$, and retained only the first order perturbative terms. Here, we have also used the unperturbed quantity ΔW_{0b}, as defined by

$$\Delta W_{0b} = [W_0(\mathbf{r}, \mathbf{q} - \mathbf{k}_b/2, t) - W_0(\mathbf{r}, \mathbf{q} + \mathbf{k}_b/2, t)] \tag{3.124}$$

It is now useful to introduce the parallel distribution $G(u)$, after integration of W over the directions perpendicular to that of the blue-detuned laser beam. We define

$$G(u) = \int W(u, \mathbf{v}_\perp) d\mathbf{v}_\perp \,, \quad \mathbf{v} = u \frac{\mathbf{k}_b}{|k_b|} + \mathbf{v}_\perp. \tag{3.125}$$

This allows us to transform equation (3.123) into

$$\frac{d}{dt} \tilde{G} = \sum_n \int g_n(u, \mathbf{v}_\perp) \Delta \tilde{W}_n(u, \mathbf{v}_\perp) d\mathbf{v}_\perp + g_b \Delta G_{0b}. \tag{3.126}$$

We should notice that the coupling strengths g_n are, in general dependent of the perpendicular velocity $\mathbf{v}_\perp$. This equation can be simplified by noting that the terms containing g_n tend to cool down the atom, and to reduce $\tilde{G}$ to zero. We can then replace it by

$$\frac{d}{dt} \tilde{G} = -\nu \tilde{G} + g_b \Delta G_{0b}, \tag{3.127}$$

where ν is a phenomenological viscosity parameter of order $|g_n|$. This equation has an obvious meaning. The perturbed distribution $\tilde{G}$ is excited by the driving blue-tune laser, as described by the term in $|g_b|$, and is damped by the cooling laser beams by decaying to zero on a time-scale of $1/\nu$. For example, by choosing

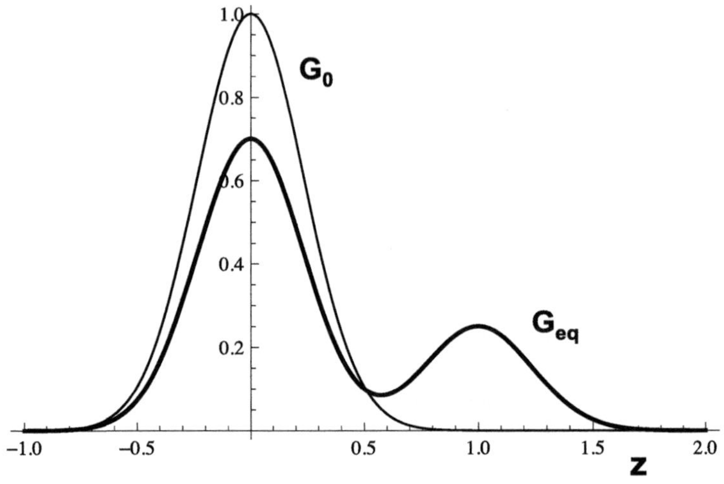

Fig. 3.6 Parallel distributions for blue-detuned excitation of a laser cooled atom: initial thermal distribution $G_0 = G(t = 0)$, and asymptotic equilibrium distribution $G_{eq} = G(t \to \infty)$, as a function of the normalized parallel velocity $z = Mu/\hbar k_b$

an equilibrium thermal distribution with temperature T, such that $G_0(u) = \sqrt{M/2\pi T}\exp(-Mu^2/2T)$, such that $\sqrt{T/M} < \hbar k_b/M$, we get the simple solution

$$\tilde{G}(u) = \frac{g_b}{\nu} G_0(u - \hbar k_b/M)\left[1 - e^{-\nu t}\right], \tag{3.128}$$

where $g_b > 0$, which is a thermal distribution displaced by in the direction of the blue-shifted laser beam. Such a displacement is due to the increase of momentum due to absorption of photons in this direction. This solution shows a linear growth at a rate $g_b G_0$, valid for short times $t \ll 1/\nu$, followed by a saturation on a time scale of $1/\nu$, when the cooling effect due to the red-shifted beams, and the excitation due to the blue-shifted beam tend to an equilibrium, $\tilde{G}(\infty) = g_b G_0(u - k_b/M)/\nu$. The resulting two energy levels in velocity space is illustrated in Fig. 3.6.

However, it should be noticed that the above description, although suggestive, is actually only experimentally relevant for sub-recoil cooled systems, for instance Raman cooled, where the initial temperature of the atoms is lower than the blue-tuned photon energy. Otherwise, the secondary atom beam distribution $\tilde{G}$ will not be able to clearly emerge from the initial distribution G_0.

References

1. E.P. Wigner, Phys. Rev. **40**, 749 (1932)
2. J.E. Moyal, Proc. Camb. Philos. Soc. **45**, 99 (1949)
3. S. Stenholm, The semiclassical theory of laser cooling. Rev. Mod. Phys. **58**, 699 (1986)

4. J. Dalibard, C. Cohen-Tannoudji, Atomic motion in laser light: connection between semiclassical and quantum description. J. Phys. B At. Mol. Phys. **18**, 1661 (1985)
5. M. Hillary, R.F. O'Connell, M.O. Scully, E.P. Wigner, Distribution functions in physics: fundamentals. Phys. Rep. **106**, 121 (1984)
6. R. Loudon, *The Quantum Theory of Light*. Oxford Science Publications (Clarendon, Oxford, 1992)
7. M.O. Scully, M.S. Zubairy, *Quantum Optics* (Cambridge University Press, Cambridge/New York, 1999)
8. M. Kasevich, S. Chu, Phys. Rev. Lett. **69**, 1741 (1992)
9. C.J. Pethick, H. Smith, *Bose-Einstein Condensation in Dilute Gases* (Cambridge University Press, Cambridge/New York, 2008)
10. A.E. Kaplan, Opt. Express **17**, 10035 (2009)
11. K. Vahala, M. Herrmann, S. Knunz, V. Batteiger, G. Saathoff, T.W. Hänsch, Th. Udem, Nat. Phys. **5**, 682 (2009)

Chapter 4
Atomic Clouds

Here, we consider atom-atom interactions. In a cold gas, for atoms confined in a magneto-optical trap, two types of interactions can be considered. First, we have the close range atom-atom collisions. Second, we have the long range interactions mediated by scattered photons. The atomic collisions play an important role in atom detrapping, in the gas thermalization during evaporative cooling, in the plasma ionization processes, and in the formation of Bose-Einstein condensates. Atom-atom collisions at very low energies have specific properties which have be studied by many authors (see for a review [1, 2]). After reviewing the basic properties of atomic collisions in the low energy limit, we discuss the Feshbach resonances, which result from the coupling with a discrete bound state. In the experiments, such resonances can be controlled by externally applied magnetic fields, and allow to control the atomic collision cross sections.

We then discuss the collective forces, mediated by secondary photons, resulting from scattering by the atoms of the incident primary photons associated with the laser cooling beams. These collective forces are therefore an intrinsic property of the laser cooled gas. Of particular interest is the influence of such forces on the atom density profile. We can distinguish two main types of density confinement regimes. First, the temperature limited regime, with a Gaussian density profile, which can be observed in clouds with a relatively small number of atoms, when the collective forces are not relevant. Second, the multiple scattering regime, where the atom density is nearly constant inside the cloud, and where the collective forces are dominant. A simple, phenomenology description, as well as a more systematic approach to these confinement regimes are given. Finally, we discuss the particularly important case of the cloud expansion in the absence of magnetic confinement. This is a Coulomb type of expansion, which results from the long range repulsive forces associated with the atomic mean field.

J.T. Mendonça and H. Terças, *Physics of Ultra-Cold Matter*, Springer Series on Atomic, Optical, and Plasma Physics 70, DOI 10.1007/978-1-4614-5413-7_4,

4.1 Atom-Atom Collisions

Here we start by reviewing the principles of collision theory, which is an essential ingredient of the cold matter physics. It is well known that collisions between two atoms with masses m_1 and m_2 can be described in the centre-of-mass frame by the wave functions $\psi(\mathbf{r}, t)$ of the relative particle with reduced mass $m_r = m_1 m_2/(m_1+m_2)$. If the interaction potential $V(\mathbf{r}_1-\mathbf{r}_2) \equiv V(\mathbf{r})$ is time independent, we can look for stationary wave solutions of the form

$$\psi(\mathbf{r}, t) = \psi(\mathbf{r}) \exp(-iEt/\hbar) \ , \quad E = p^2/2m_r, \tag{4.1}$$

where E is the energy of the particle and $\mathbf{p} = \hbar\mathbf{k} = m_r \mathbf{v}$ its momentum. In order to determine $\psi(\mathbf{r})$, we need to solve the steady-state wave equation

$$\left[\nabla^2 + k^2 + U(\mathbf{r})\right] \psi(\mathbf{r}) = 0, \tag{4.2}$$

where $U(\mathbf{r}) = 2m_r V(\mathbf{r})/\hbar^2$ is the reduced potential, to be specified later. The scattering process can then be described by asymptotic solutions of this equation, valid far away from the scattering region, of the form

$$\psi_k(\mathbf{r})|_{r\to\infty} \simeq e^{i\mathbf{k}\cdot\mathbf{r}} + \psi_{\text{scat}}(\mathbf{r}), \quad \psi_{\text{scat}} = \frac{1}{r} f_k(\theta, \phi) e^{ikr} m, \tag{4.3}$$

where have used spherical coordinates $\mathbf{r} \equiv (r, \theta, \phi)$. The first term in (4.3) represents the incident particle state, and the quantities $f_k(\theta, \phi)$ are called the scattering amplitudes. The associated current density is defined by

$$\mathbf{J}(\mathbf{r}) = \Re\left[\psi_k^* \left(\frac{\hbar}{i m_r}\right) \nabla \psi_k\right]. \tag{4.4}$$

At large distances ($r \to \infty$), the components of this current along $\mathbf{e}_\theta$ and $\mathbf{e}_\phi$ become negligible and we are left with a purely radial current, given by

$$\mathbf{J}(\mathbf{r}) \simeq J_r \mathbf{e}_r \ , \quad J_r = \frac{\hbar k}{m_r} \frac{|f_k(\theta, \phi)|^2}{r^2}. \tag{4.5}$$

This quantity is equal (or proportional) to the number of scattered particles crossing the unit area per unit time. Therefore, the number of scattered particles which cross the detecting area of a given detector, $dS = r^2 d\Omega$, where $d\Omega$ is the element of solid angle, will be $J r^2 d\Omega$. The ratio of this number to the incident flux, given by $v = \hbar k/m_r$, defines the *differential cross-section*, which reads

$$\frac{d\sigma_k}{d\Omega} = |f_k(\theta, \phi)|^2. \tag{4.6}$$

Of particular interest is the case of scattering by a central potential, $V(r)$, where the scattering amplitudes only depend on the scattering angle θ, defined as the angle between the incident and the scattered wave momenta. The ϕ dependence in Eq. (4.3) disappears, and the wavefunctions can be represent on a basis of *Legendre polynomials*, $P_l(\cos\theta)$, which form a complete set in the interval $-1 < \cos\theta < 1$. Such an expansion can be written as

$$\psi_k(\mathbf{r}) = \sum_{l=0}^{\infty} R_{kl}(r) P_l(\cos\theta). \tag{4.7}$$

Each term in this series is called a *partial wave*, which is an eigenfunction of both the L^2 and L_z angular momentum operators, with eigenvalues $l(l+1)\hbar^2$ and 1 respectively. In the usual *spectroscopic notation* the partial waves $l = 0, 1, 2, 3, \ldots$ are called the $s, p, d, f, \ldots$ waves. Replacing this in the wave equation (4.2) we get the radial equation

$$\frac{1}{r^2}\frac{d}{dr}\left(r^2 \frac{dR_{kl}}{dr}\right) + \left[k^2 - U(r) - \frac{l(l+1)}{r^2}\right] R_{kl} = 0. \tag{4.8}$$

Assuming that the interaction potential $U(r)$ falls off more rapidly than $1/r^2$ at large distances, we can write the general solution as

$$R_{kl}(r) = A_{kl}\, j_l(kr) + B_{kl}\, n_l(kr), \tag{4.9}$$

where the coefficients A_{kl} and B_{kl} are independent of r, and were $j_l(\rho)$ and $n_l(\rho)$ are the spherical Bessel and Newmann functions defined by

$$j_l(\rho) = \sqrt{\frac{\pi}{2\rho}} J_{l+1/2}(\rho)\ , \quad n_l(\rho) = -(-1)^l \sqrt{\frac{\pi}{2\rho}} J_{-l-1/2}(\rho), \tag{4.10}$$

with $J_\nu(\rho)$ denoting the ordinary ν-th order Bessel function. Using the asymptotic expressions for these functions for large distances $r \to \infty$, we get approximate solutions of the form

$$R_{kl}(r) \simeq \frac{C_{kl}}{kr} \sin\left(kr - \frac{1}{2} l\pi + \delta_{kl}\right), \tag{4.11}$$

with amplitudes $C_{kl} = \sqrt{A_{kl}^2 + B_{kl}^2}$ and *phase shifts* $\delta_{kl} = -\tan^{-1}(B_{kl}/A_{kl})$. These phase shifts in fact determine the strength of the scattered l-th partial wave by the potential $U(r)$, and become zero in its absence, if $U(r) = 0$.

Let us now relate the phase shifts with the scattering amplitudes $f_k(\theta)$, and with the scattering cross-section. This can be done by developing the incident plane wave in Legendre polynomials. For large distances, $r \to \infty$, this can be written as

$$e^{i\mathbf{k}\cdot\mathbf{r}} = \sum_{l=0}^{\infty} (2l+1) \frac{i^l}{kr} \sin\left(kr - \frac{l\pi}{2}\right) P_l(\cos\theta). \tag{4.12}$$

Using Eqs. (4.3) and (4.7) we obtain, for each partial wave, the following relation

$$R_{kl}(r) \simeq (2l+1)\frac{i^l}{kr}\sin\left(kr - \frac{l\pi}{2}\right) + \frac{e^{ikr}}{r} f_{kl}, \tag{4.13}$$

where we have performed the developments

$$f_k(\theta) = \sum_{l=0}^{\infty} f_{kl} P_l(\cos\theta). \tag{4.14}$$

Replacing Eq. (4.11) in (4.13), and equating the coefficients of $\exp(ikr)$ and $\exp(-ikr)$ on both sides of the resulting expression, we get

$$c_{kl} = (2l+1)i^l e^{i\delta_{kl}}, \quad f_{kl} = \frac{1}{2ik}(2l+1)\left(e^{2i\delta_{kl}} - 1\right). \tag{4.15}$$

Inserting the latter in (4.16), we observe that the scattering amplitudes depend only on the phase shifts δ_{kl}, as explicitly stated by

$$f_k(\theta) = \frac{1}{k}\sum_{l=0}^{\infty}(2l+1)e^{i\delta_{kl}}\sin(\delta_{kl})P_l(\cos\theta). \tag{4.16}$$

Replacing this in Eq. (4.6), and integrating over the solid angle $d\Omega = \sin\theta d\theta d\phi$, we obtain the *total cross section*

$$\sigma_k = \int |f_k(\theta)|^2 d\Omega = \frac{4\pi}{k^2}\sum_{l=0}^{\infty}(2l+1)\sin^2(\delta_{kl}). \tag{4.17}$$

It should be mentioned that, using $\theta = 0$ in Eq. (4.16), and noting that $P_l(\theta = 0) = 1$, we can also conclude that $\sigma_k = (4\pi/k)\Im[f_k(\theta = 0)]$, a result known as the *optical theorem*. In order to obtain explicit expressions for σ_k as a function of the scattering potential $U(r)$, we have to determine the phase shifts, which can be done by using a Green's function method, as discussed below.

We now focus on the concept of *scattering length*, and consider the particular case of very low energies, such that $k \to 0$, which is relevant to the study of ultra-cold atoms. We first notice that Eq. (4.17) is only useful if the series converge rapidly for large l. This is particularly true for low energies, where only the first term of the series, the so called *s-wave scattering* $l = 0$, becomes relevant. In this case, Eq. (4.11) can be written as

$$R_{k,l=0}(r) \simeq \frac{C_{k0}}{kr}\sin\left[k(r-a)\right], \tag{4.18}$$

where a is the *scattering length*, as defined by

$$a = -\lim_{k \to 0} \frac{\tan \delta_{k0}}{k}. \tag{4.19}$$

It is obvious that the s-wave phase shift is equal to $\delta_{k0} = -ka$, and that, in the same limit, the total cross section reduces to

$$\sigma_{k \to 0} = \frac{4\pi}{k^2} \delta_{k0}^2 = 4\pi a^2. \tag{4.20}$$

In order to compute δ_{kl}, and in particular δ_{k0} and a, for a given interaction potential $U(r)$, we need to go back to the steady-state wave equation (4.2), which can be written in a more convenient and equivalent form

$$\left[\nabla^2 + k^2\right] \psi_k(\mathbf{r}) = F(\mathbf{r}), \quad F(\mathbf{r}) \equiv U(\mathbf{r})\psi_k(\mathbf{r}). \tag{4.21}$$

The general solution of this equation can be written as

$$\psi_k(\mathbf{r}) = \psi_k^0(\mathbf{r}) + \int G(\mathbf{r}, \mathbf{r}') F(\mathbf{r}') d\mathbf{r}', \tag{4.22}$$

where $\psi_k^0(\mathbf{r})$ is a particular solution of the corresponding homogeneous equation, and the *Green's function* $G(\mathbf{r}, \mathbf{r}')$ satisfies the equation

$$\left[\nabla^2 + k^2\right] G(\mathbf{r}, \mathbf{r}') = \delta(\mathbf{r} - \mathbf{r}'). \tag{4.23}$$

For our scattering problem, the adequate choice for the particular solution is that of the incident plane wave $\psi_k^0(\mathbf{r}) = \exp(i\mathbf{k} \cdot \mathbf{r})$. On the other hand, solving the Green's equation (4.23), we get

$$G(\mathbf{r}, \mathbf{r}') = -\frac{\exp(ik|\mathbf{r} - \mathbf{r}'|)}{4\pi|\mathbf{r} - \mathbf{r}'|}. \tag{4.24}$$

Replacing this in (4.22), we can then write the general solution of Eq. (4.21) as

$$\psi_k(\mathbf{r}) = e^{i\mathbf{k} \cdot \mathbf{r}} - \frac{1}{4\pi} \int \frac{\exp(ik|\mathbf{r} - \mathbf{r}'|)}{|\mathbf{r} - \mathbf{r}'|} U(\mathbf{r}') \psi_k(\mathbf{r}') d\mathbf{r}'. \tag{4.25}$$

This is the integral equation for the scattering process, which is known as the *Lippmann-Schwinger equation*. We should notice that, for large distances, ($r \to \infty$, $r \gg r'$), we can use the approximation $k|\mathbf{r} - \mathbf{r}'| \simeq kr - \mathbf{k}' \cdot \mathbf{r}'$, where $\mathbf{k}'$ is a vector with module equal to k and the direction of $\mathbf{r}$, or equivalently $\mathbf{k} = k\mathbf{r}/r$. In this case, the above expression can be approximately written as

$$\psi_k(\mathbf{r}) = e^{i\mathbf{k} \cdot \mathbf{r}} - \frac{1}{4\pi} \frac{e^{ikr}}{r} \int e^{i\mathbf{k}' \cdot \mathbf{r}'} U(\mathbf{r}') \psi_k(\mathbf{r}') d\mathbf{r}'. \tag{4.26}$$

Comparing this expression with (4.3), we conclude that the scattering amplitude $f_k(\theta, \phi)$ is determined by

$$f_k = -\frac{1}{4\pi}\langle \psi_{k'}^0 | U | \psi_k \rangle \equiv -\frac{m_r}{2\pi\hbar^2} T_{k',k}, \tag{4.27}$$

where the *transition matrix elements* $T_{k',k}$ are defined as

$$T_{k',k} = \langle \psi_{k'}^0 | V | \psi_k \rangle = \int \psi_{k'}^{0*}(\mathbf{r}) V(\mathbf{r}) \psi_k(\mathbf{r}) d\mathbf{r}. \tag{4.28}$$

In order to solve this integral, we can use the *first Born approximation*, by assuming that $\psi_k(\mathbf{r}) \simeq \psi_k^0(\mathbf{r})$. The scattering amplitude is then reduced to

$$f_k \simeq -\frac{m_r}{2\pi\hbar^2} T_{k',k}^0, \quad T_{k',k}^0 = \int e^{-i\mathbf{k}'\cdot\mathbf{r}} V(\mathbf{r}) e^{i\mathbf{k}\cdot\mathbf{r}} d\mathbf{r}. \tag{4.29}$$

If the two colliding atoms have equal mass $m_1 = m_2 = M$, we have $m_r = M/2$ and, taking the low energy limit $k \to 0$, we simply obtain

$$f_k \simeq -\frac{M}{4\pi\hbar^2} \int V(\mathbf{r}) d\mathbf{r}. \tag{4.30}$$

Noticing that, in this limit, the total cross-section is $\sigma_0 = 4\pi |f_0|^2$, and comparing this with Eq. (4.20), we obtain the following expression for the scattering length

$$a = \frac{M}{4\pi\hbar^2} \int V(\mathbf{r}) d\mathbf{r}. \tag{4.31}$$

We can see that the atomic collisions in the zero energy limit can adequately be described by an effective interaction potential of the form $V_{\text{eff}}(\mathbf{r}) = V_0 \delta(\mathbf{r})$, such that $V_0 = 4\pi\hbar^2 a/M$, or in explicit terms

$$V_{\text{eff}}(\mathbf{r}) = \frac{4\pi\hbar^2}{M} a \delta(\mathbf{r}). \tag{4.32}$$

As an illustrative example, let us take the particularly useful case of a potential interaction of the form $V(r) = -\alpha/r^6$, for $r > r_a$ and $V(r \leq r_a) = \infty$, where α and r_a are given constants. This is a repulsive potential at short distances $r > r_a$, and an attractive *van der Waals potential* $1/r^6$ at large distances. It can be shown that, in this case, the scattering length is [3]

$$a = C r_0 \left[1 - \tan\left(\phi - \frac{3\pi}{8} \right) \right], \tag{4.33}$$

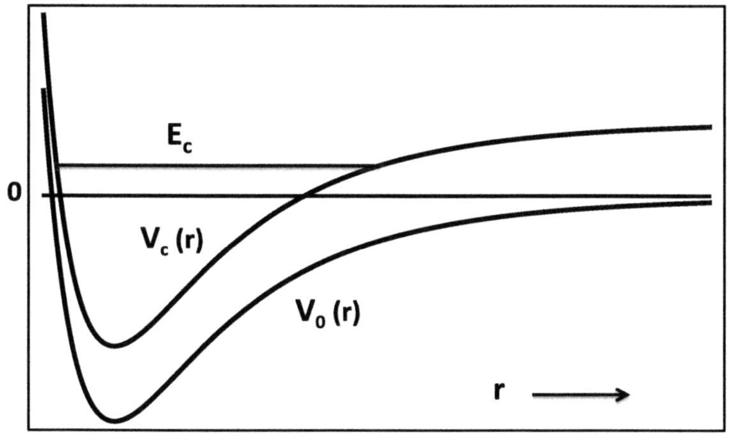

Fig. 4.1 Fashbach resonance: two potential courses, one closed $V_c(r)$ and the other open $V_0(r)$. Twp atoms colliding at energy $E \to 0$ can excite an internal state $E_c \simeq E$

where $C = \Gamma(3/2)/2\sqrt{2}\Gamma(5/4) \simeq 0.478$ is a constant, and r_0 and ϕ are determined by

$$r_0 = \left(\frac{M\alpha}{\hbar^2}\right)^{1/4}, \quad \phi = \frac{r_0^2}{2r_c^2}. \tag{4.34}$$

This result shows that the scattering length a is of order r_0. It also shows that a can be positive (which corresponds to an effective repulsive interaction) or negative (for an effective attractive interaction), depending on the value of the parameters ϕ and r_c. Such a simple van der Waals model is very useful for low energy scattering of alkali atoms.

4.2 Feshbach Resonances

Let us now focus our attention on resonant scattering processes, due to the interaction of discrete bound states with the continuum of unbound states considered above. This was first considered by Feshbach [4] in the context of nuclear physics and by Fano [5] in atomic physics. It was later introduced in the context of ultra-cold atom research [6], where they are called *Feshbach resonances*. A recent review is given by Chin et al. [7]. Feshbach resonances are used to control the strength of the atom-atom interactions, which has implications on the collision losses and well as on the mean field of condensates.

These resonant processes can be illustrated with the help of two potential energy curves (see Fig. 4.1), one open and the other closed, as described by $V_0(r)$ and $V_c(r)$, where r is the inter-atomic distance. The open channel is the entrance channel for the atomic collisions, and the closed channel allows for the existence of bound molecular states. Two atoms colliding at very low energy, $E \to 0$, can excite a bound state with energy E_c. Such a coupling can be tuned by an external magnetic field B, and the resulting scattering length takes the form

$$a(b) = a_0 \left(1 - \frac{\Delta b}{b - 1}\right), \tag{4.35}$$

where $b = B/B_0$ is the magnetic field strength normalized to the resonant value, and Δb is the resonance width. Here, a_0 is the value of the scattering length associated with the open channel $V_0(r = \infty)$, in the absence of coupling. Noting that both a_0 and Δb can be positive or negative, we realize that the scattering length can diverge to $\pm\infty$, at resonance.

Another interesting feature of Eq. (4.35) is observed for $b = 1 + \Delta b$, when the scattering length becomes equal to zero. At this point, the atom-atom interaction disappears, and we obtain a non-interacting ideal gas, which has many interesting applications in Bose-Einstein condensates and in atom interferometry. In practical terms, such a situation arises when $a(b)$ becomes of the order or smaller than the atom de Broglie wavelength.

An important simple example, for which exact analytical solutions for resonant scattering are possible, is that of scattering by a *spherical potential well.* Discrete bound states are observed for an attractive well, as defined by

$$U(r) = -U_0 H(r_c - r), \quad U_0 > 0, \tag{4.36}$$

where $H(x)$ is the *Heaviside function*, such that $H(x) = 1$ for $x > 0$, and $H(x) = 0$ otherwise. The radial wave equation (4.8) takes now a slightly different form

$$\frac{1}{r^2}\frac{d}{dr}\left(r^2 \frac{dR}{dr}\right) + \left[K^2 - \frac{l(l+1)}{r^2}\right] R = 0, \tag{4.37}$$

where $K^2 = k^2 + U_0$ and we have used $R \equiv R_{kl}$ to simplify the notation. In the external region $r > r_c$, the quantity K^2 is replaced by k^2, and the solution takes a form similar to (4.9), or

$$R^e(r) = B \left[j_l(Kr) - n_l(kr) \tan \delta_l \right]. \tag{4.38}$$

On the other hand, for the internal region $r < r_c$, the solution can be written as

$$R^i(r) = A' j_l(Kr), \tag{4.39}$$

where A' is a new constant of integration, and the term in $n_l(Kr)$ is ruled out to avoid diverge at $r \to 0$. These internal and external solutions are coupled at the boundary $r = r_c$ by the two continuity relations

$$R^i(r_c) = R^e \quad \text{and} \quad \left(\frac{dR^i}{dr}\right)_{r=r_c} = \left(\frac{dR^e}{dr}\right)_{r=r_c}. \tag{4.40}$$

By taking the ration of these two relations, we can eliminate the normalization constants A' and B, and use a single expression which allows us to determine the phase shift δ_l, which reads

$$\left(\frac{1}{R^i} \frac{dR^i}{dr} \right)_{r=r_c} = \left(\frac{1}{R^e} \frac{dR^e}{dr} \right)_{r=r_c} . \tag{4.41}$$

Using the above solutions (4.38) and (4.39), we get

$$\tan \delta_l = \frac{k j_l'(\rho^e) j_l(\rho^i) - K j_l(\rho^e) j_l'(\rho^i)}{k n_l'(\rho^e) n_l(\rho^i) - K n_l(\rho^e) n_l'(\rho^i)}, \tag{4.42}$$

where $\rho^e = k r_c$ and $\rho^i = K r_c$, and the derivatives are taken over the arguments, as $j_l' = dj_l(\rho)/d\rho$ and $n_l' = dn_l(\rho)/d\rho$. In the particular case of s-waves ($l = 0$), the Bessel and van Neumann functions are given by $j_0(\rho) = \sin \rho/\rho$ and $n_0(\rho) = -\cos \rho/\rho$, and Eq. (4.42) consequently reduces to

$$\tan \delta_0 = \frac{k \tan(K r_c) - K \tan(k r_c)}{K + k \tan(k r_c) \tan(K r_c)}. \tag{4.43}$$

The corresponding cross-section for the s-wave can then be determined by

$$\sigma_0 = \frac{4\pi}{k^2} \frac{1}{1 + \cot^2 \delta_0}. \tag{4.44}$$

For very low energies, such that $k \to 0$, we can use the expansion

$$k \cot \delta_0 \simeq -\frac{1}{a} + \frac{1}{2} r_0^2 k^2, \tag{4.45}$$

where a us the scattering length, and r_0 is the effective range of the interaction. In this limit, we have $K \to \sqrt{U_0}$, and the scattering length resulting from Eq. (4.44) is

$$a = r_c \left[1 - \frac{\tan\left(r_c \sqrt{U_0}\right)}{r_c \sqrt{U_0}} \right]. \tag{4.46}$$

For shallow potential wells, such that $r_c \sqrt{U_0} \ll 1$, this quantity is finite and negative. For increasing values of $r_c \sqrt{U_0} \to \pi/2$, the scattering length diverges, due to the coupling with the s-wave bound state. For even larger values of $r_c \sqrt{U_0}$, divergence will occur periodically at $(2n + 1)\pi/2$, with n integer.

In general, for $l \neq 0$, the cross-sections σ_l will show a similar behavior, taking the form

$$\sigma_l = \frac{4\pi}{k^2} (2l + 1) \frac{1}{1 + \cot^2 \delta_l}, \tag{4.47}$$

thus generalizing the result in (4.44). These cross-sections attain their maximum values when $\cot\delta_l \to 0$. In particular, near the resonance, we can use the expression

$$\cot\delta_l = \frac{E_R - E}{\Gamma(E_R)/2}, \tag{4.48}$$

where E_R is the resonant energy at which δ_l becomes multiple of $\pi/2$, and $\Gamma(E_R)$ determines the resonance width. Replacing this in (4.47), we obtain the *Breit-Wigner formula* for resonant scattering. As we have seen, such a resonant behavior results from the excitation of a bound state with a lifetime of the order of $\Delta t \simeq \hbar/\Delta E = \hbar\Gamma(E_R)$.

Let us now consider a more generic and formal description of resonant scattering. We assume that there are two kinds of quantum states, the bound states $|\phi\rangle$ and a continuum of unbound states $|E\rangle$ with energy E. We assume that these quantum states are orthogonal and normalized, which implies that $\langle\phi|E\rangle = 0$, and

$$\langle\phi|\phi\rangle = 1, \quad \langle E'|E\rangle = \delta(E - E'). \tag{4.49}$$

If H is the Hamiltonian of the system, then $E_\phi = \langle\phi|H|\phi\rangle$ is the energy of the bound state. Coupling between this bound state and the continuum is given by the potential $V(E) = \langle E|H|\phi\rangle$. We further assume that

$$\langle E'|H|E\rangle = E\delta(E - E'). \tag{4.50}$$

Let us now introduce the state vector $|\psi_E\rangle$ satisfying the eigenvalue equation

$$H|\psi_E\rangle = E|\psi_E\rangle. \tag{4.51}$$

We can represent this state vector as a linear combination of the orthonormal state vectors $|\phi\rangle$ and $|E\rangle$ defined above, in the form

$$|\psi_E\rangle = a_\phi|\phi\rangle + \int b_{E'}|E'\rangle dE', \tag{4.52}$$

where a_ϕ and b_E are the appropriate coefficients. Using Eq. (4.51), we can readily obtain

$$\langle\phi|H|\psi_E\rangle = a_\phi E, \quad \langle E'|H|\psi_E\rangle = b_{E'}E. \tag{4.53}$$

It is also obvious that, using the development (4.52), we can derive

$$\langle\phi|H|\psi_E\rangle = a_\phi E_\phi + \int b_{E'}V^*(E')dE'. \tag{4.54}$$

Similarly, we have

$$\langle E'|H|\psi_E\rangle = a_\phi V(E') + b_{E'}E'. \tag{4.55}$$

Now, if we compare these two equations with (4.53), we conclude that

$$a_\phi E = a_\phi E_\phi + \int b_{E'} V^*(E') dE', \tag{4.56}$$

and

$$(E - E') b_{E'} = a_\phi V(E'). \tag{4.57}$$

It is known from the theory of distributions that this last expression in equivalent to

$$b_{E'} = a_\phi V(E') \left\{ \mathcal{P} \frac{1}{E - E'} + Z(E') \delta(E - E') \right\}, \tag{4.58}$$

where $\mathcal{P}$ denotes the principal part, and $Z(E')$ is a given function of E', which determines the contribution of the singularity at $E = E'$. In order to determine $Z(E')$, we replace (4.58) in Eq. (4.56). This allows us to eliminate a_ϕ, and to write

$$E = E_\phi + \mathcal{P} \int \frac{|V(E')|^2}{E - E'} dE' + |V(E)|^2 Z(E). \tag{4.59}$$

Now, introducing the auxiliary quantities, $\Delta(E)$ and $\Gamma(E)$, defined as

$$\Delta(E) = \mathcal{P} \int \frac{\Gamma(E)}{E - E'} dE' \quad \text{and} \quad \Gamma(E) = |V(E)|^2, \tag{4.60}$$

we can finally write

$$Z(E) = \frac{E - E_\phi - \Delta(E)}{\Gamma(E)}. \tag{4.61}$$

Let us now use this result to determine the scattering length. For this purpose, we should notice that, at vary large distances, $r \to \infty$, the unbounded wave functions have the asymptotic behavior

$$|E\rangle \sim \sin[kr + \delta_0(E)], \tag{4.62}$$

where δ_0 is the s-wave phase shift in the absence of coupling with the bound state $|\phi\rangle$. On the other hand, we can also assume, at such large distances, where the bound state is absent, Eq. (4.52) can be approximately written as

$$|\psi_E\rangle \simeq \int b_{E'} |E'\rangle dE'. \tag{4.63}$$

Now, using (4.58), and integrating over E', we obtain

$$|\psi_E\rangle \simeq a_\phi V(E) \left\{ -\pi \cos[kr + \delta_0(E)] + Z(E) \sin[kr + \delta_0(E)] \right\}. \tag{4.64}$$

This can be rearranged to yield

$$|\psi_E\rangle \sim \sin[kr + \delta_0(E) + \delta_r(E)], \tag{4.65}$$

where the additional phase shift δ_r results from coupling with the bound state

$$\delta_r(E) = -\tan^{-1}\frac{\pi}{Z(E)} = -\tan^{-1}\frac{\pi\Gamma(E)}{E - E_R}, \tag{4.66}$$

with $E_R = E_\phi - \Delta(E)$. At this point, it should be noticed that, at very low energies, such that $E \to 0$ and $k \to 0$, the definition of the scattering length allows us to write $(ka_0) \to -\tan\delta_0 \simeq -\delta_0$. In what concerns the additional phase shift δ_r, it can be assumed to take a form similar to δ_0 in this limit, such that $\pi\Gamma(E) \to (ka_0)\Gamma_0$, where a_0 is the scattering length in the absence of coupling, and Γ_0 is a constant. Proceeding this way, we then compute the total scattering length, which simply reads

$$a = a_0\left(1 - \frac{\Gamma_0}{E_R}\right). \tag{4.67}$$

The importance of resonant scattering for the ultra-cold gas results from the dependence of the resonant energy E_R on external parameters, such as the static magnetic field B. Let us assume that, for a particular value $B = B_0$, we have $E_R = 0$. Then, by expanding the resonant energy around this value we get $E_R = (B - B_0)\delta\mu$, where $\delta\mu$ is the difference between the magnetic moments of the two colliding atoms μ_1 and μ_2, and the magnetic moment of the molecular bound state $|\phi\rangle$, such that $\delta\mu = (\mu_1 + \mu_2) - \mu_\phi$. This allows us to write the scattering length as a function of the static magnetic field in the form of Eq. (4.35), with $\Delta b = \Gamma_0/B_0\delta\mu$.

4.3 Collective Forces

Scattering effects associated with atom-atom collision processes dominate at short distances, but collective atomic interactions mediated by photons become dominant at scales larger than a given characteristic scale, called the *Debye length*, λ_D to be defined below. In the remaining of this chapter we will be mainly concerned with collective processes. We assume that the atomic cloud is sufficiently dense and cold, such that the Debye length becomes much shorter than the size of the cloud (in order to make the collective processes relevant) and the number of atoms inside the Debye sphere is much larger than one (in order to justify the use of a mean field approach). These two conditions are usually satisfied in current MOT experimental conditions, and will be specified below.

When a gas of cold atoms is confined in a magneto-optical trap and is submitted to the action of cooling laser beams, we can distinguish between two different kinds of forces acting on every single atom, $\mathbf{F} = \mathbf{F}_0 + \mathbf{F}_{\text{coll}}$. The first term $\mathbf{F}_0$ is due to the direct action of the laser cooling beams and the magnetic confinement field, and the second term $\mathbf{F}_{\text{coll}}$ is due to the collective interaction with all the other atoms.

Let us first consider the cooling force due to the three pairs of laser beams, as already discussed in previous chapters. It can be written as

$$\mathbf{F}_0 = \sum_{j=1}^{6} \mathbf{F}_j \,, \quad \mathbf{F}_j = \frac{\hbar \mathbf{k}_j \Gamma \Omega_j^2}{2\Omega_j^2 + 4(\Delta + \mathbf{k}_j \cdot \mathbf{v})^2 + \Gamma^2}. \tag{4.68}$$

We have seen that, in the limit of low atom velocities, such that $|v| \ll |\Delta/k|$, the resulting force can be written as

$$\mathbf{F}_0 = -\bar{\alpha} \cdot \mathbf{v}, \quad \bar{\alpha} = \sum_{j=1}^{6} \bar{\alpha}_j, \tag{4.69}$$

where $\bar{\alpha}$ is the total friction tensor, and the partial friction tensors associated with each laser beam are

$$\bar{\alpha}_j = -M \frac{\mathbf{k}_j \mathbf{k}_j}{2k^2} \beta_j \,, \quad \beta_j = \hbar k \Gamma \frac{2^4 \Omega_j^2 \Delta}{M(\Delta^2 + \Gamma^2/4)}. \tag{4.70}$$

Here we have assumed that $|\mathbf{k}_j| = k$. For simplicity, we will consider that laser beam intensities are identical ($\Omega_j^2 = \Omega_R^2$) further on.

If we now take into account the magnetic field force, due to the presence of a magnetic trap configuration, we have to make the transformation $\Delta \to \Delta - \mu \cdot \mathbf{B}$, where μ is the magnetic moment of the atom, and $\mathbf{B}$ the magnetic field. For small values of the Doppler and Zeeman effects, we can add an extra term to (4.69), and write

$$\mathbf{F}_0 = -\bar{\alpha} \cdot \mathbf{v} - \bar{\kappa} \cdot \mathbf{r}, \tag{4.71}$$

where $\bar{\kappa}$ is the spring constant tensor, with elements given by

$$\kappa_{ij} = -\frac{\partial F_{0i}}{\partial x_j}, \tag{4.72}$$

where the indices (i, j) now refer to the three spatial coordinates (x, y, z). Let us assume a typical MOT configuration, with the laser beams aligned with the axis, $\mathbf{e}_x$, $\mathbf{e}_y$ and $\mathbf{e}_z$, and the magnetic field coils along the z-axis. In the central region of the trap, we have seen that $B(\mathbf{r}) = A_0\sqrt{x^2 + y^2 + 4z^2}$, with a constant A_0. The resulting magnetic force is then

$$\mathbf{F}_B = -\hbar g_J \mu_B \frac{\partial B}{\partial \mathbf{r}} = -\kappa_i x_i, \tag{4.73}$$

with

$$\kappa_x = \kappa_y = \frac{\kappa_z}{2} = \frac{\hbar g_J \mu_B A_0}{\sqrt{r^2 + 4z^2}}. \tag{4.74}$$

Comparing with (4.71), we conclude that

$$\kappa_{ij} = k_i \delta_{ij}. \tag{4.75}$$

Let us now consider the collective force $\mathbf{F}_{\mathrm{coll}}$, resulting from the presence of the other atoms. This force results from the combined influence of two distinct processes, and can be written as

$$\mathbf{F}_{\mathrm{coll}} = \mathbf{F}_A + \mathbf{F}_R, \tag{4.76}$$

where $\mathbf{F}_A$ is a shadow force due to laser absorption, and $\mathbf{F}_R$ is a repulsive force resulting from the absorption of re-emitted photons. Let us examine them separately.

In order to understand the physical meaning of the shadow force, we first consider a one-dimensional analysis. Let us then consider the two laser beams propagating along the x-axis in opposite directions. Due to photon absorption, these two beams vary in space as

$$I_\pm(x) = I_0 \exp\left[-\sigma_L \int_{\pm\infty}^{x} n(x')dx'\right], \tag{4.77}$$

where I_0 is the initial laser intensity far away from the cloud, $I_0 = I_\pm(x \to \mp\infty)$, $n(x)$ is the density profile of the cloud, and σ_L is the laser absorption cross-section, which can be written as [8, 9]

$$\sigma_L = \frac{\sigma_0}{1 + I_0/I_s}, \tag{4.78}$$

where $I_0/I_s = 2\Omega_{\mathrm{tot}}^2/\Gamma^2$, with $\Omega_{\mathrm{tot}}^2 = 6\Omega_R^2$, and σ_0 is the laser absorption cross-section at resonance

$$\sigma_0 = \frac{3\lambda^2}{2\pi} = \frac{6\pi c^2}{\omega^2}. \tag{4.79}$$

For low intensities, $I_0 \ll I_s$, and $\Delta = \Gamma/2$, this leads to $\sigma_L = 2\sigma_0$, as used in [10]. As we have seen, the laser forces resulting from each of the two counter-propagating beams are proportional to the laser beam intensities and, for weak absorption such that $\sigma_L n_0 a \ll 1$, where n_0 us the maximum density and a the cloud radius, we can approximately write them as

$$F_\pm(x) \simeq \pm F_0 \left[1 - \sigma_L \int_{\pm\infty}^{x} n(x')dx'\right]. \tag{4.80}$$

Adding these two forces, we get the total force along the x-axis due to laser absorption. And, because this force is in the x-direction, we can write it in vectorial form, as

$$\mathbf{F}_A(x) = (F_+ + F_-)\mathbf{e}_x = -2F_0\sigma_L\mathbf{e}_x \int_{-\infty} n(x')dx'. \tag{4.81}$$

Generalizing to three dimensions and taking the divergence, we can easily arrive at

$$\nabla \cdot \mathbf{F}_A(\mathbf{r}) = -2\sigma_L^2 \frac{I_0}{c} n(\mathbf{r}), \tag{4.82}$$

where we have used $F_0 = \sigma_L I_0/c$. This result shows that $\mathbf{F}_A$ is an attractive force, which pushes the atoms towards the centre of the trap and therefore compresses the atomic cloud. This shadow force, or laser absorption force, $\mathbf{F}_A$, was first discussed by Dalibard [10]. It is clearly associated with the gradient of the incident laser intensity due to laser absorption by the atomic cloud.

A different kind of collective force, which is due to the successive emission and absorption of secondary photons, is the repulsive force $\mathbf{F}_R$. To properly describe its physical origin, let us consider the processes of photon emission and absorption. A given atom inside the trap will absorb and reemit photons at a rate $\sigma_L I$. The reemitted radiation I_R will then fall off as $1/r^2$, where r is the distance from the reemitting atom. This can be appropriately written as

$$I_R(\mathbf{r}) = \sigma_L \frac{I(\mathbf{r}')}{4\pi|\mathbf{r} - \mathbf{r}'|^2}, \tag{4.83}$$

where $\mathbf{r}'$ is the position of the absorbing atom. The force resulting from the interaction of the scattered intensity I_R with another atom at $\mathbf{r}$ will then be given by

$$F_R(\mathbf{r}) = \sigma_R \frac{I_R}{c} = \sigma_L\sigma_R \frac{I(\mathbf{r}')}{4\pi|\mathbf{r} - \mathbf{r}'|^2}, \tag{4.84}$$

where σ_R is the photon absorption cross-section [9]. This quantity differs from σ_L because the re-emitted light is different from the incident light. This is a difference in both the frequency distribution and the polarization state between the primary photons of the laser cooling beams and the secondary photons of the scattered fluorescent light.

The $1/r^2$ law is identical to that of an electrostatic force, which allows us to write for the total integrated force acting at a position $\mathbf{r}$ and resulting from the contribution of all the atoms inside the cloud, as

$$\mathbf{F}_R(\mathbf{r}) = \sigma_L\sigma_R \frac{I}{4\pi} c \int n(\mathbf{r}') \frac{(\mathbf{r} - \mathbf{r}')}{|\mathbf{r} - \mathbf{r}'|^3} d\mathbf{r}', \tag{4.85}$$

where for simplicity the intensity of the incident light $I \simeq I_0$ was assumed constant inside the cloud. Corrections due to the absorption process discussed above would lead to second order effects which can usually be neglected. In differential form, this is equivalent to the expression

$$\nabla \cdot \mathbf{F}_R(\mathbf{r}) = \sigma_L \sigma_R \frac{I}{c} n(\mathbf{r}). \tag{4.86}$$

In contrast to (4.82), this is a repulsive force which tends to push the atoms away from each other. It means that the atoms repel each other, as if they were charge particles. This collective force, $\mathbf{F}_R$, can be called a repulsive force, or a radiation trapping force, and was first considered by Sesko et al. [11]. It describes atomic repulsion, due to the radiation pressure of scattered photons on nearby atoms, and can be determined by a Poisson type of equation. The existence of an equivalent charge of the atoms will be discussed later. For the moment, let us note that the collective force which results from the superposition of the shadow and the repulsive forces is given by

$$\nabla \cdot \mathbf{F} = Q n(\mathbf{r}) \,, \quad Q = (\sigma_R - \sigma_L)\sigma_L \frac{I}{c}, \tag{4.87}$$

where we have dropped the suffix 'coll' in the force, for simplicity. In typical experimental conditions, the repulsive forces largely dominate over the shadow effect, and the quantity Q is positive [9, 12]. The physical implications of this quantity will be discussed later.

In order to establish the value of the new quantity Q, we need to determine the difference between the two cross-sections σ_R and σ_L. For that purpose, we first notice that the re-emitted secondary photons have two distinct parts: (1) the elastic Rayleigh scattering, with the same spectrum as the incident primary photons, and (2) the inelastic fluorescent part, which shows a triplet, usually called the *Mollow triplet*. This can be written as [13, 14]

$$I_R = I_{\text{el}} + I_{\text{inel}}, \quad I_{\text{el}} = \frac{I_R}{1+s}, \tag{4.88}$$

where $s = I/I_s$ is the saturation parameter. We can see that the inelastic part grows with the incident laser beam intensity. An estimate of σ_R can then be made [15] by noting that the absorption cross-section for the elastic part is the same as that for the primary photons, σ_L. We can then write

$$\sigma_R = \frac{I_{\text{el}}}{I_R}\sigma_L + \frac{I_{\text{inel}}}{I_R}\sigma_{\text{inel}}, \tag{4.89}$$

where σ_{inel} is the absorption cross-section for the inelastic scattered spectrum. It is usually larger than σ_L, because the absorption coefficient in the presence of the incident laser field is shifted towards the blue component of the Mollow triplet, which is closer to the atomic resonance. Its value can be estimated by assuming that it is equal to the usual absorption cross-section at resonance. We can write it as

$$\sigma_{\text{inel}} = \sigma_L \frac{1+s}{s + \Gamma^2/(\Gamma^2 + 4\Delta^2)}. \tag{4.90}$$

Replacing this in Eq. (4.89), and using (4.88), we finally get

$$(\sigma_R - \sigma_L) = \frac{s}{1+s} \frac{\sigma_L \Delta^2}{s(\Delta^2 + \Gamma^2/4) + \Gamma^2/4}. \tag{4.91}$$

A more accurate expression can be found in [9]. The first experimental estimate of this quantity was given in [12], as $\sigma_R \simeq 1.3\sigma_L$. The above expressions for the forces acting on the atomic clouds correspond to the simplest possible description of the laser cloud interaction, and can be used as a first approximation to model the fluid dynamics of the ultra-cold gas. We consider the basic oscillations of the gas associated with such forces by inserting these forces in the fluid equations for the ultra-cold gas, which can then be written as

$$\frac{\partial n}{\partial t} + \nabla \cdot (n\mathbf{v}) = 0 \, , \quad \frac{\partial \mathbf{v}}{\partial t} + \mathbf{v} \cdot \nabla \mathbf{v} = -\frac{\nabla P}{M} + \frac{\mathbf{F}}{M}, \tag{4.92}$$

where n and $\mathbf{v}$ are the mean density and mean velocity of the gas, and P is the gas pressure. The total force in this equation is the sum of the confining force and the collective force, $\mathbf{F} = -\nabla V_0 + \mathbf{F}_{\text{coll}}$, where V_0 is the external confining potential, and we have included the collective force $\mathbf{F}_{\text{coll}} = \mathbf{F}_A + \mathbf{F}_R$, as determined by the Poisson equation (4.87). These equations can be applied to describe the equilibrium states of the confined ultra-cold gas, as well the characteristic waves and oscillations which can be excited in such medium.

We can also refine the dynamical description of the atomic cloud by using a wave kinetic approach, similar to that used to describe the single atom cooling process. In a gas, the atom suffers the influence of the collective forces due to the other atoms. We can then write the corresponding wave kinetic equation by including the mean field potential V_{eff} associated with such forces, leading to a more detailed description of the gas, where kinetic processes such as Landau damping, kinetic instabilities and atom density correlations can be examine, as shown later.

4.4 Equilibrium Profiles

The description of the equilibrium properties of a cold atomic gas confined in a MOT is a long standing problem, as a consequence of the complexity of the optical processes. In particular, various density regimes of the non-condensed ultra-cold gas confined in a MOT have been identified, as resulting from the variation of the different trapping parameters, such as the number of atoms, the laser intensity, frequency detuning, magnetic field gradient and etc. Here we first give a phenomenological description of such regimes, along the lines of [8, 15]. We then propose a more quantitative approach based on the collective forces described above.

4.4.1 *Qualitative Discussion*

(i) – **Temperature limited regime**

For low atomic densities, the interaction between atoms is weak, and we can describe the gas as an ensemble of independent particles in the trap external fields, which include both the laser cooling beams and the static magnetic field. This is only valid when the characteristic scale of the collective interactions λ_D is much smaller than the size of the atomic cloud, a_T, as will be seen later. Anticipating the definition of λ_D, we can then write

$$\lambda_D^2 \equiv \frac{\epsilon_0 T}{n q_{\text{eff}}^2} = \frac{T}{nQ} \gg a_T^2, \tag{4.93}$$

where n is the atom density, T is the atom temperature in energy units, and q_{eff} the *atom effective charge*. Therefore, the low atomic density regime is valid for densities below a given limit, as defined by

$$n < n_{\text{lim}} \equiv \frac{T}{Q a_T^2}. \tag{4.94}$$

We can see that, for a fixed temperature, this limit scales as a_T^{-2}. In order to be more specific, we now have to determine the value of the cloud size, a_T. In steady state, the equipartition theorem states the equality of the kinetic and potential energies inside the harmonic trap, such that

$$\frac{1}{2} M v_i^2 = \frac{1}{2} \kappa_{ii} r_i^2, \quad \kappa_{ij} = -\frac{\partial F_i}{\partial r_j}, \tag{4.95}$$

where $i, j = (x, y, z)$ represent the three orthogonal components, κ_{ij} are the elements of the string tensor, and F_i the components of the confining force. In the central region of the quadrupole magnetic field of the trap, with the coil axis oriented along the z-direction, we can define the trap string constant as

$$\kappa \equiv 2\kappa_x = 2\kappa_y = \kappa_z, \tag{4.96}$$

and we can define the cloud size as

$$a_T \equiv a_x/\sqrt{2} = a_y/\sqrt{2} = a_z. \tag{4.97}$$

Assuming that the atom temperature in steady state is isotropic, which can be guaranteed by a low level of atom-atom collisions, we get from Eq. (4.95) the following simple relation between the cloud size and the temperature

$$\frac{T}{2} = \frac{1}{2} \kappa a_T^2 \rightarrow a_T = \sqrt{\frac{T}{\kappa}}, \tag{4.98}$$

where, as we have seen before, the string constant κ is proportional to the magnetic field gradient, $\kappa \propto \partial B/\partial z$. Using (4.94), we then get for the density limit of this low density regime the value $n_{\lim} = \kappa/Q$. This limit, below which the cloud behaves as a collection of independent atoms, is therefore inversely proportional to the field amplitude of the laser cooling beams, or $n_{\lim} \propto 1/\sqrt{I}$.

At thermal equilibrium, the Gaussian velocity distribution of the atoms inside the cloud implies the existence of a Gaussian density profile, according to Eqs. (4.95) and (4.98), which can be written as

$$n(r) = n(0) \exp\left[-\frac{r^2}{2a_T^2}\right], \quad n(0) = \frac{N}{2(\sqrt{2\pi}a_T)}, \tag{4.99}$$

where $n(0)$ is the peak atomic density at the cloud centre, and N is the total number of atoms in the cloud. It is clear that, in this regime, the size a_T is independent of the total number of atoms, as stated by (4.98) and, because the temperature T is also independent of N, the peak density $n(0)$ is proportional to N. This feature gives a signature of the low density regime, which was identified by the experiments. Such a regime has been observed for $N \leq 10^4$ atoms.

(ii) – **Multiple scattering regime**

For a larger number of atoms, such that inequality (4.94) is reversed, the collective forces are no longer negligible, and long range atom-atom interactions mediated by multiple photon scattering have to be taken into account. In this new regime, the trapping force is not balanced by the kinetic force as before, but by the collective repulsive force which is dominant. By taking the divergence of the trapping force $-\kappa\mathbf{r}$, and noting that $\nabla \cdot \mathbf{r} = 3$, we get

$$\nabla \cdot \kappa\mathbf{r} = \nabla \cdot F_{\mathrm{coll}} = Qn, \tag{4.100}$$

which shows the existence of an equilibrium density, $n = n_0$, such that $3\kappa = Qn_0$, or

$$n_0 = \frac{3\kappa}{Q} = \frac{3\kappa c}{I_0\sigma_L^2(\sigma_R/\sigma_L - 1)}, \tag{4.101}$$

where I_0 is here the total light intensity associated with the six trapping laser beams. This means that the atom density is nearly constant across the cloud, in contrast with the Gaussian profile of the temperature limited regime.

On the other hand, this equilibrium density n_0 is independent of the total number of atoms N, which means that by increasing this number we will increase the cloud size. This is actually an important experimental signature of the multiple scattering regime. Assuming that the cloud is approximately spherical, we obtain for the cloud size

$$a_{MS} = \left(\frac{3}{4\pi}\frac{N}{n_0}\right)^{1/3} = \left(\frac{NQ}{4\pi\kappa c}\right)^{1/3}. \tag{4.102}$$

(iii) – **Two-components regime**

For even higher number of atoms N, a third regime is eventually attained, where a dense core is surrounded by a low density halo of gas. The dense core can still be described by the above multiple scattering regime, whereas the outside region is weakly confined and can be described by a simple Gaussian model. Transition from the outer to the inner region can be established by saying that the local atom density n at some critical radius r_c is equal to the limit density n_{lim} defined by Eq. (4.94). We therefore get a multiple scattering regime for $r < r_c$, and a temperature limited regime of $r > r_c$.

An alternative definition of the transition radius r_c can be given in terms of purely atomic properties (see Ref. [8]), as the region where the Zeeman shift of the upper energy levels becomes equal to the light sift of the lower energy level of the laser cooling transition. This would lead to

$$\mu_B b r_c \simeq \hbar \frac{\Omega_R^2}{\Delta} \rightarrow r_c = \frac{\hbar \Omega_R^2}{\Delta \mu_B b}, \quad b \equiv \frac{\partial B}{\partial z}. \tag{4.103}$$

These two definitions of r_c are obviously not equivalent and, although the arguments based on a mean field theory point strongly to the validity of the first definition, only a careful experimental analysis will clarify the real nature of the transition region.

4.4.2 *Quantitative Model*

A more quantitative and unifying model of the equilibrium density profiles in a trap can be established using the mean field perspective. Our starting point is given by above fluid equations (4.92) for the atomic cloud, where both the kinetic pressure and the collective force due to multiple scattering are taken into account. This is particularly well adapted to the description of the first two regimes above, and can be extended to account for the two-components regime, as discussed next. Density equilibrium profiles can be derived from the fluid equations, by assuming the hydrostatic conditions $\partial/\partial t = 0$ and $\mathbf{v} = 0$. We are then reduced to

$$\frac{\nabla P}{n} = -\nabla V_0 + \mathbf{F}_{\text{coll}}. \tag{4.104}$$

In order to proceed further, we need to assume an equation of state relating the gas pressure P to the atom density n. For this purpose, we assume a generic polytropic equation of state of the form $P = C(T)n^\gamma$, where $C(T)$ is some unspecified function of the temperature, and γ is the usual polytropic exponent. Replacing this in Eq. (4.104), and applying the divergence operator, we get

$$C(T)\frac{\gamma}{(\gamma - 1)} \nabla^2 \left[n^{(\gamma-1)} \right] = -M\nabla^2 V_0 + Qn. \tag{4.105}$$

In what follows, we focus our attention on radially symmetric profiles, where the trapping potential can be written as

$$V_0(r) = \frac{1}{2} M \omega_0^2 r^2, \tag{4.106}$$

with ω_0 determining the restoring force of the harmonic trap. Writing the Laplacian operator ∇^2 in spherical coordinates, we can then reduce Eq. (4.105) to the following form

$$C(T) \frac{\gamma}{(\gamma - 1)} \frac{1}{r^2} \frac{\partial}{\partial r} \left[r^2 \frac{\partial}{\partial r} n^{(\gamma-1)} \right] + M\omega_0^2 - Qn = 0. \tag{4.107}$$

It is convenient to rewrite this equation using dimensionless variables, by introducing the new quantities

$$\theta = \left(\frac{n}{n_0} \right)^{(\gamma-1)} \quad \text{and} \quad \xi = \frac{r}{R_\gamma}, \tag{4.108}$$

where n_0 is some equilibrium density to be specified later, and the radius R_γ is chosen such that

$$R_\gamma = \left[\frac{P_0}{3 n_0 M \omega_0^2} \right]^{1/2}, \tag{4.109}$$

where $P_0 = C(T) n_0^\gamma$. In terms of these new variables, the equilibrium profile Eq. (4.107) can then be written in its final form, as

$$\frac{\gamma}{(\gamma - 1)} \frac{1}{\xi^2} \frac{\partial}{\partial \xi} \left(\xi^2 \frac{\partial}{\partial \xi} \theta \right) - \nu_{\text{eff}}^2 \theta^{1/(\gamma-1)} + 1 = 0, \tag{4.110}$$

where we have also used the auxiliary quantity $\nu_{\text{eff}}^2 = Q n_0 / 3 M \omega_0^2$. This quantity can be seen as a normalized atomic effective charge, describing the effect of collective atom interactions. The new Eq. (4.110) is very interesting as it closely resembles the *Lane-Emde equation*, well known in astrophysics [16]. It is however physically distinct, because it describes a repulsive interaction between the atoms in the gas, and not attractive gravitational interactions.

In order to understand its meaning, it is now useful to discuss the solutions of Eq. (4.110) in some simple but physically relevant situations, and to establish a link with the above qualitative analysis of the density equilibrium regimes in a trap. We first examine the case where the pressure effects are negligible, and we can use $\gamma = 0$. We then get a simple solution, $\theta(\xi) = 1/\nu_{\text{eff}}^2$. This is equivalent to

$$\theta(\xi) \equiv \frac{n}{n_0} = \frac{3 M \omega_0^2}{Q n_0} \tag{4.111}$$

and corresponds to the *water-bag* solution of the form

$$n(r) = n_0 \nu_{\rm eff}^2 H(a_0 - r), \tag{4.112}$$

where the cloud size is such that $n_0 \nu_{\rm eff}^2 V_{\rm ol} = N$, where $V_{\rm ol}$ is the volume of the cloud. This leads to

$$a_0 = \left(\frac{3}{4\pi} \frac{N}{n_0 \nu_{\rm eff}^2} \right)^{1/3}. \tag{4.113}$$

We can see that, by making the replacement $n_0 \nu_{\rm eff}^2 \rightarrow n_0$, this coincides with Eq. (4.101). This water-bag model therefore confirms on a more quantitative basis the above result for the multiple scattering regime, when the exchange of photons between atoms dominates over the kinetic pressure.

In the opposite case, where the cloud density is low and the collective effects are negligible, we can take $\nu_{\rm eff}^2 \simeq 0$. The equilibrium profile Eq. (4.110) now reduces to

$$\frac{\partial}{\partial \xi} \left(\xi^2 \frac{\partial}{\partial \xi} \theta \right) + \frac{\gamma - 1}{\gamma} \xi^2 = 0. \tag{4.114}$$

This has an obvious solution of the form

$$\theta(\xi) = \theta_0 - \frac{\gamma - 1}{6\gamma} \xi^2, \tag{4.115}$$

where θ_0 is a constant. Taking, for convenience, this constant equal to one, we can write the corresponding atom density profile as

$$n(r) = n_0 \left[1 - \frac{\gamma - 1}{6\gamma} \left(\frac{r}{R_\gamma} \right)^2 \right]^{1/(\gamma-1)}. \tag{4.116}$$

This solution can appropriately describe the density profiles of a cloud in the temperature limited regime. In particular, we should notice that, for $\gamma = 1$ we recover the Gaussian profile of Eq. (4.99).

A detailed numerical analysis shows that for increasing values of the parameter $\nu_{\rm eff}$, the solutions of Eq. (4.110) show a smooth transition from the Gaussian to the water-bag solution (see Fig. 4.2). This also suggests that a careful measurement of the density profiles in the trap could give information on the equation of state of the confined ultra-cold gas.

4.5 Coulomb Expansion

An interesting effect can be observed when, at some instant $t = 0$ the magnetic confinement is switched off, thus leaving the atomic cloud free to expand, due to the combined effect of the kinetic pressure and the repulsive collective forces. When the

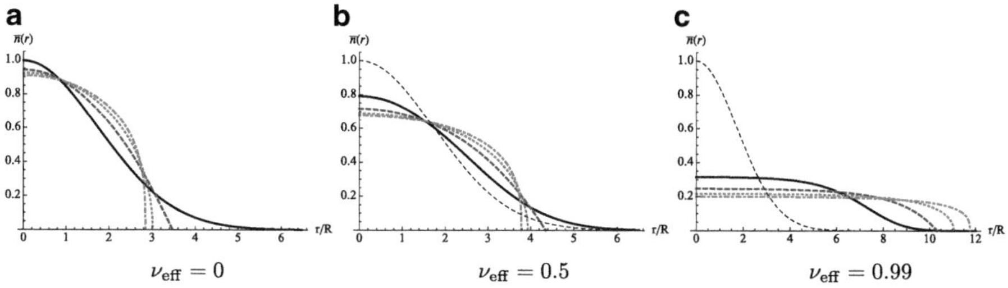

Fig. 4.2 Effect of the long-range interaction on the density profile for different polytropic exponents. The *black dashed line* is the gaussian profile obtained for the isothermal case ($\nu_{\text{eff}} = 0$ and $\gamma = 1$) and is used here to normalize the numerical solutions. *Thick lines* are obtained for $\gamma = 1$ (*black full line*), $\gamma = 2$ (*blue dashed line*), $\gamma = 3$ (*orange dotted line*) and $\gamma = 4$ (*green dash-dotted line*). (**a**) $\nu_{\text{eff}} = 0$. (**b**) $\nu_{\text{eff}} = 0.5$. (**c**) $\nu_{\text{eff}} = 0.99$

collective forces dominate this becomes a Coulomb type of expansion [9], similar to that observed in laser irradiated plasmas [17, 18]. In order to examine this situation, we consider the fluid equations for the ultra-cold gas, which can be written as

$$\frac{\partial n}{\partial t} + \nabla \cdot \Gamma = 0, \tag{4.117}$$

$$\frac{d\mathbf{v}}{dt} = -\frac{\boldsymbol{\nabla} P}{M} + \frac{\mathbf{F}}{M} - \beta \mathbf{v}, \tag{4.118}$$

where $d/dt = \partial/\partial t + \mathbf{v} \cdot \nabla$, β is the viscosity of the gas due to laser cooling, and $\Gamma = n\mathbf{v}$ is the atom flux. The force $\mathbf{F}$ is the Coulomb type of force associated with the atom effective charge, and determined by the equation $\boldsymbol{\nabla} \cdot \mathbf{F} = Qn$. On a time scale much larger than the viscosity damping time $1/\beta$, we can consider $d/dt \simeq 0$, and the momentum equation (4.118) provides the following steady state solution

$$\mathbf{v} = \frac{\mathbf{F}}{\beta} - \frac{\boldsymbol{\nabla} P}{M\beta}. \tag{4.119}$$

Let us use the polytropic equation of state $P = C(T)n^{\gamma}$, as considered before. In this case, we can write the corresponding atom flux as

$$\Gamma = \frac{n\mathbf{F}}{\beta} - D\boldsymbol{\nabla} n, \tag{4.120}$$

where we have introduced the diffusion coefficient D, as defined by

$$D = \frac{u_s^2}{\beta}, \quad \text{with} \quad u_s^2 = \frac{\gamma P}{M}. \tag{4.121}$$

Replacing this in the continuity equation (4.117), we get

$$\frac{\partial n}{\partial t} + \frac{1}{\beta}\nabla \cdot (n\mathbf{F}) - \nabla \cdot (D\nabla n) = 0. \tag{4.122}$$

This equation determines the slow time (in comparison to the single-atom relaxation time $1/\beta$) evolution of the cloud density profile in the expansion phase. In order to illustrate such an expansion, we consider the water-bag model for the density profile, with an initial spherical cloud with radius r_0 and uniform density n_0. We have seen that this water-bag model is adequate for the cloud equilibrium in the multiple scattering regime. We then have, for $t = 0$, the density profile

$$n(r, t = 0) = n_0 H(r_0 - r), \quad n_0 = \frac{3N}{4\pi r_0^3}, \tag{4.123}$$

where $H(x)$ is the Heaviside function, r is the radial coordinate, and N the total number of atoms. We can see that the diffusion term is only non-zero at the boundary, and can be ignored. We can then write the approximate evolution equation

$$\frac{\partial n}{\partial t} + \frac{Qn^2}{\beta} = 0, \tag{4.124}$$

The latter can be easily integrated, leading to

$$n(t) = \frac{n_0}{(1 + t/\tau_e)}, \quad \tau_e = \frac{\beta}{Qn_0}, \tag{4.125}$$

where τ_e is the characteristic expansion time. As a result, the cloud radius $R(t)$ increases with time, as determined by the condition $n(t) = 3N/4\pi R(t)^3$. The cloud volume increases at the same rate as the density decreases, and we get the following expansion law

$$R(t) = r_0 \left(1 + \frac{t}{\tau_e}\right)^{1/3}. \tag{4.126}$$

Such a description is of course valid in the bulk of the cloud, but not at the boundary $r = r(t)$, where diffusion effects can lead to additional expansion. However, diffusive corrections are small as long as the repulsive force at the boundary remains large as compared with the diffusive flux, as shown by Eq. (4.120), or

$$|\mathbf{F}| \gg D\beta \left|\frac{\nabla n}{n}\right| \simeq \frac{D\beta}{R(t)}. \tag{4.127}$$

Let us now estimate the cooling effect associated with the cloud expansion. For a spherically symmetric cloud, Eq. (4.120) predicts a radial velocity v_r given by

$$v_r = \frac{F}{\beta} - \frac{D}{n}\frac{dn}{dr}. \tag{4.128}$$

For the water-bag density profile, and neglecting diffusion effects at the cloud boundary, we can write the radial velocity profile simply as

$$v_r(r) = \frac{F(r)}{\beta} = \frac{Qnr}{3\beta}, \tag{4.129}$$

where the Gauss theorem was used. The maximum expansion velocity, attained at the boundary, is given by

$$v_r(r = R(t)) = \frac{QnR(t)}{3\beta} = \frac{Q}{\beta}\frac{N}{4\pi R(t)^2}. \tag{4.130}$$

Let us now define temperature of the expanding gas as $T \equiv M\langle v^2\rangle \simeq Mv_r(R)^2$. This allows us to establish the temperature evolution law as

$$T(t) = \frac{T_0}{(1 + t/\tau_e)^{4/3}}. \tag{4.131}$$

These simple expansion laws have been well verified by experiments [9]. Finally, let us go back to Eq. (4.122) and consider the case where diffusion effects dominate and condition (4.127) is reversed. Expansion is now determined by the diffusion equation

$$\frac{\partial n}{\partial t} - \nabla \cdot (D\nabla n) = 0. \tag{4.132}$$

At this point, it should be noticed that the diffusion coefficient also depends on the gas density n, as shown by Eq. (4.121), where P is determined by the polytropic equation of state. We therefore have

$$\nabla \cdot (D\nabla n) = \frac{\gamma D}{n}|\nabla n|^2 + D\nabla^2 n \tag{4.133}$$

Let us assume a spherically symmetric problem, and assume an initial Gaussian profile, as described by

$$n(r, t = 0) = C_0 \exp\left(-\frac{r^2}{r_0^2}\right). \tag{4.134}$$

We can then try a variational solution for $t > 0$ with the same Gaussian form, as given by

$$n(r, t) = C(t) \exp\left[-\frac{r^2}{R(t)^2}\right], \tag{4.135}$$

where the quantity $C(t)$ is determined by the normalization condition

$$N = \int n(r, t)dr \;\rightarrow\; C(t) = \frac{2N}{R(t)\sqrt{\pi}}. \tag{4.136}$$

Inserting (4.135) in the diffusion equation (4.132), we can confirm that this is the desired solution provided that

$$\gamma = 1, \quad R(t) = \left[2 \int_0^t D dt + r_0^2 \right]^{1/2}. \tag{4.137}$$

We notice that the restriction $\gamma = 1$ is certainly adequate, because we have seen that in this case the equation of state predicts a purely Gaussian profile for the low density regime, as assumed in (4.135). Such a solution is therefore useful near the boundaries of the expanding cloud, where the collective forces are week and the diffusion process can therefore dominate.

References

1. J. Weiner, *Cold and Ultracold Collisions in Quantum Microscopic and Mesoscopic Systems* (Cambridge University Press, Cambridge/New York, 2003)
2. J. Weiner, V. Bagnato, S. Zilio, P.S. Julienne, Experiments and theory of cold and ultra cold collisions. Rev. Mod. Phys. **71**, 1 (1999)
3. L.D. Landau, E.M. Lifshitz, *Quantum Mechanics: Nonrelativistic Theory* (Pergamon, Oxford/New York, 1977)
4. H. Feshbach, Ann. Phys. **19**, 287 (1962)
5. U. Fano, Phys. Rev. **124**, 1866 (1961)
6. W.C. Stwalley, Phys. Rev. Lett. **37**, 1628 (1976)
7. C. Chin, R. Grimm, P. Julienne, E. Tiesinga, Feshbach resonances in ultracold gases. Rev. Mod. Phys. **82**, 1225 (2010)
8. C.G. Townsend, N.H. Edwards, C.J. Cooper, K.P. Zetie, C.J. Foot, A.M. Steane, P. Szriftgiser, H. Perrin, J. Dalibard, Phys. Rev. A **52**, 1423 (1995)
9. L. Pruvost, I. Serre, H.T. Duong, J. Jortner, Phys. Rev. A **61**, 053408 (2000)
10. J. Dalibard, Opt. Commun. **68**, 203 (1988)
11. D.W. Sesko, T.G. Walker, C.E. Wieman, J. Opt. Soc. Am. B **8**, 946 (1991)
12. T. Walker, D. Sesko, C. Wieman, Phys. Rev. Lett. **64**, 408 (1990)
13. B.R. Mollow, Phys. Rev. **188**, 1969 (1969)
14. R. Loudon, *The Quantum Theory of Light*. Oxford Science Publications, 3rd edn. (Oxford University Press, Oxford/New York, 2000)
15. A.M. Steane, M. Chowdhury, C.J. Foot, J. Opt. Soc. Am. B **9**, 2142 (1992)
16. S. Chandrasekhar, *An Introduction to the Study of Stellar Structure* (Courier Dover Publications, New York, 1967)
17. I. Last, I. Schek, J. Jortner, J. Chem. Phys. **107**, 6685 (1997)
18. M. Eloy, R. Fonseca, J.T. Mendonçca, R. Bingham, J. Plasma Phys. **73**, 635 (2007)

Chapter 5
Waves and Oscillations in Clouds

In the previous chapter, we have discussed the effects of the long-range interaction in cold atomic traps and have demonstrated some of its implications on the equilibrium features of the system. In the present chapter, we discuss some of the dynamical effects associated with such Coulomb-like interactions. We will discuss the case of two kind of oscillations: small and large scale oscillations. The first are related with the elementary excitations that can take place in a homogenous system, while the second refer to collective modes of a trapped system, occurring at wavelengths comparable with the size of the system. Centre-of-mass oscillations, Tonks-Dattner modes, breathing modes and surface modes are introduced. We then discuss the nonlinear coupling between different modes in the atom cloud, with emphasis on the coupling between the centre-of-mass oscillation and an hybrid acoustic wave. We also show that the linear mode analysis can be extended to a large spectrum of waves, leading to quasi-linear diffusion in the atom velocity space. Finally, the interesting concept of a phonon laser, and its possible realization in a MOT is briefly discussed.

5.1 Hybrid Sound

5.1.1 Fluid Description

We first assume oscillations that can be excited in the cold gas with a wavelength much smaller that the size of the system (or the radius of the trap a). The medium can therefore be assumed as infinite. We then assume that the equilibrium state of the gas is perturbed by oscillations with frequency ω and wavevector $\mathbf{k}$. In the sense of linear response theory, we linearize the fluid and Poisson equations, by defining perturbations around the equilibrium quantities

$$n = n_0 + \tilde{n}, \quad \mathbf{F} = \mathbf{F}_T + \delta\mathbf{F}_{\text{coll}}, \quad \mathbf{v} = \mathbf{v_0} + \delta\mathbf{v}. \tag{5.1}$$

J.T. Mendonça and H. Terças, *Physics of Ultra-Cold Matter*, Springer Series on Atomic, Optical, and Plasma Physics 70, DOI 10.1007/978-1-4614-5413-7_5,

Since the trapping force $\mathbf{F}_T$ defines only the equilibrium quantities and plays no role in the modes we are about to describe, we drop the subscript 'coll' for the perturbation in the collective force $\delta\mathbf{F}_{\text{coll}}$ in (5.1). To describe the dynamics of the system, we make use of the fluid equations (4.92)

$$\frac{\partial n}{\partial t} + \nabla \cdot (n\mathbf{v}) = 0 \tag{5.2}$$

and

$$\frac{\partial \mathbf{v}}{\partial t} + \alpha\mathbf{v} + \mathbf{v}\cdot\nabla\mathbf{v} = -\frac{\nabla P}{M} + \frac{\mathbf{F}}{M}, \tag{5.3}$$

together with the polytropic equation of state for the pressure P, which was also considered in the previous chapter, $P(n) \sim n^{\gamma}$. After linearizing the previous equations, we may easily obtain

$$\left[\frac{\partial}{\partial t}\left(\alpha + \frac{\partial}{\partial t}\right) + \omega_p^2 - u_S^2\nabla^2\right]\tilde{n} = 0, \tag{5.4}$$

where we have defined the effective plasma frequency $\omega_p = \sqrt{Qn_0/M}$, and u_S can be identified with the sound speed

$$u_S^2 = \gamma\frac{P_0}{Mn_0}, \tag{5.5}$$

and P_0 is the equilibrium gas pressure. Assuming a space-time dependence of the perturbations $\tilde{n}$ and $\delta\mathbf{F}$ of the form $\exp(i\mathbf{k}\cdot\mathbf{r} - i\omega t)$, with a complex frequency $\omega = \omega_r + i\omega_i$, we obtain for the dispersion relation and for the corresponding damping rate, the values

$$\omega_r^2 = \omega_p^2 + k^2u_S^2 + \frac{3}{4}\alpha^2, \quad \omega_i = \frac{\alpha}{2}. \tag{5.6}$$

In the limit of very low viscosity $\alpha \ll \omega_p$, the later dispersion relation reduces to $\omega^2 = \omega_p^2 + k^2u_S^2$, see Fig. 5.1, which is formally identical to the dispersion relation of electron plasma waves in ionized media (also known as *Langmuir waves*) but where the electron thermal velocity $v_{\text{th}} = \sqrt{T_e/m_e}$, (with T_e and m_e representing the electron temperature and mass, respectively) is replaced by the sound velocity divided by a numerical factor $u_S/\sqrt{3}$. Plasma waves will be discussed later, in the this book.

The density wave mode described by Eq. (5.6) contains elements of both the electron plasma waves and acoustic waves. It possesses a lower cut-off, given by $\omega_r = \sqrt{\omega_p^2 + 3\alpha^2/4}$, which is absent in the acoustic modes, but is typical on an electron plasma wave. On the other hand, its phase velocity tends to the sound velocity u_S and becomes weakly dispersive as an acoustic wave. Its corresponding quasi-particles can therefore be seen as hybrid entities, somewhere

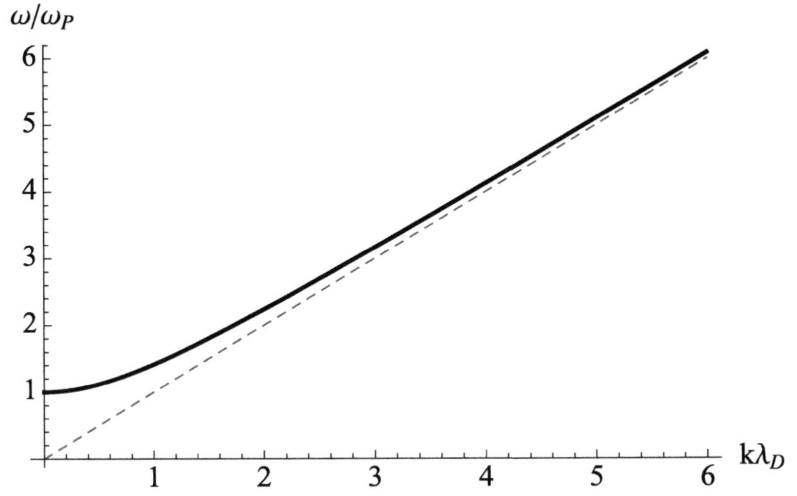

Fig. 5.1 Dispersion relation ω plotted against $k\lambda_D$. The *red dashed line* represents the acoustic asymptotic behavior of the hybrid waves

between plasmons and phonons, and for this reason we can call then *hybrid phonons*. These modes were first considered in [1].

5.1.2 Kinetic Approach

In a previous chapter we have shown that the wave equation describing the evolution of an atom inside the trap can be transformed into a wave kinetic equation of the Wigner-Moyal type, by introducing the correlation function $K(\mathbf{r}, \mathbf{s}, t) = \langle \mathbf{r} + \mathbf{s}/2 \mid \mathbf{r} - \mathbf{s}/2 \rangle$, where $\mathbf{r}$ is the center of mass the atom and $|\mathbf{r}, t\rangle$ is the atomic state vector describing its translational motion with respect to the laboratory frame. Here we use again the Wigner function, which is the Fourier transform

$$W(\mathbf{r}, \mathbf{q}, t) = \int K(\mathbf{r}, , \mathbf{s}, t) \exp(-i\mathbf{q} \cdot \mathbf{s})\, d\mathbf{s}. \tag{5.7}$$

and $\hbar\mathbf{q}$ is the momentum of the atom. As we have previously seen, the quantity W evolves according to an equation of the form

$$\left(\frac{\partial}{\partial t} + \mathbf{v} \cdot \frac{\partial}{\partial \mathbf{r}}\right) W = \frac{-i}{\hbar} \int V(\mathbf{k}') \left[W^{(-)} - W^{(+)}\right] e^{i\mathbf{k}' \cdot \mathbf{r}} \frac{d\mathbf{k}'}{(2\pi)^3}, \tag{5.8}$$

If the atom is not isolated, but is located inside an atomic cloud confined in a MOT, it suffers the influence of a mean field potential $V_{\text{coll}}(\mathbf{r})$, which results from the exchange of scattered photons with the other atoms, This means that here the quantity $V(\mathbf{k}')$ is the Fourier transform of the total potential $V(\mathbf{r}) = V_T(\mathbf{r}) + V_{\text{coll}}(\mathbf{r})$. Again, as we want to focus on the atom density fluctuations (elementary excitations) of the system, we neglect the trapping term V_T. The goal of the following calculations is to generalize the previous simple treatment, based on the fluid equations, to the kinetic quantum domain. This allows us to include the effects of phonon recoil, in formal analogy with what we have done in Chap. 3 for the case of photon recoil.

The collective potential is determined by the local atomic density $n(\mathbf{r})$, as

$$\nabla^2 V_{\text{coll}} = -Qn \equiv -Q \int W(\mathbf{r}, \mathbf{v}, t) d\mathbf{v}, \tag{5.9}$$

where we have considered the following normalization of the Wigner function

$$\int d\mathbf{r} \int d\mathbf{v}\, W(\mathbf{r}, \mathbf{v}, t) = \int n(\mathbf{r}, t) d\mathbf{r} = N. \tag{5.10}$$

In this sense, Eq. (5.8) will describe the evolution of N identical atoms, assumed as independent, except for the mean-field potential V_{coll} which creates an effective collective force associated with the exchange of photons between nearby atoms. We now consider the linear evolution of atom density perturbations around some equilibrium state defined by W_0, as determined by the confining potential V_T, and by the equilibrium collective potential. We therefore assume perturbations of the atom quasi-probability distribution W and of the collective potential V_{coll} of the form

$$\tilde{W}, \tilde{V}_{\text{coll}} \propto \exp(i\mathbf{k} \cdot \mathbf{r} - i\omega t). \tag{5.11}$$

Perturbative analysis of Eqs. (5.8) and (5.9) then lead to the following two expressions

$$\tilde{V}_{\text{coll}} = \frac{Q}{k^2} \int \tilde{W} d\mathbf{v}\,, \quad \tilde{W} = \tilde{V}_{\text{coll}} \frac{[W_0^{(-)} - W_0^{(+)}]}{\hbar(\omega - \mathbf{k} \cdot \mathbf{v})}, \tag{5.12}$$

with the quantities

$$W_0^{(\pm)} = W_0(\mathbf{v}_\pm) \equiv W_0(\mathbf{v} \pm \hbar\mathbf{k}/2M). \tag{5.13}$$

From these equations we can easily get the kinetic dispersion relation for density perturbations, which reads [2]

$$1 - \frac{Q}{\hbar k^2} \int \frac{[W_0^{(-)} - W_0^{(+)}]}{(\omega - \mathbf{k} \cdot \mathbf{v})} d\mathbf{v} = 0. \tag{5.14}$$

Before discussing the properties of the dispersion relation in this exact form, it is useful to consider its quasi-classical limit, where the momentum carried by an emitted or absorbed phonon $\hbar\mathbf{k}$ can be considered negligible with respect to the atomic translational momentum $\hbar\mathbf{q} = M\mathbf{v}$. In this case, we can use the approximation

$$W^{(\pm)} \simeq W(\mathbf{r}, \mathbf{q}, t) \pm \frac{\mathbf{k}}{2} \cdot \frac{\partial W}{\partial \mathbf{q}} + \frac{\partial}{\partial \mathbf{q}} \cdot \frac{\mathbf{kk}}{4} \cdot \frac{\partial}{\partial \mathbf{q}} W \pm \ldots, \tag{5.15}$$

which simply yields

$$\left[W^{(-)} - W^{(+)}\right] = -\mathbf{k} \cdot \frac{\partial W}{\partial \mathbf{q}}. \tag{5.16}$$

In the quasi-classical limit, the dispersion relation (5.16) reduces to [1]

$$1 - \frac{Q}{Mk^2} \int \frac{\mathbf{k} \cdot \partial W_0 / \partial \mathbf{v}}{(\omega - \mathbf{k} \cdot \mathbf{v})} d\mathbf{v} = 0. \tag{5.17}$$

Going back to the exact dispersion relation (5.14), we can consider the important case of a mono-energetic atomic beam, as determined by the equilibrium Wigner function

$$W_0(\mathbf{v}) = n_0 \delta(\mathbf{v} - \mathbf{v}_0), \tag{5.18}$$

which describes ultra-cold atoms in the $T \to 0$ limit, moving with velocity $\mathbf{v}_0$ with respect to the laboratory frame. The result is the following

$$1 - \frac{Q n_0}{M k^2} \left[\frac{1}{(\omega - \mathbf{k} \cdot \mathbf{v}_-)} - \frac{1}{(\omega - \mathbf{k} \cdot \mathbf{v}_+)} \right] = 0, \tag{5.19}$$

where $\mathbf{v}_\pm$ are defined by Eq. (5.13). Noting that $(\mathbf{v}_+ - \mathbf{v}_-) = \hbar\mathbf{k}/M$, and using the effective plasma frequency $\omega_p = \sqrt{Q n_0 / M}$, we obtain

$$1 - \frac{\omega_p^2}{(\omega - \mathbf{k} \cdot \mathbf{v}_+)(\omega - \mathbf{k} \cdot \mathbf{v}_-)} = 0. \tag{5.20}$$

In the classical limit, we can use $\mathbf{v}_\pm \simeq \mathbf{v}_0$, this would reduce to the case of a Doppler shifted plasmon oscillation: $\omega_p^2 = (\omega - \mathbf{k} \cdot \mathbf{v}_0)^2$. On the other hand, in the particular case where the atoms are at rest, Eq. (5.20) leads to

$$\omega^2 = \omega_p^2 + \frac{\hbar^2}{4} \frac{k^4}{M^2}, \tag{5.21}$$

which describes these same collective oscillations, but with a quantum dispersion term.

Let us now consider a more realistic case by plugging the effects of the temperature into the spectrum. We return to the general dispersion relation (5.17), and assume an arbitrary equilibrium distribution $W_0(\mathbf{v})$. It is useful to write this equation in the form

$$1 + \chi(\omega, \mathbf{k}) = 0, \tag{5.22}$$

where the *atomic susceptibility* is defined by

$$\chi(\omega, \mathbf{k}) = -\frac{Q}{\hbar k^2} \int \frac{[W_0^{(-)} - W_0^{(+)}]}{(\omega - \mathbf{k} \cdot \mathbf{v})} d\mathbf{v}. \tag{5.23}$$

In what follows, we separate the velocity into its parallel and perpendicular components in respect to the wave vector $\mathbf{k}$, such that

$$\mathbf{v} = u\frac{\mathbf{k}}{k} + \mathbf{v}_\perp. \tag{5.24}$$

We clearly see from Eq. (5.22) that there is a resonant velocity, $u_0 = \omega/k$, such that the atomic velocity moves in phase with the wave. We can also write the atomic susceptibility in terms of the parallel velocities, as

$$\chi(\omega, \mathbf{k}) = -\frac{Q}{\hbar k^3} \int G(u) \left[\frac{1}{u - u_0^{(-)}} - \frac{1}{u - u_0^{(+)}} \right] du, \tag{5.25}$$

where we have introduce the parallel quasi-distribution $G(u) = \int W_0(u, \mathbf{v}_\perp) d\mathbf{v}_\perp$, and used the quantities $u_0^{(\pm)} = u_0 \pm \hbar k/2M$. We should notice that the integral in (5.25) can be divided into its principal part and its resonant contribution. This leads to a complex susceptibility of the form

$$\chi(\omega, \mathbf{k}) = \chi_r(\omega, \mathbf{k}) + i\chi_i(\omega, \mathbf{k}). \tag{5.26}$$

We can easily solve the principal part of the integral, using the plausible assumption that the root mean square deviation of the atom velocity is much smaller than the phase velocity of the wave perturbation. This means that the main contribution to the integral comes from regions where $u \ll u_0$. Assuming an even function $G(u) = G(-u)$, and noting that

$$n_0 = \int G(u) du \ , \quad u_s^2 = \frac{1}{n_0} \int G(u) u^2 du, \tag{5.27}$$

where u_s can be identified with the sound speed, we arrive at the following result for the real part of the dispersion relation

$$1 - \frac{Q n_0}{\hbar k^2} \left[\frac{\mathbf{k} \cdot (\mathbf{v}_- - \mathbf{v}_+)}{(\omega - \mathbf{k} \cdot \mathbf{v}_-)(\omega - \mathbf{k} \cdot \mathbf{v}_+)} \right] \left(1 + \frac{k^2}{\omega^2} u_s^2 \right) = 0, \tag{5.28}$$

which corresponds to

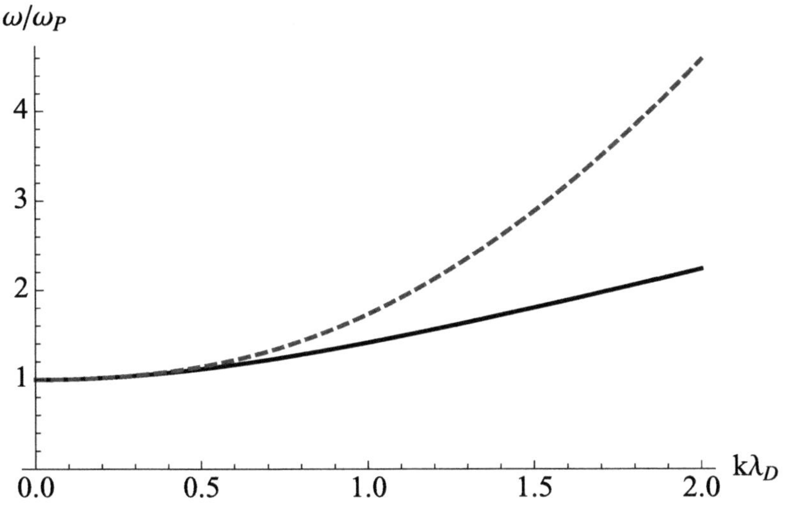

Fig. 5.2 Dispersion relation, representing the dimensionless variables, frequency ω/ω_P versus wavenumber $k\lambda_D$. The *dashed curve* was obtained for the dimensionless quantum parameter $H \equiv \hbar^2/4M^2u_s^2\lambda_D^2 = 1$. The *dashed line* was obtained for $H = 0.1$ and represents the quasi-classical regime

$$\omega^2 = \omega_p^2\left(1 + \frac{k^2}{\omega^2}u_s^2\right) + \frac{\hbar^2}{4}\frac{k^4}{M^2}. \tag{5.29}$$

Assuming oscillations close to the effective plasma frequency, $\omega^2 \simeq \omega_p^2$, the latter expression yields the following dispersion relation

$$\omega^2 = \omega_p^2\left(1 + k^2\lambda_D^2\right) + \frac{\hbar^2}{4}\frac{k^4}{M^2}. \tag{5.30}$$

where we have used $\lambda_D^2 = u_s^2/\omega_p^2$. The quantity λ_D plays the role of a characteristic length for the atom correlations inside the cloud. By analogy with a similar concept in plasma physics, we can call it the *atomic Debye length.* This dispersion relation generalizes our previous result on hybrid acoustic modes, by including a quantum dispersion term. Apart from its cut-off frequency ω_p, it also strongly resembles the dispersion relation of Bogoliubov oscillations in a Bose-Einstein condensate, as it will be shown later. Neglecting the cut-off and the quantum dispersion terms, we would get the dispersion relation for the usual acoustic waves in the gas, $\omega^2 = k^2u_s^2$ (Fig. 5.2).

Let us now turn to the imaginary part of the atomic susceptibility. By solving the resonant contribution to the integral (5.25), we get

$$\chi_i(\omega, \mathbf{k}) = \frac{\pi Q}{\hbar k^3}\left[G^{(-)}(u_0) - G^{(+)}(u_0)\right]. \tag{5.31}$$

This imaginary part implies the existence of a complex mode frequency $\omega = \omega_r + i\gamma$, where γ is the damping (or growth) rate, for a given wavevector $\mathbf{k}$. For $|\gamma| \ll \omega_r$ it is known that

$$\gamma = -\frac{\chi_i(\omega_r, \mathbf{k})}{(\partial \chi_r/\partial \omega)_r}, \tag{5.32}$$

where the derivative is taken at $\omega = \omega_r$. As a result, we obtain the expression for the Landau damping of the hybrid acoustic modes, as determined by

$$\gamma = -\frac{\pi Q}{\hbar k^3} \frac{[G(u_0 - \hbar k/2M) - G(u_0 + \hbar k/2M)]}{(\partial \chi_r/\partial \omega)_r}. \tag{5.33}$$

This result retains the exact atom recoil due to the emission or absorption of an hybrid phonon. The resulting damping rate is due to the difference in population for translational states distant by an amount of momentum $\hbar\mathbf{k}$. For an inversion of population, i.e. $G^{(-)} < G^{(+)}$, the system is dynamically unstable, as $\gamma > 0$, and the collective oscillations can start to grow out from of noise. This instability could lead to the phonon laser effect with the coherent emission of hybrid phonons, a question that we are discussing later.

To check the consistency of this result, we take the quasi-classical limit where this momentum increment can be considered negligible and where we can approximately write

$$G^{(-)} - G^{(+)} \simeq -\frac{\hbar k}{M} \left(\frac{\partial G}{\partial u} \right)_{\omega/k}. \tag{5.34}$$

By taking the derivative in the denominator of (5.33) as $\sim 1/\omega$, we can reduce this expression to [1]

$$\gamma = \frac{\pi}{\omega} \frac{Q}{Mk^2} \left(\frac{\partial G}{\partial u} \right)_{\omega/k}. \tag{5.35}$$

We can then see that the quasi-classical limit of the atomic Landau damping (or growth) rate is determined by the derivative of the parallel distribution $G(u)$, a result that is similar to the electron or ion Landau damping, well known for collective oscillations in classical plasma physics and discussed in later chapters of this book.

5.2 Tonks-Dattner Modes

The hybrid mode discussed above is only meaningful in infinite and homogeneous media. In physical terms, its dispersion relation can only be applied to waves that propagate locally, with wavelength scales much smaller than the inhomogeneity scale and the cloud dimensions. Let us now consider oscillations with a wavelength that is comparable with the size of the atomic cloud. In this case, we can no longer neglect the boundary conditions. Going back to Eq. (5.4), we observe that the density fluctuations obey the following Helmholtz equation

$$\left[\nabla^2 + k^2(\mathbf{r})\right]\tilde{n} = 0, \tag{5.36}$$

where the space dependent wavenumber $k(\mathbf{r})$ is defined by

$$k^2(\mathbf{r}) = [\omega^2 - \omega_p^2(\mathbf{r})]/u_s^2. \tag{5.37}$$

Before going into a more complex model, it is useful to consider the simple one-dimensional problem [4]. In the case of a uniform slab of cold gas, we have a constant plasma frequency ω_p, except at the boundaries $x = 0$ and $x = L$, where it vanishes. Equation (5.36) then reduce to a simple one-dimensional equation

$$\frac{d^2\tilde{n}}{dx^2} + \frac{1}{u_s^2}[\omega^2 - \omega_p^2(x)]\tilde{n} = 0. \tag{5.38}$$

Taking the boundary conditions $\tilde{n}(0) = \tilde{n}(L) = 0$, we obtain the following dispersion relation

$$\omega_\nu^2 = \omega_p^2\left[1 + \left(\pi\nu\frac{\lambda_D}{L}\right)^2\right], \tag{5.39}$$

where ν represents an integer number. This result illustrates the important role that the Debye length plays in the dispersion of the modes: in the absence of temperature ($\lambda_D = 0$), the only mode of the system would correspond to an effective plasmon oscillation $\omega = \omega_p$. It is important to remind that λ_D corresponds to a natural length above which collective effects should be expected. For instance, in a typical experiment, we estimate a plasma frequency around $\omega_p/2\pi \sim 100$ Hz and a thermal velocity of $v_{\text{th}} \sim 10$ cm/s, which yields $\lambda_D \sim v_{\text{th}}/\omega_P \sim 100\,\mu$m. However, we remark that there is no experimental evidence so far of any equation of state $P(n)$, which may compromise the definition of the sound speed u_S and consequently affect the exact definition of λ_D, at least, if we stay in the frame of a fluid description. However, in the present kinetic approach, no equation of state is necessary, and this ambiguity is solved.

The relation (5.39) shows that the finiteness of the slab implies the existence of a series of resonant modes with an integer number of half-wavelengths. The cylindrical geometry was considered, for the plasma case, in a famous paper by Parker, Nickel, Gould in 1964 [3], resulting as an explanation of the gallery of modes considered previously by Tonks [4] and Dattner [5]. In what follows, we extend the original derivation to a spherically symmetric clouds, which is more appropriate for atomic clouds.

For this purpose, we consider a gas confined in a sphere of radius a with an homogeneous profile, i.e. $\nabla n_0(r) = 0$ for $0 \leq r < a$, for which analytical solutions can be found. As we have seen in the previous chapter, the assumption of uniform density profile is valid in the multiple scattering regime. In that case, we use solutions of the form

$$\tilde{n}(\mathbf{r}) = R(r)Y(\theta, \phi), \tag{5.40}$$

where (r, θ, ϕ) are spherical coordinates. After separation of variables, we get the usual spherical harmonics for the angular part of the density perturbation

$$Y(\theta, \phi) = P_l^m(\cos\theta)\exp(im\phi), \tag{5.41}$$

where $P_l^m(\cos\theta)$ are the associated Legendre polynomials, l is a positive integer or zero, and $|m| < l$. The radial equation resulting from (5.36) can be written as

$$\frac{d}{dr}\left(r^2\frac{dR}{dr}\right) + [k^2r^2 - l(l+1)]R = 0. \tag{5.42}$$

By using a simple transformation of variables, $x = kr$, and $S(kr) = \sqrt{kr}\ R(r)$ this equation can be reduced to a Bessel equation

$$\frac{d^2S}{dx^2} + \frac{1}{x}\frac{dS}{dx} + \left[1 - \frac{(l+1/2)^2}{x^2}\right]S = 0. \tag{5.43}$$

The solutions with regular behavior at the origin $x = 0$ are therefore given by Bessel functions of the first kind, $J_{l+1/2}(x)$. From this we conclude that the Tonks-Dattner modes in a spherical homogeneous cold atom cloud are determined by

$$\tilde{n}(\mathbf{r}, t) = \sum_{l,m} \tilde{n}_l(t)\frac{J_{l+1/2}(kr)}{\sqrt{kr}}P_l^m(\cos\theta)\exp(im\phi), \tag{5.44}$$

where $\tilde{n}_l(t)$ have small amplitudes such that $|\tilde{n}_l| \ll n_0$. The mode frequencies can be obtained by remarking that $\tilde{n}$ should vanish at the border $r = a$. This implies that the allowed values for k have to obey the condition $k = z_{\nu,l}/a$, where $z_{\nu,l}$ represents the νth zero of the Bessel function of order $(l + 1/2)$. We are then led to the mode frequencies

$$\omega_{\nu,l}^2 = \omega_P^2\left\{1 + \left(z_{\nu,l}\frac{\lambda_D}{a}\right)^2\right\}. \tag{5.45}$$

Comparing with the rectangular case of Eq. (5.39) we see that the allowed eigen-frequencies for a spherical cloud now depend on two quantum numbers ν and l. But, in contrast with the similar quantum mechanical solutions for hydrogen like atoms, we have no hierarchical relation between these quantum numbers. The normalized modes for the lowest order solutions are illustrated in Fig. 5.3. In a more realistic description, the present rigid (Dirichlet) boundaries will eventually have to be replaced by soft boundaries and a generic density profiles $n_0(r)$ must be assumed.

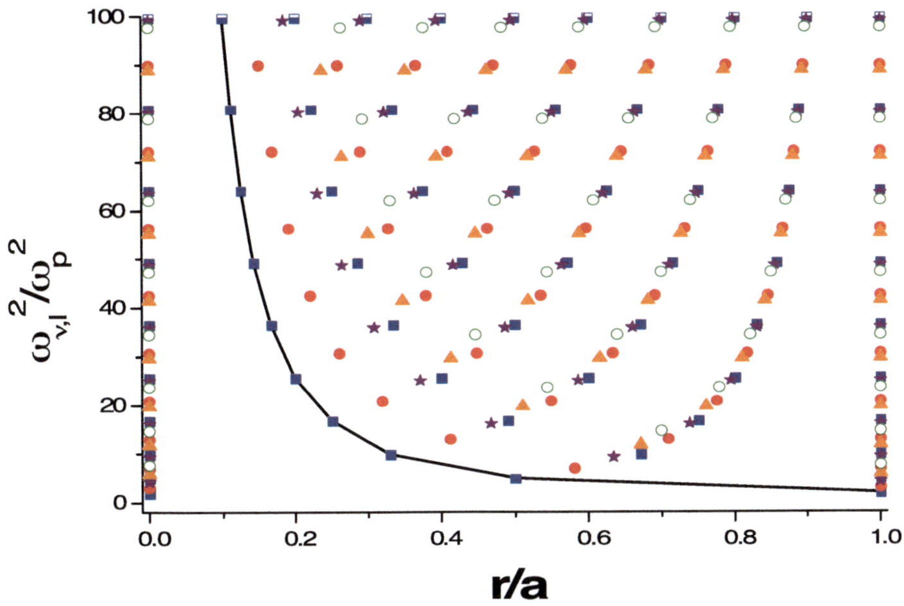

Fig. 5.3 Normalized modes $\omega_{\nu,l}^2/\omega_p^2$ plotted against the nodes $r_{\nu,l}/a$ of the radial solution for the density perturbation $\tilde{n}$, for $\lambda_D/a = 0.1$ and $1 < \nu < 10$. *Blue squares* ($l = 0$), *red circles* ($l = 1$), *violet stars* ($l = 2$), *yellow triangles* ($l = 3$), *green open circles* ($l = 4$). The *full line* is plotted at $l = 0$ and scales as $1/\nu^2$

5.3 Large Scale Oscillations

5.3.1 The Centre-of-Mass Oscillation

We now consider a different kind of oscillation, where the atomic cloud in a magneto-optical trap can oscillate as a rigid body. In contrast with the Tonks-Dattner resonances, there will be no density perturbations and the collective atomic velocity will be uniform and independent of position. In order to consider such a kind of oscillation we have to define the centre of mass position of the cloud, as

$$\mathbf{R}(t) = \frac{1}{N} \int_V \mathbf{r}\, n(\mathbf{r}, t) d\, \mathbf{r}, \tag{5.46}$$

where N is the total number of atoms in the cloud. Using the fluid equations, (5.2) and (5.3), and the Poisson-like Eq. (5.1), and neglecting the nonlinear terms, we obtain the following harmonic oscillator equation

$$\frac{d^2\mathbf{R}}{dt^2} + \omega_0^2 \mathbf{R} = 0, \tag{5.47}$$

where, for a spherically symmetric cloud, the oscillating frequency is determined by the expression

$$\omega_0^2 = \frac{Q}{MR^3} \int_0^a n(r) r^2 dr. \tag{5.48}$$

In the case of a uniform density profile $n(r) = n_0$, this reduces to

$$\omega_0 = \frac{\omega_p}{\sqrt{3}} \equiv \sqrt{\frac{Q n_0}{3M}}. \tag{5.49}$$

This frequency is commonly known as the *Mie frequency*. It characterizes the oscillation of a spherical gas with respect to its equilibrium position, similar to that of a plasma bubble. In the plasma case, the restoring force is due to the background ions, as discussed later in this book. Here, for the cold neutral atom gas, this is replaced by the magneto-optical trapping force. Due to the geometric factor of $\sqrt{3}$, this oscillation, which is also sometimes called the *sloshing mode*, cannot be confused with the hybrid and Tonks-Dattner modes discussed above. From an experimental point of view, it is possible in principle to distinguish between these modes, by analyzing the temporal and spatial structure of the oscillations.

5.3.2 Normal Modes

Another family of modes can be defined by relaxing the imposition of a vanishing density perturbation at the boundaries and including a more generic density profile. A systematic study of the large scale oscillations of the atomic cloud can then be made by linearizing the fluid equations for the ultra-cold gas, assuming a given equilibrium density profile $n_0(r)$. Using the polytropic equation of state, $P = C(T) n^\gamma$, and assuming that the perturbed quantities oscillate in time as $\exp(-i\omega t)$, we then get

$$-\omega^2 \tilde{n} - \frac{\gamma C(T)}{M} \nabla \cdot \left(n_0(r)^{\gamma-1} \nabla \tilde{n} \right) = \frac{1}{M} \nabla \cdot \left(n_0(r) \nabla \tilde{V} \right), \tag{5.50}$$

where the collective potential perturbation $\tilde{V}$ is determined by $\nabla^2 \tilde{V} = -Q\tilde{n}$. We now define an auxiliary quantity η such that,

$$\tilde{n} = \frac{1}{4\pi r^2} \frac{d\eta}{dr} \tag{5.51}$$

which allows us to transform the potential equation into

$$\frac{d}{dt} \tilde{V} = -\frac{Q}{4\pi r^2} \eta. \tag{5.52}$$

Replacing in (5.50), we get the radial equation for the density perturbation, which reads

$$-\omega^2 \frac{d\eta}{dr} - \frac{\gamma P(0)}{Mn(0)} \frac{d}{dr}\left[r^2 n_0(r)^{\gamma-1} \frac{d}{dr}\left(\frac{1}{r^2}\frac{d\eta}{dr}\right)\right] + \omega_p^2 \frac{d}{dr}\left[n_0(r)\eta\right] = 0, \quad (5.53)$$

where $n(0) = n_0(r = 0)$, $\omega_p^2 = Qn(0)/M$, and $P_0 = C(t)n(0)^\gamma$. Introducing the dimensionless variables θ and ξ defined in Sect. 4.4.2, we then get

$$-\omega^2 \frac{d\eta}{d\xi} - \frac{\gamma}{2}\omega_0^2 \frac{d}{d\xi}\left[\xi^2 \theta \frac{d}{d\xi}\left(\frac{1}{\xi^2}\frac{d\eta}{d\xi}\right)\right] + \omega_p^2 \frac{d}{d\xi}\left(\theta^{1/(\gamma-1)}\eta\right) = 0. \quad (5.54)$$

The solutions of this equation will depend on the particular equilibrium profile $n_0(r)$ that we are assuming to perturb. In general, this involves the use of numerical integration, as the equation (although linear) is non-polynomial. It is however more interesting here to consider approximate solutions, valid in the two main trap regimes, the temperature limited and the multiple scattering regime, for which some analytical solutions may be constructed.

(i) **Temperature limited regime**

We have seen that, for a small number of particles in the trap (typically $N < 10^4$) the effects of multiple scattering can be neglected. This means that we can use $\omega_p = 0$ in the above equations. Using the equilibrium profile given by Eq. (4.116), we can reduce Eq. (5.54) to a simpler expression

$$-\omega^2 \tilde{n} - \frac{\gamma}{2}\omega_0^2 \frac{1}{\zeta^2}\frac{d}{d\zeta}\left[(1-\zeta^2)\zeta^2 \tilde{n}\right] = 0, \quad (5.55)$$

where we have used a new variable $\zeta = 6\gamma/(\gamma - 1)\xi$. The solution can be formally represented in terms of hypergeometric functions. Here, instead, we use the *ansatz*

$$\tilde{n} = \sum_{jl} a_{jl}\zeta^{2j+l}, \quad (5.56)$$

where j and l are integers. Replacing this in Eq. (5.55), we can easily obtain a recurrence relation between the different coefficients a_{jl} which is found to converge, provided that

$$\omega^2 = \omega_0^2\left[2j + l + \gamma\left(2j^2 + 2jl + 1\right)\right]. \quad (5.57)$$

This result resembles the solutions obtained for the oscillations of a Bose-Einstein condensate in a spherical harmonic trap, for $\gamma = 2$, as discussed later. On the other hand, for $\gamma = 0$, we recover the result known for the free BEC in the collisionless regime, $\omega = \omega_0\sqrt{2j + l}$. Pure *surface modes* ($j = 0$), which may eventually be easier to detect experimentally, have the following frequencies

$$\omega_S = \omega_0 \sqrt{j + \gamma}. \tag{5.58}$$

On the other hand, *breathing modes* ($l = 0$) are also theoretically possible in small traps, with frequencies given by

$$\omega_B = \omega_0 \sqrt{2j + \gamma(1 + 2j^2)}. \tag{5.59}$$

Because current experimental techniques are able to measure such frequencies with a very high precision, these results could be very useful to determine the polytropic exponent γ and to confirm the assumed equation of state. In alternative, a kinetic approach can be used, where no equation of state is required, but the calculations become more involved. The breathing modes were discussed in a kinetic perspective by Olivetti et al. [6].

Multiple Scattering Regime

In contrast with the previous case, where the density profile is nearly Gaussian, in the multiple scattering regime we can use a water-bag model. In this case, one readily obtains

$$(3\omega_0^2 - \omega^2)\tilde{n} = 0, \tag{5.60}$$

which corresponds to a breathing mode of a system with long range interactions, characterized by the frequency

$$\omega_B = \omega_p = \sqrt{3}\omega_0. \tag{5.61}$$

This result corresponds to an uncompressional monopole oscillation of the system at the classical plasma frequency ω_p. It differs from the centre of mass oscillation considered above by a factor of $\sqrt{3}$. However, this solution is not unique. Manipulation of the fluid equations also yields

$$\nabla \cdot \left[\epsilon(\omega) \nabla \tilde{V}^{in}\right], \quad \epsilon(\omega) = 1 - 3\frac{\omega_0^2}{\omega^2}, \tag{5.62}$$

which holds in the interior of the atomic cloud, $r < r_0$. Outside this region, for $r > r_0$, the collective force should be a constant, thus we have $\nabla^2 V^{\text{out}} = 0$. The general solution of Eq. (5.62) is therefore given by

$$V^{\text{in}}(\mathbf{r}) = \sum_{lm} a_{lm} r^l Y_l^m(\theta, \varphi), \tag{5.63}$$

where $Y_l^m(\theta, \varphi)$ are the spherical harmonics. Similarly, for the external potential, we have

$$V^{\text{out}}(\mathbf{r}) = \sum_{lm} b_{lm} r^{-(l+1)} Y_l^m(\theta, \varphi). \tag{5.64}$$

Imposing regular continuity conditions at the boundary $r = r_0$, such that

$$V^{\text{in}}(r_0) = V^{\text{out}}(r_0), \quad \left.\frac{d}{dr} V^{\text{in}}\right|_{r=r_0} = \left.\frac{d}{dr} V^{\text{out}}\right|_{r=r_0}, \tag{5.65}$$

we obtain the frequencies for the incompressible surface modes, as

$$\omega_S = \omega_0 \sqrt{\frac{3l}{2l+1}}. \tag{5.66}$$

A similar result can be obtained from the Mie scattering theory in the context of surface plasmon-polaritons (see [12] for a review). We notice that for $l = 1$, we recover the Mie oscillations discussed at the beginning of this Section. We also notice that, for $l > 1$, the oscillations frequencies all stay below the plasma frequency $\omega_p = \sqrt{3}\omega_0$, and are independent of the size of the cloud r_0, totally differing from the Tonks-Dattner resonances.

5.4 Nonlinear Mode Coupling

In this section, we explore the possible existence of an instability process, due to the coupling between collective oscillations (hybrid waves) and center-of-mass oscillations of the cloud gas in a MOT. As we will show, the centre-of-mass oscillation (or the sloshing mode) can parametrically excite the hybrid acoustic waves [7]. Depending on the amplitude of the driving, the system can develop dynamical instabilities. We then show how this mechanism can be useful to explain the occurrence of giant oscillations in the system, as observed experimentally by Labeyrie et al. [8].

Our starting point is given by the set of fluid equations, as previously stated. For simplicity, we will use a water-bag model for the atomic density, where we assume that the density is approximately constant inside the cloud radius, $r < r_0$. In the low intensity Doppler regime, the trapping pseudo-potential is approximately given by

$$U(\mathbf{r}) = \frac{1}{2\kappa r^2}, \quad \kappa = \alpha \frac{g\mu_B}{\hbar k} b', \tag{5.67}$$

where μ_B is the Bohr magneton, g represents the Landé factor ($g \approx 2$ for the electron), κ is the spring constant, $\alpha = \alpha(\delta, I/I_s)$ is the friction coefficient, I_s is the saturation intensity and $b' = |\nabla B|$ represents the magnetic field gradient. In order to establish the balance between trapping and collective (repulsive, for red detuned MOTs) forces, the atomic density inside the cloud depends only on the

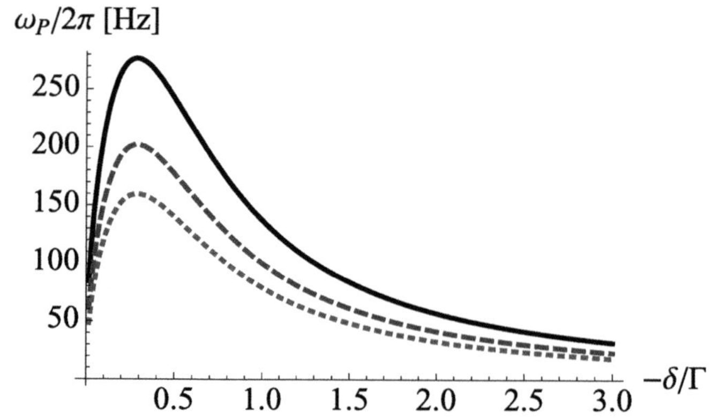

Fig. 5.4 Dependence of the plasma frequency ω_p on the detuning δ, plotted at $I/I_s = 0.3$ for different values of the magnetic field gradient. *Solid line*, $b' = 15\,\text{G/cm}$, *dashed line*, $b' = 0.8\,\text{G/cm}$, *dotted line*, $b' = 0.4\,\text{G/cm}$

MOT parameters, in such way that $\nabla \cdot (\mathbf{F}_{\text{coll}} + \mathbf{F}_T) = 0$, or equivalently

$$\frac{\kappa}{n_0} = \frac{(\sigma_R - \sigma_L)\sigma_L I_0}{c}, \tag{5.68}$$

where $\mathbf{F}_T = -\nabla U$. This condition establishes a relation between the plasma frequency and the basic parameters of the MOT, $\omega_p = \omega_p(\delta, I/I_s) = \sqrt{\kappa/M}$.

In Fig. 5.4, we represent the effective plasma frequency as a function of detuning δ, for different values of the magnetic field gradient. For typical experimental conditions of a ^{85}Rb gas, operating at $\delta = -1.5\Gamma$ with a magnetic field gradient $b' = 5\,\text{G/cm}$ and $I/I_s \approx 0.3$, the corresponding plasma frequency is $\omega_p/2\pi \approx 69\,\text{Hz}$.

We linearize the fluid equations, by separating each relevant physical quantity into its equilibrium and perturbation components, such that

$$n = n_0 + n_1 + n_2, \quad \mathbf{v} = \mathbf{v}_1 + \mathbf{v}_2, \quad \mathbf{F}_{\text{coll}} = \mathbf{F}_2. \tag{5.69}$$

Here, the subscripts 1 and 2 label the center-of-mass and the collective variables, respectively. We assume that center-of-mass dynamics is linear, which is valid for moderate oscillation amplitudes. In such limit, the velocity $\mathbf{v}_1$ is simply described by a harmonic oscillation

$$\mathbf{v}_1(t) = \mathbf{u}_1 \sin(\omega_0 t + \phi), \tag{5.70}$$

where $\omega_0 = \omega_p/\sqrt{3}$ represents the center-of-mass (sloshing) oscillation frequency, as given by Eq. (5.66) with $l = 1$, ϕ is an arbitrary phase, and $\mathbf{u}_1$ is the amplitude. We keep it undefined for the sake of generality. Because we are only considering small amplitude oscillations, the fluctuation n_1 of the bulk density due to the motion of the center-of-mass can be neglected, and the following scaling $n_1 \ll n_2 \ll n_0$ holds. Combining these approximations with the above fluid equations, and noticing that $n_0 \partial_t \nabla \cdot \mathbf{v}_2 = \omega_p^2 n_2$, one easily obtains

$$\frac{\partial^2 n_2}{\partial t^2} + \omega_p^2 n_2 + \\ +\mathbf{u}_1 \cdot \nabla \left[\frac{\partial n_2}{\partial t} \sin(\omega_0 t + \phi) + n_2 \omega_0 \cos(\omega_0 t + \phi) \right] n_2 = 0. \quad (5.71)$$

The last term on the right-hand side describes the nonlinear coupling between the center-of-mass and plasma (collective) oscillations. Assuming the separability of the solution, such that $n_2(\mathbf{r}, t) = B_2(\mathbf{r}) A_2(t)$, the later equation yields

$$\frac{d^2 A_2}{d\tau^2} + [\nu + 2\epsilon \cos(2\tau)]\, A_2 + \epsilon \sin(2\tau) \frac{dA_2}{d\tau} = 0, \quad (5.72)$$

where we have defined dimensionless variables of time $2\tau = \omega_0 t + \phi$, frequency $\nu = 4\omega_p^2/\omega_0^2$ and amplitude perturbation $\epsilon = 2\mathbf{u}_1 \cdot \nabla \ln B_2/\omega_0$. Equation (5.72) describes the dynamics of a parametrically excited system, and belongs to the family of Hill equations. It is formally similar to the Mathieu equation, which is well-known in the literature for containing unstable solutions. By using the standard Floquet theory, it is possible to verify that the solutions to Eq. (5.72) are of the form

$$A_{2,\gamma}(\tau) = e^{i\gamma\tau} P(\tau), \quad (5.73)$$

where $P(\tau) = \sum_{n=1}^{\infty} c_n \exp(2in\tau)$ represents a π-periodic function (corresponding to $2\pi/\omega_M$ in Eq. (5.71)) and $\gamma = \gamma(\nu, \epsilon)$ is the so called characteristic exponent [9]. Similarly, the solution $A_2(\tau + \pi)$ satisfies (5.72) whenever $A_2(\tau)$ does. The general solution of the system can then be written as

$$A_2(\tau) = a_1 A_{2,\gamma}(\tau) + a_2 A_{2,\gamma}(-\tau), \quad (5.74)$$

where a_1 and a_2 are constants. There is a infinite countable set of iso-γ or characteristic curves $\nu_n(\epsilon) \equiv a_n(\epsilon)$ and $\nu_n \equiv b_n(\epsilon)$, respectively yielding even and odd solutions, in agreement with the expansion

$$A_2(\tau) = \sum_n a_n \mathrm{ce}_n(\epsilon, \tau)$$

$$A_2(\tau) = \sum_n b_n \mathrm{se}_n(\epsilon, \tau), \quad (5.75)$$

where $\mathrm{ce}(\epsilon, \tau)$ and $\mathrm{se}(\epsilon, \tau)$ respectively represent the elliptical cosine and sine functions, resulting from the expansion of $A_2(\tau)$ in powers of ϵ. The functions $\mathrm{ce}_{2n}(\epsilon, \tau)$ and $\mathrm{se}_{2n+1}(\epsilon, \tau)$ are π-periodic, while the functions $\mathrm{ce}_{2n+1}(\epsilon, \tau)$ and

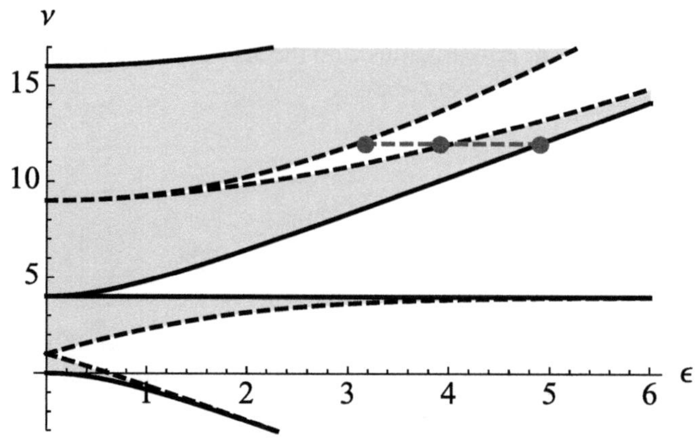

Fig. 5.5 Stability chart of Eq. (5.72) obtained numerically with a $n = 5$ Hill determinant. The *full lines* represent π-period solutions and the *dashed lines* represent the 2π-period ones. The *shadowed areas* correspond to stable (bounded) solutions. The *dots* represent the thresholds for the observation of unstable, corresponding to $\nu = 12$

$\mathrm{se}_{2n}(\epsilon, \tau)$ are 2π-periodic. For $\epsilon \geq 0$, γ is real when (ν, ϵ) lies in the regions between $a_n(\epsilon)$ and $b_{n+1}(\epsilon)$ and all solutions are stable; γ is complex in the regions between $a_n(\epsilon)$ and $b_n(\epsilon)$. In these regions all the solutions become infinite at least once. The marginal curves a_n and b_n separate the different regions of stability and are solutions of the Hill determinant, which is related to the following matrix

$$\Delta_{j,k}(n) = \left[\nu + (\gamma + 2in)^2\right]\delta_{j,k} + \epsilon\left[1 - \frac{i}{2}(\gamma + 2in)\right]\delta_{j,k+1} + \epsilon\left[1 + \frac{i}{2}(\gamma + 2in)\right]\delta_{j,k-1}, \quad (5.76)$$

where $\delta_{j,k}$ is the Kroenecker delta.

In Fig. 5.5 we represent the stability chart of Eq. (5.72), by computing the characteristic polynomial of Eq. (5.76), for $n = 5$. As $\epsilon \to 0$, $\nu \to n^2$.

The dynamics contained in Eq. (5.72) is sufficiently general and describes a family of instabilities, which mechanism is based on the nonlinear coupling between any periodic perturbation and a given collective mode. By considering the sloshing mode, we immediately set the value $\nu = 12$ in Eq. (5.72). Correspondingly, the instability is expected to occur for $3.17 < \epsilon < 3.91$ and $\epsilon > 4.90$. To unfold the parametric dependence in terms of experimentally relevant variables, we consider the low saturation limit, where the friction coefficient is approximately given by

$$\alpha = -16\hbar k_L^2 \frac{\Omega^2}{\Gamma^2} \frac{\Delta/\Gamma}{(1 + 4\Delta^2/\Gamma^2)^2}, \quad (5.77)$$

where $\Omega = \Gamma\sqrt{I/2I_s}$ is the two-level atom Rabi frequency. Inserting the later in Eq. (5.68), it is possible to derive the threshold condition in the (Δ,b')-plane, as

$$\left(\frac{\xi}{\epsilon}\right)^2 = \frac{16k_L g\mu_B b'}{M}\frac{\Omega^2}{\Gamma^2}\frac{\Delta/\Gamma}{(1+4\Delta^2/\Gamma^2)^2}. \tag{5.78}$$

For a fixed value of the magnetic field gradient b', the oscillations are unstable above a certain critical detuning δ_1. Below that value, the oscillations remain stable in the interval $[\delta_1, \delta_2]$, where δ_2 is the marginal detuning separating a new stability zone, below which the oscillations are unstable again. Finally, below the value δ_3, our model indicates that the oscillations are stable. In Fig. 5.6 we plot the marginal curves δ_1, δ_2 and δ_3 against the magnetic field gradient, obtained for a ^{85}Rb MOT operating at $I/I_s = 0.3$, matching the experimental conditions of Ref. [8]. For a coupling strength of $\xi/2\pi = 223$ Hz, the present model qualitatively reproduces the unstable threshold observed in the referred experiment, here represented by the dots in Fig. 5.6.

These results, although valid only within the low intensity Doppler model, may provide a more complete description for the so called *self-sustained* oscillations, since we can theoretically predict the existence new regions of stability in the (δ,b') phase diagram. On the other hand, this fluid dynamical approach, allows the description of different instability regimes, rooted in different driving mechanisms. In particular, one could study the case of unstable steady waves, of the Tonks-Dattner (TD) type [1], simply by replacing the frequency ω_0 by one of the corresponding (TD) modes. This fact suggests that the actual phase diagram for self-sustained oscillations may be more rich and complex than the present one, provided that different mechanisms of instability occur in the same real experiment.

A short discussion of the validity range of the present model is in order. The results considered here are only valid in the low-intensity Doppler regime, where the dynamics of single atoms remains essentially linear. We should notice that for large magnetic field gradients ($b' \sim 10$ G/cm), the matching between this theoretical model and the experimental results is less evident, as shown by Fig. 5.6. This is a clear manifestation of the deviation from the linear regime considered here. It is also known that other instability mechanisms appear for high magnetic field gradients, eventually related to stochastic diffusion of atoms in the MOT, but its description is out of the range of this simple model.

5.5 Quasi-linear Diffusion

In the study of linear wave dispersion, we have previously assumed that a given equilibrium function $W_0(\mathbf{r}, \mathbf{v})$ remains constant along the process of wave propagation and damping. This is certainly valid on a time scale larger than the wave period

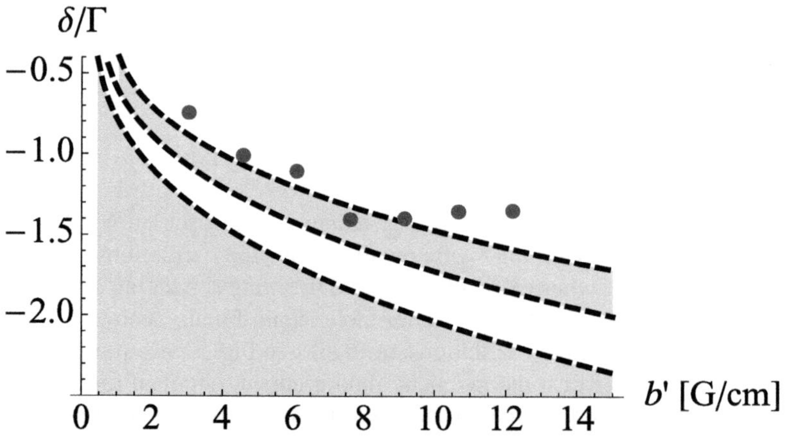

Fig. 5.6 Phase diagram of the nonlinear coupling of center-of-mass and collective oscillations, obtained for $I = 0.3 I_s$ and for a stochastic perturbation of $\xi/2\pi = 223$ Hz. The *shadowed areas* are stable

$1/\omega_r$, but is no longer valid for a much larger time scale (larger than $1/\gamma$), due to the energy exchange between the atoms and the wave modes.

In order to establish the long time evolution for W_0 we can go back to the exact wave kinetic equation (5.8) and retain the long time contribution of the nonlinear terms on a long time scale. The slowly varying part of the wave kinetic equation can then be written as

$$\left(\frac{\partial}{\partial t} + \mathbf{v} \cdot \frac{\partial}{\partial \mathbf{r},}\right) W_0(\mathbf{v}) = \frac{1}{i\hbar} \int \tilde{V}_{\text{eff}}(-\mathbf{k}) \left[\tilde{W}(\mathbf{v}_-, \mathbf{k}) + \tilde{W}(\mathbf{v}_+, \mathbf{k})\right] \frac{d\mathbf{k}}{(2\pi)^3}. \quad (5.79)$$

Now, using Eq. (5.12), which relates the potential perturbations $\tilde{V}_{\text{eff}}$ to the perturbations of the distribution function $\tilde{W}$, we obtain

$$\left(\frac{\partial}{\partial t} + \mathbf{v} \cdot \frac{\partial}{\partial \mathbf{r},}\right) W_0(\mathbf{v}) = \frac{Q^2}{\hbar^2 k^4} \int \frac{|\tilde{n}(\mathbf{k})|^2}{i(\omega - \mathbf{k} \cdot \mathbf{v})} \left[W_0(\mathbf{v} - \hbar\mathbf{k}/M) - W_0(\mathbf{v} + \hbar\mathbf{k}/M)\right] \frac{d\mathbf{k}}{(2\pi)^3}, \quad (5.80)$$

where we have used the density fluctuations $\tilde{n}(\mathbf{k}) = k^2 \tilde{V}_{\text{eff}}(\mathbf{k})/Q$. This is a kinetic equation of the Boltzmann type, associated with the inelastic collisions between the atoms and the hybrid phonons of the density fluctuation spectrum. Emission and absorption of one phonon will make the atom velocity to jump between $\mathbf{v}$ and $\mathbf{v} \pm \hbar\mathbf{k}/M$, as it would be expected.

Now, let us consider the quasi-classical limit of this equation, where the atomic recoil can be neglected. In this limit, the population difference in Eq. (5.80) is replaced by a derivative, and we can write

$$\left(\frac{\partial}{\partial t} + \mathbf{v} \cdot \frac{\partial}{\partial \mathbf{r},}\right) W_0(\mathbf{v}) = \frac{1}{i\hbar} \int \tilde{V}_{\text{eff}}(-\mathbf{k})\delta\tilde{W}(\mathbf{k}) \frac{d\mathbf{k}}{(2\pi)^3}, \tag{5.81}$$

where

$$\delta\tilde{W}(\mathbf{k}) = -\frac{\hbar\mathbf{k}}{M} \cdot \frac{\partial}{\partial \mathbf{v}} \tilde{W}(\mathbf{v}, \mathbf{k}). \tag{5.82}$$

In the same limit, we also have

$$\tilde{W}(\mathbf{v}, \mathbf{k}) = -\frac{\tilde{V}_{\text{eff}}}{M} \frac{\mathbf{k} \cdot \partial W_0 / \partial \mathbf{v}}{(\omega - \mathbf{k} \cdot \mathbf{v})} \tag{5.83}$$

from where we get

$$\delta\tilde{W}(\mathbf{k}) = \tilde{V}_{\text{eff}}(\mathbf{k}) \frac{\hbar\mathbf{k}}{M^2} \cdot \frac{\partial}{\partial \mathbf{v}} \frac{\mathbf{k}}{(\omega - \mathbf{k} \cdot \mathbf{v})} \cdot \frac{\partial}{\partial \mathbf{v}} W_0. \tag{5.84}$$

Replacing this in Eq. (5.81) we finally obtain

$$\left[\frac{\partial}{\partial t} + \mathbf{v} \cdot \frac{\partial}{\partial \mathbf{r},} - \frac{\partial}{\partial \mathbf{v}} \cdot \mathbf{D}(\mathbf{v}) \cdot \frac{\partial}{\partial \mathbf{v}}\right] W_0(\mathbf{v}) = 0 \tag{5.85}$$

with the diffusion tensor in the velocity space determined by

$$\mathbf{D} = \frac{Q^2}{M^2 k^4} \int \frac{|\tilde{n}(\mathbf{k})|^2}{i(\omega - \mathbf{k} \cdot \mathbf{v})} \mathbf{k}\mathbf{k} \frac{d\mathbf{k}}{(2\pi)^3}. \tag{5.86}$$

We can now explore the spectral symmetries, well known from the quasi-linear theory [10], by noting that $|\tilde{n}(\mathbf{k})|^2 = |\tilde{n}(-\mathbf{k})|^2$. Here we are assuming that, for each mode of the density fluctuation spectrum, there is a complex frequency $\omega \equiv \omega(\mathbf{k}) = \omega_{\mathbf{k}} + i\gamma_{\mathbf{k}}$, where $\omega_{\mathbf{k}}$ is the real part of the mode frequency and $\gamma_{\mathbf{k}}$ is the corresponding Landau damping. We also notice that $\omega_{-\mathbf{k}} = \omega_{\mathbf{k}}$. This allows us to rearrange the terms inside the integral of Eq. (5.87), leading to the following new expression for the diffusion tensor

$$\mathbf{D} = \frac{Q^2}{M^2 k^4} \int |\tilde{n}(\mathbf{k})|^2 \frac{\gamma_{\mathbf{k}} \mathbf{k}\mathbf{k}}{(\omega - \mathbf{k} \cdot \mathbf{v})^2 + \gamma_{\mathbf{k}}^2} \frac{d\mathbf{k}}{(2\pi)^3}. \tag{5.87}$$

This new expression is physically more satisfactory, because it clearly states that diffusion is a real process. Of particular interest is the case where Landau damping is a very small quantity, and where we can use the limit

$$\lim_{\gamma_{\mathbf{k}} \to 0} \frac{\gamma_{\mathbf{k}}}{(\omega - \mathbf{k} \cdot \mathbf{v})^2 + \gamma_{\mathbf{k}}^2} = \pi\delta(\omega_{\mathbf{k}} - \mathbf{k} \cdot \mathbf{v}), \tag{5.88}$$

which allows one to rewrite the diffusion tensor in a much simpler form

$$\mathbf{D} = \pi \frac{\omega_p^2}{n_0 k^4} \int |\tilde{n}(\mathbf{k})|^2 \, \mathbf{kk} \, \delta(\omega_\mathbf{k} - \mathbf{k} \cdot \mathbf{v}) \frac{d\mathbf{k}}{(2\pi)^3}. \tag{5.89}$$

The latter expression for **D** states that diffusion in velocity space is due to a succession of resonant atom wave mode interactions. At a given time, a given mode of the wave spectrum is in exact resonance with the atoms moving with velocity **v**, as shown by the Dirac delta function. On a later time, the atom velocity will change and resonance with a nearby mode will be established. The stronger the wave component energy, the faster diffusion will occur. The existence of the density fluctuation spectrum will then introduce a temperature limit for the laser cooling process.

In order to estimate this temperature limit, we have to describe the competing influence of the density fluctuations and the laser cooling force. Including the well known expression for this force, the quasi-classical kinetic equation can be written, in the form

$$\left(\frac{\partial}{\partial t} + \mathbf{v} \cdot \frac{\partial}{\partial \mathbf{r}}, \right) W_0 = \frac{\partial}{\partial \mathbf{v}} \cdot \left[\mathbf{A} + \mathbf{D} \cdot \frac{\partial}{\partial \mathbf{v}} \right] W_0, \tag{5.90}$$

where **A** is the friction coefficient associated with the cooling force, as given by

$$\mathbf{A} = \beta \mathbf{v} \,, \quad \beta = 8\hbar k^2 \Gamma \frac{\Delta |\Omega_R|^2}{M(4\Delta^2 + \Gamma^2)^2}, \tag{5.91}$$

where, was usual, Ω_R is the Rabi frequency, Γ the spontaneous decay time, and Δ the frequency detuning. This expression is valid for $|\omega_R|^2 \ll \Gamma^2/2$. Now, assuming spherical symmetry in velocity space, a steady state solution for this equation can be derived as

$$W_0(\mathbf{v}) = W_{00} \exp \left[-\frac{M v^2}{2 T_{\text{eff}}} \right], \tag{5.92}$$

where W_{00} is a constant, and the effective temperature T_{eff} is determined by

$$T_{\text{eff}} \simeq \frac{\pi^2 c^2}{\hbar \omega^2} \frac{\omega_p^2}{n_0 k^2} \int |\tilde{n}(\mathbf{k})|^2 \frac{d\mathbf{k}}{(2\pi)^2}. \tag{5.93}$$

This quantity establishes a new temperature limit for the laser cooling process, which is conceptually different, and eventually larger, than the Doppler limit associated with spontaneous emission.

5.6 Phaser, the Phonon Laser

The possibility to excite a Doppler instability of a single atom or ion trapped in a harmonic potential, has been considered by several authors [11, 12]. This could be used as the basic mechanism for the creation of a phonon laser. In this section, we show that it is possible to generalized the single trapped atom configuration to the case of a large ensemble of atoms, thus enabling collective phonon excitations. This would lead to the excitation of a phonon laser instability in the trap. Such an effect would results from a negative Landau damping of the collective density perturbations in the gas, leading to the coherent emission of phonons. It would correspond to a collective oscillation, instead of a single particle effect. In that respect, it would be closer to the paradigm of an optical laser, where the photon modes correspond to internal vibrations of the optical cavity, thus making the name *phaser* a natural denomination for such collective modes. The resulting acoustic oscillations can the, in principle, be coupled to the outside world, by mechanical or electromagnetic means, thus providing a source of coherent acoustic radiation.

The basic ingredient for the phonon laser in a MOT is provided by an inversion of population in the atom velocity states. This can be produced, for instance, by a pump laser beam, as discussed in Chap. 3, or by any other method. We use here a simplified one-dimensional model, assuming $\mathbf{v} = u(\mathbf{k}/k) + \mathbf{v}_\perp$, and integrating over the perpendicular velocities with respect to the phonon wave vector $\mathbf{k}$. Phonons are described as perturbations of the mean field potential $V(\mathbf{k})$. We define the quasi-distribution for the parallel velocities as $G(u) = \int W(u, \mathbf{v}_\perp) d\mathbf{v}_\perp$. The evolution equation for this parallel distribution can be written as

$$\frac{dG}{dt} = \tilde{g}\left[G(k_b)^{(-)} - G(k_b)^{(+)}\right] + \nu\left[G_0 - G\right], \tag{5.94}$$

where $\tilde{g}$ represents the pumping mechanism, which provides the inversion of population and ν is the viscosity associated with the laser cooling beams. We can use the average value $\nu = -(E_0/\hbar)\mathrm{Im}(d_{21}\tilde{\rho}_{21})$, where E_0 is the average magnitude (computed over the six cooling beams) of the incident electric field. The distribution $G_0(u)$ is that of the laser cooled gas, as obtained in the absence of the pumping process. This last term in (5.94) describes depopulating of the high velocity states.

Three physical processes are assumed to occur in parallel, as described in Fig. 5.7. First, the red-detuned laser cooling beams, will create a very low temperature quasi-distribution $G_0(u)$, which provide the ground state. Second, some given pumping process will excite high atomic velocity states. For clarity, we assume pumping by a blue detuned laser with wave vector $\mathbf{k}_b$, exciting the atoms in the velocity state $u + \hbar k_b/M$, due to the photon absorption. These high velocity states can be seen as an atom beam, thus creating a population inversion in the centre-of-mass states. Finally, a negative Landau damping of acoustic-like oscillations with frequency ω and wavenumber k will result in the coherent emission of phonons. Due to phonon emission, a third and intermediate velocity state will be populated. The temporal evolution of the phonon field will be dictated by a complex frequency $\omega = \omega_r + i(\gamma_k - \nu/2)$.

We assume that $G(u) = \bar{G}(u) + \tilde{G}(u)$, where $\bar{G}(u)$ is the equilibrium distribution, and $\tilde{G}(u)$ is the perturbation, which is assumed to evolve as $\exp(i\mathbf{k}\cdot\mathbf{r} - i\omega t)$. Notice that $\bar{G}(u)$ reduces to $G_0(u)$ in the absence of pumping. It is clear that we will have

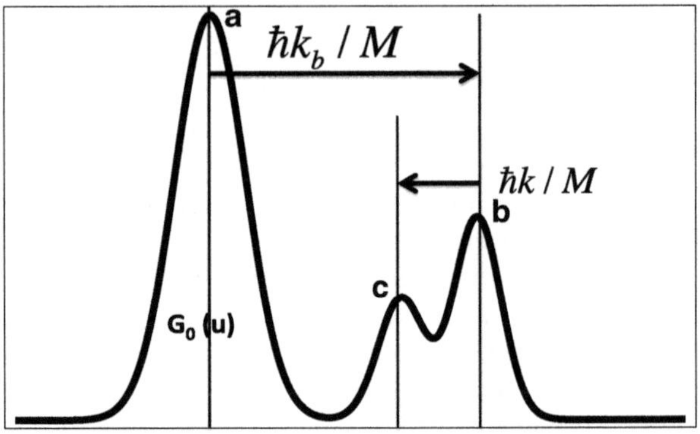

Fig. 5.7 Phonon laser scheme: (**a**) The gas is cooled down by red-detuned laser beams with frequency ω_a and wavevector $\mathbf{k}_a$ into a velocity distribution $W_0(v)$; (**b**) the ultra-cold gas is pumped, creating an inversion of population in velocity space; (**c**) Phonons with frequency ω and wavevector $\mathbf{k}$ are coherently emitted by the atoms, which decay into a lower kinetic energy state (The width of the distributions and the distance between the peaks are generic and provide only a schematic perspective of the instability mechanism. See discussion in the text)

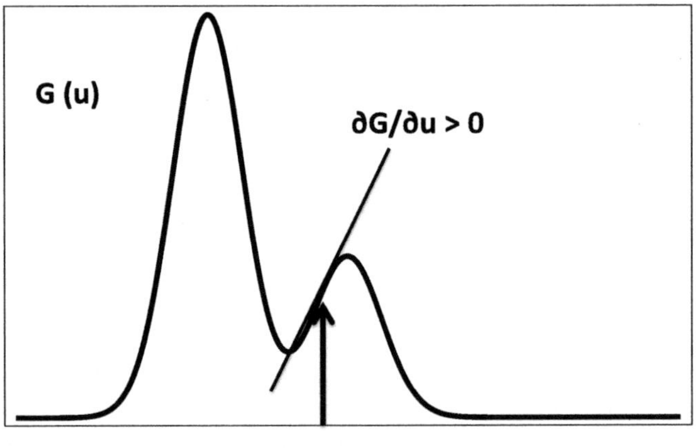

Fig. 5.8 Phonon laser scheme in the classical limit. The two upper levels are replaced by a region of negative derivative, where inverse Landau damping will occur

an acoustic wave growth if the inverse Landau damping is positive $\gamma_k > 0$, and if it is large enough to compensate for the wave losses, i.e. $\gamma_k > \nu$. We can then write the threshold condition as

$$\left[\bar{G}^{(+)} - \bar{G}^{(-)}\right]_r > \frac{2\nu}{\pi} \frac{\hbar k^3}{Q\omega_r}, \tag{5.95}$$

It is important to noticed here that this instability also exists in the classical limit, where we can develop $\bar{G}^{(\pm)}$ around $\bar{G}(u)$, as $\bar{G}^{(\pm)} \simeq \bar{G}(u) \pm (\hbar k/M)(\partial \bar{G}/\partial u)$. The threshold condition becomes

$$\left(\frac{\partial \bar{G}}{\partial u}\right)_r > \frac{2\nu}{\pi}\frac{Mk^2}{Q\omega_r}. \tag{5.96}$$

The population inversion is now represented by the derivative of the atom distribution at the resonant velocity $u_r = \omega_r/k$. The resemblance with a three level laser will be somewhat lost, but the physical principle stays the same. This is illustrated in Fig. 5.8.

Defining the population difference $\Delta G = \int [G_2(u) - G_1(u)]du$, for the atoms centered around the upper and intermediate parallel velocities u_1 and u_2, it is possible to establish the evolution equation [13]

$$\frac{d}{dt}\Delta G = -\frac{|V(\mathbf{k},t)|^2}{\hbar^2}\Delta G. \tag{5.97}$$

On the other hand, the inverse Landau damping of the phonon mode will also change and, as a consequence, the phonon square amplitude (energy) will evolve according to

$$\frac{d}{dt}|V_k(t)|^2 = \left[\frac{\pi Q\omega_r}{\hbar k^3}\delta G - \nu\right]|V_k(t)|^2, \tag{5.98}$$

where γ_k is the growth rate defined above. Introducing a new time variable, τ, and a new viscosity coefficient $\bar{\nu}$, such that $\tau = (\hbar k^3/\pi Q\omega_r)\Delta G(0)t$, and $\bar{\nu} = (\tau/t)\nu$, we can rewrite these two equations as [13]

$$\frac{dz}{d\tau} = -azy\ , \quad \frac{dy}{d\tau} = (z - \bar{\nu})y, \tag{5.99}$$

with $z \equiv \Delta G/\Delta G(0)$, $y \equiv |V_k|^2$ and $a = (\pi Q\omega_r/\hbar^3 k^3)$. These coupled evolution equations clearly exhibit the threshold condition $z(0) > 2\bar{\nu}$, already stated in Eq. (5.95), and saturation for increasing values of time. This is illustrated in Fig. 5.9. We can see that the population difference decreases along time, due to the increase of the phonon coupling between the two velocity states around u_1 and u_2. As a consequence, the growth rate slows down, leading asymptotically to saturation. The resemblance with the usual optical laser is striking.

It is useful to consider the selection mechanisms for the phonon wave vector **k**. First, we have spontaneous selection of the mode with largest growth rate. Another mechanism is imposed to the finite size and boundary conditions of the atom cloud. As we have seen, this leads to a discretization of the phonon mode spectrum, where the discrete phonon modes with a well defined internal structure are associates with the Tonks-Dattner resonances. Such modes are quantized inside the cloud, which will work as a spherical acoustic cavity. The dominant phonon mode in a MOT will then be due to the Tonks-Dattner resonance mode with the largest inverse Landau damping.

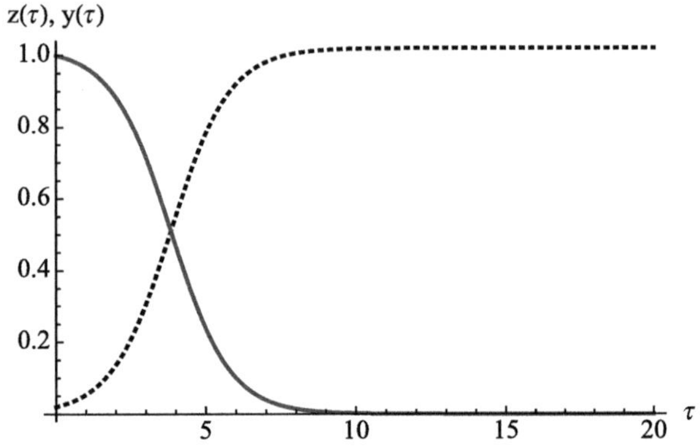

Fig. 5.9 Saturation of the laser instability, as described by the quasi-linear equations. Temporal evolution of the normalized population inversion in velocity space $z(\tau) \equiv \Delta G(u,\tau)/\Delta G(u,0)$ (*dashed line*), and the phonon amplitude square $y(\tau) \equiv |V_k(\tau)|^2$ (*full line*), as a function of the time variable τ, for $a = 1$ and $\bar{\nu} = 1/2$. For illustration purposes, we take $y(0) = 0.001$

References

1. J.T. Mendonça, R. Kaiser, H. Terças, J. Loureiro, Phys. Rev. A **78**, 013408 (2008)
2. J.T. Mendonça, Phys. Rev. A, **81**, 023421 (2010)
3. J.V. Parker, J.C. Nickel, R.W. Gould, Phys. Fluids **7**, 1489 (1964)
4. L. Tonks, Phys. Rev. **37**, 1458 (1931); ibid. **38**, 1219 (1931)
5. A. Dattner, Phys. Rev. Lett. **10**, 205 (1963)
6. A. Olivetti, J. Barré, B. Marcos, F. Bouchet, R. Kaiser, Phys. Rev. Lett., **103**, 224301 (2009)
7. H. Terças, J.T. Mendonça, R. Kaiser, Europhys. Lett. **89**, 53001 (2010)
8. G. Labeyrie, F. Michaud, R. Kaiser, Phys. Rev. Lett. **96**, 023003 (2006)
9. A. Nayfeh, D. Mook, *Nonlinear Oscillations* (Wiley, New York, 1979)
10. R.C. Davidson, *Methods in Nonlinear Plasma Theory* (Academic, New York, 1971)
11. A.E. Kaplan, Opt. Express **17**, 10035 (2009)
12. K. Vahala, M. Herrmann, S. Knunz, V. Batteiger, G. Saathoff, T.W. Hänsch, Th. Udem, Nat. Phys. **5**, 682 (2009)
13. J.T. Mendonça, H. Terças, G. Brodin, M. Marklund, Europhys. Lett. **91**, 33001 (2010)

Chapter 6
Photons in the Ultra-cold Gas

In this chapter, we discuss the properties of laser beam propagation in the ultra-cold gas, by focusing mainly our attention on the specific properties of this medium. We first consider the linear and nonlinear dispersion properties of light. The novelty here is that the laser beams used for atom cooling can couple to the low frequency oscillations of the atomic cloud. In particular they can destabilize the hybrid sound inside the gas, therefore generating small scale oscillations, which can be the seed for turbulence in the medium. Under certain conditions, the laser beam can become modulationaly unstable, and photon bubbles can eventually be formed.

The photon field associated with the laser cooling beams contains two distinct parts. One is the coherent part due to the incident laser radiation, and can be described by the usual electromagnetic wave equation. The other is an incoherent part associated with secondary photon emission, due to multiple scattering inside the gas, and can be described by a photon diffusion equation. For an optically thick gas, the diffusive part dominates. The coupling of both the coherent and the diffusion parts of the photon field with the gas are proportional to the atom density. Therefore, a density perturbation will lead to a perturbation of the local laser intensity, which can eventually become unstable. For the coherent part of the electromagnetic radiation, such instabilities can take the form of filamentation and amplitude modulation. For the diffusive part, the instability is nearly isotropic, a local photon bubble can be formed.

Another interesting aspect of photon coupling with the collective atomic forces is the formation of rotons. It is well known that rotons were introduced by Landau in the frame of his theory of superfluidity. Here we discover that rotons can also occur in a classical or non-superfluid gas, due to the correlations induced by multiple photon scattering. Such collective photon-atom interactions lead to the appearance of a roton minimum in the hybrid phonon dispersion curve, which is similar to those occurring in a superfluid. Such roton dispersive features can be limited by atomic Landau damping, and regions of undamped rotons will be revealed.

Finally, we discuss the collective Rayleigh scattering of an incident laser beam by the density perturbations in the gas. The spectrum of density perturbations can

J.T. Mendonça and H. Terças, *Physics of Ultra-Cold Matter*, Springer Series on Atomic, Optical, and Plasma Physics 70, DOI 10.1007/978-1-4614-5413-7_6,

be derived by using a test-particle method. The characteristic features of the spectrum depend on the collective properties of the ultra-cold gas, and in particular, they depend on the atomic Debye length. Observation of the scattered photon spectrum can then be used as a diagnostic tool for the collective forces inside the medium.

6.1 Optical Properties

Let us first review the basic optical properties in the gas. We consider an incident laser beam propagating in the ultra-cold medium, as described by the wave equation for the laser electric field **E**, as

$$\left(\nabla^2 - \frac{1}{c^2}\frac{\partial^2}{\partial t^2}\right)\mathbf{E} = \mu_0 \frac{\partial^2 \mathbf{P}}{\partial t^2}, \tag{6.1}$$

where the polarization **P** can be divided in two parts, $\mathbf{P} = \mathbf{P}_L + \mathbf{P}_{NL}$. The linear part is determined by

$$\mathbf{P}_L(t) = \epsilon_0 \int_0^\infty \chi^{(1)}(\tau)\mathbf{E}(t-\tau)d\tau, \tag{6.2}$$

with $\chi^{(1)}$ representing the linear optical susceptibility. Here we assume that the gas is isotropic, otherwise this scalar function would have to be replaced by a tensor. For the nonlinear part of the polarization, we can neglect the temporal dispersion, and simply write

$$\mathbf{P}_N(t) = \epsilon_0 \chi^{(3)} |E(t)|^2 \mathbf{E}(t), \tag{6.3}$$

where $\chi^{(3)}$ is the third-order susceptibility. Here we have assumed that the gas is centro-symmetric, and therefore the second order susceptibility is zero, $\chi^{(2)} = 0$ [3]. This is certainly true inside the trap, where the static magnetic field of the trap is nearly zero, and the inhomogeneity of the gas is negligible at the scale of the laser wavelength.

Let us neglect nonlinear effects, and focus on the linear laser propagation. For a given spectral component of the laser field, with frequency ω, such that

$$(\mathbf{E}, \mathbf{P}) = (\mathbf{E}_\omega, \mathbf{P}_\omega)\exp(i\mathbf{k}\cdot\mathbf{r} - i\omega t). \tag{6.4}$$

The polarization amplitude can be simply related with the electric field amplitude by the expression

$$\mathbf{P}_\omega = \epsilon_0 n_a \chi_a(\omega)\mathbf{E}_\omega, \tag{6.5}$$

where n_a is the atom density, and $\chi_a(\omega)$ is the single atom linear susceptibility. Replacing this in Eq. (6.1), we get the linear photon dispersion relation

$$k^2 = \frac{\omega^2}{c^2} \left[1 + n_a \chi_a(\omega)\right] . \tag{6.6}$$

In general, the linear susceptibility is complex, and we can write it as $\chi_a(\omega) = \chi'_a(\omega) + i\chi''_a(\omega)$. For a nearly resonant laser field with frequency $\omega \simeq \omega_a$, where ω_a is the frequency of the atomic radiative transition, the real part of the susceptibility $\chi'_a(\omega)$ can be very small, which has two important consequences. (1) The nonlinear susceptibility becomes comparable to the linear one, $\chi^{(3)} \sim \chi^{(1)}$; and (2) the resulting wave dispersion, being negligible, will facilitate the phase-matching associated with nonlinear wave mixing.

On the other hand, for nearly resonant propagation, the wave losses associated with the imaginary part $\chi''_a(\omega)$ can be significant, leading to the transfer of energy from the primary laser beam energy to secondary scattered light, propagating incoherently in other directions. We can then imagine a situation where such scattering losses are dominant, and the incident coherent beam of photons is transformed into an incoherent spectrum of diffused light. This case will be considered in Sect. 6.3. It is known that the atomic susceptibility for laser interaction with a two-level atom in steady state is given by [3]

$$\chi(\omega) = \frac{1}{3}(\rho_{22} - \rho_{11})|p_{21}|^2 \frac{\Delta + i\Gamma/2}{\Delta^2/4 + s}, \tag{6.7}$$

where ρ_{11} and ρ_{22} are the population probabilities of the internal energy levels, and $s = |E|^2/|E_s|^2$ is the saturation parameter, with $|E_s|^2 = \hbar^2/4|p_{21}|^2$. By developing this expression as a power series in $|E|^2$, we get

$$\chi(\omega) = \chi^{(1)}(\omega) + 3\chi^{(3)}|E|^2 + \ldots \tag{6.8}$$

This leads to well known linear and nonlinear optical properties. In particular, the third order susceptibility, is responsible for the occurrence of four-wave mixing processes, obeying the energy and momentum conservation rules

$$\omega_1 + \omega_2 = \omega_3 + \omega_4 \, , \quad \mathbf{k}_1 + \mathbf{k}_2 = \mathbf{k}_3 + \mathbf{k}_4, \tag{6.9}$$

where ω_j and $\mathbf{k}_j$ are the frequencies and wavenumbers of the interacting photons. Four-wave mixing in a three-level Λ atomic configuration has been considered in theory [2]. We should add that higher order effects have also been studied. In particular, it has been noticed that the nonlinear effects associated with the fifth-order susceptibility $\chi^{(5)}$, and involving six-wave mixing, can be of the same order as those associated with the $\chi^{(3)}$ effects [4].

6.2 Modulational Instability

Of particular interest are processes specific to the ultra-cold gas in a MOT, as those directly involving the laser cooling beams. These processes can couple the temporal variations of the atom density n_a, and the low frequency oscillations of the gas, with the amplitude modulations of the beam. We first examine the case when multiple scattering is negligible and the laser radiation is mainly associated with the incident laser cooling beams. The opposite case where multiple scattering dominates will be discussed in the next section.

Starting from the wave equation which describes the evolution of an incident laser beam (6.1), and assuming a single frequency component such that Eqs. (6.4) and (6.5) are valid, we can derive an envelope equation for the electric field amplitude, of the form

$$\left(\frac{\partial}{\partial t} - \mathbf{v}_\omega \cdot \nabla \right) \mathbf{E}_\omega = -\alpha_\omega \mathbf{E}_\omega \frac{\partial n_a}{\partial t}, \tag{6.10}$$

where the photon group velocity $\mathbf{v}_\omega$, and the coupling parameter α_ω, are determined by

$$\mathbf{v}_\omega = \frac{c^2}{\omega} \frac{\mathbf{k}}{[1 + n_a \chi_a(\omega)]} = \frac{\omega \mathbf{k}}{k^2} \tag{6.11}$$

and

$$\alpha_\omega = \frac{\chi_a(\omega)}{[1 + n_a \chi_a(\omega)]} = \frac{\omega^2}{k^2 c^2} \chi_a(\omega). \tag{6.12}$$

Notice that the quantity α_ω is in general complex, such that

$$\alpha_\omega = \alpha' + i\alpha'' = \frac{\omega^2}{k^2 c^2} \left[\chi'(\omega) + i\chi''_\omega \right]. \tag{6.13}$$

Let us now assume an equilibrium state characterized by $\mathbf{E}_\omega = \mathbf{E}_0$, and $n_a = n_0$. If this equilibrium is perturbed, we can then use $\mathbf{E}_\omega = \mathbf{E}_0 + \tilde{\mathbf{E}}$, and $n_a = n_0 + \tilde{n}$, where $\tilde{\mathbf{E}}$ and $\tilde{n}$ are the perturbations. Linearizing the envelope equation (6.10) with respect to such perturbed quantities, we get

$$\left(\frac{\partial}{\partial t} + \mathbf{v}_\omega \cdot \nabla \right) \tilde{\mathbf{E}} = -\alpha_\omega \mathbf{E}_0 \frac{\partial \tilde{n}}{\partial t}. \tag{6.14}$$

On the other hand, we know from the mean field fluid theory that the atom density n_a can be described by the fluid wave equation

$$\left(\frac{\partial^2}{\partial t^2} - u_s^2 \nabla^2 \right) n_a = -\omega_p^2 n_a, \tag{6.15}$$

where u_s and ω_p are the sound velocity and the effective plasma frequency of the ultra-cold gas. Linearizing this equation, we get

$$\left(\frac{\partial^2}{\partial t^2} + \omega_{p0}^2 - u_s^2 \nabla^2 \right) \tilde{n} = -\beta \left[(\mathbf{E}_0 \cdot \tilde{\mathbf{E}}^*) + (\mathbf{E}_0^* \cdot \tilde{\mathbf{E}}) \right], \tag{6.16}$$

where ω_{p0}^2 is associated with the unperturbed density n_0, and $\beta_a = \epsilon_0(\sigma_R - \sigma_L) \sigma_L n_0^2/2$ is the coupling coefficient, depending on the effective charge of the atom. The quantities σ_R and σ_L are the atom radiation scattering and atom laser absorption cross sections, as previously considered. In order to solve the two coupled equations for the laser and the atom density perturbations, $\tilde{\mathbf{E}}$ and $\tilde{n}$, (6.14) and (6.16), we assume that they both evolve in space an time as $\exp(i\mathbf{q} \cdot \mathbf{r} - i\Omega t)$. Noticing that the term in $|\mathbf{E}_0 \cdot \tilde{\mathbf{E}}^*|$ will not contribute to the same Fourier component and can be ignored, we get a couple of algebraic equations relating the two perturbed quantities $\tilde{\mathbf{E}}$ and $\tilde{n}$, as

$$-i(\Omega - \mathbf{v}_\omega \cdot \mathbf{q})\tilde{\mathbf{E}} = i\Omega\alpha_\omega \mathbf{E}_0 \tilde{n} \, , \quad \left(\Omega^2 - \omega_{p0}^2 - u_s^2 q^2 \right) \tilde{n} = \beta(\mathbf{E}_0^* \cdot \tilde{\mathbf{E}}). \tag{6.17}$$

From here we can easily get the dispersion relation for the laser and density perturbations, as

$$(\Omega - \mathbf{v}_\omega \cdot \mathbf{q}) \left(\Omega^2 - \omega_{p0}^2 - u_s^2 q^2 \right) = -\beta\epsilon\Omega, \tag{6.18}$$

where we have used $\epsilon = \alpha_\omega |E_0|^2$. Of particular interest is the case of perturbations in a direction perpendicular to that of the incident laser beam, such that $(\mathbf{v}_\omega \cdot \mathbf{q}) = 0$. If unstable, these perturbations will lead to *laser filamentation*. In this case, Eq. (6.18) reduces to

$$(\Omega^2 - \omega_s^2) = -\beta\epsilon \, , \quad \omega_s^2 \equiv \omega_{p0}^2 + u_s^2 q^2. \tag{6.19}$$

In the absence of coupling $\beta = \epsilon = 0$, this would reduce to the dispersion relation of the hybrid acoustic modes, $\Omega = \omega_s$, as discussed in previous chapters. Noting that the quantity ϵ is complex, and using $\Omega = \Omega_r + i\gamma$, we can separately write the real and imaginary parts of the above equation, as

$$\Omega_r^2 = \omega_s^2 + \gamma^2 - \beta\epsilon', \quad \gamma = -\frac{\beta\epsilon''}{2\Omega_r}, \tag{6.20}$$

where ϵ' and ϵ'' are the real and imaginary parts of ϵ. We can see that a filamentation instability occurs for $\gamma > 0$, or equivalently for $\beta\epsilon'' < 0$. This implies, $\beta > 0$, which is the usual case, and $\chi''(\omega) < 0$. This means that filamentation can only be observed for a blue shifted laser beam, and not for the usual laser cooling beams.

Let us now turn to the general case of perturbations propagating in arbitrary directions, where $(\mathbf{v}_\omega \cdot \mathbf{q}) \neq 0$. Going back to the dispersion relation (6.18), we assume a double resonant condition

$$\mathbf{v}_\omega \cdot \mathbf{q} = \omega_s = \Omega - \delta, \tag{6.21}$$

with $|\delta| \ll \omega_s$. Retaining the dominant terms, we get

$$\delta = \pm i \sqrt{\frac{\beta}{2}\epsilon}. \tag{6.22}$$

We therefore have and instability with growth rate $\gamma \equiv \Im(\delta)$ with

$$\gamma = \frac{\omega_{p0}}{\sqrt{2}} \left[\chi'^2(\omega) + \chi''^2(\omega) \right]^{1/4} \cos\varphi, \tag{6.23}$$

where $2\varphi = \tan^{-1}(\epsilon''/\epsilon')$. For $\epsilon' \sim \epsilon''$, we have $\varphi \sim \pi/8$. We can see that instability can occur for the favorable situation of a double resonant condition (6.21), with significant growth rates. Notice however, that we always have $\gamma \ll \omega_{p0}$, which is compatible with the condition $|\delta| \ll \omega_s$. When this condition brakes down, we need to solve directly the dispersion relation (6.18).

This analysis shows that a *modulational instability*, with oscillating frequency $\Omega \simeq \omega_s$ and $(\mathbf{v}_\omega \cdot \mathbf{q}) \neq 0$ satisfying the double resonance condition (6.21) can occur for the incident laser cooling beams. In contrast, transverse filamentation instabilities with $(\mathbf{v}_\omega \cdot \mathbf{q}) = 0$ are unlikely to occur, because they would imply the use of blue detuned incident radiation. On the other hand, the length scale of the unstable modulations is determined by $q \simeq \omega_{p0}/c \cos\theta$, where θ is the angle between $\mathbf{q}$ and $\mathbf{k}$. This implies that $\cos\theta \gg (u_s/c)$, which is a very small quantity, showing that a wide range of wavenumbers q can become unstable, with nearly identical growth rates.

6.3 Photon Bubbles

In contrast with the previous section, we now discuss the case where photon multiple scattering dominates over direct field propagation [10]. This is a relevant situation for current MOT conditions, where the incident laser cooling beams are deflected by multiple scattering, randomizing the photon propagation direction, and diffusion effects become dominant. In order to describe this new situation we use the energy transport equation for the photon field which, in a region where radiation sources are absent, can be generally stated as

$$\frac{\partial}{\partial t} I_\omega + \nabla \cdot \mathbf{S}_\omega = -\gamma_\omega I_\omega, \tag{6.24}$$

Here, γ_ω describes photon absorption, and I_ω is the photon energy density corresponding to the frequency ω. This quantity can be defined as the spectral energy density of the electromagnetic radiation $W(\omega, \mathbf{k})$, integrated over all the possible directions of propagation, as

$$I_\omega = \int W(\omega, \mathbf{k}) \delta(k - k_\omega) \frac{d\mathbf{k}}{(2\pi)^3}, \tag{6.25}$$

where k_ω is the solution of the photon dispersion equation, $k^2c^2 = \omega^2[1+n_a \chi_a(\omega)]$. In the following, such a solution is assumed, as it will not play any explicit role in the calculations. If radiation is made isotropic by multiple scattering, the energy flux is determined by a diffusive process, characterized by (see e.g. [7, 12])

$$\mathbf{S}_\omega = -D\nabla I_\omega \,, \quad \gamma_\omega = Dk_a^2, \tag{6.26}$$

where k_a is the inverse of the energy absorption length. The diffusion coefficient is determined by $D = l^2/\tau$, where the photon mean free path is $l = 1/n_a\sigma_R$, where n_a is the atom number density, and σ_L the laser atom scattering cross section. The photon diffusion time τ can be considered as nearly independent from the atom density, as shown by cold atom experiments [9]. We therefore get $D \propto n_a^{-2}$.

On the other hand, the atom density n_a can be determined in the mean field approximation by the fluid equations, which determine the atom density n_a and mean velocity $\mathbf{v}$. In the momentum equation it is useful to introduce a damping force term, $-\nu\mathbf{v}$, where ν is the damping rate resulting from the viscosity of the gas. As we have previously discussed, the collective force $\mathbf{F}$, resulting from the exchange of photons between nearby atoms is determined by a Poisson type of equation, $\nabla \cdot \mathbf{F} = Qn_a$. We notice that the charge parameter Q is proportional to the laser intensity (see Eq. (4.87)), thus providing a coupling between the atom density n_a and the photon intensity I_ω.

The diffusion equation (6.24), coupled with the fluid equations, can now be studied, using a perturbation analysis. We assume that the equilibrium state of the atomic density and radiation intensity is perturbed, as described by the decomposition $I_\omega = I_0 + \tilde{I}$, and $n_a = n_0 + \tilde{n}$. The linearized photon transport equation can be written as

$$\left(\frac{\partial}{\partial t} - \nabla \cdot D_0 \nabla \right) \tilde{I} = -g(\nabla^2 I_0 + \nabla I_0 \cdot \nabla)\tilde{n} - \gamma_0 \tilde{I}, \tag{6.27}$$

where $D_0 = 1/n_0^2\sigma_L^2\tau$ is the unperturbed diffusion coefficient, $\gamma_0 = D_0 k_a^2$, and $g = D_0/2n_0$. Similarly, from the linearized fluid equations we obtain

$$\left[\frac{\partial}{\partial t} \left(\frac{\partial}{\partial t} + \nu \right) + \omega_p^2 - u_s^2 \nabla^2 \right] \tilde{n} = -\beta \tilde{I}, \tag{6.28}$$

where u_s is the sound speed, and ω_p is the effective plasma frequency, defined by $\omega_p^2 = Qn_0/M$. Assuming that the gas of ultra-cold atoms follows an equation of state of the form $P \propto n_a^{\gamma_a}$, where γ_a is the adiabatic constant, the sound speed will be determined by $u_s = \gamma_a P/M$. We have also introduced the coupling coefficient $\beta = \omega_p^2 n_0/I_0$.

In order to proceed with the stability analysis, we assume a space and time dependence of the perturbed quantities of the form $\exp(i\mathbf{q}\cdot\mathbf{r} - i\Omega t)$. Replacing this in Eqs. (6.27) and (6.28), we arrive at the dispersion relation

$$(-i\Omega + D_0 q^2 + \gamma_0)(\Omega^2 + i\omega\nu - \omega_s^2) = -\beta(\epsilon + ia), \Omega, \qquad (6.29)$$

where we have used the quantities $\epsilon = g\nabla^2 I_0$, $\quad a = g(\mathbf{q}\cdot\nabla I_0)$, and $\omega_s^2 \equiv \omega_p^2 + u_s^2 q^2$. Notice that, in the absence of coupling between the photons and the atomic gas ($\beta = 0$), this dispersion relation would describe two independent modes: (1) – a purely decaying photon mode, such that $\Omega = -i(D_0 q^2 + \gamma_0)$, where damping results from both diffusion and absorption; and (2) – an oscillating fluid mode, determined by $\Omega = \omega_s - i\nu$, which is damped by viscosity. It is now interesting to consider the modes resulting from the coupling between these two basic solutions of both photon intensity and atom density.

Let us first examine oscillating perturbations with frequency of order ω_s. Using $\Omega = \omega_s + \delta$, and assuming $|\delta| \ll \omega_s$, we obtain solutions of the form

$$\delta = -\frac{B}{2}\left[1 \pm \sqrt{1 - 2i\left(\frac{\nu}{B} + \frac{\beta\epsilon}{\omega_s B^2}\right)}\,\right], \qquad (6.30)$$

where we have used $B = \omega_s + i(D_0 q^2 + \gamma_0)$. The choice of the positive sign would lead to $|\delta| \sim \omega_s$, thus violating our initial assumption. We therefore focus on the solution with the negative sign. After expansion of the expression inside the square rooth, we get

$$\delta = -\frac{i}{2}\nu - \frac{\beta(\epsilon + ia)}{2\omega_s^2}\left(i + \frac{D_0 q^2}{\omega_s} + \frac{\gamma_0}{\omega_s}\right). \qquad (6.31)$$

This can be considered as the perturbed oscillating fluid mode. We can see that, apart from a small correction in the real part of the frequency, we can have a growth or damping rate, determined by the quantity $\Gamma = \Im(\delta) = \Im(\Omega)$. We analyze two distinct situations were $\Gamma > 0$ and instability becomes possible. One is the case when the quantity a can be neglected. In explicit terms, this occurs when the following inequality is satisfied

$$|\mathbf{q}\cdot\nabla I_0| \ll \frac{D_0\omega_p^2}{I_0\omega_s} q^2 \nabla^2 I_0. \qquad (6.32)$$

In this case, instability will occur if the following two conditions are simultaneously satisfied

$$\beta\epsilon < 0 \,, \quad \frac{|\beta\epsilon|}{\omega_s^2} > \nu. \tag{6.33}$$

The first condition implies that, for $\beta > 0$ (or, equivalently, for $\sigma_R > \sigma_L$) we should have $\nabla^2 I_0 < 0$. Notice that a positive β corresponds to the most natural experimental scenario in a MOT. But we could also have $\beta < 0$, (or $\sigma_R < \sigma_L$), which would then imply $\nabla^2 I_0 > 0$, for MOT conditions where the atom effective charge would become imaginary. On the other hand, the second condition in (6.33) determines the instability threshold, where the growth rate has to compensate for the losses due to viscosity. In the absence of viscosity, or for conditions well above the threshold, the growth rate attains its maximum value $\Gamma_{\max} = |\beta\epsilon|/2\omega_s^2$. This will lead to the formation of photon bubbles, as discussed below.

Let us now consider the situation where a becomes the dominant term in (6.31) and the inequality (6.32) is reversed. In this case, we can have mode instability if $\mathbf{q} \cdot \nabla I_0 < 0$, and at the same time we verify

$$|\mathbf{q} \cdot \nabla I_0| \frac{g\beta}{2\omega_s^2} \left(\frac{D_0 q^2}{\omega_s} + \frac{\gamma_0}{\omega_s} \right) > \nu. \tag{6.34}$$

In contrast with the previous case specified by (6.33), where the instability was isotropic, here the instability conditions will strongly dependent on the direction of $\mathbf{q}$, and the growth rate will maximize along the gradient of the photon intensity.

Let us now turn to purely damped (or growing) perturbations, with $\Re(\Omega) \simeq 0$. This can now be considered as the perturbed photon decay mode. Assuming that $|\Omega|^2 \ll \omega_s^2$, Eq. (6.29) leads to

$$\Gamma = \left[\frac{\beta\epsilon}{\omega_s^2} - (D_0 q^2 + \gamma_0) \right]. \tag{6.35}$$

This shows that unstable modes can indeed exist, for $\beta\epsilon > 0$. It should also be noticed that these modes are nearly (but not exactly) purely growing modes, because a residual value of the mode frequency still exists, $\Re(\Omega) \simeq (a D_0 / 2 I_0)$. This quantity goes to zero with $a \equiv \mathbf{q} \cdot \nabla I_0$, and therefore vanishes in the spherically symmetric case.

It is useful to compare these modes with the above oscillating modes. First, they occur for opposite signs of the quantity $\beta\epsilon$, the oscillating modes for a negative sign, and the purely growing modes for a positive sign. The maximum growth rate for the purely growing modes are two times larger than for the oscillating modes, $\Gamma_{\max} = \beta\epsilon/\omega_s^2$. On the other hand, the threshold conditions are different: (1) For the oscillating modes, Eq. (6.33) shows that the instability is limited by the viscosity of the atomic gas. (2) In contrast, for purely growing modes, Eq. (6.35) shows that the instability is limited by photon diffusion and losses.

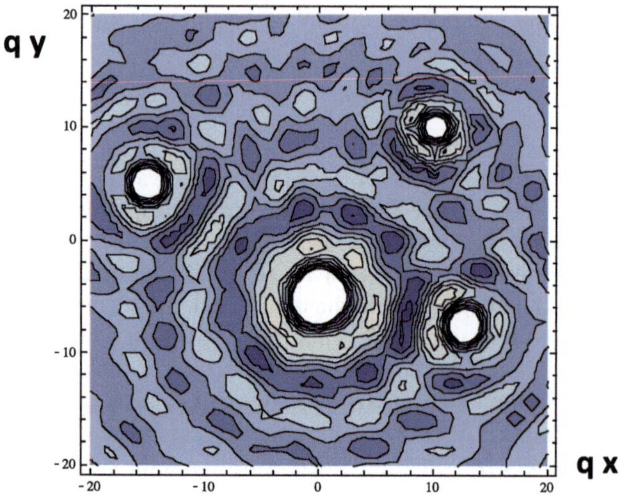

Fig. 6.1 Schematic representation of spherical bubbles resulting from the unstable coupling of laser light with the atom mean field

In this comparison we have ignored the mode (6.34), and assumed a negligible a, when the instabilities can be seen as isotropic. In this situation, we can associate such unstable modes with the formation of photon bubbles. In order to describe more explicitly the formation of such bubbles, we replace the plane wave modes by spherically symmetric perturbations described by $(\tilde{I},\ \tilde{n}) \propto F(r)\exp(-i\Omega t)$, where $F(r)$ satisfies the equation

$$\nabla^2 F(r) \equiv \frac{1}{r^2}\frac{\partial}{\partial r}\left(r^2 \frac{\partial}{\partial r}\right) F(r) = -q^2 F(r), \tag{6.36}$$

where q^2 is a positive real quantity. Going back to the perturbed equations, but ignoring the quantity $a \equiv \mathbf{q} \cdot \nabla I_0$ which would brake the spherical symmetry, we arrive again at the dispersion relation (6.29), but with $a = 0$. Therefore, our previous analysis for the oscillating and purely growing modes remains valid for such spherically symmetric perturbations. Equation (6.36) can be written as a spherical Bessel function, with a well known non-singular solution, given by

$$F(r) = \sqrt{\frac{\pi}{2qr}} J_{1/2}(qr) = \frac{\sin(qr)}{qr} \tag{6.37}$$

A superposition of such solutions is represented in Fig. 6.1, for illustration. Of course, superposition is only valid for linear solutions. Nonlinear terms will give rise to instability saturation. For purely growing modes, such the saturation level can be easily estimated by replacing the equilibrium diffusion coefficient D_0 by its perturbed value $D_0 n_0^2/(n_0 - |\tilde{n}|)^2$ in the threshold condition (6.35). Saturation will then occur for

$$|\tilde{n}|_{\text{sat}} \simeq n_0 \left[1 - \left(\frac{D_0 q^2}{\Gamma_{\text{max}} - \gamma_0} \right)^{1/2} \right]. \tag{6.38}$$

Saturation can then be understood as a result of an increase of diffusion losses, due to a local density depletion. As a consequence of the coupled dynamics, a local decrease in the atomic number density $\tilde{n}$ also leads to an increase of the local photon intensity. Near the instability saturation, $\Im(\Omega) \sim 0$, and the static bubbles will satisfy the following relation $\tilde{I} = -\tilde{n}/(D_0 q^2 + \gamma_0)$. Therefore, the perturbations $\tilde{I}$ and $\tilde{n}$ have opposite signs.

Finally, it is useful to write the threshold conditions and growth rates in a more explicit form, in terms of the typical time and scale lengths. Introducing the photon intensity scale length L, such that $L^{-2} \equiv I_0^{-1} |\nabla^2 I_0|$, we can write, threshold and growth rate for purely growing bubbles as

$$L < \frac{1}{\sqrt{2}} \frac{\omega_p}{\omega_s} \frac{1}{(q^2 + k_a^2)^{1/2}}, \quad \Gamma_{\text{max}} = \frac{1}{\tau} \frac{l^2}{L^2}. \tag{6.39}$$

Given that $\omega_p \simeq \omega_s$, and typically we have $k_a^2 \gg q^2$, we get the threshold condition $L < 1/k_a$, which can be easily satisfied. For instance, taking the experimental conditions of Ref. [9], we have $l \simeq 300\,\mu\text{m}$, $D \simeq 0.66\,\text{m}^2/\text{s}$, corresponding to $\tau \simeq 0.1\,\mu$. Thus, even for large $L \gg l$, the maximum growth rate Γ_{max} can be much larger than $1\,\text{s}^{-1}$. A similar analysis can be made for the oscillating bubbles, and the result is

$$L < \frac{\omega_p}{\omega_s} \frac{l}{\sqrt{2\tau\nu}}, \quad \Gamma_{max} = \frac{1}{2\tau} \frac{l^2}{L^2}. \tag{6.40}$$

This can also be satisfied in current experimental conditions, but for a different sign of $\beta\epsilon$. The two threshold conditions are represented in Fig. 6.2.

6.4 Roton Instability

In the previous section, we have discussed the mechanism of photon bubble formation, if the system is dominated by diffusive light. This is a consequence of the coupled dynamics between the atoms and the photons. Although this coupling is not strong enough to effectively form atomic polaritons, the combined dynamic competition between the long-range and stochastic forces between the atoms (where the later results from the ballistic behavior of the light) this atom + photon "soup" can exhibit polariton-like features in the spectrum of the elementary excitations. In what follows, we show that it may eventually result on the emergence of a roton minimum.

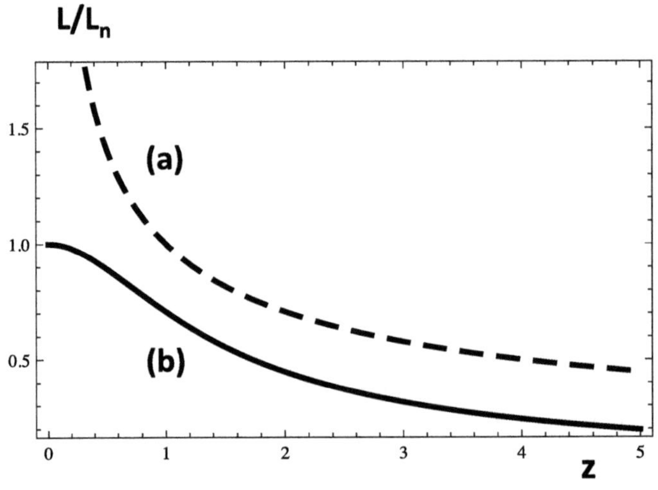

Fig. 6.2 Threshold curves for photon bubble formation in the ultra-cold gas: (*a*) Oscillating bubbles, photon characteristic scale L/L_n, versus normalized wavenumber $z = q/k_a$ with $L_n = (\omega_p/\omega_s)/\sqrt{2}k_a$; (*b*) Purely growing bubbles, L/L_n versus $z = \sqrt{\tau \nu}$, with normalization factor $L_n = (\omega_p/\omega_s)(l/\sqrt{2})$. Instability occurs below theses curves

We consider that the dynamics of cold atoms in MOTs is described by the Vlasov equation and the collective force can be described by the Poisson equation

$$\nabla \cdot \mathbf{F}(\mathbf{r}, t) = Q \int d\mathbf{v} W(\mathbf{r}, v, t), \tag{6.41}$$

where $W(\mathbf{r}, v, t)$ is the atom velocity distribution. To describe the dynamics of light in optically thick traps, we make use of the photon diffusion equation (6.24)[1]

$$\frac{\partial I}{\partial t} - \nabla \cdot D \nabla I = 0, \tag{6.42}$$

where the photon absorption term γ_a is neglected. The diffusion coefficient is again determined by $D = l^2/\tau$, which can be explicitly written in terms of the atom distribution as

$$D(\mathbf{r}, t) = \frac{1}{\sigma_L^2 \tau^2 n^2} = \frac{1}{\sigma_L^2 \tau^2 n_0^2} \left[\int W(\mathbf{r}, \mathbf{v}, t) d\mathbf{v} \right]^{-2}. \tag{6.43}$$

We now linearize the Vlasov and photon diffusion equation with respect to the fluctuations, such that $W = W_0 + \delta W$, $I = I_0 + \delta I$ and $D = D_0 + \delta D$,

[1]In this section, we have dropped the subscript ω for simplicity.

and we obtain

$$\left(\frac{\partial}{\partial t} + \mathbf{v} \cdot \nabla\right) \delta W + \frac{1}{m} \delta \mathbf{F} \cdot \nabla_{\mathbf{v}} W_0 = 0 \tag{6.44}$$

$$\frac{\partial}{\partial t} \delta I - D_0 \nabla^2 \delta I - \delta D \nabla^2 I_0 = 0 \tag{6.45}$$

$$\nabla \cdot \delta \mathbf{F} = Q_0 n_0 \int \delta W d\mathbf{v} + Q_0 n_0 \frac{\delta I}{I_0}, \tag{6.46}$$

where $Q_0 = \sigma_L(\sigma_R - \sigma_L) I_0 / c$. Assuming periodic perturbations in space and time, such that $(\delta f, \delta I, \delta D) \propto \exp(i \mathbf{q} \cdot \mathbf{r} - i \Omega t)$, Eqs. (6.43) and (6.45) yield

$$\delta I = \frac{\beta}{i \Omega - D_0 q^2} \int d\mathbf{v} \delta W, \tag{6.47}$$

where

$$\beta = \frac{2 \nabla^2 I_0}{n_0^3 \sigma_L^2 \tau} = \frac{2 \nabla^2 I_0}{n_0} D_0 \tag{6.48}$$

now represents the photon inhomogeneity parameter. Combining the latter results, we can derive a kinetic dispersion relation of the form

$$1 = \frac{\omega_p^2}{q^2} \left(1 + \frac{\omega_d}{i \Omega - D_0 q^2}\right) \int \frac{1}{v_z - \Omega / q} \frac{\partial W_0}{\partial v_a} d\mathbf{v}, \tag{6.49}$$

where we have considered propagation along the z-direction, $\mathbf{q} = q e_{\mathbf{z}}$, for definiteness. Here, we have defined two typical frequencies of the system. The first one is plasma frequency $\omega_p = \sqrt{Q_0 n_0 / M}$, and the second one is the photon scattering rate, or simply the *diffusion frequency*

$$\omega_d = \frac{\beta n_0}{I_0} = \frac{2 \nabla^2 I_0}{I_0} D_0. \tag{6.50}$$

We notice that this frequency depends on the scale at which the diffusive processes occur (micro-, meso- or macroscopic), as it depends upon the spatial scale L at which the light intensity varies. We will discuss the macroscopic case below.

The integral in Eq. (6.49) can be evaluated using the Landau prescription, according to which the full information about the initial conditions is cast if the integration path is set to pass below the pole $\Omega = v_z q$. We split the integral into two parts, as discussed in Chap. 4. We can therefore write

$$\int \frac{1}{v_z - \Omega/q} \frac{\partial W_0}{\partial v_z} d\mathbf{v} = \mathrm{P} \int \frac{1}{v_z - \Omega/q} \frac{\partial W_0}{\partial v_z} d\mathbf{v} + i\pi \left(\frac{\partial W_0}{\partial v_z} \right) \bigg|_{v_z=\Omega/k} d\mathbf{v}, \tag{6.51}$$

where P stands for the Cauchy principal value. Assuming a phase speed $v_{\text{ph}} = \omega/q$ much greater than the width of the distribution, such that W_0 and its derivatives get small as v_z gets large, we may expand the denominator in (6.51) which, together with the relation

$$\int \frac{1}{v_z - \Omega/q} \frac{\partial W_0}{\partial v_z} d\mathbf{v} = \int \frac{W_0}{(v_z - \Omega/q)^2} d\mathbf{v}, \tag{6.52}$$

simply yields

$$\mathrm{P} \int \frac{1}{v_z - \Omega/q} \frac{\partial W_0}{\partial v_z} d\mathbf{v} \simeq \int W_0 \left(1 + 3\frac{q^2 v_z^2}{\Omega^2} \right) dv_x dv_y dv_z. \tag{6.53}$$

Assuming the atomic equilibrium to be described by a Maxwell distribution

$$W_0(v) = \frac{1}{(2\pi v_{\text{th}})^{3/2}} e^{-v^2/2v_{\text{th}}^2}, \tag{6.54}$$

with $v_{\text{th}} = \sqrt{T/m}$ standing for the thermal speed, we obtain

$$1 = \frac{\omega_p^2}{\Omega^2} \left[\left(1 + \frac{u_s^2 q^2}{\Omega^2} \right) \left(1 + \frac{\omega_d}{i\Omega - D_0 q^2} \right) \right] + i\pi \frac{\omega_p^2 \Omega^2}{q^2} \frac{\partial W_0}{\partial v_z} \bigg|_{v_z=\Omega/k}, \tag{6.55}$$

where we have defined the sound atomic speed as $u_s = \sqrt{3} v_{\text{th}}$. Separating the frequency into its real and imaginary parts, $\Omega = \omega + i\gamma$, with $\gamma \ll \omega$, we may finally write

$$\omega^2 = \left(\omega_p^2 + u_s^2 q^2 \right) \left(1 - \frac{\omega_d D_0 q^2}{\omega_p^2 + D_0^2 q^4} \right) \tag{6.56}$$

and

$$\gamma = \frac{\omega_d}{2} \frac{\omega_p^2 + u_s^2 q^2}{\omega_p^2 + D_0^2 q^4} - \frac{3}{\sqrt{8\pi}} \frac{1}{q^3 \lambda_D^3} e^{-3/(2q^2 \lambda_D^2)}, \tag{6.57}$$

This dispersion relation describes a quasi-polariton resulting from the coupling between the hybrid phonons and the photons.

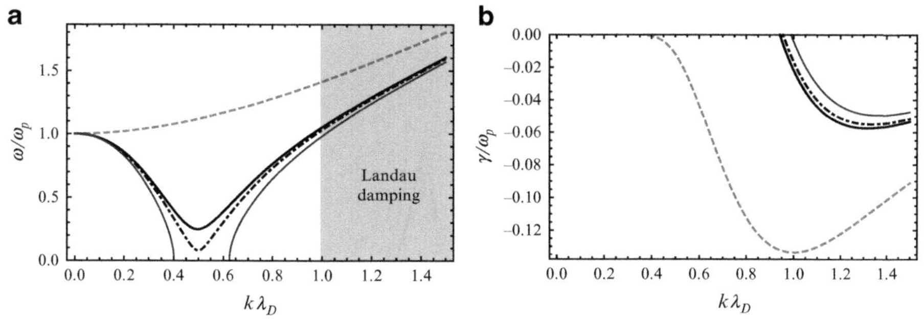

Fig. 6.3 Illustration of the real (*left panel*) and imaginary (*right panel*) parts of the polariton dispersion relation in the macroscopic regime for $D_0 = 2.0\lambda_D^2\omega_p$. We can observe the emergence of a roton minimum for $\omega_d = 1.9\omega_p$ (*full black line*) and $\omega_d = 1.99\omega_p$. The rotons softens the frequency to $\Omega(q_{\mathrm{rot}}) = 0$ at the critical value $\omega_d^{(c)} = 2.0\omega_p$. Roton instability is illustrated here for $\Omega = 2.2\omega_p$. The short-wavelength oscillations corresponding to $q\lambda_D > 1$ are kinematically damped. The usual plasma dispersion relation of hybrid phonons is presented here for comparison (*dashed gray line*)

The diffusive description of light is the result of a macroscopic approximation which is known to hold if absorption takes place at scales much larger than the mean free path ℓ [12]. This is true provided the following hierarchy for the relevant length scales

$$\lambda \ll \ell \ll a \ll L, \tag{6.58}$$

where λ is the light wavelength and a is the size of the system. According to typical experimental conditions [9], the mean-free path is found to value $\ell \sim 300\,\mu\mathrm{m}$ and the diffusion coefficient $D_0 \simeq 0.66\,\mathrm{m}^2\,\mathrm{s}^{-1}$. Based on our previous estimates, the effective plasma frequency and Debye length respectively value $\omega_p \sim 2\pi \times 100\,\mathrm{Hz}$ and $\lambda_D \sim 100\,\mu\mathrm{m}$. Therefore, provided the identification $L = 2\nabla^2 I_0/I_0$ in (6.50) and using the inequalities in Eq. (6.58) with $a \sim 1\,\mathrm{mm}$ [9], the diffusive approximation is valid provided that the diffusion and plasma frequencies are of the same order, $\omega_d \sim \omega_p$. This is achieved if $\ell_d = \sqrt{D_0/\omega_p} \sim 1\,\mathrm{cm}$ is of the same order of the intensity variation length L, which may be possibly achieved for typical experimental conditions.

In Fig. 6.3, it is shown that a roton minimum emerges in the dispersion relation (6.56) in the diffusive regime. As the values of ω_d increase (i.e., for stronger diffusion), the polariton frequency decreases (mode softening). At the critical value $\omega_d^{(c)} = 2\omega_p$, the mode softens towards zero, which is a clear manifestation of a roton instability mechanism. For $\omega_d > \omega_d^{(c)}$, the system enters a crystallization phase. This mechanism has been recently discussed in the literature as it can lead to the formation of supersolids [6]. An important remark is related to the Landau damping at short wavelengths. Modes in the region $q\lambda_D \gtrsim 1$ will undergo a kinematic damping and cannot be observed. Fortunately, rotons are possible to be excited at longer wavelengths ($q_{\mathrm{rot}}\lambda_D < 1$), thus avoiding the Landau damping mechanism.

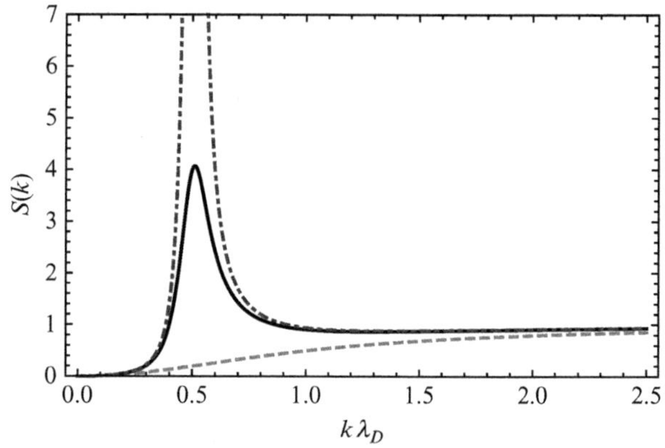

Fig. 6.4 Static structure factor $S(k)$ depicted for $D_0 = 2.0\lambda_D^2\omega_p$. $\omega_d = 0$ (*gray dashed line*), $\omega_d = 1.9\omega_p$ (*black full line*) and $\omega_d = 1.99\omega_p$ (*black dashed line*). A pick emerges in the static structure factor around $k_{\text{rot}} = \lambda_D/\ell_d^2 = 0.5\lambda_D^{-1}$

Moreover, the onset of diffusion tends to decrease the damping rate (see Fig. 6.3). This nourishes hope for rotons to be experimentally observable.

A remarkable feature of the polariton spectrum in (6.56) is that it exhibits a roton minimum in a three dimensional system even in the absence of strong interactions. It is clear that the present spectrum does not correspond to that of a superfluid, as the mode is gapped at the origin. Here it is a consequence of the long-range interaction between the cold atoms. Another important property of these classical rotons is that they carry information about the long-range correlations of the system. From the dissipation-fluctuation theorem [5], we know that the *dynamic structure factor* is given by

$$S(q,\Omega) = (Tq^2/\pi\Omega)\text{Im}\,\epsilon(q,\Omega)^{-1}, \tag{6.59}$$

where $\epsilon(q,\Omega) \equiv 1 + \chi(q,\Omega)$ and $\chi(q,\Omega)$ is the susceptibility. As a result, the dispersion relation (6.55) is defined as the root of the function $\epsilon(q,\Omega) \equiv 1 + \chi(q,\Omega)$. In the absence of hydrodynamic damping, the *static structure factor*, $S(q)$ is actually given by the classical expression [13]

$$S(q) = \pi^{-1}S(q,\Omega)\text{Re}\,\frac{i\Omega}{\epsilon(q,\Omega)^{-1}} = \frac{v_{\text{th}}^2 q^2}{\Omega(q)^2}. \tag{6.60}$$

In Fig. 6.4, we illustrate the behavior of $S(q)$ for the same parameters of Fig. 6.3.

The static two-point correlation function

$$g(r) = \frac{\langle n(r)n(0)\rangle}{\langle n(r)\rangle\langle n(0)\rangle} = \frac{\langle n(r)n(r')\rangle}{n_0^2}, \tag{6.61}$$

can then be easily calculated provided the relation $g(r) = 1 + \mathcal{F}^{-1}[S(q) - 1]$ [1], where $\mathcal{F}^{-1}$ represents the inverse Fourier transformation. After integrating out the angular variables, we obtain

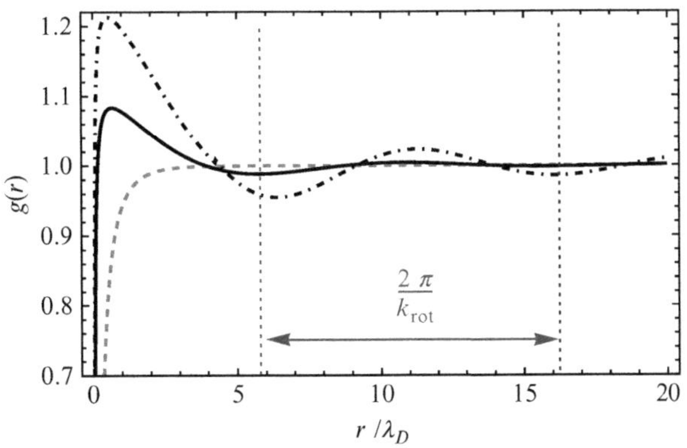

Fig. 6.5 Two-point correlation function $g(r)$ depicted for $D_0 = 2.0\lambda_D^2\omega_p$. $\omega_d = 0$ *(gray dashed line)*, $\omega_d = 1.9\omega_p$ *(black full line)* and $\omega_d = 1.99\omega_p$ *(black dashed line)*. The correlation function oscillates with the period a of $2\pi/q_{\text{rot}} \simeq 4\pi\lambda_D^{-1}$

$$g(r) = 1 + \frac{1}{\pi^2}\int_0^{\infty} \frac{q\sin(qr)}{r}\left[S(q) - 1\right]. \tag{6.62}$$

Computing the integral in the complex plane, we can immediately reconstruct the correlation function. It is observed that the appearance of a minimum in the excitation spectrum (6.56) is associated with the occurrence of long-range correlation in the system. By inspection, one founds that the roton minimum occurs at $k_{\text{rot}} \simeq \lambda_D/\ell_d^2$, which is exactly the period at which the correlation function oscillates (see Fig. 6.5). This feature can be qualitatively understood in the context of Percus-Yevick theory [11, 14], where the correlation function is approximated by

$$g(r)^{\text{PV}} \simeq 1 + c_0 r^{-1}\cos(q_0 + \delta_0)e^{-\kappa_0 r}, \tag{6.63}$$

where c_0 is a constant and $z_0 = \kappa_0 + iq_0$ is the pole of the function $S(q) - 1$. We remark, however, that this theory was originally developed for hard-sphere potentials, and therefore does not apply to systems with long-range interactions. For that reason, we have not used it to compute $g(r)$ in (6.62).

6.5 Density Fluctuations

We now propose to study the spectrum of density fluctuations in magneto-optical traps, by using collective laser scattering techniques. First, in this section, we determine the scattered fields, and the expected average scattered power. Second, in the next section, we determine the spectrum of density fluctuations, in the quasi-classical limit. We make use of the test particle method, where the field and density fluctuations associated with each atom is individually studied, by considering it as a text particle [8]. We then characterize the main properties of the expected scattered signal.

As previously stated, in the classical approximation, a trapped alkali gas in a MOT can be described by a kinetic Vlasov equation and the Poisson equation for the mean field potential. They determine the evolution of the atomic distribution function, $W \equiv W(\mathbf{r}, \mathbf{v}, t)$ and the collective force $\mathbf{F}$. The effects of trapping will be ignored, as we are interested in fluctuations that occur on a very short scale, when compared to the size of the system. We use the test particle method, by considering a single atom, moving with speed $\mathbf{v}'$ as the perturbation source. The presence of such a test particle will induce perturbations in the gas distribution function $\tilde{W}$, as determined by the linearized kinetic equation

$$\left(\frac{\partial}{\partial t} + \mathbf{v} \cdot \nabla\right) \tilde{W} = -\frac{\delta \mathbf{F}}{M} \cdot \frac{\partial}{\partial \mathbf{v}} W_0. \tag{6.64}$$

Here, $W_0 = W - \tilde{W}$ is the unperturbed (or the equilibrium) distribution function. The collective force perturbation $\delta \mathbf{F}$, also induced by the test particle, is determined by

$$\nabla \cdot \delta \mathbf{F} = Q \int \left[\tilde{W} + \delta(\mathbf{r} - \mathbf{r}')\delta(\mathbf{v} - \mathbf{v}')\right] d\mathbf{v}. \tag{6.65}$$

The test particle position is assumed given by $\mathbf{r}' \equiv \mathbf{r}'(t) = \mathbf{r}_0 + \mathbf{v}' t$. Let us multiply Eq. (6.65) by $\exp(-i\mathbf{k} \cdot \mathbf{r} + i\omega t)$ and perform integrations over both space and time. Defining the Fourier components $\delta \mathbf{F}_{\omega\mathbf{k}}$ and $\tilde{W}_{\omega\mathbf{k}}$ for the perturbed force and distribution, respectively, we can easily obtain

$$i\mathbf{k} \cdot \delta \mathbf{F}_{\omega\mathbf{k}} = 2\pi Q \delta(\omega - \mathbf{k} \cdot \mathbf{v}') + Q \int \tilde{W}_{\omega\mathbf{k}}(\mathbf{v}) d\mathbf{v}. \tag{6.66}$$

Similarly, the Fourier transformation of Eq. (6.64) leads to

$$\tilde{W}_{\omega\mathbf{k}} = -\frac{i}{M} \frac{\delta \mathbf{F}_{\omega\mathbf{k}} \cdot \partial W_0 / \partial \mathbf{v}}{(\omega - \mathbf{k} \cdot \mathbf{v})}. \tag{6.67}$$

Replacing this in Eq. (6.66), and noting that the collective force is purely longitudinal, in such a way that $\delta \mathbf{F}_{\omega\mathbf{k}} = \mathbf{k} \delta F_{\omega\mathbf{k}} / k$, we can write

$$\delta \mathbf{F}_{\omega\mathbf{k}} = -i \frac{2\pi Q}{k^2} \frac{\delta(\omega - \mathbf{k} \cdot \mathbf{v}')}{\epsilon(\omega, \mathbf{k})} \mathbf{k}, \tag{6.68}$$

where we have used the dielectric function of the atomic gas $\epsilon(\omega, \mathbf{k}) = 1 + \chi(\omega, \mathbf{k})$, as previously stated. Replacing Eq. (6.68) in the expression of the perturbed distribution, we then get

$$\tilde{W}_{\omega\mathbf{k}} = -\frac{2\pi Q}{M k^2} \frac{\delta(\omega - \mathbf{k} \cdot \mathbf{v}')}{\epsilon(\omega, \mathbf{k})} \frac{\mathbf{k} \cdot \partial W_0 / \partial \mathbf{v}}{(\omega - \mathbf{k} \cdot \mathbf{v})}. \tag{6.69}$$

After inverse Fourier transforming, we can compute the perturbed force due to the test particle, as

$$\delta\mathbf{F}(\mathbf{r},t) = -i\frac{2Q\mathbf{k}}{k^2}\int \frac{\exp[i\mathbf{k}\cdot(\mathbf{r}-\mathbf{r}_0-\mathbf{v}'t)]}{\epsilon(\mathbf{k}\cdot\mathbf{v}',\mathbf{k})}\frac{d\mathbf{k}}{(2\pi)^3} \tag{6.70}$$

and the associated perturbation of the distribution function

$$\tilde{W}(\mathbf{r},\mathbf{v},t) = -\frac{Q}{Mk^2}\int \frac{\exp[i\mathbf{k}\cdot(\mathbf{r}-\mathbf{r}_0-\mathbf{v}'t)]}{\epsilon(\mathbf{k}\cdot\mathbf{v}',\mathbf{k})}\frac{\mathbf{k}\cdot\partial W_0/\partial\mathbf{v}}{\mathbf{k}\cdot(\mathbf{v}'-\mathbf{v})}\frac{d\mathbf{k}}{(2\pi)^3}. \tag{6.71}$$

Finally, the density perturbation created at a position $\mathbf{r}$ and instant t by a test particle located at position $\mathbf{r}_0$ at $t = 0$ moving with velocity $\mathbf{v}'$, can now be determined by

$$\tilde{n}(\mathbf{r},t) = \delta(\mathbf{r}-\mathbf{r}') + \int \tilde{W}(\mathbf{r},\mathbf{v},t)d\mathbf{v}. \tag{6.72}$$

At this point, it should be noticed that all the atoms in the gas can be considered as possible test particles. This means that the total averaged perturbations can be calculated by integrating the relevant quantities defined above over the equilibrium distribution function $W_0(\mathbf{r}_0,\mathbf{v}')$. In particular, the averaged mean force perturbation can be determined by

$$\langle\delta\mathbf{F}(\mathbf{r},t)\rangle = \int d\mathbf{r}_0\int d\mathbf{v}'\delta\mathbf{F}(\mathbf{r},t)W_0(\mathbf{v}). \tag{6.73}$$

This quantity is a linear superposition of purely oscillating quantities, and it can easily be found using Eq. (6.70) that it is identically zero, $\langle\delta\mathbf{F}(\mathbf{r},t)\rangle \equiv 0$, while the averaged quadratic mean force $\langle|\delta\mathbf{F}(\mathbf{r},t)|^2\rangle$, on the other hand, generally is not. It will be rather determined by

$$\langle|\delta\mathbf{F}(\mathbf{r},t)|^2\rangle = \frac{Q^2}{V}\int d\mathbf{r}_0\int d\mathbf{v}'\int \frac{d\mathbf{k}_1}{(2q\pi)^3}\frac{\mathbf{k}_1}{k_1^2}\frac{\exp[i\mathbf{k}_1\cdot(\mathbf{r}-\mathbf{r}_0-\mathbf{v}'t)]}{\epsilon(\mathbf{k}_1\cdot\mathbf{v}',\mathbf{k}_1)}$$
$$\times\int \frac{d\mathbf{k}_2}{(2q\pi)^3}\frac{\mathbf{k}_2}{k_2^2}\frac{\exp[i\mathbf{k}_2\cdot(\mathbf{r}-\mathbf{r}_0-\mathbf{v}'t)]}{\epsilon(\mathbf{k}_2\cdot\mathbf{v}',\mathbf{k}_2)}. \tag{6.74}$$

We next express the fluctuations inside the trap in terms of the statistical average over all test particles in the system. Let us introduce $W_0(\mathbf{v}) = n_0 f_0(\mathbf{v})$, where the new distribution function $f_0(\mathbf{v})$ is the normalized distribution function, such that

$$\int W_0(\mathbf{v})d\mathbf{v} = n_0\int f_0(\mathbf{v})d\mathbf{v} = n_0. \tag{6.75}$$

Using Eqs. (6.71) and (6.72), it is possible to express the atom density fluctuations due to a given test particle initially located at $\mathbf{r}(t=0) = \mathbf{r}_0$ as

$$n(\mathbf{r},t) = \delta(\mathbf{r}-\mathbf{r}_0-\mathbf{v}'t) + \int \tilde{W}(\mathbf{r},\mathbf{v},t)d\mathbf{v}, \tag{6.76}$$

which can be represented in a more suggestive fashion

$$n(\mathbf{r},t) = \int \frac{d\mathbf{k}}{(2\pi)^3}\left\{1-\omega_p^2\int \frac{\exp[i\mathbf{k}\cdot(\mathbf{r}-\mathbf{r}_0-\mathbf{v}'t)]}{\epsilon(\mathbf{k}\cdot\mathbf{v}',\mathbf{k})}\frac{\mathbf{k}\cdot\partial W_0/\partial\mathbf{v}}{\mathbf{k}\cdot(\mathbf{v}'-\mathbf{v})}\right\}. \tag{6.77}$$

Here, we have introduced the definition of the plasma frequency for the neutral gas, $\omega_p^2 = Qn_0/M$. Using the average, as in Eq. (6.73), we can then state the density correlations as

$$\langle n(\mathbf{r}_1,t_1)n(\mathbf{r}_2,t_2)\rangle = n_0\int d\mathbf{r}_0\int d\mathbf{v}'W_0(\mathbf{v}')n(\mathbf{r}_1,t_1)n(\mathbf{r}_2,t_2). \tag{6.78}$$

More explicitly, this can be written in the following way

$$\langle n(\mathbf{r}_1,t_1)n(\mathbf{r}_2,t_2)\rangle = n_0\int d\mathbf{r}_0\int d\mathbf{v}'W_0(\mathbf{v}')\int \frac{d\mathbf{k}_1}{(2\pi)^3}e^{i\varphi_1}$$
$$\times\int \frac{d\mathbf{k}_2}{(2\pi)^3}e^{i\varphi_2}g(\mathbf{k}_1,\mathbf{v}')g(\mathbf{k}_2,\mathbf{v}'), \tag{6.79}$$

where we have introduced the new auxiliary quantities

$$\varphi_j = i\mathbf{k}_j\cdot(\mathbf{r}_j-\mathbf{r}_0-\mathbf{v}'t) \tag{6.80}$$

and

$$g(\mathbf{k}_j,\mathbf{v}') = \left[1-\omega_p^2\int \frac{\mathbf{k}_j\cdot\partial W_0/\partial\mathbf{v}}{k_j^2\epsilon(\mathbf{k}_j\cdot\mathbf{v}',\mathbf{k}_j)\mathbf{k}_j\cdot(\mathbf{v}'-\mathbf{v})}\right]. \tag{6.81}$$

Integration over the initial position for the generic test particle $\mathbf{r}_0$ leads to the appearance of a delta function $\delta(\mathbf{k}_1+\mathbf{k}_2)$ which, in its turn, allows us to integrate over $\mathbf{k}_2$. We then replace $\mathbf{k}_1$ by $\mathbf{k}$, in order to simplify the notation. By noting that the velocity $\mathbf{v}$ only appears in the parallel direction with respect to the wavevector $\mathbf{k}$, we can write

$$\mathbf{v} = u\frac{\mathbf{k}}{k}+\mathbf{v}_\perp \tag{6.82}$$

and introducing the parallel distribution function

$$G_0(u) = \int W_0(\mathbf{v})d\mathbf{v}_\perp, \tag{6.83}$$

we can finally write

$$\langle n(\mathbf{r}_1,t_1)n(\mathbf{r}_2,t_2)\rangle = n_0\int du'G_0(u')\int \frac{d\mathbf{k}}{(2\pi)^3}e^{i\varphi}|I(k,u')|^2, \tag{6.84}$$

where we have used the phase

$$\varphi = i\mathbf{k} \cdot (\mathbf{r}_1 - \mathbf{r}_2) - iku'(t_1 - t_2) \tag{6.85}$$

and the integral

$$I(\mathbf{k}, u') = \left[1 + \omega_p^2 \int \frac{\partial G_0/\partial u}{k^2 \epsilon(ku', \mathbf{k})(u' - u)} du \right]. \tag{6.86}$$

This is formally very similar to the results obtained for an electron-ion plasma [8]. The main difference relies on the fact that we only have one particle species here, the neutral atoms, which suggests that the system can be regarded as a one-component plasma. We can now use these results to determine the structure factor. Putting equations (6.84)–(6.86) together, and successively integrating over $\mathbf{r}_1$, $\mathbf{r}_2$, t_1, t_2, $\mathbf{k}$, and u', we finally obtain

$$S(k', \omega') = \frac{G_0(\omega'/k')}{k'} |I(k', \omega')|^2, \tag{6.87}$$

where we have used

$$I(k', \omega') = \left[1 + \omega_p^2 \int \frac{(\partial G_0/\partial u) du}{k'^2 \epsilon(k', \omega')(u - \omega'/k')} \right]. \tag{6.88}$$

6.6 Collective Laser Scattering

The propagation of a laser pulse along the ultra-cold gas is governed by the wave equation (6.1). relating $\mathbf{E}$, the laser electric field, with the polarization field in the gas, $\mathbf{P}$, as determined by the general expression

$$\mathbf{P}(\mathbf{r}, t) = \epsilon_0 \int_0^\infty \Pi(\mathbf{r}, \tau) \cdot \mathbf{E}(\mathbf{r}, t - \tau) d\tau, \tag{6.89}$$

where $\Pi(\mathbf{r}, t)$ is the atomic susceptibility tensor. Using a temporal Fourier transformation, one obtains

$$\mathbf{P}(\mathbf{r}, \omega) = \epsilon_0 \Pi(\mathbf{r}, \omega) \cdot \mathbf{E}(\mathbf{r}, \omega). \tag{6.90}$$

Because we can neglect the anisotropies in the majority of the experiments, the atomic susceptibility tensor Π can be safely replaced by the linear susceptibility function χ which, for the two-level atom, is given by

$$\chi(\mathbf{r}, \omega) = n(\mathbf{r}) \chi_a(\omega) \, , \quad \chi_a(\omega) = \frac{\Omega_R^2}{3\epsilon_0 \hbar} \frac{\Delta + i\Gamma}{\Delta^2 + \Gamma^2}. \tag{6.91}$$

Here, $n(\mathbf{r})$ is the number of atoms per unit volume, and Ω_R represents the Rabi frequency. The local density can fluctuate in time, at frequencies much lower than those of the radiation field frequency ω, thus leading to the replacement of $n(\mathbf{r})$ by $n(\mathbf{r}, t)$ in the above expression for the susceptibility. The total radiation field $\mathbf{E}$ will be determined by the sum of two parts, the incident laser field $\mathbf{E}_0$, and the scattered field $\mathbf{E}_s$. For a real incident laser field, we can use

$$\mathbf{E}_0(\mathbf{r}, t) = \frac{1}{2}\mathbf{E}_0 \exp(i\mathbf{k}_0 \cdot \mathbf{r} - i\omega_0 t) + c.c., \tag{6.92}$$

where the wave-vector $\mathbf{k}_0$ follows the dispersion relation $k_0^2 = (\omega_0^2/c^2)\,[1 + \chi(\omega_0)]$, with $\chi(\omega_0) = n_0\chi_a(\omega_0)$, and n_0 is the atomic mean density. On the other hand, the scattered field can be generically written as

$$\mathbf{E}_s(\mathbf{r}, t) = \int \mathbf{E}_s(\mathbf{r}, \omega) \exp(-i\omega t)\frac{d\omega}{2\pi} \tag{6.93}$$

and the field components can be determined by solving the corresponding wave equation. This can be written as

$$\left(\nabla^2 + k^2\right) \mathbf{E}_s(\mathbf{r}, \omega) = \frac{\omega^2}{2c^2}\mathbf{E}_0 e^{i\mathbf{k}_0\cdot\mathbf{r}} n(\mathbf{r}, \omega') + \text{c.c.}, \tag{6.94}$$

where

$$n(\mathbf{r}, \omega') = \int n(\mathbf{r}, t) e^{i\omega' t} dt \tag{6.95}$$

and $\omega' = \omega - \omega_0$ is the frequency of the density fluctuations. In its turn, the scattered wave number $\mathbf{k}$ will be determined by the same dispersion relation as that of the incident wave, with k_0 and ω_0 respectively replaced by k and ω. The solution to Eq. (6.94) can therefore be written as

$$E_s(\mathbf{r}, \omega) = -\frac{i\omega^2}{2kc^2}\chi_a(\omega_0)(\mathbf{e}_\omega \cdot \mathbf{e}_0) E_0 e^{i\mathbf{k}\cdot\mathbf{r}} \int d\mathbf{r} \int \frac{d\mathbf{k}'}{(2\pi)^3} n(\mathbf{k}', \omega') e^{i(\mathbf{k}_0 + \mathbf{k}' - \mathbf{k})\cdot\mathbf{r}} + \text{c.c.}, \tag{6.96}$$

where we have used the unit polarization vectors for the incident and the scattered fields, $\mathbf{e}_0 = \mathbf{E}_0/E_0$ and $\mathbf{e}_\omega = \mathbf{E}_s(\omega)/E_s(\omega)$. The integration in volume leads to the appearance of a delta function $\delta(\mathbf{k}_0 + \mathbf{k}' - \mathbf{k})$, and subsequent integration over $\mathbf{k}'$ finally leads to

$$E_s(\mathbf{r}, \omega) = -iA_s \exp(i\mathbf{k} \cdot \mathbf{r}) + \text{c.c.}, \tag{6.97}$$

where the amplitude of the scattered field is determined by

$$A_s(\omega) = \frac{\omega^2}{2kc^2}\chi_a(\omega_0)(\mathbf{e}_\omega \cdot \mathbf{e}_0) E_0 n(\omega', \mathbf{k}') \tag{6.98}$$

with $\omega = \omega_0 + \omega'$ and $\mathbf{k} = \mathbf{k}_0 + \mathbf{k}'$. The relative scattered intensity along the direction of $\mathbf{k}$ is therefore determined by the dynamical structure factor

$$S(\mathbf{k}, \omega) = \frac{|E_s(\mathbf{k}, \omega)|^2}{|E_0|^2} = \frac{i\omega^4}{4k^2c^4}|\chi_a(\omega_0)|^2(\mathbf{e}_\omega \cdot \mathbf{e}_0)^2|n(\mathbf{k}', \omega')|^2. \tag{6.99}$$

This quantity can easily be related with the density correlation function, by noting that

$$|n(\mathbf{k}', \omega')|^2 = \int d\mathbf{r}_1 dt_1 \int d\mathbf{r}_2 dt_2 n(\mathbf{r}_1, t_1) e^{-i\mathbf{k}'\cdot\mathbf{r}_1 + i\omega' t_1}\ n(\mathbf{r}_2, t_2) e^{i\mathbf{k}'\cdot\mathbf{r}_2 - i\omega' t_2}. \tag{6.100}$$

Performing an average in both space and time, the latter equation reads

$$\langle |n(\mathbf{k}', \omega')|^2 \rangle = \lim_{V,T\to\infty} \frac{1}{VT} \int d\mathbf{r}_1 dt_1 \times \int d\mathbf{r}_2 dt_2 \langle n(\mathbf{r}_1, t_1) n(\mathbf{r}_2, t_2)\rangle e^{i\mathbf{k}'\cdot(\mathbf{r}_2 - \mathbf{r}_1)} e^{-i\omega'(t_2 - t_1)}. \tag{6.101}$$

Equation (6.99) establishes the relation between the scattered signal and the fluctuations inside the cloud, which contain the basic features of the laser cooled gas. We should therefore proceed by understanding how to compute such fluctuations by taking into account the microscopic distribution of the atomic states.

In order to illustrate the collective behavior of the scattered signal, we assume that the atoms in the trap approximately follow a Maxwell-Boltzmann distribution

$$G_0(v') = \frac{1}{\sqrt{\pi} v_{th}} e^{-v'^2/v_{\rm th}^2}, \tag{6.102}$$

where $v_{\rm th} = \sqrt{2T/m}$ represents the thermal velocity of the atoms. Defining the scattering parameter $\alpha = k'\lambda_D$, where $\lambda_D = v_{\rm th}/\omega_P$ is the Debye length, the dynamical structure factor can be explicitly given by

$$S(\alpha, \omega') = \frac{e^{-\alpha\omega'/\omega_p}}{\sqrt{\pi}\alpha} I(\alpha, \omega'/\omega_p), \tag{6.103}$$

where $\alpha = k'\lambda_D$ represents the scattering parameter, and $I(\alpha, \omega'/\omega_p)$, is determined by Eq. (6.88). The latter defines the nature of the scattering processes. Moreover, for $\alpha \lesssim 1$, the scattering signal results from a coherent process, as the system is essentially composed of interacting particles. It means that the Debye length λ_D is larger or comparable to the perturbation wavelength $2\pi/k'$ and therefore the correlations between the atoms play an important role on the scattering signal (correlated atoms). On the contrary, if $\alpha \gg 1$, the scattering is incoherent,

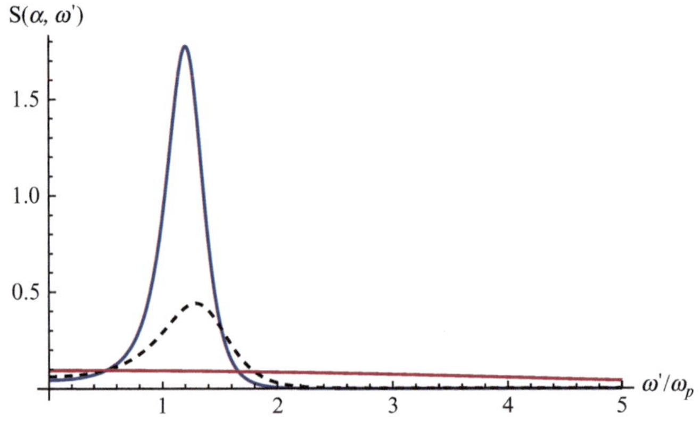

Fig. 6.6 Dynamical structure factor $S(\alpha, \omega')$ for different values of the scattering parameter α: *blue full line* ($\alpha = 0.85$), *black dashed line* ($\alpha = 1.0$), and *red full line* ($\alpha = 5.0$). Both coherent and incoherent scattering processes are represented

and the Debye length is much smaller than the typical size of the perturbation (uncorrelated atoms).

In Fig. 6.6, we have plotted the normalized structure factor in Eq. (6.103) for three different values of α. We observe a strong resonance near $\omega' = \omega_p$ for the case of coherent scattering ($\alpha = 0.85$ and $\alpha = 1.0$), while a very broad spectrum is obtained for the case of incoherent scattering ($\alpha = 5.0$). The physical reason for such results are related with the interference mechanism occurring at scattering: the excitation of collective perturbations of the atom density will result on the constructive interference of the scattered signal around the natural frequency characterizing the long-range order of the interaction, i.e., ω_p. On the contrary, if uncorrelated, single-atoms fluctuations are excited in the systems, the interference will be destructive and, therefore, no resonance is observed in the spectrum $S(\alpha, \omega')$. This could be used as an experimental method to study the collective processes in a trap.

References

1. S.S.Z. Ashraf, A.C. Sharma, K.N. Vyas, J. Phys. Condens. Matter **19**, 306201 (2007)
2. P.R. Berman, X. Xu, Phys. Rev. A **78**, 053407 (2008)
3. R.W. Boyd, *Nonlinear Optics* (Academic, san Diego, 1992)
4. J.A. Greenberg, B.L. Schmittberger, D.J. Gauthier, Opt. Express **19**, 22535 (2011)
5. J.P. Hansen, I.R. McDonald, *Theory of Simple Liquids* (Academic, London, 1976)
6. N. Henkel, R. Nath, T. Pohl, Phys. Rev. Lett. **104**, 195302 (2010)
7. A. Ishimaru, *Wave Propagation and Scattering in Random Media* (Academic, New York, 1978). Chap.9
8. N.A. Krall, A.W. Trivelpiece, *Principles of Plasma Physics* (McGraw Hill, New York, 1973)

9. G. Labeyrie, E. Vaujour, C.A. Müller, D. Delande, C. Miniatura, D. Wilkowski, R. Kaiser, Phys. Rev. Lett. **91**, 223904 (2003)
10. J.T. Mendonça, R. Kaiser, Phys. Rev. Lett. **108**, 033001 (2012)
11. J.K. Percus, G.J. Yevick, Phys. Rev. **110**, 1 (1958)
12. M.C.W. van Rossum, Th.M. Nieuwenhuizen, Rev. Mod. Phys. **71**, 313 (1999)
13. X. Wang, A. Bhattacharjee, Phys. Plasmas **4**, 3759 (1997)
14. M.S. Wertheim, Phys. Rev. Lett. **10**, 321 (1963)

Part II
The Physics of Bose-Einstein Condensates

Chapter 7
Bose Einstein Condensates

We now turn to the condensate phase of ultra-cold matter. It is known that condensation in cold atomic clouds occurs when the cooling lasers are switched off and for this reason the atom-atom interactions associated with an exchange of scattered photons are no longer present. What remains is short range atom collisions, which are responsible for the occurrence of a new type of mean field potential. Even in a dilute gas, where the collision frequency is very low, this mean field potential is very a important ingredient of the condensate physics, as shown in the next chapters.

We first start by reviewing the conditions for Bose Einstein condensation, showing the differences between condensation of a uniform gas (which only occurs in momentum space), and condensation of a trapped gas (which also occurs in real space and leads to a density spike). This introduction is completed by a discussion of the atom correlations.

We then consider the basic evolution equations for a condensate, starting with the celebrated Gross Pitaevskii (GP) equation. This is in fact a nonlinear Schrödinger equation, where the nonlinearity is due to the atomic mean field created by collisions. From this equation we can derive quantum fluid equations, which although approximate can be used to describe in a simple way, many important properties of a condensate. We also introduce the wave kinetic formulation, which is exactly equivalent to the GP equation but allows us to consider the condensate properties in a different perspective. The wave kinetic equation for a condensate is formally identical to that considered before for the non-condensed gas. The only difference is the physical origin of the mean field potential. This difference and similarity between the condensed and the non-condensed gas are very stimulating conceptually, and one of the main characteristic features of the physics of ultra-cold matter.

J.T. Mendonça and H. Terças, *Physics of Ultra-Cold Matter*, Springer Series on Atomic, Optical, and Plasma Physics 70, DOI 10.1007/978-1-4614-5413-7_7,

7.1 Uniform Gas

The condensation process of a boson gas in free space is well described in many textbooks [1, 2], and we briefly review it here. This will be useful for future comparison with the trapped case, to be considered next. Let us consider an unconfined gas of noninteracting bosons. It is well known that, in the low temperature limit, its behavior is very different from that of a fermion gas, because its minimum energy state for $T = 0$ is zero. This happens because, in contrast with fermions, all the bosons are can be simultaneously in a zero-energy state. The quantum state of a bosonic atom in a given kinetic energy state ϵ, is determined by the operators

$$\hat{\psi}^\dagger(\mathbf{r},t) = \int \psi_{\mathbf{p}}^*(\mathbf{r},t)\hat{a}_{\mathbf{p}}^\dagger \frac{d\mathbf{p}}{(2\pi\hbar)^3}, \quad \hat{\psi}(\mathbf{r},t) = \int \psi_{\mathbf{p}}(\mathbf{r},t)\hat{a}_{\mathbf{p}} \frac{d\mathbf{p}}{(2\pi\hbar)^3}, \tag{7.1}$$

where the wavefuntion $\psi_p(\mathbf{r},t)$ is simply given by a plane wave

$$\psi_{\mathbf{p}}(\mathbf{r},t) = \exp(i\mathbf{p}\cdot\mathbf{r}/\hbar - i\epsilon_{\mathbf{p}}t/\hbar), \tag{7.2}$$

with $\epsilon_{\mathbf{p}} = p^2/2M$, and the operator $\hat{a}^\dagger$ and $\hat{a}$ are the usual bosonic creation an destruction operator, obeying the commutation relations $[\hat{a}_{\mathbf{p}}^\dagger, \hat{a}_{\mathbf{p}'}] = \delta_{\mathbf{p},\mathbf{p}'}$ and $[\hat{a}_{\mathbf{p}}^\dagger, \hat{a}_{\mathbf{p}'}^\dagger] = [\hat{a}_{\mathbf{p}}, \hat{a}_{\mathbf{p}'}] = 0$. We can also write $\langle \hat{a}_{\mathbf{p}}^\dagger \hat{a}_{\mathbf{p}'}\rangle = \delta_{\mathbf{p},\mathbf{p}'}\langle n_{\mathbf{p}}\rangle$, where, in thermal equilibrium at a temperature T, the occupation number $\langle n_{\mathbf{p}}\rangle$ is determined by the Bose Einstein distribution

$$\langle n_{\mathbf{p}}\rangle = \frac{1}{\mathcal{Z}^{-1}\exp(\beta\epsilon_{\mathbf{p}}) - 1}, \quad \mathcal{Z} = e^{\beta\mu}, \tag{7.3}$$

where $\beta = 1/k_B T$, (or simply $1/T$, if we write T in energy units), μ is the chemical potential and $\mathcal{Z}$ is the fugacity. To discuss the features of an ideal Bose gas it is necessary to have in mind that its chemical potential is always negative (or eventually zero), $\mu \leq 0$. The exact value of $\mu(T)$ can be obtained by integrating the distribution (7.3) over all the possible energy states $\epsilon_{\mathbf{p}}$, noting that these energy states are independent of the direction of the atom momentum $\mathbf{p}$. This leads to the total number of atoms in the volume V,

$$N = \frac{2\pi}{\lambda_T^3} V \int_0^\infty \frac{\sqrt{x}dx}{\mathcal{Z}^{-1}e^x - 1}, \tag{7.4}$$

with $x = \beta\epsilon_{\mathbf{p}}$, and $\lambda_T = h(2\pi M T)^{-1/2}$. Using the derivative of an implicit function, it can easily be shown that, for a Bose gas, we have $(\partial\mu\ \partial T) < 0$. This means that the chemical potential decreases (or its absolute value increases) when the temperature increases. When we cool down a Bose gas, the chemical potential μ will increase until it reaches its maximum value $\mu = 0$, occurring at a critical temperature $T = T_c$. It is possible to determine the value of T_c, by taking the explicit dependence of λ with respect to the temperature, and making $\mu = 0$ in (7.4).

The integral in this equation can be expressed in terms of the Gamma function $\Gamma(z)$, and of the zeta Riemann function $\zeta(z)$, as

$$\int_0^\infty \frac{\sqrt{x}dx}{e^x - 1} = \Gamma\left(\frac{3}{2}\right)\zeta\left(\frac{3}{2}\right) = \frac{\sqrt{\pi}}{2}2.612 \simeq 2.31. \quad (7.5)$$

This leads to the explicit the value of the critical temperature $T = T_c$, for which the chemical potential μ becomes zero, as

$$T_c = 3.31\frac{\hbar^2}{M}\left(\frac{N}{V}\right)^{2/3}. \quad (7.6)$$

Notice that this temperature grows with the atom density to the power of $2/3$, and decreases for increasing mass. Now, let us see what happens for temperatures below critical value $T < T_c$. In this region we will always have $\mu = 0$, because the chemical potential can only increase with temperature, and it cannot become positive. In this case, Eq. (7.4) cannot remain valid, unless we replace the total number of particles N by a the number of atoms with non-zero energy $N' \equiv N(\epsilon > 0)$, such that $N = N' + N_0$, with

$$\frac{N'}{N} = \left(\frac{T}{T_c}\right)^{3/2}, \quad \frac{N_0}{N} = \left[1 - \left(\frac{T}{T_c}\right)^{3/2}\right]. \quad (7.7)$$

In order to understand this result (Fig. 7.1) we should note that in the integral of Eq. (7.4) the contribution from the atoms with zero-energy is not retained, because of the factor $\sqrt{x}$. Therefore, the number N' refers only to the positive energy states, $N' = N(\epsilon > 0)$. We can then conclude that, for $T < T_c$, the number of particles N_0 in the zero-energy state $\epsilon = 0$, is $N(\epsilon = 0) \equiv N_0$. This phenomenon of the accumulation of particles in the lowest energy state $\epsilon = 0$ for temperatures below T_c is the *Bose Einstein condensation*. It corresponds to a separation of the gas in two distinct components: one with a population N' and positive energies, another with a population $N_0 = N - N'$ at zero-energy (see Fig. 7.1). Note that for the present unbounded uniform gas, condensation takes place in the energy (or in the momentum) space, and not necessarily in the real configuration space. The situation changes with trapped atoms, as discussed next.

7.2 Trapped Gas

Let us now consider a system of N atoms trapped in a quadratic potential of the form

$$V(\mathbf{r}) = \frac{M}{2}\sum_{j=x,y,z}\omega_j^2 r_j^2, \quad (7.8)$$

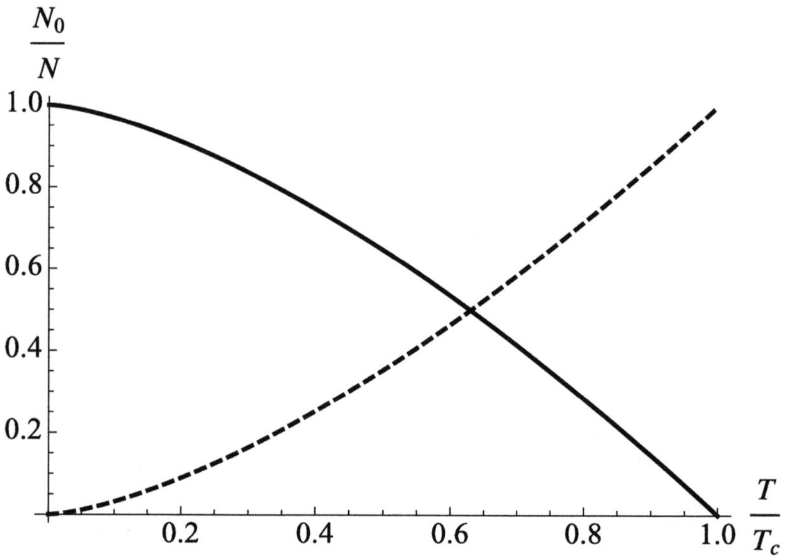

Fig. 7.1 Bose condensation: number of particles in the zero energy state N_0, in bold, and number of atoms in positive energy states N', as a function of the temperature T, below the critical temperature T_c

where $r_j = (x, y, z)$, and ω_j are the respective oscillating frequencies. The Hamiltonian of the system is the sum of single particle Hamiltonians, with eigenvalues

$$\epsilon_{\mathbf{n}} = \sum_j \left(n_j + \frac{1}{2} \right) \hbar \omega_j, \tag{7.9}$$

where $\mathbf{n} \equiv (n_x, n_y, n_z)$, and n_j are zero or positive integers. The quantum state of the system is now represented by

$$\hat{\psi}^\dagger(\mathbf{r}, t) = \sum_{\mathbf{n}} \psi^*_{\mathbf{n}}(\mathbf{r}, t) \hat{a}^\dagger_{\mathbf{n}}, \quad \hat{\psi}(\mathbf{r}, t) = \sum_{\mathbf{n}} \psi_{\mathbf{n}}(\mathbf{r}, t) \hat{a}_{\mathbf{n}}, \tag{7.10}$$

where the plane wave functions (7.2) of the unbounded system are now replaced by functions of the form

$$\psi_{\mathbf{n}}(\mathbf{r}, t) = \psi(0) \mathcal{H}(\mathbf{r}) \exp(-i \epsilon_{\mathbf{n}} t / \hbar), \tag{7.11}$$

with $\psi(0)$ representing a normalization constant and $\mathcal{H}(\mathbf{r})$ are the solutions of the harmonic oscillator in terms of the Hermite polynomials. In the extreme case of zero temperature, the atoms are all in the lowest energy state $\mathbf{n} = 0$, and we can write

$$\psi_0(\mathbf{r}, t) = \psi(0) \exp\left(-\frac{M}{2\hbar} \sum_j \omega_j r_j^2 \right) \exp\left(-\frac{i}{2} \sum_j \omega_j t \right). \tag{7.12}$$

The density distribution inside the confining potential is now given by $n(\mathbf{r}) = N|\psi_0(\mathbf{r},t)|^2$. For a spherical trap, such that $\omega_j = \omega_0$, we get

$$n(\mathbf{r}) = N\left(\frac{M\omega_0}{\pi\hbar}\right)^{3/2} \exp\left(-\frac{M}{\hbar}\omega_0 r^2\right), \tag{7.13}$$

where $r = \sqrt{x^2+y^2+z^2}$. We can see that the size of the condensate is independent of the number of atoms, and only depends on the frequency of the trap, which is related to the harmonic oscillator length

$$a_0 = \left(\frac{\hbar}{M\omega_0}\right)^{1/2}. \tag{7.14}$$

This somewhat counterintuitive result is a consequence of neglecting the interactions between the atoms, remaining valid for the ideal gas only. We will see later that a dependence of the size of the condensed cloud with N will appear when such interactions are taken into account.

It is now useful to compare this with the density profiles of a non-condensed, or *thermal*, gas. If $T \gg \hbar\omega_0$, we can use the Boltzmann distribution to characterize $n(\mathbf{r})$. In this case, the equilibrium density profile is given by

$$n(\mathbf{r}) \propto \exp[-\beta V(\mathbf{r})] = \exp\left(-\beta\frac{M}{2}\omega_0^2 r^2\right), \tag{7.15}$$

and the size if the cloud is now given by

$$a_T = a_0\sqrt{\frac{T}{\hbar\omega_0}} > a_0. \tag{7.16}$$

The difference between the condensate and the thermal gas is illustrated in Fig. 7.2. The occurrence of a sharp peak in the condensed cloud is a major characteristic feature of Bose-Einstein condensation in atomic traps, and strongly contrasts with the case of the unbounded medium considered before.

The case of an axially symmetric trap can be discussed in similar grounds. If we define the axial and the radial variables as z and $r_\perp = \sqrt{x^2+y^2}$, and define the axial and radial frequencies as ω_z and $\omega_\perp = \omega_x = \omega_y$, we obtain for the condensed gas at zero temperature the density profile

$$n(\mathbf{r}) = \frac{\lambda^{1/2}}{\pi^{3/2}a_\perp^3} \exp\left[-\frac{1}{2a_\perp^2}\left(r_\perp^2 + \lambda z^2\right)\right] \tag{7.17}$$

where now the harmonic oscillator strength in the perpendicular direction is $a_\perp = \sqrt{\hbar/M\omega_\perp}$, and the parameter $\lambda = \omega_z/\omega_\perp$ characterizes the trap asymmetry. We have a disk shaped trap for $\lambda > 1$, and a cigar shaped one for $\lambda < 1$.

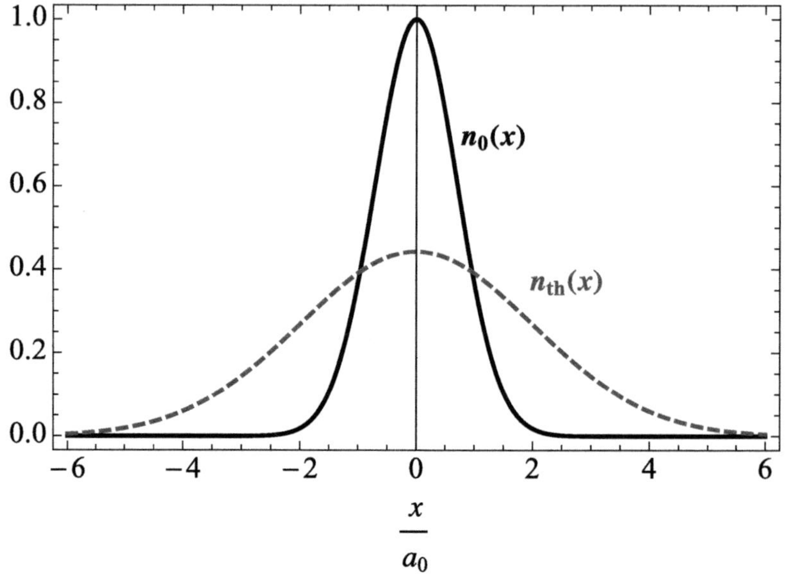

Fig. 7.2 Normalized density profile of a condensed ($n_0(x)$) and a thermal gas ($n_{th}(x)$) over a given direction x, for a spherical trap. We have used $T/\hbar\omega = 4$

Let us now consider the computation of the critical temperature in the case of a trapped ideal Bose gas. According to the grand canonical distribution (7.3), we can determine the total number of atoms in the trap as

$$N = \sum_{\mathbf{n}} \frac{1}{\mathcal{Z}^{-1} \exp(\beta\epsilon_{\mathbf{n}}) - 1}. \tag{7.18}$$

As before, we have to separate the number of atoms in the lowest energy state, N_0, from the atoms with are not in the condensate phase $N' = N - N_0$. But now, in contrast with the unbounded case, the chemical potential at the critical temperature cannot be zero, but it has to be equal to the lowest energy state, $\mu_0 = (1/2)\hbar \sum_j \omega_j$, or for a spherical trap, simply $\mu_0 = (3/2)\hbar\omega_0$. Replacing this value in (7.18), we get for the trapped gas

$$N' = \sum_{\mathbf{n}\neq 0} \frac{1}{\exp(\beta\epsilon_{\mathbf{n}}) - 1} \simeq \int_0^\infty \frac{d\mathbf{n}}{\exp(\beta\epsilon_{\mathbf{n}}) - 1}, \tag{7.19}$$

where $d\mathbf{n} \equiv \Pi_j dn_j = dn_x dn_y dn_z$. The last integration is, indeed, an approximation, which is valid when the successive energy states can be considered as a continuum, or when the spacing between adjacent energy levels is much smaller than the thermal energy of the cloud, or $T \gg \hbar\omega_j$. Integration can be easily performed with auxiliary variables $x_j = \beta\hbar\omega_j n_j$, and the result is

$$N' = \zeta(3)\left(\frac{T}{\hbar\bar{\omega}}\right)^3, \qquad \bar{\omega} = (\omega_x\omega_y\omega_z)^{1/3}. \tag{7.20}$$

For a spherical trap we have $\bar{\omega} = \omega_0$. By defining the critical temperature as that corresponding to the case $N' \to N$, we get

$$T_0 = \hbar\bar{\omega}\left(\frac{N}{\zeta(3)}\right)^{1/3} = 0.94\hbar\bar{\omega}N^{1/3}. \tag{7.21}$$

Replacing this in Eq. (7.20), we obtain, for $T < T_0$, the number of condensed atoms

$$\frac{N_0}{N} = \left[1 - \left(\frac{T}{T_0}\right)^3\right]. \tag{7.22}$$

which is clearly different from the homogeneous case of Eq. (7.7).

7.3 Atom Correlations

In order to complete the discussion of the previous section, let us now discuss atom correlations, following the lines of Ref. [3]. We can use the atom field operators (7.1), and define the first order atom correlation function as

$$G^{(1)}(\mathbf{r}_1, t_1; \mathbf{r}_2, t_2) = \langle\hat{\psi}^\dagger(\mathbf{r}_1, t_1)\hat{\psi}(\mathbf{r}_2, t_2)\rangle. \tag{7.23}$$

This definition of $G^{(1)}$ is formally similar to that used for the photon field in quantum optics [4]. By making $\mathbf{r}_1 = \mathbf{r}_2 = \mathbf{r}$ and $t_1 = t_2 = t$, this expression gives the atom density of the cloud, as

$$n(\mathbf{r}, t) = G^{(1)}(\mathbf{r}, t; \mathbf{r}, t) = \langle\hat{\psi}^\dagger(\mathbf{r}, t)\hat{\psi}(\mathbf{r}, t)\rangle. \tag{7.24}$$

Notice that, in thermal equilibrium, we can determined the atom density by

$$n(\mathbf{r}) = \langle\mathbf{r}|\hat{\rho}|\mathbf{r}\rangle = \sum_{\mathbf{n}} \psi_{\mathbf{n}}(\mathbf{r})\psi_{\mathbf{n}}^*\langle n_n\rangle, \tag{7.25}$$

where the sum is over all the atom energy states in the trap. The density operator here is

$$\hat{\rho} = \sum_{\mathbf{n}} |\psi_{\mathbf{n}}\rangle\langle\psi_{\mathbf{n}}|\langle n_{\mathbf{n}}\rangle, \tag{7.26}$$

where $\langle n_{\mathbf{n}}\rangle$ can be described by a grand-canonical distribution as in Eq. (7.3) for the uniform gas,

$$\langle n_{\mathbf{n}}\rangle = \frac{1}{\mathcal{Z}^{-1}\exp(\beta\epsilon_{\mathbf{n}}) - 1}. \tag{7.27}$$

The only difference is that here the sum is taken over an infinite set of discrete energy states $\epsilon_{\mathbf{n}}$. It is clear that, in the zero temperature limit, $T \rightarrow 0$, where all the atoms are in the ground state, we have

$$N_0 = N = \int n(\mathbf{r}) d\mathbf{r} = \frac{1}{\mathcal{Z}^{-1} \exp(\beta \epsilon_0) - 1}, \tag{7.28}$$

where $\epsilon_0 = (3/2)\hbar\tilde{\omega}$ is the lowest energy state, and $\tilde{\omega} = \sum_j \omega_j/3$ is the arithmetic mean of the trap frequencies. In this limit the chemical potential equals this minimum energy, $\mu = \epsilon_0$, and Eq. (7.28) defines a maximum for the fugacity $\mathcal{Z}$, given by

$$\mathcal{Z}_{\max} = \frac{N_0}{N_0 + 1} \exp(-\beta \epsilon_0) \simeq \exp(-\beta \epsilon_0), \tag{7.29}$$

where we assumed a large number of atoms $N_0 \gg 1$. By a simple generalization of Eq. (7.25), we can also define the first order correlation function in terms of the density operator $\hat{\rho}$, as

$$G^{(1)}(\mathbf{r}_1, \mathbf{r}_2) = \langle \mathbf{r}_1 | \hat{\rho} | \mathbf{r}_2 \rangle, \tag{7.30}$$

where time dependence was ignored. It is now useful to give a more explicit form to Eq. (7.30), which can be done by developing the expression (7.27) as

$$\langle n_{\mathbf{n}} \rangle = \sum_{l=1}^{\infty} \mathcal{Z}^l \exp(l \beta \epsilon_{\mathbf{n}}). \tag{7.31}$$

The latter simply yields

$$G^{(1)}(\mathbf{r}_1, \mathbf{r}_2) = \sum_{l=1}^{\infty} \mathcal{Z}^l \mathcal{G}(\mathbf{r}_1, \mathbf{r}_2, l\beta), \tag{7.32}$$

where the quantity

$$\mathcal{G}(\mathbf{r}_1, \mathbf{r}_2, \beta) = \sum_{\mathbf{n}} \psi_{\mathbf{n}}^*(\mathbf{r}_2) \exp(-\beta \epsilon_{\mathbf{n}}) \psi_{\mathbf{n}}(\mathbf{r}_1) \tag{7.33}$$

is sometimes called the finite-temperature Green function [5]. This quantity can be evaluated by using the atom wave function in a harmonic trap, leading to

$$\mathcal{G}_l(\mathbf{r}_1, \mathbf{r}_2) = \frac{1}{(\sqrt{\pi}\bar{\sigma})^3} \Pi_j \frac{1}{\sqrt{1 - \exp(-2\tau_j)}} \exp\left[-\left(\frac{r_{1j} + r_{2j}}{2\sigma_j} \right)^2 \tan(\tau_j/2) \right.$$
$$\left. - \left(\frac{r_{1j} - r_{2j}}{2\sigma_j} \right)^2 \operatorname{cotan}(\tau_j/2) \right], \tag{7.34}$$

where $\bar{\sigma} = \Pi_j \sigma_j^{2/3}$ is the geometric mean trap size, and $\tau_j = \beta\hbar\omega_j$. Replacing this in (7.32) and using $\mathbf{r}_1 = \mathbf{r}_2$, we get for the atomic density the profile

$$n(\mathbf{r}) \equiv G^{(1)}(\mathbf{r}_1, \mathbf{r}_2) = \lambda_T^{-3} \sum_l Z^l \Pi_j \sqrt{\frac{2\tau_j}{1 - \exp(-2\tau_j)}} \exp\left[-\left(\frac{r_j}{\sigma_j}\right)^2 \tan(\tau_j/2) \right], \tag{7.35}$$

where we have rescaled the fugacity, in order for its maximum value to be one, by introducing $Z = \mathcal{Z} \exp(-3\beta\hbar\bar{\omega}/2)$. In the limit of hight temperatures, $\tau_j \ll 1$, the first term of the sum in this expression will dominate. In this limit, the atomic cloud profile is given by a the Boltzmann distribution

$$n(\mathbf{r}) = \frac{Z}{\lambda_T^3} \Pi_j \exp\left[-\frac{r_j^2}{2\sigma_j^2} \right] = \frac{Z}{\lambda_T^3} \exp[-\beta V(\mathbf{r})], \tag{7.36}$$

where the size of the thermal cloud is $\sigma_j = v_T/\omega_j$, with $v_T = \sqrt{T/M}$. The maximum density of this cloud, at the center $\mathbf{r} = 0$ is equal to Z/λ_T^3. In the opposite limit of zero temperature $T \to 0$, the above density profile reduces to the ground state density

$$n_0(\mathbf{r}) = \frac{Z}{1 - Z} |\psi_0(\mathbf{r})|^2. \tag{7.37}$$

This result is only valid, of course, for the ideal gas. It will be shown later that atom-atom interactions will lead to a different profile. In the intermediate case of moderate temperatures, we can still use $\tau_j \ll 1$ but retain all the terms in the sum of Eq. (7.35). We now get

$$n(\mathbf{r}) = \lambda_T^{-3} g_{3/2}(x) \ , \quad x = Z \exp[-\beta V(\mathbf{r})], \tag{7.38}$$

where $g_{3/2}(x)$ is the polylogarithmic function of order $3/2$, defined by

$$g_u(x) = \sum_{l=1}^{\infty} \frac{x^l}{l^u}. \tag{7.39}$$

This represents a refinement of the semi-classical density profile discussed in the previous section. By integrating the general expression (7.35) in space, we get the total number of atoms in the trap. It simply provides

$$N = \sum_{l=1}^{\infty} \frac{Z^l}{\Pi_j [1 - \exp(-l\tau_j)]}. \tag{7.40}$$

In the semi-classical approximation, this reduces to

$$N \simeq \frac{1}{\bar{\tau}^3} g_3(Z). \tag{7.41}$$

From here, we can easily get the critical temperature and the number of condensed atoms as given by the above Eqs. (7.21) and (7.22). If we retain the first order term with respect to the small parameters τ_j we get

$$N \simeq \frac{1}{\bar{\tau}^3}\left[g_3(Z) + \frac{3}{2}\tilde{\tau} g_2(Z)\right] + \frac{Z}{1-Z}, \tag{7.42}$$

where the last term was introduced in order to account for the contribution from the ground state density. Now we can determine the critical fugacity. Disregarding size effects, we get

$$Z_c \simeq 1 - \bar{\tau}^{3/2}\zeta(2)^{-1/2}. \tag{7.43}$$

We see a small deviation with respect to the value 1, corresponding to $\mathcal{Z}_{\max}$. This small correction will modify the critical temperature and, as a result, we get the following condition for the onset of condensation

$$\lambda_T^3 n(0) = \zeta(3/2) + 2\sqrt{2\zeta(2)} \simeq 6.24. \tag{7.44}$$

This result shows the contribution from the lower energy state which was neglected in our previous estimate.

Let us now consider the second order correlation function. This can be defined as

$$G^{(2)}(\mathbf{r}_1, t_1; \mathbf{r}_2, t_2) = \langle \hat{\psi}^\dagger(\mathbf{r}_1, t_1)\hat{\psi}^\dagger(\mathbf{r}_2, t_2)\hat{\psi}(\mathbf{r}_2, t_2)\hat{\psi}(\mathbf{r}_1, t_1)\rangle. \tag{7.45}$$

It is known that for a multi-mode thermal gas, only the pairwise ordering operators survive the statistical average over the grand canonical ensemble. In this case, we have for first order correlation averages $\langle \hat{a}_{\mathbf{p}}^\dagger \hat{a}_{\mathbf{p}'}\rangle = \langle \hat{a}_{\mathbf{p}}^\dagger \hat{a}_{\mathbf{p}'}\rangle \delta_{\mathbf{p},\mathbf{p}'}$, and for the second order correlation the average operator contributions take the form

$$\langle \hat{a}_{\mathbf{p}}^\dagger \hat{a}_{\mathbf{p}'}^\dagger \hat{a}_{\mathbf{p}''} \hat{a}_{\mathbf{p}'''}\rangle = \langle \hat{a}_{\mathbf{p}}^\dagger \hat{a}_{\mathbf{p}''}\rangle \langle \hat{a}_{\mathbf{p}'}^\dagger \hat{a}_{\mathbf{p}'}\rangle \delta_{\mathbf{p},\mathbf{p}''}\delta_{\mathbf{p}',\mathbf{p}'''}. \tag{7.46}$$

It immediately follows that, at thermal equilibrium, we must have

$$G^{(2)}(\mathbf{r}_1, t_1; \mathbf{r}_2, t_2) = n(\mathbf{r}_1, t_1)n(\mathbf{r}_2, t_2) + |G^{(1)}(\mathbf{r}_1, t_1; \mathbf{r}_2, t_2)|^2. \tag{7.47}$$

For a finite degenerate gas, such as a cloud of atoms confined in a trap, this result is no longer valid [6], and we have to introduced a correction [3, 7], by subtracting the contribution from the ground state, $\langle \hat{a}_0^\dagger \hat{a}_0\rangle^2$. For $t_1 = t_2$, and for the steady state situation of a trapped gas of boson atoms in equilibrium, the corrected expression for the second order correlation is then given by

$$G^{(2)}(\mathbf{r}_1, \mathbf{r}_2) = n(\mathbf{r}_1)n(\mathbf{r}_2) + |G^{(1)}(\mathbf{r}_1, \mathbf{r}_2)|^2 - n_0(\mathbf{r}_1)n_0(\mathbf{r}_2), \tag{7.48}$$

where we have defined $n_0(\mathbf{r})$ as the density of atoms in the ground state, or equivalently, the density of condensed atoms. It is also useful to consider the normalized correlation function, defined as

$$g^{(2)}(\mathbf{r}_1, \mathbf{r}_2) = \frac{G^{(2)}(\mathbf{r}_1, \mathbf{r}_2)}{n(\mathbf{r}_1)n(\mathbf{r}_2)}. \tag{7.49}$$

The measurement of this quantity is very important to characterize the statistical properties of the atoms. Being in a bosonic state, the condensed atoms should demonstrate the expected bunching properties. By changing the internal atomic state, and changing from a boson to a fermion gas, we can then observe the suppression of the bunching phenomenon [8].

7.4 Mean Field Approximation

The mean field approximation for Bose-Einstein condensates can be described in the following simple terms. The quantum state described by the operators (7.10) can be decomposed into its condensed and fluctuating parts, as

$$\hat{\psi}(\mathbf{r}, t) = \Phi(\mathbf{r}, t) + \delta\hat{\psi}(\mathbf{r}, t), \quad \Phi(\mathbf{r}, t) \equiv \langle \hat{\psi}(\mathbf{r}, t) \rangle, \tag{7.50}$$

where the average part is a complex function, such that its module square corresponds to the density of the condensate, $n_0(\mathbf{r}, t) = |\Phi(\mathbf{r}, t)|^2$, and can be treated as a classical field. Because condensation corresponds to a very large number of atoms $N_0 \gg 1$ in the ground state, the average term in (7.50) grows with $\sqrt{N_0/V}$, and the fluctuating part can be considered negligible for very large $N_0 \gg 1$.

The evolution equation for the condensate wave function $\Phi(\mathbf{r}, t)$ can be derived from the Heisenberg equation for the field operator $\hat{\psi}(\mathbf{r}, t)$, which is written in the standard form as

$$\frac{\partial}{\partial t}\hat{\psi}(\mathbf{r}, t) = -\frac{i}{\hbar}\left[\hat{\psi}(\mathbf{r}, t), \hat{H}\right], \tag{7.51}$$

where the Hamiltonian operator of the system is

$$\hat{H} = \int \hat{\psi}^{\dagger}(\mathbf{r}, t)\left[\frac{\mathbf{p}^2}{2M} + V_{\text{ext}}(\mathbf{r}, t)\right]\hat{\psi}(\mathbf{r}, t)d\mathbf{r} + \hat{H}_{\text{int}}. \tag{7.52}$$

Here $V_{\text{ext}}(\mathbf{r}, t)$ a generic external potential, and the interaction term $\hat{H}_{\text{int}}$ accounts for the atom-atom interactions which have been neglected until now in this chapter, and takes the form

$$\hat{H}_{\text{int}} = \frac{1}{2}\int d\mathbf{r} \int d\mathbf{r}' \hat{\psi}^{\dagger}(\mathbf{r}, t)\hat{\psi}^{\dagger}(\mathbf{r}', t)V(\mathbf{r} - \mathbf{r}')\hat{\psi}(\mathbf{r}, t)\hat{\psi}(\mathbf{r}', t), \tag{7.53}$$

where $V(\mathbf{r}-\mathbf{r}')$ is the interaction potential. In the $T \rightarrow 0$ limit, we can replace the operator $\hat{\psi}(\mathbf{r},t)$ by its average $\Phi(\mathbf{r},t)$, and we get

$$\frac{\partial}{\partial t}\Phi(\mathbf{r},t) = -\frac{i}{\hbar}\left[\frac{\mathbf{p}^2}{2M} + V_{\text{ext}}(\mathbf{r},t) + \int \Phi^\dagger(\mathbf{r}',t)V(\mathbf{r}-\mathbf{r}')\Phi(\mathbf{r},t)d\mathbf{r}'\right]\Phi(\mathbf{r},t). \tag{7.54}$$

A further simplifying assumption can now be made, by noting that in a cold and dilute gas, only low energy binary collisions are important. As seen before, these collisions depend on a single parameter a, the s-wave scattering length, such that

$$V(\mathbf{r}-\mathbf{r}') = g\,\delta(\mathbf{r}-\mathbf{r}')\,, \quad g = 4\pi\hbar^2\frac{a}{M}. \tag{7.55}$$

We are finally led to

$$i\hbar\frac{\partial}{\partial t}\Phi(\mathbf{r},t) = H\Phi(\mathbf{r},t)\,, \quad H \equiv \left[-\frac{\hbar^2\nabla^2}{2M} + V_{\text{ext}}(\mathbf{r},t) + g\,|\Phi(\mathbf{r},t)|^2\right]. \tag{7.56}$$

This is the Gross-Pitaevskii equation [9, 10], which very efficiently describes many observed phenomena in Bose-Einstein condensates. Notice that its derivation implied two successive approximations. First, the replacement of the exact atom operator by a mean field $\Phi(\mathbf{r},t)$ and the neglect of fluctuations with respect to this mean field. It is valid only for very large number of atoms in the condensed phase, $N_0 \gg 1$. Second, the use of a simplified interaction term, which is only valid for a dilute gas, when the mean distance between particles is much larger than the s-wave scattering length a. This can be stated as $n_0|a|^3 \ll 1$, where $|a|^3$ is the scattering volume. Notice that, for a dense medium, the neglect of fluctuations in the collision integral can be questioned.

The GP equation is obviously of the form of a nonlinear Shrödinger equation, where the nonlinear potential term results from binary collisions between atoms. This equation is well known in many areas of Physics, from quantum physics and nonlinear optics to fluid mechanics and plasma physics [11, 12], and its mathematical properties are well understood. Here it is important to notice that the s-wave scattering length can be positive, $a > 0$, for repulsive interactions between atoms, and negative, $a < 0$, for attractive interactions. The equilibrium properties of the condensate are very different in these two cases, as discussed later.

The Gross-Pitaevskii equation is strictly valid in the limit of zero temperature, where all the atoms are in the ground state, but can be generalized to a finite temperature gas. In this case we consider single atom wave functions ψ_i, where $i = 0$ represents the lower energy state, each of them occupied by n_i atoms, such that $N = \sum_i N_i$, where these wave functions are normalized to unity, $\int \psi_i(\mathbf{r})d\mathbf{r} = 1$. By minimizing the expectation value for the energy of the system, we can derive the coupled equations [14], for the condensed phase

$$i\hbar\frac{\partial\psi_0}{\partial t} = \left[V(\mathbf{r}) + g(N_0|\psi_0|^2 + 2N') - \frac{\hbar^2}{2M}\nabla^2\right]\psi_0 \tag{7.57}$$

and for the non condensed states

$$i\hbar\frac{\partial\psi_i}{\partial t} = \left[V(\mathbf{r}) + g(2N_0|\psi_0|^2 + 2N_t) - \frac{\hbar^2}{2M}\nabla^2\right]\psi_i, \tag{7.58}$$

where the non condensed or thermal number of atoms is $N' = N - N_0$. Notice a factor of 2 appearing in the coupling terms. The condensed and thermal number densities are

$$n_0(\mathbf{r},t) = N_0|\psi_0(\mathbf{r},t)|^2\ , \quad n_T(\mathbf{r},t) = \sum_{i\neq 0} N_i|\psi_i(\mathbf{r},t)|^2. \tag{7.59}$$

It has been noted [13] that the finite temperature Gross-Pitaevskii equations are not completely consistent, due to the dependence of the effective potential on the densities n_0 and n_T. The range of validity of this theory will discussed later and the physics of the thermal cloud will be considered further on.

7.5 Thomas-Fermi Approximation

Let us now consider the equilibrium configurations, as predicted by the mean field theory. This can be done by assuming that the condensate wave function is given by

$$\Phi(\mathbf{r},t) = \psi(\mathbf{r})\exp(-i\epsilon_0 t/\hbar), \tag{7.60}$$

where ϵ_0 is the ground state energy of the confined atoms. By replacing this in the Gross-Pitaevskii equation (7.56), and noticing that $\epsilon_0 = \mu$ at $T = 0$, where $\mu = \mu_{\text{int}} > 0$ the chemical potential for the interacting condensate, we obtain

$$\mu\ \psi(\mathbf{r},t) = \left[-\frac{\hbar^2\nabla^2}{2M} + V_{\text{ext}}(\mathbf{r}) + g\ \psi(\mathbf{r})^2\right] \tag{7.61}$$

where the time independent wave function $\psi(\mathbf{r})$ was assumed real, and the confining potential was assumed static. For the harmonic potential (7.8), and if the collision term was neglected, $g = 0$, we would then get (apart from a normalization factor) the ground state solution for the harmonic oscillator, in accordance with Eq. (7.12). Here, we want to retain the contribution of the interaction term and to inquire about its influence on the density profiles of the condensate. Multiplying the static equation by $\psi(\mathbf{r})$ and integrating in volume, we get

$$N\mu = E_{\text{kin}} + E_{\text{pot}} + 2E_{\text{int}}, \tag{7.62}$$

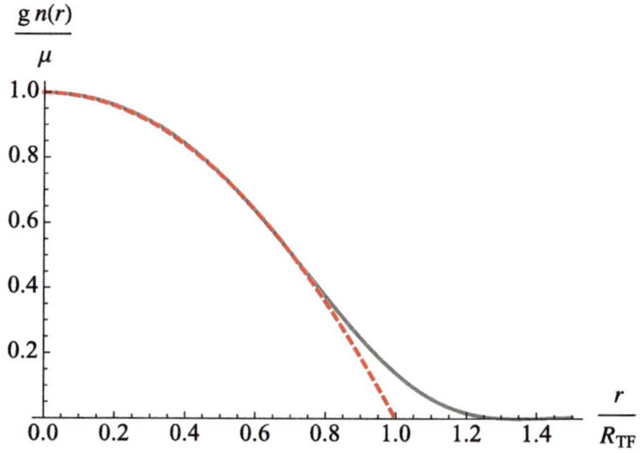

Fig. 7.3 Comparison between the Thomas-Fermi profile (*dashed line*) and the numerical solution (*full line*) of time-independent Gross-Piatevskii equation (7.61), obtained for $\mu = 10\hbar\omega_0$ and $g = 9.5\mu/n(0)$. The small difference at the vicinity of the Thomas-Fermi radius R is of the order of few ξ

where the kinetic, potential and interaction energies are defined as

$$E_{\text{kin}} = \frac{\hbar^2}{2M}\int \left|\nabla\sqrt{n(\mathbf{r})}\right|^2 d\mathbf{r}, \quad E_{\text{pot}} = \int n(\mathbf{r})V_{\text{ext}}(\mathbf{r})d\mathbf{r}, \quad E_{\text{int}} = \frac{gN}{2}\int n(\mathbf{r})d\,d\mathbf{r}. \tag{7.63}$$

It is obvious that the kinetic energy is of the order of $E_{\text{kin}} \sim N\hbar\omega_0 \sim N/a_0^2$, where a_0 is the scale length defined by (7.14), and the interaction energy is of order $E_{\text{int}} \sim |a|N^2/a_0^3$. This means that we have the following scaling $E_{\text{int}}/E_{\text{kin}} \sim N|a|/a_0$. In the case of repulsive interactions, with $a > 0$, and for the interaction dominated case where $Na/a_0 \gg 1$, analytical solutions of the static equation (7.61) can be found [15].

These solutions show that, changing this parameter only slightly changes the profile at the border of the condensate, leaving the main profile invariant. In fact, that quantum pressure associated with E_{kin}, only becomes relevant at the border, and can be neglected everywhere else. This suggests the neglect of the term in $\nabla^2\sqrt{n(\mathbf{r})}$ in Eq. (7.61), which directly gives the following density profile

$$n(\mathbf{r}) \equiv \psi^2(\mathbf{r}) = \frac{1}{g}\left[\mu - V_{\text{ext}}(\mathbf{r})\right], \tag{7.64}$$

with $n = 0$ in the outside region where μ becomes smaller than $V_{\text{ext}}(\mathbf{r})$. This is usually called the *Thomas-Fermi approximation*. By using the explicit form of the harmonic potential, as given by Eq. (7.8), we can see that the density profile of the condensate takes the form of an inverted parabola, as illustrated in Fig. 7.3.

The size of the condensate is therefore determined by the condition $\mu = V_{\text{ext}}(R)$. For a spherical trap, this implies that $\mu = M\omega_0^2R^2/2$. On the other hand, the normalization condition for the density profile (7.64) allows us to relate the chemical potential with the total number of condensed atoms $N_0 = N$, leading to

$$\mu = \frac{1}{2}\hbar\omega_0\left(\frac{15Na}{a_0}\right)^{2/5}. \tag{7.65}$$

From these two relations we can then determine the external radius of the condensate, as a function of the number of atoms, reading

$$R = a_0 \left(\frac{15Na}{a_0} \right)^{1/5} . \tag{7.66}$$

If, instead of a spherical trap we had an axial trap, the radial and axial sizes of the condensate, $R_\perp$ and Z, would be determined by

$$\mu = \frac{1}{2} M \left(\omega_\perp^2 R_\perp^2 + \omega_z^2 Z^2 \right) . \tag{7.67}$$

It is now useful to compare these predictions with those derived previously for non-interacting particles. The Thomas-Fermi density profile (7.64) predicts a density at the centre of the trap $n_{\text{TF}}(0) = \mu/g$. In contrast, the harmonic oscillator solution (7.13) gives $n_{\text{ho}}(0) = N/(\pi^{3/2} a_0^3)$. Using (7.65), we can then write the ratio between there two values for the maximum density as

$$\frac{n_{TF}(0)}{n_{ho}(0)} = \sqrt{\pi} \frac{15^{2/5}}{8} \left(\frac{Na}{a_0} \right)^{-3/5} . \tag{7.68}$$

For typical experimental conditions, where the parameter Na/a_0 is usually much larger than one, this leads to a reduction of the density of the condensed cloud with respect to that of a non-interacting gas. This effect can be significant even for a very dilute gas.

The Thomas-Fermi approximation is very useful to describe the equilibrium of the condensed cloud. But, as referred above, it breakes down near the critical radius $r = R$, in a layer with dimensions of the order of $d \simeq R(a_0/R)^{4/3}$. The approximation will better hold in the region $R \gg a_0$. In order to improve the description of the density profile in this outer region of the cloud, we can linearize the trapping potential around the critical radius, by using $V_{\text{ext}}(r) = M\omega_0^2 R(r - R)$, and solve the corresponding Gross-Pitaevskii equation in this potential [14].

In the above discussion, we have only considered the equilibrium for a repulsive interaction between atoms, $a > 0$. Let us now consider the opposite case of an attractive interaction, with $a < 0$. In this case, the density profile of the condensate can become unstable, above a given number of condensed atoms. This make the experimental study of condensates in this regime much more difficult, but also leads to very interesting new problems, because such a collapse for attractive forces can be seen as the analogue of the gravitational collapse of a star. Such a collapse is supposed to occur above a critical density $N_0 > N_c \sim a_0/|a|$. This can be seen, for a spherical trap, by using a Gaussian *ansatz* of the form [15]

$$\psi(r) = \left(\frac{N}{w^3 a_0^3 \pi^{3/2}} \right)^{1/2} \exp\left(-\frac{r^2}{2w^2 a_0^2} \right), \tag{7.69}$$

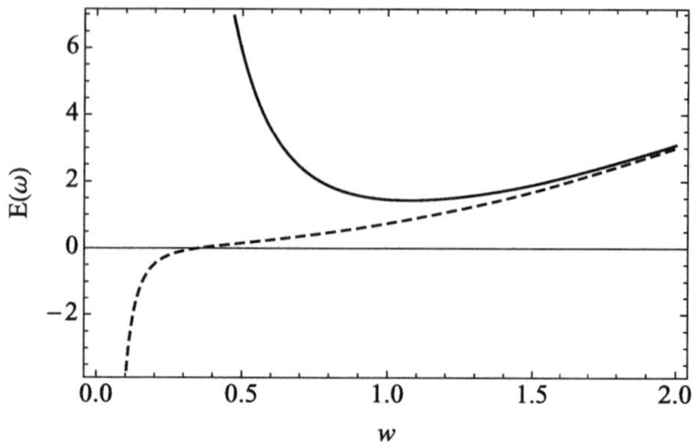

Fig. 7.4 Variational Thomas-Fermi energy in (7.70) as a function of the width of the cloud. Stable (*full line*) and unstable (*dashed line*) regimes are observed as a function of the condensed number of particles N

where the dimensionless parameter w determines the width of the condensate in units of a_0. By using a variational approach and looking for the value of w which minimizes the total energy of the condensate E, we get

$$E(w) = N\hbar\omega_0 \left[\frac{3}{4}\left(w^2 + \frac{1}{w^3} \right) - \frac{N|a|}{a_0} \frac{1}{\sqrt{2\pi w^3}} \right]. \tag{7.70}$$

As we can see from the Fig. 7.4, for low values of N, this energy curve has a minimum near 1, which shifts towards lower values and vanishes at $w = 1/5^{1/4}$, for a critical density of $N_c a_0/|a|$. Above this density we expect collapse of the condensate, due to the attractive forces which tend to accumulate the atoms near the centre of the trap. At this point, we should notice that such a collapse could be affected by several processes which are not retained in the above mean field model, such as three-body collisions, inelastic collisions or quantum tunneling.

7.6 Fluid and Kinetic Formulations

7.6.1 Quantum Fluid Equations

We now discuss alternative formulations of the mean field theory of a condensate. Starting from the GP equation, we first derive the quantum fluid equations for the condensate. They can be very helpful in the discussion of the waves and oscillations of the condensate, as shown in the next chapter. Let us introduce the *Madelung transformation* [16], defined by

$$\Phi(\mathbf{r}, t) = \sqrt{n(\mathbf{r}, t)} \exp\left[i\varphi(\mathbf{r}, t) \right], \tag{7.71}$$

where $n \equiv n(\mathbf{r}, t) = |\psi|^2$ is the density of the condensate, and $\varphi(\mathbf{r}, t)$ is a phase function. The gradient of this function can be identified with the fluid momentum, allowing us to define the fluid velocity by

$$\mathbf{v} = \frac{\hbar}{M}\nabla\varphi(\mathbf{r}, t). \tag{7.72}$$

The space and time derivatives of (7.71), are

$$\frac{\partial \Phi}{\partial t} = \psi \left[i\frac{\partial \varphi}{\partial t} + \frac{1}{2n}\frac{\partial n}{\partial t} \right] \tag{7.73}$$

and

$$\nabla\Phi = \psi \left[i\nabla\varphi + \frac{1}{2n}\nabla n \right]. \tag{7.74}$$

Replacing this in the GP equation (7.56), and equating to zero separately the real and imaginary parts of the resulting expression, we get, from the imaginary part

$$\frac{\partial n}{\partial t} + \nabla \cdot (n\mathbf{v}) = 0. \tag{7.75}$$

This is nothing but the equation of conservation of mass for the condensate. On the other hand, from the imaginary part, we obtain

$$\frac{\partial \varphi}{\partial t} = -\frac{\hbar}{2M}\left[((\nabla\varphi)^2 - \frac{1}{2n}\nabla^2 n + \frac{1}{4n^2}(\nabla n)^2 \right] - \frac{1}{\hbar}\left[V_{\text{ext}} + g\, n \right]. \tag{7.76}$$

Multiplying this by $(\hbar/M)\nabla$, we can then obtain

$$\frac{\partial \mathbf{v}}{\partial t} + \mathbf{v} \cdot \nabla\mathbf{v} = -\frac{1}{M}\nabla\left(V_{\text{ext}} + gn \right) + \frac{\hbar^2}{2M^2}\nabla\left(\frac{\nabla^2\sqrt{n}}{\sqrt{n}} \right) \tag{7.77}$$

This is the momentum conservation equation for the quantum fluid. Here we can clearly identify a specific quantum potential, known as the *Bohm potential*, as defined by

$$V_B = -\frac{\hbar^2}{2M^2}\frac{\nabla^2\sqrt{n}}{\sqrt{n}}. \tag{7.78}$$

which is a direct consequence of quantum non-locality. Such a definition allow us to re-write the momentum equation in a more familiar form

$$\frac{\partial \mathbf{v}}{\partial t} + \mathbf{v} \cdot \nabla\mathbf{v} = -\frac{\nabla P}{Mn} - \frac{1}{M}\nabla\left(V_{\text{ext}} + V_B \right) \tag{7.79}$$

where we have identified P with the pressure, as determined by

$$\nabla P = n\nabla(gn) \longrightarrow P = \frac{g}{2}n^2. \tag{7.80}$$

One final remark concerns the properties of the quantum fluid. Looking at the definition of the fluid mean velocity, as given by Eq. (7.72), we can easily conclude that

$$\nabla \times \mathbf{v} = \frac{\hbar}{M} \nabla \times (\nabla \varphi) = 0. \tag{7.81}$$

which means that the condensate flow is irrotational, as long as the phase function $\varphi(\mathbf{r}, t)$ stays non-singular. This has important physical consequences, as discussed later in connection with the existence of vortices in rotating condensates.

7.6.2 Wave Kinetic Equation

We show now that a wave kinetic equation can be derived, which is exactly equivalent to the GP equation, thus providing a different approach to the mean field theory. We then show that this equation can be reduced to a kinetic equation of the Vlasov type, if we take the quasi-classical limit. The quantum and quasi-classical versions of the kinetic equation provide a very useful description of the kinetic properties of the condensate, and in particular of the resonant interaction between the atoms and the elementary excitations, leading to Landau damping. These effects will be discussed in the next chapter.

We have seen that the condensate is described by a nonlinear Scrödinger equation, the GP equation (7.56). In order to built up our alternative kinetic approach we follow the Wigner-Moyal procedure of Chap. 3. We first define the auto-correlation function for the condensate wave function, $K \equiv \Phi(\mathbf{r}_1, t_1)\Phi^*(\mathbf{r}_2, t_2)$. As we have seen, this can also be written as $K(\mathbf{r}, \mathbf{s}) = \Phi(\mathbf{r}+\mathbf{s}/2, t+\tau/2)\Phi^*(\mathbf{r}-\mathbf{s}/2, t-\tau/2)$, where two vector positions are $\mathbf{r} = (\mathbf{r}_1 + \mathbf{r}_2)/2$ and $\mathbf{s} = \mathbf{r}_1 - \mathbf{r}_2$, and the two time variables are $t = (t_1 + t_2)/2$ and $\tau = t_1 - t_2$. We then take its double Fourier transformation, by integrating over the fast (or short scale) variables $\mathbf{s}$ and τ, as

$$W(\omega, \mathbf{k}; \mathbf{r}, t) = \int d\mathbf{s} \int d\tau \; K(\mathbf{r}, \mathbf{s}) \exp(-i\mathbf{k} \cdot \mathbf{s} + i\omega\tau). \tag{7.82}$$

This is a straightforward extension of the formalism employed in Chap. 3 onto the space-time domain, which becomes necessary here. Going back to the original GP equation (7.56), we can then derive the exact evolution equation for the quasi-distribution $W(\omega, \mathbf{k}; \mathbf{r}, t)$, in the following form

$$i\hbar\left(\partial_t + \frac{\hbar}{m}\mathbf{k} \cdot \nabla\right) W = \int \frac{d\mathbf{q}}{(2\pi)^3} \int \frac{d\Omega}{2\pi} \, V(\mathbf{q}, \Omega) \, [W_- - W_+] \exp(i\mathbf{q} \cdot \mathbf{r} - i\Omega t). \tag{7.83}$$

Here $V(\mathbf{q}, \Omega)$ is the double Fourier transformation of the total potential $V = V_{\text{ext}} + V_{NL}$, with $V_{NL} = g|\Phi(\mathbf{r}, t)|^2$,

$$V(\mathbf{r}, t) = \int \frac{d\mathbf{q}}{(2\pi)^3} \int \frac{d\Omega}{d2\pi} V(\mathbf{q}, \Omega) \exp(i\mathbf{q} \cdot \mathbf{r} - i\Omega t) \tag{7.84}$$

and $W_\pm$ are determined in the usual way, $W_\pm \equiv W(\omega \pm \Omega/2, \mathbf{k} \pm \mathbf{q}/2)$. If we consider a single Fourier component of the potential,

$$V(\mathbf{r}, t) = \tilde{V} \exp(i\mathbf{q} \cdot \mathbf{r} - i\Omega t) \equiv \tilde{V} e^{i\theta}, \tag{7.85}$$

we can write the evolution equation in the following simple form

$$\left(\partial_t + \frac{\hbar}{m}\mathbf{k} \cdot \nabla\right) W = \frac{\tilde{V}}{i\hbar} e^{i\theta} \left[W_- - W_+\right]. \tag{7.86}$$

Moreover, in the case of a periodic potential of the form

$$V(\mathbf{r}, t) = \tilde{V} \cos(i\mathbf{q} \cdot \mathbf{r} - i\Omega t) \equiv \frac{\tilde{V}}{2} \left(e^{i\theta} + e^{-i\theta}\right), \tag{7.87}$$

we would get

$$\left(\partial_t + \frac{\hbar}{m}\mathbf{k} \cdot \nabla\right) W = \frac{\tilde{V}}{\hbar} \sin\theta \left[W_- - W_+\right]. \tag{7.88}$$

These various forms of the wave kinetic equation are exactly equivalent to the initial Gross-Pitaevskii equation, for different kinds of potentials, and can be used to solve specific problems in BECs, in particular, their kinetic stability.

References

1. K. Huang, *Bose-Einstein Condensation and Superfluidity* (Cambridge University Press, Cambridge, 1995)
2. R. Balescu, *Equilibrium and Nonequilibrium Statistical Mechanics* (Wiley, NewYork, 1975)
3. J.V. Gomes, M. Belsley, D. Boiron, Phys. Rev. A **77**, 026101 (2008)
4. R. Loudon, *The Quantum Theory of Light* (Oxford Science Publications, Oxford, 1992)
5. R.P. Feynman, *Progress in Low Temperature Physics*, vol. 1 (North-Holland, Amsterdam, 1955)
6. H.D. Politzer, Phys. Rev. A **54**, 5048 (1996)
7. N. Naraschewsli, R. Glauber, Phys. Rev. A **59**, 4595 (1999)
8. S. Ritter, A. Ottl, T. Donner, T. Bourdel, M. Köhl, T. Esslinger, Phys. Rev. Lett. **98**, 090402 (2007)
9. E.P. Gross, il Nuovo Cimento **20**, 454 (1961); J. Math. Phys. **4**, 195 (1963)
10. L.P. Pitaevskii, Zh. Eksp. Teor. Fiz **40**, 646 (1961); Sov. Phys. JETP **13**, 431 (1961)

11. T.J.M. Boyd, J.J. Sanderson, *The Physics of Plasmas* (Cambridge University Press, Cambridge, 2003)
12. R.K. Dodd, J.C. Eilbeck, J.D. Gibbon, H.C. Morris, *Solitons and Nonlinear Wave Equations* (Academic, New York, 1982)
13. A.J. Legget, Rev. Mod. Phys. **73**, 307 (2001)
14. F. Dalfovo, S. Giorgini, L.P. Pitaevskii, S. Stringari, Rev. Mod. Phys. **71**, 463 (1999)
15. G. Baym, C. Pethick, Phys. Rev. Lett. **76**, 6 (1996)
16. E. Madelung, Zeit. für Phys. **40**, 322 (1926)

Chapter 8
Elementary Excitations in BECs

We now discuss the elementary waves and oscillations that can take place in a Bose Einstein condensate. We first deal with the elementary excitations as described by the quantum fluid equations. This allows us to study the sound waves in the condensate, sometimes also called Bogoliubov waves, as well as the global modes of the condensed cloud. We then approach the study of kinetic effects, based on the use of the wave kinetic equation. This will allow us to understand the microscopical wave-atom interactions taking place inside the condensate, as well as more refined processes such as the atomic Landau damping, fluid and kinetic instabilities such as beam and modulational instabilities. We show that the sound waves can become unstable, and determine the instability growth rates in both the quantum and the quasi-classical regime. Two specific examples of unstable oscillations are considered in detail. First, the two-stream instability, associated with two counter-propagating condensate beams. Then, the wakefield excitation by a moving condensate in a background thermal gas. Such instabilities can eventually be the seed for the excitation of a broadband turbulent spectrum.

8.1 Sound Waves

Let us assume that the condensate is in some equilibrium state, with a density profile $n_0(\mathbf{r})$, and globally at rest $\mathbf{v}_0 = 0$. We address the problem of the elementary excitations that can be excited when, for some reason, such an equilibrium is perturbed. We then define the density as $n = n_0 + \tilde{n}$, where it is assumed that $|\tilde{n}| \ll n_0$. Linearizing the fluid equations with respect to these infinitesimal perturbations, and ignoring the external confinement potential $V_{\text{ext}}(\mathbf{r})$, we obtain

$$\frac{\partial \tilde{n}}{\partial t} + \nabla \cdot (n_0 \mathbf{v}) = 0 \, , \quad \frac{\partial \mathbf{v}}{\partial t} = -\frac{\nabla \tilde{P}}{M n_0} - \frac{1}{M} \nabla \tilde{V}_B . \tag{8.1}$$

J.T. Mendonça and H. Terças, *Physics of Ultra-Cold Matter*, Springer Series on Atomic, Optical, and Plasma Physics 70, DOI 10.1007/978-1-4614-5413-7_8,

Let us restrict our analysis to short wavelength perturbations, much shorter than the scale length of the density variations of the equilibrium profile. This means, in practice, that they are much shorter than the size of the condensate. In this case, we can consider the gas as locally uniform, $\nabla n_0 \simeq 0$, and we can derive from these equations a density wave equation, of the form

$$\frac{\partial^2 \tilde{n}}{\partial t^2} - \frac{n_0}{M}\nabla^2\left(g\tilde{n} + \tilde{V}_B\right) = 0. \tag{8.2}$$

where the perturbation in the Bohm potential is determined by

$$\tilde{V}_B = -\frac{\hbar^2}{4M}\frac{\nabla^2 \tilde{n}}{n_0}. \tag{8.3}$$

Assuming that the perturbations evolve in space and time as $\exp(i\mathbf{k}\cdot\mathbf{r} - i\omega t)$ we then easily obtain the dispersion relation for waves with frequency ω and wavevector $\mathbf{k}$, of the form

$$\omega^2 = c_s^2 k^2 + \frac{\hbar^2 k^4}{4M^2}, \tag{8.4}$$

where we have introduced the *Bogoliubov sound speed*, as

$$c_s = \sqrt{\frac{g n_0}{M}}. \tag{8.5}$$

This dispersion relation characterizes the propagation of sound waves in the condensed gas. Two differences with respect with sound waves in ordinary classical fluids are, first a different definition of the sound speed, and second a quantum dispersive term proportional to k^4. We notice that, for large wavelengths, it reduces to the typical non-dispersive sound waves, such that $\omega \simeq c_s k$. In contrast, for short wavelengths, the quantum dispersive term becomes dominant, and we have

$$\omega \sim \frac{\hbar k^2}{2M}, \quad \epsilon_k \equiv \hbar\omega \sim \frac{\hbar^2 k^2}{2M}, \tag{8.6}$$

where ϵ_k is the phonon quantum of energy. These expressions show that, in this limit, phonons behave very much like free particles, with mass equal to the mass of the individual condensed atoms, and momentum equal to $\hbar k$. The dispersion relation (8.4) is valid in the zero temperature limit, $T = 0$. It will be seen later in this chapter that thermal effects will introduce an additional dispersive term.

In the above discussion we have implicitly assumed that $g > 0$, which corresponds to atoms with repulsive interactions. The situation changes in the presence of attractive interactions, such that $g < 0$. The sound speed (8.5) becomes imaginary, and the dispersion relation (8.4) is transformed into

$$\omega^2 = -\frac{n_0}{M}|g|k^2 + \frac{\hbar^2 k^4}{4M^2}. \tag{8.7}$$

For a given critical wavenumber k_c, the mode frequency reduces to zero, $\omega = 0$, as defined by

$$k_c^2 = n_0|g|\frac{4M}{\hbar^2}. \quad (8.8)$$

For $k < k_c$, the frequency is imaginary and the condensate becomes unstable to longer wavelength perturbations. We can also see that, by increasing the density n_0, the critical number decreases as $n_0^{-1/2}$, which shortens the spectral range of stable oscillations. Therefore, dense condensates with attractive interactions tend to be more unstable, as confirmed by experiments.

8.2 Global Oscillations

The above discussion is valid for short wavelength or local oscillations associated with phonon modes. Let us now consider global oscillations of the condensed cloud, with a scale length of the order of the size of the BEC. These global modes were first discussed by Stringari [1]. For that purpose, we go back to the fluid equations of the condensed gas, which can be rewritten as

$$\frac{\partial n}{\partial t} + \nabla \cdot (n\mathbf{v}) = 0\,, \quad \frac{\partial \mathbf{v}}{\partial t} + (\mathbf{v}\cdot\nabla)\mathbf{v} = -\frac{1}{M}\nabla\mu, \quad (8.9)$$

where we have introduced the *chemical potential*

$$\mu = V_0(\mathbf{r}) + gn - \frac{\hbar^2}{2M}\frac{\nabla^2\sqrt{n}}{\sqrt{n}}. \quad (8.10)$$

The equilibrium profile $n = n_0(\mathbf{r})$ is determined by the condition $\partial/\partial t = 0,\ \mathbf{v} = 0$, and corresponds to

$$n_0(\mathbf{r}) = \frac{1}{g}\left[\mu - V_0(\mathbf{r})\right]. \quad (8.11)$$

Here we have neglected the contribution of the Bohm potential, therefore making use of the *Thomas-Fermi* approximation described in the previous chapter. For a spherical harmonic potential, such that $V_0(\mathbf{r}) = M\omega_0^2 r^2/2$, we know that the chemical potential μ has to be equal to the trapping potential at the boundary of the condensate, $r = R$, or $\mu = V_0(R)$. This leads to the following parabolic profile

$$n_0(r) = \frac{1}{2}M\omega_0^2\left(R^2 - r^2\right). \quad (8.12)$$

Let us now consider global oscillations around such an equilibrium density profile. By using $n = n_0(r) + \tilde{n}$, and consistently neglecting V_B, we get from (8.9) the linearized equations

$$\frac{\partial \tilde{n}}{\partial t} + \nabla \cdot (n_0 \mathbf{v}) = 0 \,, \qquad \frac{\partial \mathbf{v}}{\partial t} = -\frac{g}{M} \nabla \tilde{n}. \tag{8.13}$$

From this we can easily get an equation for the perturbed density, as

$$\frac{\partial^2 \tilde{n}}{\partial t^2} - \frac{g}{M} \boldsymbol{\nabla} \cdot (n_0 \boldsymbol{\nabla} \tilde{n}) = 0. \tag{8.14}$$

For harmonic oscillations at a frequency ω, we can write $\tilde{n}(\mathbf{r}, t) = \tilde{n}(\mathbf{r}) \exp(-i\omega t)$. Using spherical coordinates $\mathbf{r} \equiv (r, \theta, \varphi)$, and noting that the equilibrium density is spherically symmetric and only depends on r, we obtain

$$\tilde{n} = -\frac{g}{M\omega^2} \left[\left(\frac{\partial n_0}{\partial r} \right) \frac{\partial \tilde{n}}{\partial r} + \nabla^2 \tilde{n} \right]. \tag{8.15}$$

Now, using the Thomas-Fermi equilibrium profile (8.11), this equation becomes

$$\tilde{n} = \frac{\omega_0^2}{\omega^2} \left[r \frac{\partial \tilde{n}}{\partial r} - \frac{1}{2} (R^2 - r^2) \nabla^2 \tilde{n} \right]. \tag{8.16}$$

It is now useful to write the Laplacian operator as

$$\nabla^2 = \frac{1}{r^2} \frac{\partial}{\partial r} \left(r^2 \frac{\partial}{\partial r} \right) - \frac{\mathcal{L}^2}{r^2}, \tag{8.17}$$

with the angular operator

$$\mathcal{L}^2 = -\frac{1}{\sin\theta} \frac{\partial}{\partial \theta} \left(\sin\theta \frac{\partial}{\partial \theta} \right) - \frac{1}{\sin^2\theta} \frac{\partial^2}{\partial \varphi^2}. \tag{8.18}$$

We know that this angular operator satisfies the eigenvalue equation

$$\mathcal{L}^2 Y_{lm}(\theta, \varphi) = l(l+1) Y_{lm}(\theta, \varphi), \tag{8.19}$$

where l and m are integer and $Y_{lm}(\theta, \varphi)$ are spherical harmonics. We can therefore try density perturbation solutions of the form

$$\tilde{n}(\mathbf{r}) \equiv \tilde{n}(r, \theta, \varphi) = r^l F(r) Y_{lm}(\theta, \varphi), \tag{8.20}$$

where $F(r)$ is a radial function to be determined. Replacing this in Eq. (8.16), and using $\epsilon = \omega^2/\omega_0^2$, we obtain the radial equation

$$(\epsilon - l) F = r \frac{dF}{dr} - \frac{1}{2} (R^2 - r^2) \left[2(l+1) \frac{1}{r} \frac{dF}{dr} + \frac{d^2 F}{dr^2} \right]. \tag{8.21}$$

Introducing the new variable $x = r^2/R^2$, we can transform this into the *hypergeometric equation*, which can be written in its canonical form as

$$x(1-x)F''(x) + [c - (a+b+1)x]\,F'(x) - abF(x) = 0, \tag{8.22}$$

where the parameters a, b and c, are determined by the relations

$$c = \frac{1}{2}(2l+3)\,, \quad a+b = c\,, \quad ab = -\frac{1}{2}(\epsilon - l). \tag{8.23}$$

Solutions of this equation can be identified with the hypergeometric function

$$F(x) \equiv {}_2F_1(a,b,c;x) = \sum_{n=0}^{\infty} \frac{(a)_n (b)_n}{(c)_n} \frac{x^n}{n!}, \tag{8.24}$$

where the notation $(a)_n = (a+n-1)!/(a-1)!$ has been used. It is known that the series in this expression is reduced to a simple polynomial if either a or b are zero, or are equal to a negative integer. Let us then assume that $a = -n$, with $n = 0, 1, 2, \ldots$ According to (8.23) this implies that $b = n + l + 3/2$, and finally to $(\epsilon - l) = n(2l+3) + 2n^2$. Using the explicit expression of ϵ, we therefore conclude that the global mode frequencies are given by

$$\omega = \omega_0 \sqrt{2n^2 + n(2l+3) + l}. \tag{8.25}$$

A particularly interesting family of modes corresponds to $l = 0$ and $n \neq 0$. These are purely radial modes, with no angular structure. They are called *breathing modes*, with the frequency spectrum given by

$$\omega = \omega_0 \sqrt{2n^2 + 3n}. \tag{8.26}$$

Another interesting family of modes is defined by the conditions $l \neq 0$ and $n = 0$, the so called *surface modes*, which obey the following frequency spectrum

$$\omega = \omega_0 \sqrt{l}. \tag{8.27}$$

Comparing with Eq. (8.21) we realize that these modes are such that $(\epsilon - l) = 0$, which means that the radial function $F(r)$ reduces to a constant. The mode structure is then described by

$$\tilde{n}(r, \theta, \varphi) \propto r^l Y_{lm}(\theta, \varphi). \tag{8.28}$$

We can see that the amplitude of the oscillation increases for large r, and attains its maximum at the boundary $r = R$. For this reason they are called *surface modes*.

8.3 Kinetic Processes

We now turn to the wave kinetic description of elementary excitations, centering our analysis on the short wavelength modes [2]. Our starting point will be given by Eq. (7.85). Let us then consider a perturbation $\tilde{V}$ of the mean field potential, and study its stability. If the confining potential V_0 is static, this can only result from a modulation of the density distribution of the condensate, such that $\tilde{V}(\mathbf{r}, t) = g\tilde{I}(\mathbf{r}, t)$ with

$$\tilde{I}(\mathbf{r}, t) = \int \tilde{W}(\omega, \mathbf{k}, \mathbf{r}, t) \frac{d\mathbf{k}}{(2\pi)^3} \frac{d\omega}{2\pi}. \tag{8.29}$$

Notice that ω and $\mathbf{k}$ refer here to the kinetic energy and momentum of the condensed atoms. Linearizing the wave kinetic equation (7.85) with respect to these perturbations, we get

$$(\partial_t + \mathbf{v} \cdot \nabla) \tilde{W} = \frac{g}{i\hbar} \tilde{I}(\mathbf{q}, \Omega) \left[W_{0-} - W_{0+}\right] \exp(i\mathbf{q} \cdot \mathbf{r} - i\Omega t), \tag{8.30}$$

where $\mathbf{v} = d\mathbf{r}/dt = \hbar\mathbf{k}/M$, and $W_{0\pm}$ are the equilibrium values. Here, Ω and $\mathbf{q}$ refer to the low frequency oscillations of the acoustic type. For perturbations with a scale length $2\pi/q$ much shorter than the length of the confinement region, we can again neglect the influence of the confining potential on the evolution along the axial direction. We can focus our attention on the longitudinal perturbations propagating along the z-axis, of the form $\tilde{W}(\mathbf{r}, t) = \tilde{W} \exp(iqz - i\Omega t)$. Fourier analysis of Eq. (8.30) then leeds to

$$\tilde{W} = \frac{g}{\hbar} \tilde{I} \frac{\left[W_{0-} - W_{0+}\right]}{(\Omega - qv_z)}. \tag{8.31}$$

Integrating over the spectrum of the condensate particles, k and ω, and using of the definition of $\tilde{I}$, given by Eq. (8.29), then yields

$$1 - \frac{g}{\hbar} \int \frac{d\omega}{2\pi} \int \frac{d\mathbf{k}}{(2\pi)^3} \frac{W_0(\mathbf{k} - q\mathbf{e}_z/2) - W_0(\mathbf{k} + q\mathbf{e}_z/2)}{(\Omega - qv_z)} = 0. \tag{8.32}$$

This expression can be simplified by noting that the kinetic energy of the atoms are related with their momentum state, by the well known constraint, sometimes called the quasi-particle approximation,

$$W(\omega, \mathbf{k}; \mathbf{r}, t) = W(\mathbf{k}; \mathbf{r}, t)\, \delta(\omega - \omega_k), \tag{8.33}$$

where $\omega_k = \hbar k^2/2M$ represents the atom kinetic energy. This allows us to integrate over ω and obtain an equation depending only on the usual (reduced) Wigner distribution $W(\mathbf{k}; \mathbf{r}, t)$. We can also neglect the influence of the perpendicular motion, by further assuming that $W(\mathbf{k}; \mathbf{r}, t) = W(k_z; \mathbf{r}, t)\, \delta(\mathbf{k}_\perp)$. Equation (8.32) can then be reduced to a much simpler form

$$1 + \frac{g}{\hbar}\int W_0(k_z)\left[\frac{1}{(\Omega_+ - qv_z)} - \frac{1}{(\Omega_- - qv_z)}\right]\frac{dk_z}{2\pi} = 0, \tag{8.34}$$

where we have used the following quantities

$$\Omega_\pm = \Omega \pm \frac{\hbar q^2}{2M}. \tag{8.35}$$

This expression resultes from the substitution of k_z by $k_z \pm q/2$, in the denominator, and the use of the axial velocity $v_z = \hbar k/M$. Equation (8.34) is the kinetic dispersion relation for low frequency oscillations of the condensate.

Let us consider first the simplest case of a condensate beam with no translational temperature. We can use $W_0(k_z) = 2\pi W_0 \delta(k_z - k_0)$. Integration of Eq. (8.34) leads to

$$1 - \frac{qW_0}{\hbar}\frac{q^2}{(\Omega - qv_0)^2 - (\hbar q^2/2M)^2} = 0, \tag{8.36}$$

where $v_0 = \hbar k_0/M$. This simple dispersion relation can also be written as

$$(\Omega - qv_0)^2 = q^2 c_s^2 + q^4 \frac{\hbar^2}{4M^2}, \tag{8.37}$$

where $c_s = \sqrt{gW_0/m}$ is the Bogoliubov sound speed. Apart from the obvious change of notation, it coincides with Eq. (8.4), derived from the quantum fluid equations, for a condensate at rest $v_0 = 0$.

8.4 Landau Damping

Let us now consider a more general case where the kinetic effects are not neglected. Going back to the dispersion relation (8.34), we note that we have to solve integrals of the form

$$\int_{-\infty}^{\infty} \frac{W_0(z)}{(z - \Omega_\pm)} dz = \mathcal{P}\int_{-\infty}^{\infty} \frac{W_0(z)}{(z - \Omega_\pm)} dz + i\pi W_0(z = \Omega_\pm), \tag{8.38}$$

where $\mathcal{P}$ represents the principal part of the integral, and the variable of integration is $z = qv_z$. The resulting kinetic dispersion equation can then be written in the following general form, already used for the non-condensed gas

$$\epsilon(\Omega, q) \equiv 1 + \chi(\Omega, q) = 0, \tag{8.39}$$

where $\chi(\Omega, q)$ is the BEC susceptibility. Due to the decomposition (8.38) this quantity is always complex, with $\chi = \chi' + i\chi''$. The real part of this function is

$$\chi'(\Omega, q) = \frac{g}{\hbar}\frac{M}{2\pi q}\mathcal{P}\int W_0(z)\left[\frac{1}{(z-\Omega_+)} - \frac{1}{(z-\Omega_-)}\right]dz \tag{8.40}$$

and the imaginary part

$$\chi''(\Omega, q) = -\frac{gM}{2\hbar^2 q}\left[W_0(z=\Omega_+) - W_0(z=\Omega_-)\right]. \tag{8.41}$$

For low translational temperatures, we can approximate Eq. (8.40) by writing

$$\chi'(\Omega, q) \simeq \frac{g}{\hbar}\frac{M}{2\pi q}W_0\left[\frac{1}{\Omega_+} - \frac{1}{\Omega_-}\right], \tag{8.42}$$

where $W_0 = \int W_0(k_z)dk_z/2\pi$. Equating to zero the quantity $1 + \chi'$, we are led to the dispersion relation

$$1 + \frac{gW_0}{M}\frac{q^2}{\Omega^2 - (\hbar q^2/2M)^2} = 0, \tag{8.43}$$

which corresponds to Eq. (8.37) with $v_0 = 0$.

On the other hand, the term $\chi''(\Omega, q)$ will lead to wave attenuation. In order to study this wave attenuation we introduce $\Omega = \Omega_r + i\gamma$, where wave damping is determined by

$$\gamma = -\frac{\chi''(\Omega, q)}{(\partial\chi'/\partial\Omega)_{\Omega=\Omega_r}}. \tag{8.44}$$

Performing the calculation, we obtain the following explicit expression for the damping coefficient

$$\gamma = \frac{g^2 W_0}{4\hbar^2}\frac{q}{\Omega}\left[W_0(z=\Omega_+) - W_0(z=\Omega_-)\right], \tag{8.45}$$

where

$$W_0(z=\Omega_\pm) = W_0(v_z = \Omega_\pm/q) = W_0\left(v_z = \Omega/q \pm \frac{\hbar q}{2M}\right). \tag{8.46}$$

Equation (8.45) describes the kinetic non-dissipation wave attenuation known as Landau damping, and already discuss for the case of hybrid waves in the non-condensed gas. We see that, if the lower translactional energy level $\hbar\Omega_-$ is more populated than the higher energy one $\hbar\Omega_+$, we will have $\gamma < 0$ and wave attenuation occurs. However, if there is inversion of population in the kinetic energy states, such that

$$W_0\left(\Omega/q + \frac{\hbar q}{2M}\right) > W_0\left(\Omega/q - \frac{\hbar q}{2M}\right), \tag{8.47}$$

we will have wave growth, or wave instability.

It should also be noticed that Eq. (8.45) takes a form similar to the Landau damping coefficient associated with discrete transitions between two excited states of the condensate, as directly derived from the GP equation [3]. In this case $W_\pm$ will be replaced by the thermal occupations numbers of the two discrete states. Here, however, transitions induced by sound waves occur between two arbitrary states of the translational continuum, associated with the axial motion of the bosons.

Let us then consider the case where the increment $\delta v_z = \hbar q/M$, corresponding to the absorption or emission of a phonon by a cold atom of the condensate is negligible with respect to the sound speed c_s. This is equivalent to take $\hbar q/M \ll \Omega/q$, and corresponds to an infinitesimal atom recoil. In this case, we can develop the Wigner distribution around the value $v_z = \Omega/q$, as

$$W_0\left(\Omega \pm \frac{\hbar q^2}{2M}\right) \simeq W_0(\Omega) \pm \frac{\hbar q^2}{2M}\left(\frac{\partial W_0}{\partial z}\right)_{z=\Omega}. \tag{8.48}$$

Replacing this in the expression for the damping coefficient (8.45) leads to

$$\gamma \simeq \frac{g^2 W_0}{4\hbar^2}\frac{q}{\Omega}\frac{\hbar q^2}{M}\left(\frac{\partial W_0}{\partial z}\right)_{z=\Omega} = \frac{g^2 W_0}{4\hbar^2}\frac{q^2}{\Omega}\left(\frac{\partial W_0}{\partial k_z}\right)_{\text{res}}, \tag{8.49}$$

where the derivative is calculated at the resonant velocity defined by $v_z = \hbar k/M$. Noting that $q^2 \simeq \Omega^2 c_s^2$ and using the definition of the Bogoliubov sound speed, we can also write

$$\gamma \simeq \frac{g\Omega W_0}{4\hbar^2}\left(\frac{\partial W_0}{\partial k_z}\right)_{\text{res}}. \tag{8.50}$$

This quasi-classical expression for the condensate Landau damping was first derived in reference [4]. Inversion of population will lead to wave excitation, as predicted by Eqs. (8.45) and (8.50) for the quantum and classical regimes, in the same way as described for the hybrid sound in the ultra-cold non-condensed gas. Such an instability can eventually decay, due to nonlinear coupling, into other modes, eventually generating a broad turbulent spectrum, similar to that observed in the experiments [5]. A simple situation where instability can occur is that of two counter-streaming condensates, as discussed next.

8.5 Dynamical Instabilities

In this section, we apply the quantum kinetic model to study the instability of Bogoliubov waves in two counter-propagating BECs [6]. Such an instability is similar to the two-stream instability in plasma physics [7–9]. In the two-stream instability problem, the energy of the streams is transferred to the wave, increasing its amplitude in time. We show that the quantum nature of the Bogoliubov waves will decrease the instability growth rate and establishes restrictions to the unstable modes.

To address this problem, we go back to the kinetic dispersion relation (8.34). In the case of a single condensate beam at $T = 0$ propagating along the axial direction, the corresponding equilibrium distribution function is $W_0(k_z) = 2\pi n_0 \delta (k_z - k_0)$, and Eq. (8.34) provides quantum Bogoliubov waves with the dispersion relation $(\Omega - q v_0)^2 = c_s^2 q^2 + \hbar^2 q^4 / 4M^2$, as we have seen. Doppler shifted Bogoliubov waves are stable, since Ω is real. Notice that the classical limit $\hbar \to 0$, is obtained if the resonant phonon energy $\hbar\omega_0$ is much less than the beam kinetic energy $m v_0^2$, where $\omega_0 = c_s k_0$ is the frequency corresponding to a wave mode number equal to the beam wave vector $k_0 = v_0/\hbar M$. In practice, this limit is achieved either if the beam is highly energetic (compared to the sound speed c_s) or if the BEC is dilute.

We now devote our attention to the case of two condensate beams, propagating in opposite directions with the same velocity v_0. This equilibrium configuration can be theoretically represented by the Wigner function $W_0(v) = \pi n_0 \left[\delta(v - v_0) + \delta(v + v_0)\right]$ and its dynamics of is now governed by the following dispersion relation

$$1 - \frac{K^2}{2\beta^2} \left\{ \frac{1}{(\tilde{\Omega} + K)^2 - H^2 K^4} + \frac{1}{(\tilde{\Omega} - K)^2 - H^2 K^4} \right\} = 0, \tag{8.51}$$

where we defined the normalized quantities $\tilde{\Omega} = \Omega/\omega_0$, $K = q v_0/\omega_0$ and introduced two dimensionless parameters, $H = \hbar\omega_0/m v_0^2$ and $\beta = v_0/c_S$. The parameter H is the dimensionless Planck constant $\hbar$ and characterizes the influence of the phonon recoil, measuring the importance of the quantum effects. Hence, for $H \geq 1$, quantum effects are important, while for $H \ll 1$, the dynamics is essentially classical. The quantity β defines a sonic parameter, the Mach number at $T = 0$. It indicates when the beam flow is either subsonic ($\beta < 1$) or supersonic ($\beta > 1$). A similar definition of this parameter can be found in the case of finite temperature systems, where the problem of the analog of Hawking radiation in BEC is addressed [10]. Factorization of the above dispersion relation provides two possible wave branches

$$\tilde{\Omega}^2 = \frac{1}{2\beta^2} \left\{ K^2 + 2\beta^2 K^2 (1 + H^2 K^2) \pm K^2 \sqrt{1 + 8\beta^2 + 16\beta^4 H^2 K^2)} \right\}, \tag{8.52}$$

one of which ($\tilde{\Omega}_+^2$) is always positive and describes stable oscillations. However, the solution $\tilde{\Omega}_-^2$ is not positive defined and negative solutions are found for the modes K that verify the condition

$$1 + 2\beta^2 (1 + H^2 K^2) \leq (1 + 8\beta^2 + 16\beta^4 H^2 K^2)^{1/2}. \tag{8.53}$$

This condition defines two different unstable regions in the (K, H) plan, as illustrated in Fig. 8.1. The two-stream instability regions are distinguished by the

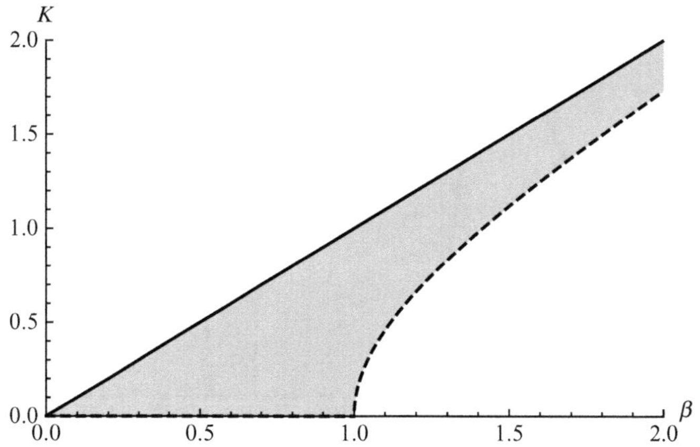

Fig. 8.1 Stability diagram for the two-stream BEC. The shadowed are is unstable. The *upper curve* corresponds to $KH = 1$. The *dashed lower curve* corresponds to the case $K = (\beta^2 - 1/H^2\beta^2)^{1/2}$, for $\beta = 1.01$

Mach number β: in the subsonic regime, $\beta < 1$, all the modes satisfying $K \leq 1/H$ are unstable; in the supersonic regime, $\beta > 1$, unstable modes are obtained for

$$\left[\frac{(\beta^2 - 1)}{H^2\beta^2}\right]^{1/2} \leq K \leq \frac{1}{H}. \tag{8.54}$$

We split the frequency into its real and imaginary part, $\Omega = \Omega_r + i\Gamma$, and define the dimensionless growth rate $\tilde{\Gamma} = \Gamma/\omega_0 = \mathrm{Im}(\tilde{\Omega})$. The stability of the oscillations is determined in terms of the sign of $\tilde{\Gamma}$: if $\tilde{\Gamma} < 0$, the waves are damped and the energy is transferred from the waves to the condensate; if $\tilde{\Gamma} > 0$, instability occurs. Numerical inspection of $\tilde{\Gamma}(K, H)$ shows that the oscillations are always unstable, for both supersonic and subsonic regimes.

The Mach number plays an important role in the nature of the instability. In the subsonic regime, all wave modes K are unstable, up to the cut-off given by the condition $KH = 1$. This means that every mode q is unstable until it reaches the cut-off value $q_c = Mv_0/\hbar$, i.e., until the recoil phonon possesses the same energy of the beam atoms. In that case, all the wave energy is transferred to the BEC, and the instability vanishes.

In the supersonic regime, $\beta > 1$, the picture changes. First, to become unstable, the phonon must have a momentum higher than the resonant value $q_{\mathrm{res}} = \gamma^{-1}Mc_S/\hbar$, where $\gamma = [(v_0/c_S)^2 - 1]^{-1/2}$. It means that the phonon energy must be resonant with that of the supersonic beam atoms, so energy can be transferred from the BEC to the wave. The cut-off limit has the same meaning and has exactly the same value of the subsonic regime.

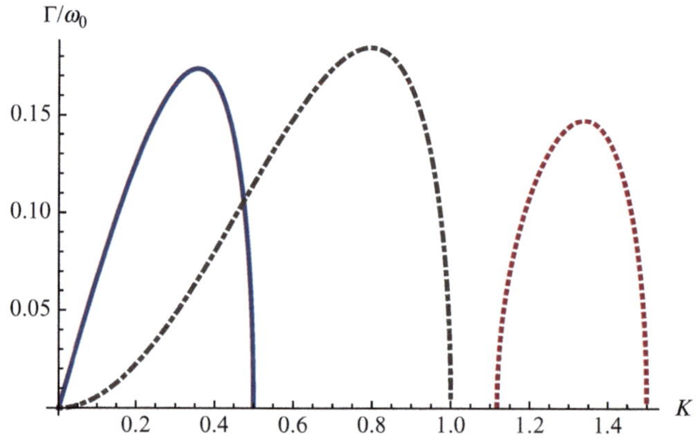

Fig. 8.2 Normalized wave growth rate $\tilde{\Gamma} = \mathrm{Im}(\tilde{\Omega})/\omega_0$ for different values of the parameters β and H. *Blue lines* ($\beta = 0.5$), *black lines* ($\beta = 1.0$) and *red lines* ($\beta = 1.5$). *Full lines* ($H = 0.5$), *dashed lines* ($H = 1.0$) and *dotted lines* ($H = 1.5$). The maximum growth rate, corresponding to the most unstable mode, occurs for the lowest values of both β and H

Generally speaking, an increase of H reduces the value of the cut-off mode q_c and the value of the maximum growth rate, associated with the most unstable mode. In the deeply quantum limit, $H \to \infty$, no instability is expected to occur and, in particular, in the supersonic regime, we have $q_c \approx q_{res} \to 0$. This means that the quantum effects tend to prevent instability. In the classical case $H \to 0$, we get $q_{res} \approx q_c \to \infty$. For a finite value of H, we find the behavior $q_{res} \approx q_c \to 0$, in the infrasonic limit $\beta \ll 1$, and $q_{res} \approx q_c = M v_0/\hbar$ in the ultrasonic limit $\beta \gg 1$, which shows that only the resonant mode becomes unstable for highly energetic beams. We then observe that the Mach number does not enhances the instability, but it rather changes the value of the resonant mode q_{res} (Fig. 8.2).

The application of a quantum kinetic equation to address the problem of unstable Bogoliubov waves in two counter-propagating elongated BECs revealed important features of the system. First, phonon recoil processes play an important role in the stability criteria of the two counter-streaming condensates. Second, the resonance condition is different in the subsonic and supersonic limits. Finally, the quantum recoil effects prevent the instability to occur, bringing the cut-off mode to zero, and decreasing the value of the maximum growth rate.

8.6 Wakefields in Bose-Einstein Condensates

In what follows, we consider a different kind of instability, leading to the excitation of a wakefield by BEC moving along a background thermal gas [11]. The non-condensate gas will be described by fluid equations, as used in previous work on trapped BE gas at finite temperature [12, 13]

For the present purpose, we describe the dynamics of the condensate with the kinetic equation (7.83) in the quasi-classical limit

$$\left(\frac{\partial}{\partial t} + \mathbf{v} \cdot \nabla + \mathbf{F} \cdot \frac{\partial}{\partial \mathbf{p}}\right) W(\mathbf{r}, \mathbf{p}, t) = 0, \tag{8.55}$$

where $\mathbf{F} = -\nabla(V_0 + V_{\mathrm{NL}})$ is the force acting on the BEC. Such a description is very well suited for the present case as we are not interested in describing the quantum features of the condensate itself, rather its interaction with the thermal gas. In order to determine the density of the background thermal gas $n(\mathbf{r}, t)$, we can use the corresponding fluid equations, including an additional force term that is associated with the presence of the condensate. These equations, derived by Zaremba, Nikuni and Griffin (ZNG) [12], pretty much in the spirit of the two-fluid model by Landau [14] and Khalatnikov [15], relate the velocity $\mathbf{v}_n$ and the pressure P of the thermal gas within the self-consistent Hartree-Fock approximation. They can thus be coupled to the above kinetic equation describing the evolution of the BEC. We can then assume that the relative motion of the condensate and the thermal gas will produce density perturbations $\tilde{N}$ and $\tilde{n}$ around their respective equilibrium values N_0 and n_0. Similarly, we take the pressure as $P = P_0 + \tilde{P}$. The gas background is assumed initially at rest, which means that the equilibrium value of its velocity $\mathbf{v}_n$ is equal to zero. The fluid equations for the non-condensate gas can be written, in linearized form, as

$$\frac{\partial \tilde{n}}{\partial t} + \nabla \cdot (n_0 \mathbf{v}_n) = 0, \qquad \frac{\partial \mathbf{v}_n}{\partial t} = -\frac{1}{M n_0} \nabla \tilde{P} + \mathbf{F}_n, \tag{8.56}$$

$$\frac{\partial \tilde{P}}{\partial t} + \frac{5}{3} \nabla \cdot (P_0 \mathbf{v}_n) = \frac{2}{3} \mathbf{v}_n \cdot \nabla P_0, \tag{8.57}$$

with the force per unti mass $\mathbf{F}_n$ defined as $\mathbf{F}_n = -(1/M)\nabla[V_0 + 2g(\tilde{n} + N_0 + \tilde{N})]$. In order to simplify the calculations and to concentrate on our main purpose here, we neglect the gradients of n_0, P_0 and V_0, which can easily be included in the solutions and will not change their physical features. However, we will retain the gradient of N_0 for reasons that will become apparent in the following discussion. We can then recognize that $\partial \tilde{P}/\partial t = (5/3)(P_0/n_0)\partial \tilde{n}/\partial t$, and the above equations can be reduced to the following wave equation for the density perturbations driven by the condensate motion in the thermal background

$$\left(\frac{\partial^2}{\partial t^2} - u_s^2 \nabla^2\right) \tilde{n} = 2n_0 \frac{g}{M} \nabla^2 (N_0 + \tilde{N}), \tag{8.58}$$

where we have used the following expression for the sound speed u_s in the thermal gas: $u_s^2 = 5P_0/3Mn_0 + 2gn/M$. This is equivalent to Eq. (65) of reference [13], if we neglect the condensate contribution. Such a contribution is here included in the term containing $\tilde{N}$, as it will become apparent below. Equation (8.58) is our basic equation for the description of the wakefield phenomena.

Let us first consider the case of a very short condensate beam moving with velocity $\mathbf{u}_0 = u_0\mathbf{e}_x$ across the non-condensed background gas. In such conditions, the gradient of N_0 in Eq. (8.58) is larger then that of $\tilde{N}$. We can then use $\tilde{N} = 0$, and we are left with a driving force of the acoustic oscillations that is determined by $\nabla^2 N_0$. We retain the possible existence of a standing wave structure of the sound waves in the perpendicular direction. The perpendicular part of the ∇^2 operator will then lead to the appearance of a cut-off frequency ω_0, which will be determined by the transverse wave structure. In order to study the evolution of the perturbation in the parallel direction, we can thus add a term $\omega_0^2\tilde{n}$ to the left hand side of Eq. (8.58), while replacing ∇^2 by $\partial^2/\partial x^2$. Neglecting the distortion of the beam envelope, on the time scale of the instability growth rate, we can also describe the condensate beam by a function of the type $N_0(x - u_0 t)$. The evolution Eq. (8.58) will then be transformed into

$$\left[\frac{\partial^2}{\partial\tau^2} + (u_0^2 - u_s^2)\frac{\partial^2}{\partial\xi^2} - 2u_0u_s\frac{\partial^2}{\partial\xi\partial\tau} + \omega_0^2\right]\tilde{n} = 2n_0\frac{g}{M}\frac{\partial^2}{\partial\xi^2}N_0(\xi), \tag{8.59}$$

where we have used a coordinate transformation from the variables (x, t) to (ξ, τ), such that $\xi = x - u_0 t$ and $\tau = t$. The interest of this transformation is obvious. The time variable τ describes now the variations occurring in the frame of the moving condensate. It is useful to take here the quasi-static approximation, $\partial/\partial\tau \simeq 0$. This is appropriate to describe stationary wakefield solutions. In this approximation Eq. (8.59) reduces to

$$\frac{\partial^2\tilde{n}}{\partial\xi^2} + k_0^2\tilde{n} = \alpha_0\frac{\partial^2 N_0(\xi)}{\partial\xi^2}, \tag{8.60}$$

where we have introduced the characteristic wavenumber k_0, defined by $k_0^2 = \omega_0^2/(u_0^2 - u_s^2)$, and used the auxiliary quantity $\alpha_0 = 2n_0 g/M(u_0^2 - u_s^2)$. Let us now use the dimensionless space variable $\eta = k_0\xi$. We get then the equation of a driven harmonic oscillator with unit frequency $(d^2/d\eta^2 + 1)\tilde{n} = f(\eta)$, where the driving force term is determined by $f(\eta) = \alpha_0\partial^2 N_0/\partial\eta^2$. The particular solution of this equation adequate to the wakefield problem (such that, at a given position x no oscillations of the thermal gas can exist before arrival of the condensate) can be written in the form

$$\tilde{n}(\eta) = \int_\infty^\eta f(\eta')\sin(\eta - \eta')d\eta'. \tag{8.61}$$

Noting that $df/d\eta = 0$ for $\eta \to \infty$, and integrating by parts twice, we can write this particular solution in its final form

$$\tilde{n}(\eta) = 2n_0\frac{g}{m}\frac{1}{(u_0^2 - u_s^2)}\left[N_0(\eta) - \int_\infty^\eta N_0(\eta')\sin(\eta - \eta')d\eta'\right]. \tag{8.62}$$

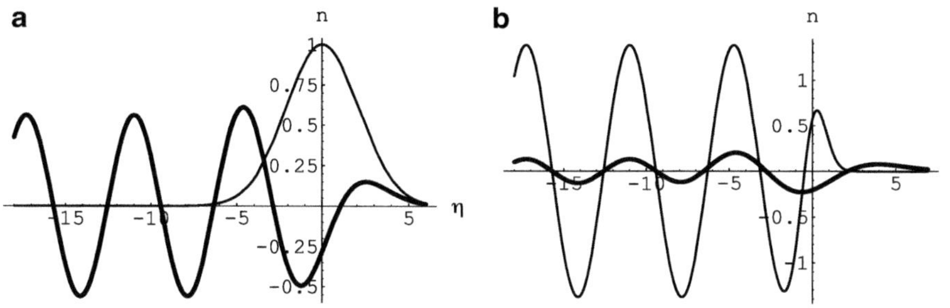

Fig. 8.3 Wakefield excitation: (**a**) Wakefield oscillation (*bold curve*) excited by a condensate beam with a Gaussian shape $N_0(\eta) = \exp[-(\eta/\sigma)^2]$ (*thin curve*), for $\sigma = 3$; (**b**) Two cases of wakefield excitation, for two different widths of the Gaussian beam, $\sigma = 4$ (*bold curve*) and $\sigma = 1$ (*thin curve*)

This solution contains two terms with distinct physical meanings. The first one is just a local perturbation of the background gas, occurring where the condensate is located at a given time. The second term is much more interesting physically, because it corresponds to an acoustic wake, left behind the condensate beam and remaining long after its passage. See Fig. 8.3 for illustration. It can easily be realized that this wakefield is in fact an acoustic wave propagating in the background gas with a phase velocity equal to the translational velocity of the condensate beam. So, by using the linear dispersion relation of the acoustic waves with frequency ω and parallel wavenumber k, as stated by $\omega^2 = \omega_0^2 + k^2 u_s^2$, we can determine the frequency of the wake in the laboratory frame as $\omega = \omega_0 (u_0^2 - u_s^2)^{1/2}/u_0$.

This shows that the wakefields can only be generated if the acoustic wave has some transverse structure, which means $\omega_0 \neq 0$, and also if the condensate moves faster than the sound speed, $u_0 > u_s$. Otherwise, the frequency ω will be zero or imaginary and the second term in the solution (8.62) will not exist. We also notice that this solution will diverge when the resonant condition is attained, $u_0 \to u_s$. This means that, in this case, the system will be unstable to infinitesimal perturbations and that, in the wakefield equation (6) we cannot neglect the perturbation of the condensate $\tilde{N}$ anymore.

In order to discuss the stability of the condensate in these resonant conditions we go back to Eq. (8.58) but neglect the driving term in N_0, which was considered before but is irrelevant for the present stability analysis. In fact, even in the absence of a driving term, the system will eventually be unstable to infinitesimal perturbations. In order to solve this equation we have to establish a relation of $\tilde{N}$ with respect to $\tilde{n}$. This can be done by linearizing the kinetic equation (8.55) with respect to the perturbations, which leads to

$$\left(\frac{\partial}{\partial t} + \mathbf{v} \cdot \nabla \right) \tilde{W} = g \nabla (\tilde{N} + 2\tilde{n}) \cdot \frac{\partial W_0}{\partial \mathbf{p}}. \tag{8.63}$$

Here, W_0 and $\tilde{W}$ are the equilibrium and the perturbation of the Wigner function which, after integration in $\mathbf{p}$ give the quantities N_0 and $\tilde{N}$ respectively. We then have two coupled equations for the three perturbed quantities $\tilde{n}$, $\tilde{W}$ and $\tilde{N}$. If we assume that these perturbations evolve in space and time as $\exp(i\mathbf{k}\cdot\mathbf{r} - i\omega t)$, we obtain

$$
\begin{aligned}
(\omega^2 - u_s^2 k^2)\tilde{n} &= 2n_0 \frac{g}{m} k^2 \tilde{N} \\
(\omega - \mathbf{k}\cdot\mathbf{v})\tilde{W} &= -g(\tilde{N} + 2\tilde{n})\mathbf{k}\cdot\frac{\partial W_0}{\partial \mathbf{p}}.
\end{aligned} \tag{8.64}
$$

It can easily be seen that, by integration of this last equation over the momentum space $\mathbf{p}$, we obtain $\tilde{N} = -G(\tilde{N} + 2\tilde{n})$, or $\tilde{N} = -2G/(1+G)\,\tilde{n}$, where G represents the integral

$$
G = g \int \frac{\mathbf{k}\cdot\partial W_0/\partial\mathbf{p}}{(\omega - \mathbf{k}\cdot\mathbf{p}/M)} \frac{d\mathbf{p}}{(2\pi\hbar)^3}. \tag{8.65}
$$

This leads to the following kinetic dispersion relation of the system "condensate coupled with the thermal gas background"

$$
(\omega^2 - u_s^2 k^2) = -4n_0 (g/m) k^2 G/(1+G). \tag{8.66}
$$

On the left hand side of this equation we recognize the dispersion relation of the sound waves in the thermal gas alone, $\omega = u_s k$, and on the right hand side we have the contribution of the condensate to these oscillations. This will lead to a change in the sound speed and, more significantly, to an instability of the sound oscillations. On the other hand, in the absence of a non-condensate phase, this kinetic approach would lead to a simple description of Bogoliubov oscillations of the condensate, where a kinetic Landau damping is included.

To simplify, we will limit our discussion to the case $\tilde{N} \ll 2\tilde{n}$, for which we can neglect G in the denominator. For a nearly mono-kinetic condensate beam, such that $W_0(\mathbf{p}) = (2\pi\hbar)^3 N_0 \delta(\mathbf{p} - \mathbf{p}_0)$, we can easily derive the dispersion relation for wave propagating in the parallel direction. This dispersion relation takes the form

$$
(\omega^2 - u_s^2 k^2)(\omega - k u_0)^2 = \frac{4n_0}{N_0}(k c_s)^4. \tag{8.67}
$$

Here, we have used the BEC beam velocity $u_0 = p_0/M$, and the Bogoliubov sound speed in the condensate gas, as defined by $c_s = \sqrt{g N_0/M}$.

Notice that, in the absence of coupling between the two types of gas, the above expression will lead two separate dispersion relation, $\omega = k u_s$ for sound waves in the thermal gas, and $\omega - k u_0 = k c_s$, for Doppler shifted sound waves in the condensate. The coupling between the two fluids, resulting from the expression of G, will imply the occurrence of beam-like instabilities. To illustrate this important question, let us consider the resonant condition $u_s = u_0$. The instability growth rates can be obtained from the previous dispersion relation by considering $\omega = \omega_r + \eta$, with $\omega_r = k u_0 = k u_s$, which then leads to a growth rate $\gamma = \Im(\eta)$ given by

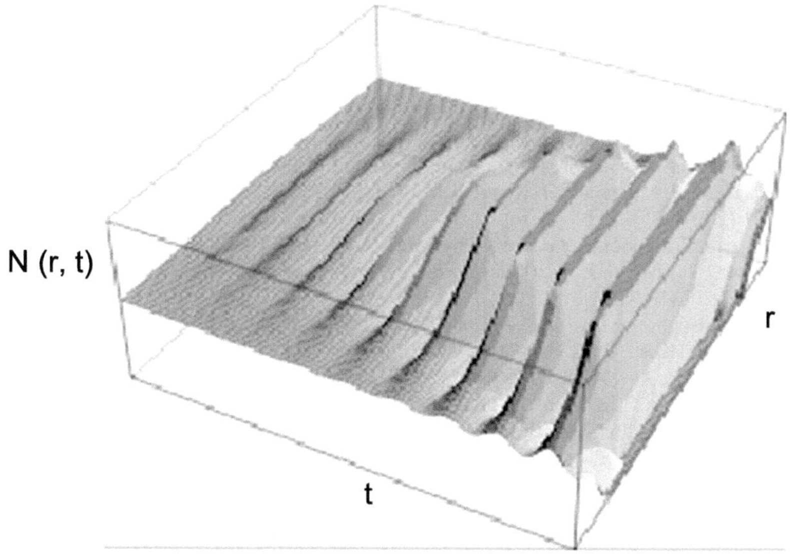

Fig. 8.4 Modulational instability of a condensate moving in a background non-condensate gas, with a velocity equal to the sound speed of the non-condensate gas, $u_0 = u_s$, as determined by equations. Equation (8.63) for a transverse Gaussian density profile

$$\gamma = \sqrt{3}\left(\frac{n_0}{4N_0}\right)^{1/3}\left(\frac{c_s}{u_0}\right)^{4/3}\omega_r . \tag{8.68}$$

Let us now compare this case of a cold condensate beam with that of a beam with a small translational velocity spread around the mean value $\mathbf{u}_0$. Splitting the integral of G in its principal part and its singular part, we can establish the following kinetic dispersion relation

$$(\omega^2 - u_s^2k^2) \simeq 4\frac{n_0}{N_0}\frac{(kc_s)^4}{\omega^2} + i4\pi n_0\frac{g^2k^2}{M}\left(\frac{\partial I_0}{\partial p}\right)_{\text{res}} , \tag{8.69}$$

where the derivative is calculated at the resonant value of the momentum such that $p = \omega m/k$, and we have used the axial beam distribution as determined $I_0 = \int W_0 d\mathbf{p}_\perp/(2\pi\hbar)^2$. Using $\omega = \omega_r + i\gamma$ we obtain $\omega_r \simeq ku_s$. For a Gaussian beam, of the form $I_0(p) = I_0 \exp[-(p - p_0)^2/\sigma^2]$, we will have a maximum value for the growth rate given by $\gamma_{\max} = 2\sqrt{2}n_0g^2k^2e^{-1/2}I_0/\sigma m\omega_r$. This kinetic growth rate can be seen as a negative Landau damping of sound waves by a resonant condensate beam. These two instability regimes (corresponding to a cold and a warm beam) confirm the previous analysis of the wakefield excitation, where a singularity occurring at the resonant conditions $u_s = u_0$ suggested the possible excitation of growing waves from infinitesimal perturbations. This is illustrated in Fig. 8.4, where the modulational instability of a condensate with a transverse Gaussian profile is represented.

References

1. S. Stringari, Phys. Rev. A **77**, 2360 (1996)
2. J.T. Mendonça, A. Serbeto, P.K. Shukla, Phys. Lett. A **372**, 2311 (2008)
3. M. Guilleuma, L.P. Pitaevskii, Phys. Rev. A **67**, 053607 (2003)
4. J.T. Mendonça, P.K. Shukla, R. Bingham, Phys. Lett. A **340**, 355 (2005)
5. E.A.L. Henn, J.A. Seman, G. Roati, K.M.F. Magalhães, V.S. Bagnato, Phys. Rev. Lett. **103**, 045301 (2009)
6. H. Terças, J.T. Mendonça, G. Robb, Phys. Rev. A **79**, 065601 (2009)
7. F.F. Chen, *Introduction to Plasma Physics and Controlled Fusion*, 2nd edn. (Plenum, New York, 1984)
8. D.R. Nicholson, *Introduction to Plasma Theory* (Wiley, New York, 1983)
9. H. Saleem, R. Khan, Phys. Scr. **71**, 314 (2005)
10. S. Wüster, Phys. Rev. A **78**, 021601(R) (2008)
11. J.T. Mendonça, R. Bingham, P.K. Shukla, J. Exp. Theor. Phys. (JETP) **201**, 942 (2005)
12. E. Zaremba, A. Griffin, T. Nikuni, Phys. Rev. A **57**, 4695 (1998)
13. T. Nikuni, A. Griffin, Phys. Rev. A **69**, 023604 (2004)
14. L. Landau, The theory of superfluidity of Helium II. J. Phys. USSR **5**, 71 (1941)
15. I.M. Khalatnikov, *An Introduction to the Theory of Superfluidity* (Westview, New York, 2000)

Chapter 9
Solitons

In this chapter, we discuss the main properties of the solitary structures in Bose-Einstein condensates. As we are about to see, they occur as a natural consequence of the description of the condensate, at the mean-field level, in terms of the Gross-Pitaevskii equation, which is nonlinear in the condensate wave function ϕ. In general terms, a soliton is an "envelope", very often being bell-shaped, which propagates with very little dispersion. They usually arise as a unique solution of a widespread class of nonlinear field equations, where the effects of the dispersion balance with the nonlinearity of the system. According to the accepted historical facts, the soliton phenomenon was first observed by by John Scott Russell, who observed a solitary wave in the Union Canal in Scotland. He reproduced the phenomenon in a wave tank and named it the "wave of translation" [1].

The first theoretical description of solitary waves was due to Kortweg and de Vries back in the nineteenth-century [2], who derived an equation for the shallow-water surface $\phi(x,t)$

$$\frac{\partial \phi}{\partial t} + \frac{\partial^3 \phi}{\partial x^3} + 6\phi \frac{\partial \phi}{\partial x} = 0, \tag{9.1}$$

which was originally obtained from the Navier-Stokes equation by applying the so-called reductive perturbation method [3]. The latter is known under the name of Kortweg-de Vries (KdV) equation and the corresponding solitary soliton is given by

$$\phi(x,t) = \frac{1}{2} v \text{sech}^2 \left[\frac{\sqrt{v}}{2}(x - vt + \varphi) \right], \tag{9.2}$$

where v is the soliton (dimensionless) velocity and φ is an arbitrary phase. This solution corresponds to a bump of density which propagates without deformation and is often referred to as *bright soliton* (see Fig. 9.1).

In the case of Bose-Einstein condensates, solitary waves can also be excited. However, the corresponding solution significantly differs from that of a bright soliton in (9.2) and often corresponds to a 'deep' propagating along the background

J.T. Mendonça and H. Terças, *Physics of Ultra-Cold Matter*, Springer Series on Atomic, Optical, and Plasma Physics 70, DOI 10.1007/978-1-4614-5413-7_9,

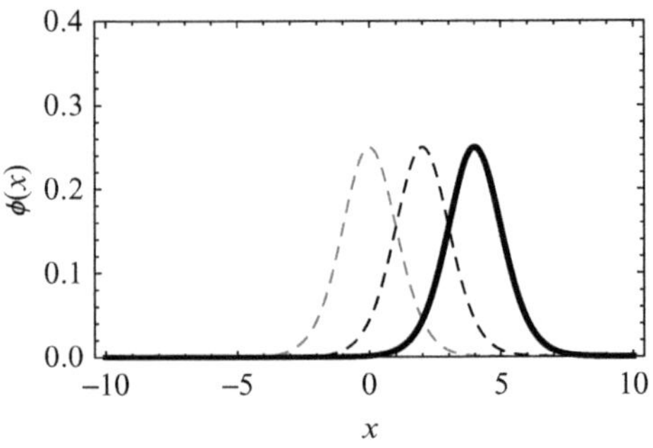

Fig. 9.1 Solitary solution (9.2) propagating from *left* to the *right* ($v > 0$)

density. For that reason, they are classified as *dark solitons*. In fact, dark solitons are the most fundamental excitations of a very general nonlinear model, the nonlinear Schrödinger equation (NSL), where the nonlinearity is associated to a defocusing term. For that reason, they have been extensively studied, both theoretically and experimentally, in a large variety of systems and can be found in different physical forms, such as temporal pulses in optical fibers [4, 5], spatial structures in waveguides [6], non-propagating kinks ins in parametrically-driven shallow fluids [7], standing waves in mechanical systems [8] and dissipative nonlinear waves in complex plasmas [9].

In the next sections, we present a summarized review of the mains aspects of dark-solitons in Bose-Einstein condensates. We start with the one-dimensional case and then enclose a short discussion on their dynamical properties in higher dimensions. A detailed and comprehensive review on the main properties of dark-solitons can be found in Ref. [10].

9.1 Effective One-Dimensional Gross-Pitaevskii Equation

First of all, it should be noticed that Bose-Einstein condensation of an ideal gas can not occur in $d \leq 2$, since the number of the particles in the excited states (i.e. above the condensate) does not converge. However low-dimensions BECs are, possible in trapped systems. As a consequence, the effective interaction must differ from the three-dimensional case, where the later is characterized (at the mean-field level) by the coupling strength g. Let us again consider the condensate confined in a asymmetric harmonic trap

$$V_{\text{trap}}(x, y, z) = \frac{1}{2} M \left(\omega_x^2 x^2 + \omega_y^2 y^2 + \omega_z^2 z^2 \right), \tag{9.3}$$

where ω_j ($j = 1, 2, 3$) represent the trap frequencies along the three directions. This sets the characteristic length scales through the hamornic oscillatior lengths $a_j = (\hbar/M\omega_j)^{1/2}$, thus generalizing the spherical case in Eq. (7.14). Another important length scale is associated with the so-called *healing length*, defined as $\xi = \hbar/\sqrt{2Mgn}$ (where N/V is the particle number density). The latter defines the scale over which the condensate wavefunction "heals" over defects and is thus connected with the width of the dark solitons.

As we can see through the Thomas-Fermi profile (7.64), it is clear that the shape of the condensate is controlled by the relative value of the trap frequencies. For example, the isotropic trap, $\omega_x = \omega_y = \omega_z \equiv \omega_0$, corresponds to a spherical BEC, while for the anisotropic case $\omega_\perp > \omega_z$, where $\omega_\perp = \sqrt{\omega_x^2 + \omega_z^2}$, the condensates is cigar-shaped. Such an elongated condensate is, in general, a three-dimensional (3D) object, but can acquire an quasi- 1D character for strong transverse confinement, $\omega_\perp \gg \omega_z$. This dimensionality 3D-1D crossover can be described by the following dimensionless parameter [11]

$$d = \lambda \frac{Na}{a_\perp}, \tag{9.4}$$

where $\lambda = \omega_z/\omega_\perp$ represents the *aspect ratio* of the trap. In particular, if $d \gg 1$, the condensate may still be regarded as a 3D object and the corresponding ground-state is given by the Thomas-Fermi profile

$$n_\perp(r) = n_0\left(1 - \frac{r^2}{R_\perp^2}\right), \quad n_z(z) = n_0\left(1 - \frac{z^2}{L^2}\right), \tag{9.5}$$

where $R_\perp = [2\mu/M\omega_\perp^2]^{1/2}$ and $Z = [2\mu/M\omega_\perp^2]^{1/2}$ are the Thomas-Fermi lengths. On the other hand, if $d \ll 1$, excited states along the transverse direction are no longer energetically accessible and the system is effectively 1D. Finally, for $d \simeq 1$, the BEC is in the crossover regime, for which some recent experimental activity has been performed [12].

For very elongated, quasi-1D condensates (corresponding to $d \ll 1$), the wave function can be factorized into its transverse and longitudinal components [13]

$$\Phi(\mathbf{r}; t) = \varphi(r, t)\psi(z, t). \tag{9.6}$$

In that case, the transverse state can not be described in terms of a Thomas-Fermi approximation, and rather corresponds to the Gaussian ground state of the harmonic oscillator, such that the transverse radius equals the transverse oscillator length, $R_\perp \simeq a_\perp$. On the other hand, the longitudinal component is described by the 1D Gross-Pitaevskii equation

$$i\hbar\frac{\partial\psi(z,t)}{\partial t} = \left\{-\frac{\hbar^2}{2M}\frac{\partial^2}{\partial z^2} + V(z) + g_{1D}|\psi(z,t)|^2\right\}\psi(z,t), \tag{9.7}$$

where $V(z) = M\omega_z^2 z^2/2$ and g_{1D} is the renormalized interaction parameter

$$g_{1D} = \frac{g}{2\pi a_\perp} = 2a_s\hbar\omega_\perp. \tag{9.8}$$

We should remark at this point that the factorization performed in (9.6) is very restrictive. In general, there are deviations from the pure 1D-case and the factorization must then correspond to an 'effective' decomposition, whereas the transversal component of the condensate is also a function of z, i.e. $\varphi = \varphi(r,z;t)$. Several decomposition schemes can be found in the literature [14–16]. By means of variational methods, the transverse wavefunction $\varphi(r,z,t)$ can still be described by a Gaussian ansatz, where the transverse radius $R_\perp(z)$ is treated as a variational quantity. These variational schemes usually lead to the following effective 1D nonlinear Schrödinger (NLS) equation

$$i\hbar\frac{\partial\psi(z,t)}{\partial t} = \left\{-\frac{\hbar^2}{2M}\frac{\partial^2}{\partial z^2} + V(z) + f\left[|\psi(z,t)|^2\right]\right\}\psi(z,t). \tag{9.9}$$

The details about the functional dependence of the nonlinear function $f[n]$ are out of the scope of the present discussion, but we remark that they strongly depend on the variational scheme that is employed. In particular, following Ref. [15], where the variation of the action functional is used, $f[n]$ is given by

$$f[n] = \frac{g}{2\pi a_\perp^2}\frac{n}{\sqrt{1+2na_s}} + \frac{\hbar\omega_\perp}{2}\left(\frac{1}{\sqrt{1+2na_s}} + \sqrt{1+2na_s}\right), \tag{9.10}$$

and the respective NLS equation is known in the literature as the non-polynomial Schödinger equation. If the method of Ref. [16] is considered instead, the latter would simply reduce to $f[n] = \hbar\omega_\perp\sqrt{1+4na_s}$. It should be noticed, however, that both expressions lead to the same result in the weakly-interacting limit $na_s \ll 1$. In that case, the description based on Eq. (9.7) again holds, as $R_\perp \to a_\perp$. Another interesting case corresponds to the limiting situation of impenetrable bosons [17], the so-called *Tonks-Girardeau gas*, where $f[n] = \pi^2\hbar^2 n^2/2M$.

9.2 One-Dimensional Dark and Grey Solitons

We now restrict the discussion to the solitary solutions of Eq. (9.7), which can therefore be treated as one-dimensional solitons. In what follows, we consider the case of a very elongated homogeneous BEC, so that $V(z)$ can be ignored. Let us consider the case of a soliton traveling with speed v along the condensate. In that case, we should parametrize it as follows

$$\psi(z,t) = \psi(z)e^{i\theta} = \sqrt{n_0}f(z-vt)e^{-i\mu t/\hbar}. \tag{9.11}$$

Inserting the latter in Eq. (9.7), we obtain the following equation for f

$$-\frac{\hbar^2}{2M}\frac{d^2 f}{d\zeta^2} + gn_0(1-|f|^2)f + i\hbar v\frac{df}{d\zeta} = 0, \tag{9.12}$$

where $\zeta = z - vt$ is the Euler variable and using the fact that homogeneous background density ($f = 1$) imposes $\mu = gn_0$. Separating f into its real and imaginary parts, $f(\zeta) = f_1(\zeta)+if_2(\zeta)$, with the real function f_1 and f_2 obeying the normalization condition $f_1^2(\pm\infty) + f_2^2(\pm\infty) = 1$, Eq. (9.12) may be rewritten as

$$f_1'' + (1 - f_1^2 - f_2^2)f_1 + \sqrt{2}\beta f_2' = 0, \quad (1 - f_1^2 - f_2^2)f_2 - \sqrt{2}\beta f_1' = 0, \tag{9.13}$$

where differentiation was taken in order to $\tilde{\zeta} = \zeta/\xi$ and $\beta = v/c_s$. Let us try a solution for which the imaginary part is constant, such that $f_2 = f_0$. In that case, the latter equations read

$$f_1'' + (1 - f_1^2 - f_2^2)f_1 = 0 \tag{9.14}$$

and

$$(1 - f_1^2 - f_0^2)f_0 = \sqrt{2}\beta f_1'. \tag{9.15}$$

Multiplying equation (9.14) by f_1' and integrating, we obtain

$$2(f_1')^2 = (1 - f_1^2 - f_0^2)^2, \tag{9.16}$$

where the boundary condition $f_1^2 + f_0^2 = 1$ is used to rule out the integration constant. We immediately notice that this result is consistent with Eq. (9.15) only if $f_0 = \beta$. In that case, the solution to Eq. (9.16) is

$$f_1(\tilde{\xi}) = \sqrt{1-\beta^2}\tanh\left[\frac{\sqrt{1-\beta^2}\tilde{\zeta}}{\sqrt{2}}\right]. \tag{9.17}$$

The traveling soliton solution can then be given in terms of the initial parameters

$$\psi(z,t) = \sqrt{n_0}\left[i\frac{v}{c_s} + \sqrt{1-\frac{v^2}{c_s^2}}\tanh\left(\frac{z-vt}{\sqrt{2}\xi}\right)\right]e^{-i\mu t/\hbar}, \tag{9.18}$$

associated with the density profile

$$n(z,t)=\psi(z,t)^*\psi(z,t)=n_0(f_1^2+f_0^2)=n_0\left[\frac{u^2}{c_s^2}+\left(1-\frac{u^2}{c_s^2}\right)\tanh^2\left(\frac{z-vt}{\sqrt{2}\alpha}\right)\right], \tag{9.19}$$

where $\alpha = (1 - \beta^2)^{-1/2}\xi$ represents the soliton width. A more compact and convenient way of writing the soliton density is

$$n_{\rm ds} = n_0 - (n_0 - n_{\rm min})\mathrm{sech}^2\left[\frac{z-vt}{\sqrt{2}\alpha}\right], \tag{9.20}$$

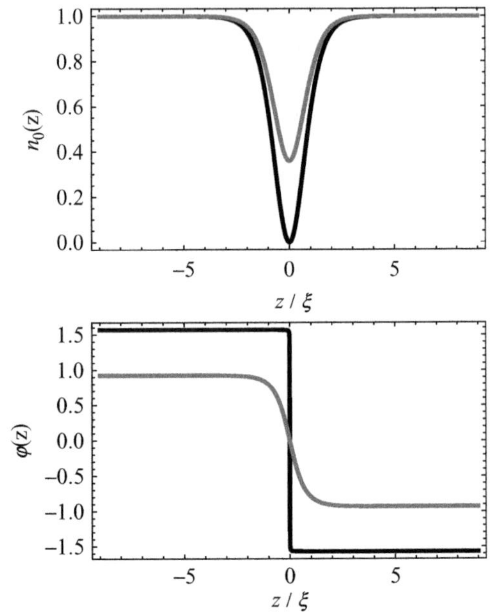

Fig. 9.2 *Dark* ($v = 0$) and *grey* ($v = 0.3c_s$) solitary solutions obtained from (9.21) (*left panel*) and (9.22) (*right panel*)

where

$$n_{\min} = n_0 \frac{v^2}{c_s^2}. \tag{9.21}$$

The later solution is known in the literature as a *dark soliton*. The reason for that name is very evident: for $v = 0$, the minimum density (the 'deep' in the density) reaches zero. So, the slowest the soliton, the deepest is the perturbation. This feature is totally contrary to the case of bright solitons described by KdV equation (9.2), for which the faster the speed, the highest is the perturbation 'hill'. For consistency, solutions to (9.7) in the intermediate regime $0 \ll v \lesssim c_s$ are usually referred to as *grey solitons* (see Fig. 9.2a)). The phase of the condensate corresponds to the argument of the wavefunction,

$$\varphi(z,t) = \arctan\left(\frac{f_0}{f_1}\right). \tag{9.22}$$

The total phase shift $\Delta\varphi = \varphi(+\infty) - \varphi(-\infty)$ across the soliton is then given as

$$\Delta\varphi = \arctan\left(\frac{\beta}{\sqrt{1-\beta^2}}\right) - \left[\pi - \arctan\left(\frac{\beta}{\sqrt{1-\beta^2}}\right)\right] = -2\arccos\left(\frac{v}{c_s}\right). \tag{9.23}$$

For a dark soliton moving to the right (i.e. in the positive z direction), the phase change is negative. The reason for that is because a dark soliton is accompanied by a density depression and the fluid velocity associated to it,

$$u_{\rm ds} = \frac{\hbar}{M}\frac{d\varphi}{dz} = v\left(1 - \frac{n_0}{n_{\rm ds}}\right), \tag{9.24}$$

is in the negative direction. The first experimental observations of dark solitons in condensates were due to the National Institute of Standards and Technology (NIST) [18] and to Hannover groups [19], where the solitons where produced with the so-called phase-imprinting method. From Eq. (9.24) it is quite easy to understand how the phase can be used to effectively control the dynamics of the condensate.

9.2.1 Energy of the Soliton

The energy of the soliton can be easily extracted from its definition in terms of the wavefunction (9.18)

$$E_{\rm ds} = \int_{-\infty}^{\infty} dz \left[\frac{\hbar^2}{2M}\frac{d\psi}{dz}\frac{d\psi^*}{dz} + \frac{1}{2}g\left(|\psi|^2 - n_0\right)^2\right], \tag{9.25}$$

which is the result of taking the difference between the grand canonical energies in the presence and in the absence of the soliton [20]. Using the fact that $\int_{-\infty}^{\infty} dx \cosh^{-4}(x) = 4/3$, together with Eq. (9.20), we obtain

$$E_{\rm ds} = \frac{4}{3} n_0 \hbar c_s \left(1 - \frac{v^2}{s^2}\right)^{3/2}. \tag{9.26}$$

As we can see, the energy decreases with the velocity, which simply means that the effective mass of the soliton is negative. In order to estimate it, we Taylor expand the energy in terms of v/c_s, such that

$$E_{\rm ds} \simeq \frac{4}{3} n_0 \hbar c_s - 2 n_0 \hbar c_s \frac{2 n_0 \hbar}{c_s} v^2. \tag{9.27}$$

Rewriting the latter as $E_{\rm ds} = E_0 - M_{\rm ds} v^2/2$, we immediatly conclude that the soliton effective mass is

$$M_{\rm ds} = -\frac{4 n_0 \hbar}{c_s}. \tag{9.28}$$

On the other hand, the momentum carried by the soliton can be determined starting from the canonical relationship $v = \partial E_{\rm ds}/\partial P_c$, which can be easily integrated to yield

$$P_c = \int_0^v \frac{dv'}{v'}\frac{\partial E}{\partial v'} = -2\hbar n_0 \left[\frac{v}{c_s}\sqrt{1-\frac{v^2}{c_s^2}} + \arcsin\left(\frac{v}{c_s}\right)\right]. \tag{9.29}$$

For small values of v, $P_c \simeq -4\hbar n_0 v/c_s$, while for $v \sim c$, the latter results on

$$P_c \simeq \hbar n_0 \left[\frac{4}{3}\left(1-\frac{v^2}{c_s^2}\right)^{3/2} - \pi\right] = -\pi\hbar n_0 + \frac{E_{\rm ds}}{c_s}. \tag{9.30}$$

We immediately notice that the canonical moment P_c, associated with the particle character of the soliton, does not correspond with the 'physical' momentum $P = M\int J_z dz$ related to the fluid current $J_z = n_0 u_{ds}$. In fact, from Eq. (9.24), it is easily given by

$$P = -2\hbar n_0 \frac{v}{c_s}\sqrt{1-\frac{v^2}{c_s^2}} = P_c + \hbar n_0 (\pi - \Delta\varphi). \tag{9.31}$$

The difference between the two momenta P_c and P is due to the fact that the soliton exhibits unusual phase discontinuities at infinity, even if it can still be regarded as a localized structure from the point of view of its density profile (see Ref. [21] for a more detailed discussion on this issue).

9.3 The Inverse Scattering Transform

The dark soliton solution (9.18) can be derived by means of a powerful method of the *Inverse Scattering Transform* (IST). This procedure consists in mapping the Gross-Pitaevskii equation into an auxiliary eigenvalue problem. The IST method is due to Zakharov and Shabat (ZS) [22] in an attempt of developing a spectral representation (roughly speaking, defining a nonlinear Fourier transform) for a wide class of integrable nonlinear equations.

Let us consider the time-independent Gross-Pitaevskii equation (9.7) for the initial wave function $\psi(z, 0)$

$$\lambda^2 \psi(z,0) = \left\{-\frac{\hbar^2}{2M}\frac{\partial^2}{\partial z^2} + g_{1D}|\psi(z,0)|^2\right\}\psi(z,0), \tag{9.32}$$

where λ is the eigenvalue associated with the soliton (i.e. the square of its energy). Let us also consider an auxiliary function $\Psi = (\Psi_1, \Psi_2)$ compatible with (9.32)

$$\hat{L}\Psi \equiv \begin{bmatrix} i\dfrac{\hbar}{\sqrt{2M}}\dfrac{\partial}{\partial z} & \sqrt{g_{1D}}\psi(z,0) \\ \sqrt{g_{1D}}\psi(z,0)^* & -i\dfrac{\hbar}{\sqrt{2M}}\dfrac{\partial}{\partial z} \end{bmatrix} \begin{bmatrix} \Psi_1 \\ \Psi_2 \end{bmatrix} = \lambda \begin{bmatrix} \Psi_1 \\ \Psi_2 \end{bmatrix}, \tag{9.33}$$

such that nontrivial solutions of Ψ can only exist if the GP equation holds. Since the operator $\hat{L}$ is self-adjoint, the ZS problem has a discrete set of real-valued eigenvalues λ_j. It is not easy to obtain the spectrum of the ZS problem in general, since exact analytical solutions to Eq. (9.33) are, in general, not available, regardless its linearity. However, an important feature of the IST method is that it is very useful to provide a class of solutions, if at least one of them is known. To illustrate that, let us consider the "most" dark soliton solution

$$\psi(z,0) = \sqrt{n_0}\tanh\left(\frac{z}{\sqrt{2}\xi}\right). \tag{9.34}$$

In that case, Eq. (9.33) possesses exact analytical solutions, providing the following eigenvalues [23]

$$\lambda_1 = 0, \quad \text{and} \quad \lambda_{2j} = -\lambda_{2j+1} = \frac{\hbar}{\sqrt{M}\xi}\sqrt{1-\left(1-\frac{j}{2^{1/4}\sqrt{n_0\xi}}\right)^2}, \tag{9.35}$$

and the corresponding soliton velocities are "darknesses" given by

$$v_j = \sqrt{\frac{2}{M}}\lambda_j, \quad n_0 - n_{\min,j} = n_0\left(1-\frac{v_j^2}{c_s^2}\right). \tag{9.36}$$

The number N_s of soliton pairs, propagating to the left and to the right of the dark soliton (9.34), is immediately determined by the condition

$$N_s = \left[\sqrt{\sqrt{2}n_0\xi}\right], \tag{9.37}$$

where [.] denotes the integer part. Notice that the total number of eigenvalues and, thus, the total number of solitons, is $2N_s + 1$ and depends on the soliton width, which in the present case is $\sqrt{2}\xi$. The generation of multiple solitons has been studied for a large number of initial configurations apart from Eq. (9.34), such as box-like pulses, phase steps, etc. [24]. Generally, for initial solutions $\psi(z,0)$ with odd (even) symmetry, will produce an odd (even) number of solitons, a fact that was experimentally verified in Ref. [25].

9.4 Interaction Between Two Dark Solitons

Another interesting aspect regarding the dynamics of solitons has to deal with their interactions. To shed some light into this problem, we should consider the case of two-interacting solitons.

In the previous section, we have considered the spectral properties of the dark-soliton solution (9.34). However, the IST method also describes the total evolution $\psi(z,t)$. This is achieved within the second part of the method, known as the spectral evolution problem, which consists on the study of trajectories of the eigenvalues λ_j as time evolves [22] (and, consequently, also the wavefunction $\psi(z,t)$, which we have seen to act as a time-dependent potential in the auxiliary linear problem in the space spanned by Ψ). In the asymptotic limit $t \rightarrow \pm\infty$, the resulting solutions can describe the elastic collision between dark solitons, since the multiple-soliton solutions can be written as a superpositon of individual single-soliton solutions. In order to better understand the physical meaning of these solutions, let us consider the two-soliton wave function $\psi(z,t)$, which takes the following asymptotic form [10]

$$\psi_{1,2}(z,t) \simeq \psi(z - v_1 t, z_1^+) + \psi(z - v_2 t, z_2^+), \quad \text{for} \quad t \rightarrow +\infty, \tag{9.38}$$

$$\psi_{1,2}(z,t) \simeq \psi(z - v_1 t, z_1^-) + \psi(z - v_2 t, z_2^-), \quad \text{for} \quad t \rightarrow -\infty, \tag{9.39}$$

where $z_j^{\pm}$ represents the soliton positions, before and after the collision. Although the shape of each single solution is preserved, their phase apparently is not, and the corresponding shifts read

$$\Delta z_1 = z_1^+ - z_1^- = \frac{\sqrt{2}\xi}{2} \gamma_1 \ln \left[\frac{(\beta_1 - \beta_2)^2 + \left(\gamma_1^{-1} + \gamma_2^{-1}\right)^2}{(\beta_1 - \beta_2)^2 + \left(\gamma_1^{-1} - \gamma_2^{-1}\right)^2} \right], \tag{9.40}$$

$$\Delta z_2 = z_2^+ - z_2^- = -\frac{\sqrt{2}\xi}{2} \gamma_2 \ln \left[\frac{(\beta_1 - \beta_2)^2 + \left(\gamma_1^{-1} + \gamma_2^{-1}\right)^2}{(\beta_1 - \beta_2)^2 + \left(\gamma_1^{-1} - \gamma_2^{-1}\right)^2} \right], \tag{9.41}$$

where we have used $\beta_j = v_j / c_s$ and

$$\gamma_j = \frac{1}{\sqrt{1 - \beta_j^2}}, \tag{9.42}$$

importing the usual notation employed in relativistic mechanics. In the special case of two counter-propagating solitons with the same velocity, such that $\beta_1 = -\beta_2 = \beta$ and $\gamma_1 = \gamma_2 = \gamma$, the phase shift takes the simple form

$$\Delta z = -\sqrt{2}\xi\gamma \ln \beta. \tag{9.43}$$

This shows that the shift on each soliton trajectory is in the same propagation direction of each individual soliton, which means that they repel each other. To study the dynamics of the solitons near the collision points, rather than its asymptotic properties described in (9.38) and (9.39), we should refer to the full form of a two-soliton wavefunction. For the case of solitons with same speeds, but propagating in opposite directions, the IST problem provides [26]

$$\psi(z,t) = \frac{A(z,t)}{B(z,t)} e^{-i\mu t/\hbar}, \tag{9.44}$$

where

$$A(z,t) = 2(n_0 - 2n_{\min})\cosh\left(\sqrt{1-\frac{v^2}{c_s^2}}\frac{\sqrt{2}vt}{\xi}\right) - 2n_0\frac{v}{c_s}\cosh\left(\sqrt{1-\frac{v^2}{c_s^2}}\frac{\sqrt{2}z}{\xi}\right)$$
$$+ i\sinh\left(\sqrt{1-\frac{v^2}{c_s^2}}\frac{\sqrt{2}vt}{\xi}\right), \tag{9.45}$$

$$B(z,t) = 2\sqrt{n_0}\cosh\left(\sqrt{1-\frac{v^2}{c_s^2}}\frac{\sqrt{2}vt}{\xi}\right) + 2\sqrt{n_{\min}}\cosh\left(\sqrt{1-\frac{v^2}{c_s^2}}\frac{\sqrt{2}z}{\xi}\right). \tag{9.46}$$

To determine the soliton minima $z_0(t)$, we make use of the fact that $\partial_z|\psi(z,t)|_{z=z_0} = 0$, which provides the following condition

$$\cosh\left(\frac{\sqrt{2}z_0}{\gamma\xi}\right) = \frac{1}{\beta}\cosh\left(\frac{\sqrt{2}vt}{\gamma\xi}\right) - \frac{2\beta}{\cosh\left(\frac{\sqrt{2}vt}{\gamma\xi}\right)}. \tag{9.47}$$

From the latter, we can then compute the distance $2\delta z$ between the two solitons at the collision time $t = 0$,

$$\delta z = \frac{\sqrt{2}\xi}{2}\gamma\operatorname{arccosh}\left(\frac{1}{\beta} - 2\beta\right), \tag{9.48}$$

which is valid for $v < 0.5c_s$ ($\beta < 0.5$). For higher speeds, the two solitons cross each other and therefore the present definition for the collision distance has no physical meaning. In Fig. 9.3, we illustrate the dynamics for different values of the velocity v. There are three different regimes: for $v < 0.5c_s$, the solution do not cross each other, inverting their motion; for the critical velocity $v = 0.5c_s$, the solution collide and overlap, but their motion is still reverted. Finally, for $v > 0.5c_s$, the soliton kinetic energy is enough for them to cross each other.

It is possible to show that the specific form of the solution (9.44) encodes the information about the interaction properties. To extract such an information, let us come back to the relative coordinate trajectory described by Eq. (9.47). Far away from the collision point, the second term on the right-hand side can be neglected and the distance between the solitons may be written as

$$\delta z(t) \simeq \frac{1}{2}\gamma\beta^{-1}\cosh^{-1}\left(2\beta\sqrt{1-\beta^2}t\right). \tag{9.49}$$

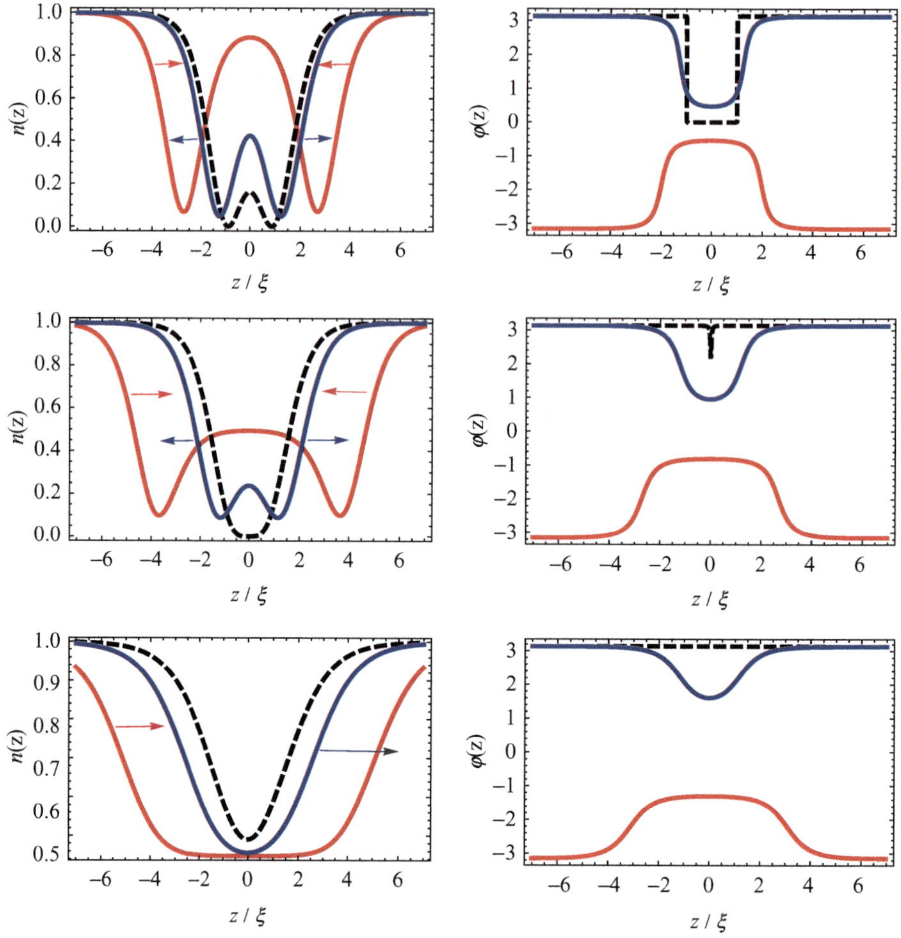

Fig. 9.3 Collision dynamics of two identical dark solitons. The panels on the *left* correspond to the density $n_{12}(z,t) = |\psi_{1,2}(z,t)|^2$, while the respective phase portraits are located in the *right* side. From *top* to *bottom*: $v = 0.3c_s$, $v = 0.5c_s$ and $v = 0.8c_s$. The *arrows* indicate the direction of propagation before (*red*) and after (*blue*) the collision. The *black dashed line* indicates both the density and phase profiles at the collision instant $t = 0$

Taking the derivative with respect to time, we obtain [10, 23]

$$\frac{d\delta z}{dt} = \frac{\sinh\left(2\beta^2\sqrt{1-\beta^2}t\right)}{\sqrt{\beta^{-1}\cosh^2\left(2\beta^2\sqrt{1-\beta^2}t\right)-1}}, \tag{9.50}$$

which, at the collision instant $t = 0$, yields $\delta z = 0$. Thus, as the solitons approach each other, their velocity (and, consequently, their depth) is increased (decreased) and becomes completely dark – or black – at the collision point. Afterwards, the solitons are reflected by each other and start moving in opposite directions, with asymptotic velocities given by $d\delta z/dt \simeq \pm v$ as $t \to \infty$.

Moreover, by taking the second time derivative of Eq. (9.50), and making use of Newton's law in the form

$$\frac{d^2\delta z}{dt^2} = -\frac{\partial V_{\rm int}}{\partial z}, \tag{9.51}$$

we can derive the pseudo-potential (divided by the mass) associated with the interaction between the two solitons

$$V_{\rm int}(z) = \frac{1}{2}c_s^2 \frac{1-\beta^2}{\sinh^2\left(\frac{\sqrt{2}\sqrt{1-\beta^2}z}{\xi}\right)}. \tag{9.52}$$

Is is easy to conclude that $V_{\rm int}(z)$ is indeed a repulsive potential, which is a clear signature that dark solitons tend to repeal each other. If the separation between the solitons is sufficiently large when compared to its core (i.e. $z \gg \xi$), the potential takes the asymptotic form as $V_{\rm int}(z) \sim \exp(-2\sqrt{2}\sqrt{1-\beta^2}z/\xi)$ (see Ref. [27] for a Lagrangian derivation). Due to the dependence on the soliton velocity v, Eq. (9.52) describes a pseudo-potential, as it is non-conservative. In fact, it reduces to an effective potential for the case of black soliton ($v \sim 0$), which reads

$$V_{\rm eff}(z) = \frac{1}{2}c_s^2 \mathrm{cosech}^2\left(\frac{\sqrt{2}z}{\xi}\right). \tag{9.53}$$

9.5 Bright Solitons

There is another sort of nonlinear excitations in Bose-Einstein condensates beyond the dark soliton solution. For attractive interactions, $g < 0$, the soliton solutions exhibits a maximum rather than a minimum, being therefore referred to as *bright solitons*. Coming back to Eq. (9.16), the change in the sign $g \to -|g|$ yields

$$2(f_1')^2 = (1 + f_1^2 - f_0^2)^2, \tag{9.54}$$

to which the solution can be found to be like

$$f_1 = a\mathrm{sech}(bx), \tag{9.55}$$

where a and b are given by

$$a^2 = \frac{m\mu}{\hbar^2}, \quad b^2 = \frac{2\mu}{g}, \tag{9.56}$$

which is valid provided that

$$\mu = -\frac{1}{2}|g|n_0. \tag{9.57}$$

The corresponding solitary solution, similar to (9.18), can be written as

$$\psi(z,t) = \sqrt{n_0}\,\mathrm{sech}\left(\frac{z-vt}{\sqrt{2}\xi}\right)e^{-i\mu t/\hbar}, \quad \text{with} \quad \xi = \frac{\hbar}{\sqrt{2m|g|n_0}}. \tag{9.58}$$

Bright solitons differ from dark (or gray) solitons in many aspects. As it can be seen directly from the solution, the soliton speed and amplitude are not connected, as it is the case of the former. This simply means that the energy of the soliton is not dispersive. Also, there is no phase shift as $z \to \pm\infty$, as it can be easily observed

$$\Delta\varphi = \varphi(+\infty) - \varphi(-\infty) = 0. \tag{9.59}$$

This result has important implications in what concerns the dynamics of bright soliton trains [28].

9.6 Dark Solitons in Harmonic Traps

In the previous sections, we observe that dark solitons exhibits particle-like behavior. It is, therefore, very instructive to understand how far we can go with this "classical" description and understand what happens if we put a single soliton in a trap. For the case of a harmonic trap, Kohn's theorem assures that a particle should oscillate at the trapping frequency ω_z. But does it hold, in general, for the case of a soliton? The answer is no and it may be instructive to understand why.

Let us come back to Eq. 9.7 and incorporate the external potential $V(z) = M\omega_z^2 z^2/2$ as follows

$$i\hbar\frac{\partial\psi(z,t)}{\partial t} = \left\{-\frac{\hbar^2}{2M}\frac{\partial^2}{\partial z^2} + V(z) + g_{1D}|\psi(z,t)|^2\right\}\psi(z,t). \tag{9.60}$$

If the external potential varies slowly on the soliton scale, as it is the case of a harmonic trap, one may assume that a perturbed soliton solution can be written in the following manner

$$\psi(z,t) = \psi_s(z,t) + \delta\psi(z,t), \tag{9.61}$$

where $\psi_s(z,t)$ is the soliton solution given by (9.18) and $\delta\psi(z,t)$ represents the phonon radiation emitted by the soliton. It turns out that $\delta\psi(z,t) \ll \psi_s(z,t)$, as radiation only takes place for strong perturbations. Therefore, the first order approximation (also known as adiabatic approximation [29]) consists in neglecting the emission of such sound waves. There are many techniques to describe the adiabatic dynamics of solitons, based on extensions of Hamiltonian and Lagrangian methods originally constructed for a constant density profile. Here, we will employ a perturbative approach to the Gross-Pitaevskii equation, by seeking a solution of the form $\psi(z,t) = \psi(z,t)\psi_s(z,t)$

$$\psi(z,t) = \Psi(z)e^{-i\mu t/\hbar}, \tag{9.62}$$

where $\Psi(z)$ is the constant (real-valued) background governed by the time-independent problem

$$\mu\Psi + \frac{\hbar^2}{2M}\frac{\partial^2}{\partial z^2}\Psi - g\Psi^3 = V(z)\Psi(z). \tag{9.63}$$

On the other hand, the dark soliton wave function reads

$$i\hbar\frac{\partial\psi_s}{\partial t} + \frac{\hbar^2}{2M}\frac{\partial^2}{\partial z^2}\psi - g\Psi^2\left(|\psi_s|^2 - 1\right)\psi_s = -\frac{\hbar^2}{2M}\frac{d}{dz}\left[\ln(\Psi)\frac{\partial\psi_s}{\partial z}\right]. \tag{9.64}$$

For the case of a harmonic trap, which is smooth, the corresponding Thomas-Fermi equilibrium profile is $\Psi(z)_{\mathrm{TF}} = \sqrt{(\mu - V(z))g} = \sqrt{1 - r^2/R_{\mathrm{TF}}^2}$ and both the logarithmic and nonlinear terms can be treated as perturbations. Indeed, the right-hand-side of Eq. (9.64) can be approximated as

$$-\frac{d}{dz}\left[\ln(\Psi)\frac{\partial\psi_s}{\partial z}\right] \simeq \frac{1}{2}\frac{dV}{dz}\left(1 + V + V^2 + \ldots\right), \tag{9.65}$$

which allows us to write the approximate evolution equation for the soliton as

$$i\hbar\frac{\partial\psi_s}{\partial t} + \frac{\hbar^2}{2M}\frac{\partial^2}{\partial z^2}\psi - g\left(|\psi_s|^2 - 1\right)\psi_s = \mathcal{Q}\left[\psi_s, V\right], \tag{9.66}$$

where the perturbation potential is given by

$$\mathcal{Q}[\psi_s, V] = \left(1 - |\psi_s|^2\right)\psi_s V + \frac{1}{2}\frac{\partial\psi_s}{\partial z}\frac{dV}{dz}\left(1 + V + V^2\right). \tag{9.67}$$

In order to proceed further, we now postulate a variational solution inspired in the homogeneous solution (9.18),

$$\psi_s(z,t) = \sqrt{n0}\left[\cos(\phi)\tanh(\zeta) + i\,\sin(\phi)\right], \tag{9.68}$$

where $\sin[\phi(t)] = v(t)/c_s(0)$, $\cos[\phi(t)] = \gamma^{-1}(t)$ and

$$\zeta(t) = \cos[\phi(t)]\frac{z - z_0(t)}{\sqrt{2}\xi} \tag{9.69}$$

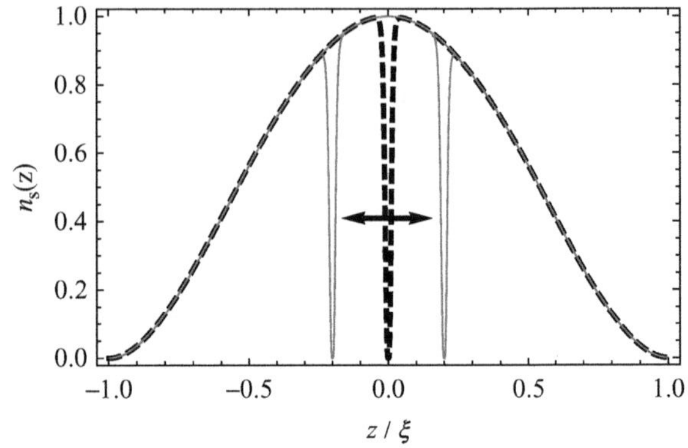

Fig. 9.4 Illustration of the dynamics of a dark soliton performing slow oscillations in a harmonic trap

are now time-dependent, but slowly varying, adiabatic parameters. The velocity of the centroid is then simply given by

$$\frac{dz_0}{dt} = \sin(\phi). \tag{9.70}$$

To properly determine the evolution of the phase angle $\phi(t)$, we make use of the energy Eq. (9.25) and write

$$\frac{dE_{\rm ds}}{dt} = -4\mu \cos^2(\phi) \sin(\phi) \frac{d\phi}{dt}. \tag{9.71}$$

On the other hand, from Eq. (9.67), we can express the energy evolution as follows

$$\frac{dE_{\rm ds}}{dt} = -\int_{-\infty}^{+\infty} \left\{ \mathcal{Q}[\psi_s, V] \frac{\partial \psi_s^*}{\partial t} + \mathcal{Q}^*[\psi_s, V] \frac{\partial \psi_s}{\partial t} \right\} dz. \tag{9.72}$$

Putting the latter results together, one can easily obtain

$$\frac{d\phi}{dt} = \frac{1}{2\cos^2\phi \sin\phi} \mathrm{Re} \left\{ \int_{-\infty}^{+\infty} \mathcal{Q}[\psi_s, v] \frac{\partial \psi_s^*}{\partial t} dz \right\}. \tag{9.73}$$

By Taylor expanding the trapping potential $V(z)$ around the soliton centroid $z_0(t)$ and, assuming small oscillations around the center of the trap (see Fig. 9.4), such that $\mu = g n_0(0) \gg V$, we can safely neglect the terms proportional to V and V^2 in Eq. (9.67), which in its turn yields

$$\frac{d\phi}{dt} \simeq -\frac{1}{2} \cos(\phi) \left. \frac{\partial V}{\partial z} \right|_{z=z_0}. \tag{9.74}$$

Using the same procedure as in the case of two-interacting solitons, we take the black soliton approximation, $\cos(\phi) \approx 1$ to finally obtain

$$\frac{d^2 z_0}{dt^2} = -\frac{1}{2M}\frac{\partial V}{\partial z_0}. \tag{9.75}$$

This result contains the remarkable feature that a soliton in an external potential behaves (in the adiabatic approximation) as a point-like particle of effective mass $m_{\text{eff}} = 2M$. Notice that this is not in contradiction with the previous statement that a dark soliton has a negative mass (see Eq. (9.28)), which is only physically meaningful within the wave description and is associated with its dispersion relation (and not the dynamics of its centroid). This is a consequence of the vanishing boundary conditions of the wave function ($|\psi_s| \to 0$ as $z \to \pm\infty$) due to confinement, whereas in the homogeneous case the assympotic behaviour of the wave function is $|\psi_s| \to 1$ as $z \to \pm\infty$. For the special case of the harmonic potential, Eq. (9.75) describes an oscillatory motion at the frequency

$$\omega_A = \frac{\omega_z}{\sqrt{2}}, \tag{9.76}$$

which is known in the literature as the *anomalous mode* [30]. The reason for this name can be better understood in the context of the Bogolyubov (or a similar linear) analysis, which reveals the emergence of this mode, occurring below the Kohn mode $\omega = \omega_z$, only in the presence of topological excitations [31].

To establish the parallelism with the homogeneous case, we now compute the total energy of the trapped soliton. Making use of the local density approximation, which allows us to replace c_s by its slowly-varying equivalent $c_s(z) = \sqrt{gn_0(z)/M}$, together with Eq. (9.67), we obtain

$$E_{\text{ds}} = \frac{4}{3}\hbar n_0(0)c_s(0) + \frac{1}{2}M_{\text{ds}}v^2 + \frac{1}{4}M_{\text{ds}}\omega_z z^2, \tag{9.77}$$

where $M_{\text{ds}} = -4\hbar n_0(0)/c_s(0)$ is in agreement with the previous definition. Due to the negative mass of the soliton,[1] the application of an external potential indeed decreases the energy of the soliton. Again, it is possible to observe that the particle-like behaviour of the soliton is associated with an effective mass of $m_{\text{eff}} = 2m$, which can be verified by calculation the ratio of the soliton mass M_{ds} to the potential energy, $4/\omega_z^2 z^2$, which is twice as bigger as in the case of a trapped atomic particle.

9.7 The Soliton Gas

A natural question that may arise in the context of one-dimensional solitons is related with the dynamics of a collection of dark solitons disposed in an array. Such a configuration can be understood as a soliton lattice, or even as a soliton gas. It is

[1] Notice that we are now considering the soliton as a wave, described by the ansatz in (9.68)

intuitive to predict that the fact of solitons carrying a negative mass, the dynamics of such a system is not totally stable even if the effective potential is repulsive, as we have previously stated.

To describe the dynamics of a dark soliton gas, we postulate that its phase-space distribution $f(x, v, t)$ is governed by following kinetic equation

$$\frac{df}{dt} \equiv \frac{\partial f}{\partial t} + v\frac{\partial f}{\partial x} + \dot{v}\frac{\partial f}{\partial x} = 0, \tag{9.78}$$

where the collision integral is neglected assuming that only elastic processes are involved in the system. The acceleration term can be defined by

$$\dot{v} = -\frac{\partial U}{\partial x}, \tag{9.79}$$

where the potential is simply given in terms of the two-soliton effective potential (9.53)

$$U(x,t) = \int_{-\infty}^{\infty} V_{\text{eff}}(x - x')n(x',t)dx' = \int_{-\infty}^{\infty}\int_{-\infty}^{\infty} V_{\text{eff}}(x - x')f(x',v,t)dx'dv. \tag{9.80}$$

Equation (9.78) generalizes the continuity equation presented in Ref. [32] by the introduction of the soliton interaction potential. We now linearize the system around its equilibrium, $f = f_0 + \tilde{f}$, and rewrite the kinetic equation as follows

$$\left(\frac{\partial}{\partial t} + v\frac{\partial}{\partial x}\right)\tilde{f}(x,v,t) - \frac{\partial V_{\text{eff}}}{\partial x}\int_{-\infty}^{\infty}\frac{\partial f_0}{\partial v}\tilde{f}\, dv = 0. \tag{9.81}$$

After Fourier transforming, the latter equation can be rewritten as

$$1 = k\tilde{V}(k)\int_{-\infty}^{\infty}\frac{1}{\omega - vk}\frac{\partial f_0}{\partial v}\, dv = k^2\tilde{V}(k)\int_{-\infty}^{\infty}\frac{f_0(x,v)}{(\omega - vk)^2}\, dv, \tag{9.82}$$

where $\tilde{V}(k)$ is the Fourier transform of the effective potential. Let us now exemplify with a special configuration. The most simple and straightforward one is to assume that all the solitons are seated in the lowest energy configuration, which corresponds then to the behavior of a boson gas

$$f_0(x,t) = N_0\delta(v), \tag{9.83}$$

where N_0 is the one-dimensional soliton density. Inserting it in the dispersion relation (9.82), we have

$$\omega^2 = -\frac{1}{4}N_0\xi c_s^2 k^2\left[2\sqrt{2} + \pi k\xi \coth\left(\frac{\pi k\xi}{2\sqrt{2}}\right)\right]. \tag{9.84}$$

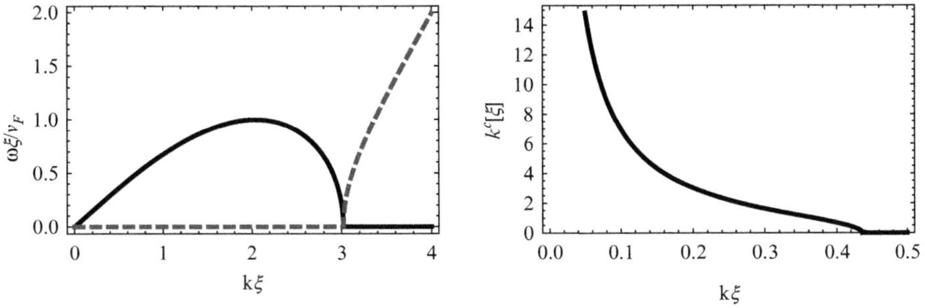

Fig. 9.5 Real (*full*) and imaginary (*dashed*) parts of the dispersion relation of a dark soliton gas with $N_0 = 0.15 n_0$. The *right panel* illustrates the dependence of k^* on α

We can immediately see that the equilibrium configuration (9.83) is unstable, as the frequency only possesses purely imaginary roots. Fortunately, due to the peculiarity of low-dimensional systems, such a configuration does not correspond, indeed, to a physical one. In fact, it is known that the excitations in one-dimensional Bose gases below the critical velocity c_s follow a fermionic statistics. This important remark was established in the famous Lieb-Liniger theory [33, 34] and it is also considered in Ref. [35] in the context of excitations of a superfluid flow passing an impurity. Actually, this result is quite easy to understand by simply looking at the phase profiles in Fig. 9.3: by exchanging solitons, there is a overall phase shift of $\pm\pi$, which implies that the wave function of two solitons located at positions z_1 and z_2 transforms as

$$\psi(z_2, z_1) = e^{\pm i\pi}\psi(z_1, z_2) = -\psi(z_1, z_2), \tag{9.85}$$

which is the signature of a fermionic statistics. Incorporating this fact, a new (and more physical) equilibrium configuration can be written as

$$f_0(z, v) = \frac{N_0}{2v_F}\Theta(v_F - |v|), \tag{9.86}$$

where $v_F = \hbar\pi N_0/(2|M_{\rm ds}|)$ is the one-dimensional Fermi velocity. In that case, the dispersion relation reads [36]

$$\omega^2 = v_F^2 k^2 - \frac{\alpha}{4}c_s^2 k^2\left[2\sqrt{2} + \pi k\xi \coth\left(\frac{\pi k\xi}{2\sqrt{2}}\right)\right], \tag{9.87}$$

where $\alpha = N_0\xi$ is the soliton concentration parameter. The latter result contains very interesting features, which are illustrated in Fig. 9.5. For low values of α, the system displays a stable oscillation up to a certain critical wavevector $k^* = k^*(\alpha)$. For $k > k^*$, the oscillations are unstable. The system becames totally unstable for $\alpha \gtrsim 0.423$. For that reason, we notice that Eq. (9.87) describes oscillations at frequencies which are much lower than for typical Bogoliubov oscillations, which can be verified by the following ratio

$$\frac{\omega}{\omega_{\text{Bog}}} \sim \frac{v_F}{c_s} = \frac{\pi}{4}\frac{N_0}{n_0} = \frac{\pi}{4}\frac{\alpha}{n_0\xi} \sim \frac{\pi}{4}\alpha, \tag{9.88}$$

using the fact that $\xi n_0 \sim 1$. The reasoning for this dynamical behavior is quite intuitive: because of the negative mass of the system, the interaction between solitons in a gas is effectively attractive. This attraction is balanced by the Fermi pressure, which can only overcome the collapse up to a certain critical density.

9.8 Solitons in Two Dimensions

Dark solitons can also be excited in Bose-Einstein condensates with dimensions higher than one. For example, as we have seen, there is a smooth transition between "effective" one- and three dimensional settings in cigar-shaped BECs, being the latter the most common experimental situation to produce the quasi one-dimensional solitons discussed in the previous sections. Furthermore, quasi-1D solitons may also exist in *disk-shaped* BECs, which is achieved if $\omega_z \gg \omega_\perp$, therefore being a solution of the following Gross-Pitaevskii

$$i\hbar\frac{\partial\psi(x,y,t)}{\partial t} = \left\{-\frac{\hbar^2}{2M}\left(\frac{\partial^2}{\partial x^2} + \frac{\partial^2}{\partial y^2}\right) + V(r) + g_{2D}|\psi(x,y,t)|^2\right\}\psi(x,y,t), \tag{9.89}$$

where $g_2D = 2\sqrt{2}aa_z$ is the effective 2D interaction strength and

$$V(r) = \frac{1}{2}\omega_\perp r^2, \quad r^2 = x^2 + y^2, \quad \text{with} \quad a_z = \frac{\hbar}{\sqrt{m\omega_z}}. \tag{9.90}$$

In the homogeneous case ($V(r) = 0$), and if we choose the direction of propagation along the $x-$axis, the soliton solution is simply given by (9.18)

$$\psi_s(x,t) = \sqrt{n_0}\left[i\frac{v}{c_s} + \sqrt{1 - \frac{v^2}{c_s^2}}\tanh\left(\frac{x - vt}{\sqrt{2}\xi}\right)\right]e^{-i\mu t/\hbar}. \tag{9.91}$$

In the case of a trapped system, the soliton corresponds to a dark "stripe" on top of the Thomas-Fermi equilibrium, such that the total wave function is given by

$$\psi(r,t) = \Psi_{\text{TF}}(r)\psi_s(x,t)e^{-i\mu\hbar/t}, \tag{9.92}$$

similarly to what was discussed for the one-dimensional case (see Eq. (9.62)) (Fig. 9.6).

An important question associated with the soliton solution (9.92) is related with its dynamic stability. As it was first demonstrated in Ref. [37], the soliton undergoes

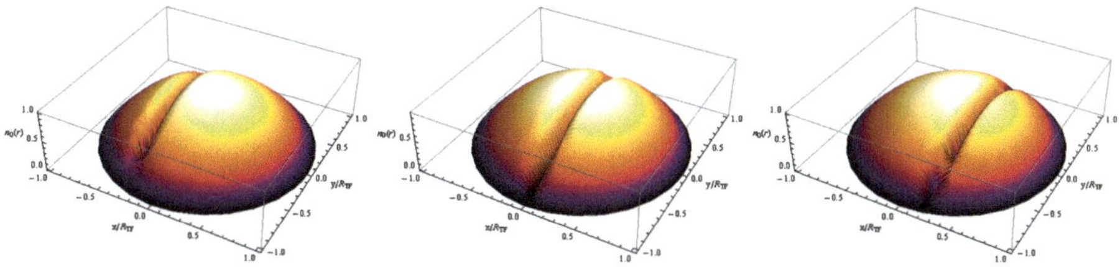

Fig. 9.6 Snapshots of a rectilinear dark-soliton propagating in a two-dimensional BEC

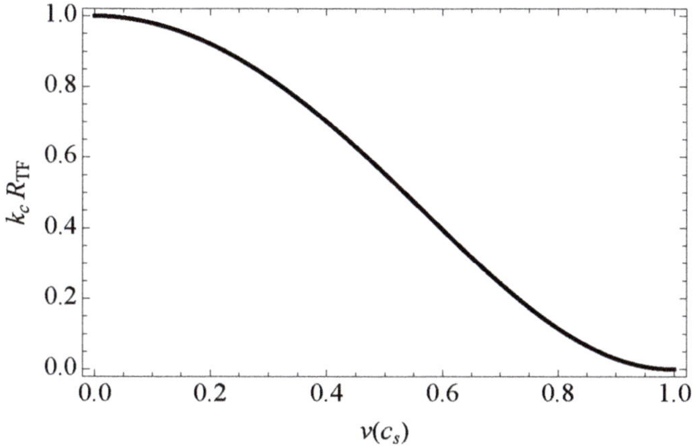

Fig. 9.7 Dependence of the critical wavevector k_c on the soliton speed

a modulation of the form $\delta\psi \sim \cos(ky)$, where k is a function of its velocity v/c_s. Below the critical value k_c given by

$$k_c = \frac{1}{R_{TF}}\left[1 - \frac{v^2}{c_s^2} - 2 + 2\sqrt{\left(1 - \frac{v^2}{c_s^2}\right)^2 + \frac{v^2}{c_s^2}}\,\right], \tag{9.93}$$

the soliton becomes modulationally unstable and grows exponentially along the transverse direction. This mechanism is known in the literature as "snaking instability". Such an instability was extensively studied in the field of nonlinear optics (see e.g. [38]) and was found to be in the root of the decay of the soliton into vortex–anti-vortex pairs. Notice that the darker the soliton, the most stable it is (see Fig. 9.7). In fact, when the snaking instability sets in, a plane 'black' soliton decays into vortex pairs; on the other hand, unstable gray solitons may not decay into vortices, but rather undergo long-lived oscillations associated with the emission of sound waves [39].

It is, however, possible to prevent the transverse modulational instability of dark solitons by using the finite-size of the system where it propagates. Such a suppression can be found by means of a scale competition criterion [40]. In

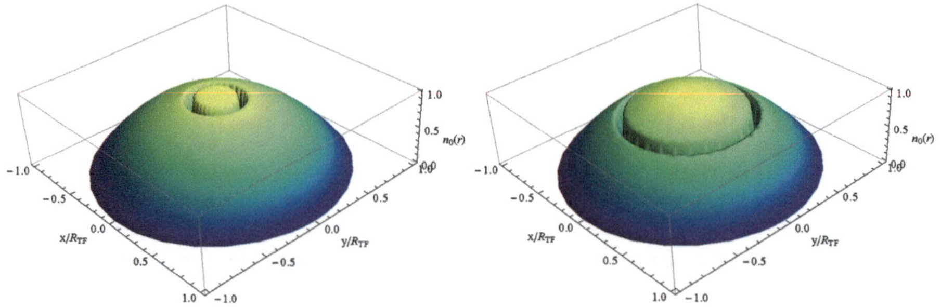

Fig. 9.8 Snapshots of a ring dark-soliton propagating in a two-dimensional BEC

particular, if the critical wavelength $\lambda_c = 2\pi/k_c$ is greater than the Thomas-Fermi radius R_{TF}, then the snaking instability will not occur. For example, for a black soliton with $v \sim 0$, $\lambda_c = 2\pi R_{TF}$. In that case, the above scale competition argument, $R_{\mathrm{TF}} < \lambda_c$, implies that the use of a sufficiently strong confinement along the transverse direction, such that

$$\omega_\perp > \omega_\perp^* = \frac{\sqrt{2}}{\pi}\omega_z =\simeq 0.45\omega_z, \tag{9.94}$$

can suppress the modulational instability. According to numerical simulations performed by Kevredikis et al. [40], the actual critical trapping frequency is less than the theoretically predicted in (9.94), namely $\omega_\perp^* \simeq 0.31\omega_z$. The origin of such a discrepancy stems form the fact that, for small BECs, the presence of a dark soliton significantly changes the value of the chemical potential μ, and the rescaling factor is $\tilde{\mu} = 0.35\mu$. Beyond the rectilinear dark soliton, two-dimensional systems can support another kind of quasi-one1D dark solitons, called *ring solitons* (see Fig. 9.8). An important feature of these objects is connected with the fact that they do not exhibit snaking instability.

References

1. J.S. Russell, Report on waves, in *14th Meeting of the British Association for the Advancement of Science*, London (1844)
2. D.J. Korteweg, G. de Vries, On the change of form of long waves advancing in a rectangular canal and on a new type of long stationary waves. Philos. Mag. **39** 422 (1895)
3. A. Jeffrey, T. Kawara, *Assympthotic Methods in Nonlinear Wave Theory* (Pitman, Boston, 1982)
4. P. Emplit, J.P. Haimade, F. Reynaud, C. Froehly, A. Barthelemey, Opt. Commun. **62**, 374 (1987)
5. D. Krökel, N.J. Halas, G. Giuliani, D. Grischkowsky, Phys. Rev. Lett. **60**, 29 (1988)
6. G.R. Allan, S.R. Skinner, D.R. Andersen, A.L. Smirl, Opt. Lett. **16**, 156 (1991)
7. B. Denardo, W. Wright, S. Putterman, A. Larraza, Phys. Rev. Lett. **64**, 1518 (1990)

8. B. Denardo, B. Galvin, A. Greenfield, A. Larraza, S. Putterman, W. Wright, Phys. Rev. Lett. **68**, 1730 (1992)
9. R. Heidemann, S. Zhdanov, R. Stütterlin, H.M. Thomas, G.E. Morfill, Phys. Rev. Lett. **102**, 135002 (2009)
10. D.J. Frantzeskakis, Dark solitons in atomic Bose-Einstein condensates: from theory to experiments. J. Phys. A Math. Theor. **43**, 213001 (2010)
11. C. Menotti and S. Stringari, Phys. Rev. A **66**, 043610 (2002).
12. A. Weller, J.P. Ronzheimer, C. Gross, D.J. Frantzeskakis, G. Theocharis, P.G. Kevrekidis, J. Esteve, M.K. Oberthaler, Phys. Rev. Lett. **101**, 120406 (2008)
13. V.M. Pérez-García, H. Michinel, H. Herrero, Phys. Rev. A **57**, 3837 (1998)
14. A.E. Muryshev, G.V. Shlyapnikov, W. Ertmer, K. Sengstock, M. Lewenstein, Phys. Rev. Lett. **89**, 110401 (2002)
15. L. Salanisch, A. Parola, L. Reatto, Phys. Rev. A **65**, 043614 (2002)
16. F. Gerbier, Europhys. Lett. **66**, 771 (2004)
17. T. Kinoshita, T. Wenger, D.S. Weiss, Science **305**, 1125 (2004)
18. J. Denschlag, J.E. Simsarian, D.L. Feder, C.W. Clark, L.A. Collins, J. Cubizolles, L. Deng, E.W. Hagley, K. Helmerson, W.P. Reinhardt, S.L. Rolston, B.I. Schneider, W.D. Phillips, Science **287**, 97 (2000)
19. S. Burger, K. Bongs, S. Dettmer, W. Ertmer, K. Sengstock, A. Sanpera, G.V. Shlyapnikov, M. Lewenstein, Phys. Rev. Lett. **92**, 230401 (2004)
20. C.J. Pethick, H. Smith, *Bose-Einstein Condensation in Dilute Gases*, 2nd edn. (Cambridge University Press, Cambridge, 2008)
21. S.I. Shevchenko, Sov. J. Low Temp. Phys. **14**, 553 (1988)
22. V.E. Zakharov, A.S. Shabat, Sov. Phys. JETP **37**, 823 (1973); ibid. Sov. Phys. JETP **34**, 62 (1971)
23. V.V. Konotop, V.E. Vekslerchik, J. Phys. A Math. Gen. **24**, 767 (1991)
24. S.A Gredeskul, Yu. S. Kivshar, M.V. Yanovskaya, Phys. Rev. A **41**, 3994 (1990)
25. A.M. Weiner, J.P. Heritage, R.J. Hawkings, R.N. Thurston, E.M. Kirschner, D.E. Learid, W.J. Tomlinson, Phys. Rev. Lett. **61**, 2445 (1988)
26. N.N. Akhmediev, A. Ankiewich, Phys. Rev. A **47**, 3213 (1993)
27. Yu S. Kivshar, W. Królikowski, Opt. Commun. **114**, 353 (1994)
28. K.E. Strecker, G.B. Partridge, A.G. Truscott, R.G. Hulet, Nature **417**, 150 (2002)
29. Y.S. Kivshar, B.A. Malomed, Rev. Mod. Phys. **61**, 763 (1989)
30. T. Bush, J.R. Anglin, Phys. Rev. Lett. **84**, 2298 (2000)
31. A.L. Fetter, A.A. Svidzinksy, J. Phys. Condens Matter **13**, R135 (2001)
32. G.A. El, A.M. Kamchatnov, Phys. Rev. Lett. **95**, 204101 (2005)
33. E.H. Lieb, W. Liniger, Phys. Rev. **130**, 1605 (1963)
34. E.H. Lieb, Phys. Rev. **130**, 1616 (1963)
35. G.E. Astrakharchik, L.P. Pitaevskii, Phys. Rev. A **70**, 013608 (2004)
36. H. Terças, D.D. Solnyshkov, and G. Malpuech, *Topological Wigner Crystals of Half-Solitons in Spinor Bose-Einstein Condensates*, arXiv:1209.6197 (2012)
37. E.A. Kuznetsov, S.K. Turitsyn, Sov. Phys. JETP **67**, 1583 (1988)
38. A.V. Mamaev, M. Saffman, A.A. Zozulya, Phys. Rev. Lett. **76**, 2262 (1996)
39. Y. Kivshar, D.E Pelinovsky, Phys. Rep. **331**, 117 (2000)
40. P.G. Kevredikis, G. Theocharis, D.J. Frantzeskakis, A. Trombettoni, Phys. Rev. A **70**, 043616 (2004)

Chapter 10
Quantum Field Theory of BECs

We now consider the Bogoliubov theory of a Bose Einstein condensate, which describes the phonon field as a quantum field, and allows to study quantum field phenomena at low energies, including vacuum fluctuations and phonon emission. Mechanisms for phonon pair creation from vacuum are similar to those known for photons in quantum electrodynamics, and will be discussed in this chapter. They include: time refraction, the dynamical Casimir effect and Hawking radiation. This last radiation process also allows us to discuss condensate analogues of a gravitational field and the equivalent space-time metric.

10.1 Bogoliubov Theory

Our starting point is the mean field Hamiltonian, as defined in the GP equation, H. The expectation value of this Hamiltonian gives the energy of the system, for a given quantum state $|\Phi\rangle$, as $E = \langle\Phi|H|\Phi\rangle$, or in more explicit terms

$$E = \int \left[-\Phi^*(\mathbf{r}) \frac{\hbar^2}{2M} \nabla^2 \Phi(\mathbf{r}) + V_0(\mathbf{r}) |\Phi(\mathbf{r})|^2 + \frac{g}{2} |\Phi(\mathbf{r})|^4 \right] d\mathbf{r}. \tag{10.1}$$

In quantum field theory, the state vector $|\Phi\rangle$ is replaced by an operator $\hat{\psi}$, and the mean energy E by the new Hamiltonian operator $\hat{H}$, defined by

$$\hat{H} = \int \left[-\hat{\psi}^\dagger(\mathbf{r}) \frac{\hbar^2}{2M} \nabla^2 \hat{\psi}(\mathbf{r}) + V_0(\mathbf{r}) \hat{\psi}^\dagger(\mathbf{r}) \hat{\psi}(\mathbf{r}) + \frac{g}{2} \hat{\psi}^\dagger(\mathbf{r})^2 \hat{\psi}(\mathbf{r})^2 \right] d\mathbf{r}. \tag{10.2}$$

This is obtained from (10.1) by making the replacements $|\Phi\rangle \to \hat{\psi}$, and $E \to \hat{H}$. By using a Fourier decomposition of the field operators $\hat{\psi}$ and $\hat{\psi}^\dagger$, we can write

$$\hat{\psi}(\mathbf{r}) = \frac{1}{\sqrt{V}} \sum_{\mathbf{k}} a_{\mathbf{k}} e^{i\mathbf{k}\cdot\mathbf{r}} = \sqrt{V} \int a_{\mathbf{k}} e^{i\mathbf{k}\cdot\mathbf{r}} \frac{d\mathbf{k}}{(2\pi)^3}, \tag{10.3}$$

J.T. Mendonça and H. Terças, *Physics of Ultra-Cold Matter*, Springer Series on Atomic, Optical, and Plasma Physics 70, DOI 10.1007/978-1-4614-5413-7_10,

where V is the volume of the system, and the operators $a_{\mathbf{k}}$ can be defined by the inverse transformation

$$a_{\mathbf{k}} = \frac{1}{\sqrt{V}} \int \hat{\psi}(\mathbf{r}) e^{-i\mathbf{k}\cdot\mathbf{r}} d\mathbf{r}. \tag{10.4}$$

Let us restrict our discussion to the case of a free system, where the influence of the external confining potential $V_0(\mathbf{r})$ is assumed negligible. By replacing this Fourier decomposition in Eq. (10.2), we obtain

$$\hat{H} = \sum_{\mathbf{k}} \frac{\hbar^2 k^2}{2M} a_{\mathbf{k}}^\dagger a_{\mathbf{k}} + \frac{g}{2V} \sum_{\mathbf{k},\mathbf{k}',\mathbf{k}''} \left(a_{\mathbf{k}}^\dagger a_{\mathbf{k}+\mathbf{k}''}^\dagger a_{\mathbf{k}'-\mathbf{k}''} a_{\mathbf{k}} \right). \tag{10.5}$$

The operators $a_{\mathbf{k}}$ and $a_{\mathbf{k}}^\dagger$ are the destruction and creation operators of boson states with momentum $\hbar\mathbf{k}$. We assume that they satisfy the bosonic commutation relations

$$\left[a_{\mathbf{k}}, a_{\mathbf{k}'}^\dagger \right] = \delta_{\mathbf{k}\mathbf{k}'}, \quad [a_{\mathbf{k}}, a_{\mathbf{k}'}] = 0, \quad \left[a_{\mathbf{k}}^\dagger, a_{\mathbf{k}'}^\dagger \right] = 0. \tag{10.6}$$

We can define occupation number operators $\hat{N}_{\mathbf{k}} = a_{\mathbf{k}}^\dagger a_{\mathbf{k}}$, as well as the total number operator $\hat{N}$, such that

$$\hat{N} = \sum_{\mathbf{k}} \hat{N}_{\mathbf{k}} = \sum_{\mathbf{k}} a_{\mathbf{k}}^\dagger a_{\mathbf{k}}. \tag{10.7}$$

In the spirit of the Bogoliubov theory, we now assume that most of the atoms are in the condensed state $\mathbf{k} = 0$. This allows us to use the approximation

$$\hat{N} \simeq N_0 + \sum_{\mathbf{k}\neq 0} a_{\mathbf{k}}^\dagger a_{\mathbf{k}}, \tag{10.8}$$

where the first term is dominant and N_0 is a c-number representing the total number of condensed bosons. In the same spirit, we can use a similar approximation in the last term of the operator (10.5), by writing

$$\sum_{\mathbf{k},\mathbf{k}',\mathbf{k}''} \left(a_{\mathbf{k}}^\dagger a_{\mathbf{k}+\mathbf{k}''}^\dagger a_{\mathbf{k}'-\mathbf{k}''} a_{\mathbf{k}} \right) \simeq a_0^{\dagger 2} a_0^2 + \sum_{\mathbf{k}\neq 0} \left(a_0^{\dagger 2} a_{\mathbf{k}} a_{-\mathbf{k}} + a_{\mathbf{k}}^\dagger a_{-\mathbf{k}}^\dagger a_0^2 + 4 a_0^\dagger a_0 a_{\mathbf{k}}^\dagger a_{\mathbf{k}} \right). \tag{10.9}$$

Replacing the operators $a_0^{\dagger 2}$, a_0^2 and $a_0^\dagger a_0$ by the c-number N_0, we get

$$\sum_{\mathbf{k},\mathbf{k}',\mathbf{k}''} \left(a_{\mathbf{k}}^\dagger a_{\mathbf{k}+\mathbf{k}''}^\dagger a_{\mathbf{k}'-\mathbf{k}''} a_{\mathbf{k}} \right) \simeq N_0 \left[N_0 + \sum_{\mathbf{k}\neq 0} \left(a_{\mathbf{k}} a_{-\mathbf{k}} + a_{\mathbf{k}}^\dagger a_{-\mathbf{k}}^\dagger + 4 a_{\mathbf{k}}^\dagger a_{\mathbf{k}} \right) \right]. \tag{10.10}$$

On the other hand, Eq. (10.8) allows us to use

$$N_0^2 + 2N_0 \sum_{\mathbf{k}\neq 0} a_{\mathbf{k}}^\dagger a_{\mathbf{k}} \simeq \left(N_0 + \sum_{\mathbf{k}\neq 0} a_{\mathbf{k}}^\dagger a_{\mathbf{k}} \right)^2 \simeq N^2, \tag{10.11}$$

where N is the total number of particles in the system. Of course, these approximations are only valid for $N_0 \simeq N$. Replacing this in Eq. (10.5), we obtain

$$\hat{H} = \frac{N}{2} g n_0 + \sum_{\mathbf{k}\neq 0} \left\{ (\epsilon_k + g n_0) a_{\mathbf{k}}^\dagger a_{\mathbf{k}} + \frac{g}{2} n_0 \left(a_{\mathbf{k}}^\dagger a_{-\mathbf{k}}^\dagger + a_{\mathbf{k}} a_{-\mathbf{k}} \right) \right\} \tag{10.12}$$

where we have used the density of the condensed atoms $n_0 = N_0/V$, and the free energy of the atoms $\epsilon_k = \hbar^2 k^2/2M$. Let is now introduce the *Bogoliubov transformations*

$$a_{\mathbf{k}} = u b_{\mathbf{k}} - v b_{-\mathbf{k}}^\dagger , \quad a_{-\mathbf{k}}^\dagger = u b_{-\mathbf{k}}^\dagger - v b_{\mathbf{k}}, \tag{10.13}$$

where the coefficients u are v are assumed real, and will be specified later, and the new operators $b_{\mathbf{k}}$ and $b_{-\mathbf{k}}^\dagger$ satisfy bosonic commutation relations, similar to (10.6), as defined by

$$\left[b_{\mathbf{k}}, b_{\mathbf{k}'}^\dagger \right] = \delta_{\mathbf{k}\mathbf{k}'} , \quad [b_{\mathbf{k}}, b_{\mathbf{k}'}] = 0 , \quad \left[b_{\mathbf{k}}^\dagger, b_{\mathbf{k}'}^\dagger \right] = 0. \tag{10.14}$$

By inserting Eq. (10.13) in the commutation relations (10.6), we can conclude that the coefficients u and v have to satisfy the following condition

$$u^2 - v^2 = 1. \tag{10.15}$$

The inverse transformations can then be easily established, as

$$b_{\mathbf{k}} = u a_{\mathbf{k}} + v a_{-\mathbf{k}}^\dagger , \quad b_{-\mathbf{k}}^\dagger = u a_{-\mathbf{k}}^\dagger + v a_{\mathbf{k}}. \tag{10.16}$$

Replacing the Bogoliubov transformations (10.13) in Eq. (10.12), we obtain an expression in terms of the new operators

$$\begin{aligned} \hat{H} = \frac{N}{2} g n_0 + \sum_{\mathbf{k}\neq 0} \Big\{ & b_{\mathbf{k}}^\dagger b_{\mathbf{k}} \left[\xi_k u^2 - \xi_0 u v \right] + b_{-\mathbf{k}} b_{-\mathbf{k}}^\dagger \left[\xi_k v^2 - \xi_0 u v \right] \\ & + \left(b_{\mathbf{k}}^\dagger b_{-\mathbf{k}} + b_{-\mathbf{k}} b_{\mathbf{k}} \right) \left[-\xi_k u v + \frac{1}{2} \xi_0 \left(u^2 + v^2 \right) \right] \Big\}. \end{aligned} \tag{10.17}$$

Here we have used the auxiliary quantities $\xi_k = \epsilon_k + g n_0$, and $\xi_0 = g n_0$. Let us now choose the coefficients u and v such that the off-diagonal terms in this Hamiltonian vanish, in order to diagonalize it. This implies that

$$\xi_k u v = \frac{1}{2} \xi_0 \left(u^2 + v^2 \right). \tag{10.18}$$

Motivated by condition (10.15), we can now define a single transformation parameter, s, such that

$$u = \cosh s, \quad v = \sinh s. \tag{10.19}$$

Replacing this in (10.18), we get an equation for s, as

$$\xi_0 \left(\cosh^2 s + \sinh^2 s\right) - 2\xi_k \sinh s \cosh s = 0, \tag{10.20}$$

from where we get the solution $\tanh 2s = \xi_0/\xi_k$, or more explicitly

$$s = \frac{1}{2} \tanh^{-1} \left(\frac{g n_0}{\epsilon_k + g n_0} \right). \tag{10.21}$$

This completely defines the Bogoliubov transformation. We can also determine the quantities appearing in (10.17), such as

$$u^2 = \frac{1}{2}\left(\frac{\xi_k}{\nu} - 1\right), \quad u^2 = \frac{1}{2}\left(\frac{\xi_k}{\nu} - 1\right), \quad uv = \frac{1}{2}\frac{\xi_0}{\nu} \tag{10.22}$$

with $\nu^2 = \xi_k^2 - \xi_0^2$. Noting that the coefficients in the summation of (10.17) are independent of the direction of $\mathbf{k}$, and therefore equal for the terms $b_{\mathbf{k}}^\dagger b_{\mathbf{k}}$ and $b_{-\mathbf{k}}^\dagger b_{-\mathbf{k}}$, we can symmetrize the Hamiltonian and finally write

$$\hat{H} = \frac{N}{2} g n_0 + \sum_{\mathbf{k} \neq 0} \hbar\omega_k b_{\mathbf{k}}^\dagger b_{\mathbf{k}} - \frac{1}{2} \sum_{\mathbf{k} \neq 0} [(\epsilon_k + g n_0) - \hbar\omega_k] \tag{10.23}$$

where we have defined

$$\hbar\omega_k = \sqrt{\epsilon_k (\epsilon_k + 2 g n_0)}. \tag{10.24}$$

By using the explicit expression for ϵ_k and reminding the definition of the Bogoliubov sound speed, $c_s = \sqrt{g n_0/M}$, we rewrite this result as

$$\omega_k^2 = c_s^2 k^2 + \frac{\hbar^2 k^4}{4M^2}, \tag{10.25}$$

which coincides with the dispersion relation for sound waves in a BEC previously derived within the fluid description. The Hamiltonian operator $\hat{H}$, as given by Eq. (10.23), is then made of the sum of the zero point energy of the condensate E_0, plus the energy associated with the spectrum of phonons, as

$$\hat{H} = E_0 + \sum_{\mathbf{k} \neq 0} \hbar\omega_k b_{\mathbf{k}}^\dagger b_{\mathbf{k}}. \tag{10.26}$$

with

$$E_0 = \frac{N}{2} g n_0 - \frac{1}{2} \sum_{\mathbf{k} \neq 0} [(\epsilon_k + g n_0) - \hbar \omega_k] \tag{10.27}$$

This equation explicitly states the contributions of both the condensed part, and of the quantum vacuum fluctuations associated with higher level energy states.

10.2 BEC Depletion

Let us now consider the ground state energy in more detail. We first notice that the summation over **k** in Eq. (10.27) diverges in the long wavelength limit. Let us transform for commodity this summation into an integral, by using the auxiliary variable $x = \sqrt{\epsilon_k / g n_0}$. We then get

$$\sum_{\mathbf{k} \neq 0} [(\epsilon_k + g n_0) - \hbar \omega_k] \to \int_0^\infty \left(1 + z^2 - z\sqrt{z^2 + 2} - \frac{1}{2z^2} \right) z^2 dz = -\frac{8}{15}\sqrt{2}, \tag{10.28}$$

where the last term in the integrand was introduced to eliminate the unphysical divergence. We then get, for the energy of the condensed phase, the following result

$$E_0 = \frac{g N^2}{2V} \left[1 + \frac{128}{15} \sqrt{\frac{n a^2}{\pi}} \right], \tag{10.29}$$

where we have used $n_0 \simeq n = N/V$, and the definition of $g = 4\pi\hbar^2 a/M$. This is a well known result, which was first derived by Lee and Yang [1]. Notice that the introduction of the correction term $1/z^2$ in the the integral of (10.28) is not a simple mathematical trick. It indeed represents a correction (valid for finite momentum) to the interaction potential g, which is only valid for zero momentum. The inclusion of higher order corrections (see [2]) would lead to

$$E_0 = \frac{g N^2}{2V} \left[1 + \frac{128}{15} \sqrt{\frac{n a^2}{\pi}} + 8 \left(\frac{4\pi}{3} - \sqrt{3} \right) n a^3 \ln \left(n a^3 \right) \right]. \tag{10.30}$$

As we can see from the Hamiltonian operator (10.26), that the condensate state vector $|\psi\rangle$ can be represented in terms of the Fock states $|n_{\mathbf{k}_1}, n_{\mathbf{k}_2}, \ldots\rangle$, where $n_{\mathbf{k}_j}$ are the phonon occupation numbers with momentum $\mathbf{k}_j$. Using the number operators $b_{\mathbf{k}}^\dagger b_{\mathbf{k}}$, we have

$$b_{\mathbf{k}}^\dagger b_{\mathbf{k}} |\ldots n_{\mathbf{k}} \ldots\rangle = n_{\mathbf{k}} |\ldots n_{\mathbf{k}} \ldots\rangle. \tag{10.31}$$

The condensate ground state $|\psi_0\rangle \equiv |0,0,\ldots\rangle$ is the state where no phonons are present, and satisfy the condition $b_{\mathbf{k}}|\psi_0\rangle = 0$ for every $\mathbf{k}$.

Let us now rewrite the particle number operator (10.8) in terms of the operators $b_{\mathbf{k}}^\dagger$ and $b_{\mathbf{k}}$, as determined by the Bogoliubov transformations (10.13). Noting that the coefficients u and v depend on the value of k, we get

$$\hat{N} = N_0 + \sum_{\mathbf{k}\neq 0} \left[v_k^2 + \left(u_k^2 + v_k^2\right) b_{\mathbf{k}}^\dagger b_{\mathbf{k}} - u_k v_k \left(b_{\mathbf{k}}^\dagger b_{-\mathbf{k}}^\dagger + b_{-\mathbf{k}} b_{\mathbf{k}} \right) \right], \tag{10.32}$$

We can see that the last term in this expression is equal to zero, for eigenstates of the Hamiltonian (10.26). This allows us to write, using the explicit expressions for the coefficients u_k and v_k as given by equations (10.22), this operator can be written as

$$\hat{N} = N_0 + \sum_{\mathbf{k}\neq 0} v_k^2 + \sum_{\mathbf{k}\neq 0} \frac{\xi_k}{\sqrt{\xi_k^2 - \xi_0^2}} b_{\mathbf{k}}^\dagger b_{\mathbf{k}}. \tag{10.33}$$

This shows that, in the presence of phonon modes, the number of condensed particles has to change, for a constant number of particles in the system. We then have to make the change $N_0 \to N_0 + \delta N_0$, where the variation in the number of condensed particles is given by

$$\delta N_0 = -\sum_{\mathbf{k}\neq 0} v_k^2 = -\frac{1}{2} \sum_{\mathbf{k}\neq 0} \left(\frac{\xi_k}{\sqrt{\xi_k^2 - \xi_0^2}} - 1 \right). \tag{10.34}$$

By transforming the sum into an integral, and solving the integral, we then get

$$N_0 + \delta N_0 = N_0 \left(1 - \frac{8}{3} \sqrt{\frac{na^3}{\pi}} \right). \tag{10.35}$$

Such a reduction in the number of condensed atoms is called the *condensate depletion*. This shows that, for a low density gas, such that $(an^3) \ll 1$, the Bogoliubov theory which assumes that most of the atoms are in the condensed state remains valid, which turns out to be the case for the majority of experiments with alkaline atoms. However, with the use of Feshbach resonances, the value of the scattering length a (consequently the interaction strength) can be increased, and the condensate depletion becomes important. This is a signature that a mean-field description of BECs is valid as long the interactions are not very strong.

10.3 Phonon Pair Creation

10.3.1 Time Refraction

Due to its extremely low temperature, which implies a negligible thermal emission, the condensate offers remarkable conditions for the study of quantum vacuum processes. This was proposed recently, for detection of Hawking and Unruh radiation, and the identification of horizon where particle emission of radiation, in this case sound waves, could take place [3]. Here we consider a closely related problem of quantum vacuum, that of time refraction of phonons in a condensate.

It is well known that three different processes of quantum vacuum can be considered: time refraction, dynamical Casimir effect and Unruh radiation. They all lead to the emission of particle pairs [4]. The concept of time refraction [5], is an extension of the usual (space) refraction into the temporal domain. It was first developed for photons in optical media, and then extended to plasmons. It is associated to the temporal change of the optical properties of a dielectric medium. This concept has both been explored in the classical and quantum domains. The similarities between time refraction and the dynamical Casimir effect have also been explored [4], where the empty cavity with a vibrating mirror was replaced by a static cavity with a time dependent optical medium. The equivalence between the two configurations was demonstrated, and, in the absence of losses, an exponential growth of the photon number inside the cavity was predicted.

The dynamical Casimir effects is an instability of the electrodynamical vacuum, which has received considerable attention in recent years (see for a recent review[6]). For an empty optical cavity with a vibrating mirror, with no losses, it leads to the exponential growth of trapped photons, if the oscillating frequency of the mirror is equal to twice the photon frequency. Its main interest is related with the fact that it can occur for arbitrarily low photon energies, thus stimulating the hope of observing quantum vacuum properties in small scale experiments.

In this section, we discuss first time refraction, and then the dynamical Casimir effect, for phonon modes in Bose Einstein condensates. They both lead to the emission of phonons propagating in opposite directions. For the present purpose, we consider a condensate where the trapping potential is cylindrically symmetric, as given by $V(\mathbf{r}) = m(\omega_\perp^2 r_\perp^2 + \omega_z^2 z^2)/2$, providing a strong confinement along the transverse direction, $\omega_\perp^2 \gg \omega_z^2$. This allows us to neglect, in a first step, the axial confinement by assuming $\omega_z^2 \simeq 0$.

We assume that some equilibrium state of the condensate has been achieved, such that $n = n_0$. By suddenly changing the transverse trapping frequency $\omega_\perp$, we change the equilibrium density on a very fast time scale. This means that we can assume a time dependent quasi-equilibrium density $n_0(t) = n_0 f(t)$, where $|f(t)| \simeq 1$. We then consider $n = n_0(t) + \tilde{n}$, where $\tilde{n}$ describes a perturbation propagating in the axial direction. Linearizing the quantum fluid equations, we can easily derive an evolution equation for the density perturbations, which takes the form

$$\left(\frac{\partial^2}{\partial t^2} - c_s(t)\frac{\partial^2}{\partial z}\right)\tilde{n} = \nu(t)\frac{\partial \tilde{n}}{\partial t} - \frac{\hbar^2 f(t)}{4m^2}\frac{\partial^4}{\partial z^4}\tilde{n}, \tag{10.36}$$

where we have defined

$$\nu(t) = \frac{d\ln f(t)}{dt}, \quad c_s^2 = \frac{gn_0}{m}f(t). \tag{10.37}$$

Notice that in the absence of any temporal variation of the background density, we set $f(t) = 1$ and $\nu = 0$, and for wave solutions of the form $\tilde{n} \propto \exp(-i\omega t + ikz)$ we obtain the dispersion relation for sound waves in the condensate $\omega^2 = c_s^2k^2 + (\hbar^2/4m^2)k^4$. For very slow temporal changes, such that $\nu \sim 0$ and the first term on the r.h.s. of Eq. (10.36) is negligible, we can try solutions of the form

$$\tilde{n}(z,t) = Ae^{\pm ikz}\exp\left[-i\int^t \omega(t')dt'\right], \tag{10.38}$$

which are a WKB-like approximation. Replacing this in Eq. (10.36), we obtain the time dependent dispersion relation

$$\omega^2(t) = c_s^2(t)k^2 + \frac{\hbar^2 f(t)}{4m^2}k^4. \tag{10.39}$$

This shows that, in a region of uniform density larger than $2\pi/k$, the temporal change in the background medium will imply a frequency shift for the sound wave propagating in the condensate. This is one of the main characteristic feature of time refraction, well suited for experimental purposes.

We now consider the case where the term in $\nu(t)$ cannot be neglected. An exact solution of Eq. (10.36) can still be found, where two distinct solutions $\tilde{n}_k$ and $\tilde{n}_{-k}$, similar to (10.39), with the same axial wavenumber k, but propagating in opposite directions will become coupled by the time variation of the medium, as shown next.

The sound speed c_s determines the characteristic time for changes in the density of a Bose Einstein condensate. For changes along the transverse direction, the time scale for the mean density variations is of order $R_\perp/c_s$, where $R_\perp$ is the transverse Thomas-Fermi radius of the condensate. This can be of the order or shorter than the period of the phonon oscillations with frequency ω if the following condition $R_\perp k \leq 1$ is satisfied. An alternative way, eventually more interesting, of changing the mean density, is to act on the scattering length, by using a Feshbach resonance.

Let us start by considering a sudden change in the medium, which can be assumed as instantaneous on the time scale of a phonon period $1/\omega$. In order for the fluid wave equation (10.36) to stay valid for all times, including the time of discontinuity $t = 0$, the following continuity conditions have to be satisfied

$$\tilde{n}(z,t=0^+) = \tilde{n}(z,t=0^-) \quad \text{and} \quad \tilde{v}(z,t=0^+) = \tilde{v}(z,t=0^-). \tag{10.40}$$

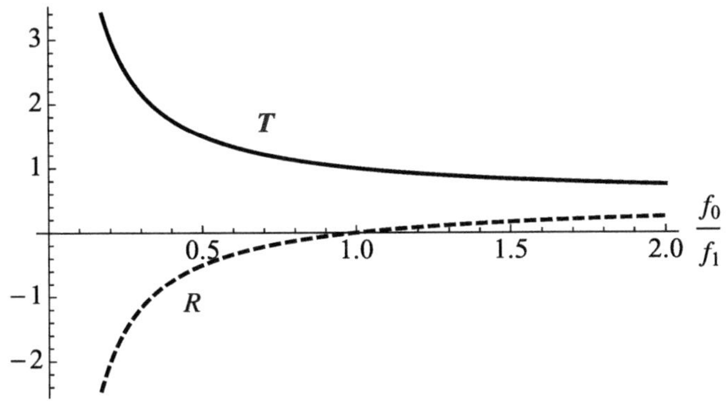

Fig. 10.1 Time transmission T, and reflection R coefficients, as a function of the temporal discontinuities of the atom density, represented by the ratio f_0/f_1

If, for $t < 0$, we had a single sound wave propagating along the condensate, as described by

$$\tilde{n}(z, t < 0) = A_0 \exp(ikz - i\omega_0 t) + \text{c.c.}, \tag{10.41}$$

then, for $t > 0$, we will have

$$\tilde{n}(z, t\ 0) = A_1 \exp(ikz - i\omega_1 t) + A_1' \exp(ikz + i\omega_1 t) + \text{c.c.} \tag{10.42}$$

In these solutions, the phonon frequencies ω_0 and ω_1 have to obey the dispersion relation (10.39), with $f(t < 0) = f_0$, and $f(t > 0) = f_1$, with the corresponding sound velocities $c_{s,j}^2 = gn_0 f_j/m$, for $j = 0, 1$. The term in A_1' has to be included in the wave solution (10.42) in order to satisfy the continuity conditions (10.40). Making use of these conditions, we can easily obtain a relation between the different phonon amplitudes, such that

$$A_0 = A_1 + A_1' \ , \quad \alpha_0 A_0 = \alpha_1 (A_1 - A_1'), \tag{10.43}$$

with $\alpha_j = \omega_j / n_0 f_j k$, for $j = 0, 1$. This allows us to calculate the reflected and transmitted coefficients associates with a temporal discontinuity, as given by

$$T \equiv \frac{A_1}{A_0} = \frac{1+\alpha}{2\alpha}, \quad R \equiv \frac{A_1'}{A_0} = -\frac{(1-\alpha)}{2\alpha}, \tag{10.44}$$

where we have used $\alpha = \alpha_1/\alpha_0 = (\omega_1/\omega_0)(f_0/f_1)$. This shows that a temporal discontinuity, not only leads to a frequency shift, as noticed before, but also leads to the appearance of a counter-propagating phonon (as in ordinary reflection). These two features are qualitatively analogous to those of a photon field [7]. The results are illustrated in Fig. 10.1, where it is clearly shown that these temporal coefficients are such that the sum $|T|^2 + |R|^2 \neq 1$, and can be larger than one. This results from the fact that time refraction conserves wave momentum, but not wave energy (in contrast with the usual refraction, which conserves wave energy but not momentum). The additional energy is provided by the external source, responsible for the temporal change in the condensate.

Let us finally consider the general case of an arbitrary temporal change in the background density, as described by $n(t) = n_0 f(t)$. This can be treated as a succession of infinitesimal discontinuities of the kind just described. A more direct method, leading to the same final result, is the following. We assume a solution of the form

$$\tilde{n}(z,t) = A(t)\exp[ikz - i\varphi(t)] + A'(t)\exp[ikz + i\varphi(t)] + \text{c.c.}, \quad (10.45)$$

with the phase given by $\varphi(t) = \int^t \omega(t')dt'$. Replacing this in Eq. (10.36), and equating the terms with the same space dependence, $\exp(ikz)$, we get

$$\frac{dA'}{dt} = \eta(t)A\,, \quad \frac{dA}{dt} = \eta^*(t)A', \quad (10.46)$$

with

$$\eta(t) = \frac{1}{2}\left(\frac{1}{\omega}\frac{d\omega}{dt} - \nu(t)\right)\exp[-2i\varphi(t)]. \quad (10.47)$$

Here, we have assumed that the wave amplitudes vary slowly with respect with the wave period, such that $|dA/dt| \ll |\omega A|$, and $|dA'/dt| \ll |\omega A'|$. In the simple but important case where initially a single wave mode is excited in the medium, the effect of time refraction effect is not very strong. We can then assume that $|A'| \ll |A|$ for all times, and $A(t) \simeq$ const. This allows us to integrate the first of equations (10.46), leading to

$$A'(t) = R(t)A(t), \quad (10.48)$$

where the reflection coefficient is now given by

$$R(t) \simeq 1 - \frac{1}{2}\int^t \eta(t')dt'. \quad (10.49)$$

This is actually very similar to the reflection coefficient $R(z)$ associated with the usual (space) refraction, with the space variable z replaced by the temporal one t. However, the physical meaning is clearly distinct, because there is no physical boundary or discontinuity from where the initial phonon wave could be reflected, if the medium is assumed as axially homogeneous. The above discussion is purely classical, but it can easily be translated into the language of quantum field theory. The corresponding phonon field operators will be introduced next.

10.3.2 Dynamical Casimir Effect

We have seen above that, a given k phonon mode can be described by the oscillator equation

$$\left[\frac{d^2}{dt^2} + \omega^2(t)\right]\tilde{n} = \nu(t)\frac{d\tilde{n}}{dt}, \tag{10.50}$$

where $\omega(t)$ is determined by the instantaneous dispersion relation (10.39). In the absence of temporal perturbations of the condensate, $\nu(t) = 0$, this oscillator can be quantized in the usual way. And, generalizing this quantization to a time varying medium, we can define the quantum operator

$$\tilde{n}(z,t) = A_k(t)\exp[ikz - i\varphi(t)] + A_{-k}^{\dagger}(t)\exp[ikz + i\varphi(t)] + \text{h.c.} \tag{10.51}$$

This is formally identical to Eq. (10.45), where the amplitude A is replaced by A_k, the destruction operator corresponding to the phonon state k, and A' by $A_{-k}^{\dagger}$, the creation operator corresponding to the phonon state $-k$, which propagates in the opposite direction. By following a procedure similar to that described above, we can then get the coupled equations describing the temporal evolution of these two operators

$$\frac{dA_{-k}^{\dagger}}{dt} = \eta(t)A_k, \quad \frac{dA_k}{dt} = \eta^*(t)A_{-k}^{\dagger}. \tag{10.52}$$

It is now useful to consider the perturbations of the phonon vacuum, induced by the temporal changes in the condensate. For that purpose, we focus on the paradigmatic case os a sinusoidal perturbation of the form

$$f(t) = 1 + \epsilon\sin(\Omega t) \quad \text{and} \quad \nu(t) \simeq \epsilon\Omega\cos(\Omega t), \tag{10.53}$$

where the amplitude of the modulation is assumed small $\epsilon \ll 1$. This allows us to derive, to the lowest order in ϵ,

$$\eta(t) \simeq -\frac{\epsilon\Omega}{2^3}\left(e^{i\omega_+ t} + e^{i\omega_- t}\right), \quad \omega_\pm = 2\omega_0 \pm \Omega \tag{10.54}$$

where ω_0 satisfies the phonon dispersion relation for the unperturbed case $\epsilon = 0$. It is important to notice that only the constant or the slowly varying part of $\eta(t)$ will contribute significantly to the evolution of the operators, as seen from the above equations. The oscillating part will average to zero, It is then obvious that a strong effect only occurs for a driving frequency such that $\Omega = 2\omega_0$. We then have

$$\eta(t) = \eta_0\left(1 + e^{3i\omega_0 t}\right), \quad \text{with} \quad \eta_0 = -\frac{\epsilon\omega_0}{4}. \tag{10.55}$$

We can then approximate the relations in (10.52) by

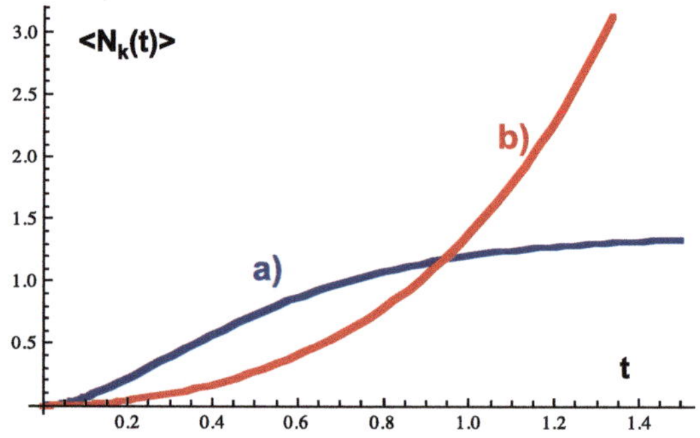

Fig. 10.2 Number of phonon pair created from vacuum, (*a*) for a sudden change in the atom density, as described by $n(t) - n_0 = [1 + \tanh(2t - 1)]/2$; (*b*) for a sinusoidal perturbation such that $\eta_0 = 1$

$$\frac{dA_{-k}^{\dagger}}{dt} = \eta_0 A_k, \quad \frac{dA_k}{dt} = \eta_0 A_{-k}^{\dagger}. \tag{10.56}$$

The relevant solution is therefore given by

$$\begin{aligned} A_k(t) &= \cosh(\eta_0 t) A_k(0) + \sinh(\eta_0 t) A_{-k}^{\dagger}(0), \\ A_{-k}^{\dagger}(t) &= \cosh(\eta_0 t) A_{-k}^{\dagger}(0) + \sinh(\eta_0 t) A_k(0). \end{aligned} \tag{10.57}$$

We can now calculate the mean number of phonons generated from vacuum at time t. The result is

$$\langle N_k(t) \rangle \equiv \langle A_k^{\dagger}(t) A_k(t) \rangle = \sinh^2(\eta_0 t). \tag{10.58}$$

The same number is generated for phonons propagating in the opposite direction, in order to preserve the total momentum of vacuum, $\langle N_{-k}(t) \rangle = \langle N_k(t) \rangle$. We can see that, for short times, we have a quadratic growth, $\langle N_k(t) \rangle \simeq \eta_0^2 t^2$, whereas for long times, we observe an exponential growth, $\langle N_k(t) \rangle \simeq \exp(2|\eta_0|t)$. This instability of the phonon vacuum will eventually be prevented due to a finite coherence time, or finite life time, of the phonons inside the condensate, as discussed recently for the case of photons [8]. The cases of a sudden change and a sinusoidal perturbation of the condensate density are illustrated in Fig. 10.2.

The instability that we have just described is formally analogous to the dynamical Casimir effect for the photon field inside an empty optical cavity with a modulated length. Here, the photons were replaced by phonons, and the optical medium by the Bose-Einstein condensate. The phonon version of time refraction in a Bose Einstein condensate can be seen as an interesting alternative for the observation of such vacuum processes.

But an oscillating length could also be conceived for a condensate. We known that the Thomas-Fermi length L in the axial direction is determined by the trapping potential, and the phonon wavenumber could be quantized due to these axial boundary conditions, which were ignored above. Let us focus on some phonon mode with wavenumber $k = k_l \equiv 2\pi l/L$, where l is an integer. It is obvious that a temporal variation of the axial length will introduce a shift in the mode frequency. Assuming $L(t) = L_0[1 - \epsilon(\Omega t)]$, we get, for $\epsilon \ll 1$,

$$\omega^2(t) = \omega_0^2[1 + \epsilon' \sin(\Omega t)] \ , \quad \epsilon' = \left(2 + \frac{\hbar^2 k_{l0}^4}{4m^2\omega_0^2}\right)\epsilon, \tag{10.59}$$

where $k_{l0} = 2\pi l/L_0$, and ω_0 is the corresponding mode frequency. This is similar to the situation considered before, where the length was fixed and only the background density was assumed to change. Now we have $\nu(t) = 0$, because the density is assumed fixed, but the frequency shift will lead to a new value of the parameter $\eta(t)$, which is now given by

$$\eta(t) = \frac{1}{2}\epsilon' \Omega \cos(\Omega t) e^{-2i\varphi}. \tag{10.60}$$

Phonon pair creation is also expected to occur here, as determined by Eq. (10.58), with η_0 replaced by η_{DC} (standing for dynamical Casimir), which is the constant part of this new quantity. A relation with the previous physical configuration is very easy to establish and can be stated as $\eta_{DC} = 2|\eta_0|\epsilon/\epsilon'$. This shows the equivalence, in terms of phonon vacuum properties, between the two distinct physical configurations: (1) *time refraction*, characterized by a constant axial length of the condensate L_0, but a varying background density $n(t)$, as imposed by a time variation of the transverse confining potential; (2) the *dynamical Casimir effect*, associated with a variable length $L(t)$, in a constant density n_0. The difference with respect to the case of photon vacuum is that, in a Bose-Einstein condensate, these two configurations can eventually coexist.

10.4 Acoustic Black Holes

Analogue models for general relativity provide specific and clear examples in which effective space-time structures emerge from condensed-matter systems [9]. For low levels of resolution, and under certain circumstances, one can appropriately describe the physical properties of the system by means of a classical (or quantum) field theory in a curved Lorentzian space-time. At higher resolution levels, however, it is necessary to proceed to a full quantum theory [10]. Therefore, although analogue models cannot be considered as complete models of quantum gravity (since they do not lead to the Einstein equations in any regime or approximation), they provide specific and tractable models that reproduce many aspects of the overall scenario expected in the realm of real gravity (an extended discussion can be found in Ref. [11]).

Here, we summarize the main properties of gravity-like configurations with Bose-Einstein condensates, revealing the emergence of an effective acoustic metric.

An interesting feature of these systems is that their theoretical description in terms of the GP equation naturally contains some quantum features, which in their turn are the result of the quantum coherence of the condensation phenomenon itself.

10.4.1 Hawking Radiation

According to Einstein's general theory of relativity, the space-time is deformed in the presence of massive objects, like a star or a planet, which are responsible for the curvature of space-time. If the source of matter is dense enough, the space-time metric can show a singularity, which is identified as a *black hole.* The black hole is characterized by a radius, known as *horizon,* beyond which nothing (not even light) can scape.

Classically, a black hole is indeed black, i.e. does not radiate. However, some quantum features associated with the black hole suggest that it may not be the case. Due to the amount of available gravitational energy, pairs of photons are formed, as a consequence of electromagnetic fluctuations. At the horizon, some of those photons are captured by the hole, while some of them can in fact scape. The latter are responsible for the so-called *Hawking radiation.*

In order to illustrate the physics underlying the Hawking radiation mechanism, we consider a spherically symmetric Schwarzchild metric (in astrophysical units $G = c = \hbar = 1$) [12]

$$ds^2 = g_{\mu\nu} dx_\mu dx_\nu = -\left(1 - \frac{2M}{r}\right) dt^2 + \left(1 - \frac{2M}{r}\right)^{-1} dr^2 + r^2 d\Omega^2, \quad (10.61)$$

where M represents the mass of the black hole. The horizon is located at the distance where the metric is singular, i.e. $r_{\text{hor}} = 2M$. Electromagnetic fluctuations allow photon pairs of energy δE to be produced, which coexist for a short period of time $\delta t = \hbar/\Delta E$. If such a pair of photons is created at the black hole horizon, the one with negative energy $-E$ can cross the horizon within a time $\hbar/E$ without recombining with that of positive energy and simply falls inside the hole. An observer located inside the horizon measures the energy of an incoming photon to be

$$E = E_- - \left(\frac{2M}{r} - 1\right)^{-1/2} p_r, \quad (10.62)$$

where p_r is the radial component of the photon four-momentum. On the other hand, the other photon is allowed to scape to infinity. The energy of the photon at infinity as measured by an observer freely falling toward the horizon can be shown to be given by [15]

$$E = E_+ = \frac{\hbar}{4M}. \quad (10.63)$$

Accordingly, the black hole radiation spectrum is characteristic of a black body with a temperature, in energy units, of

$$T_H = \frac{\hbar}{8\pi M}. \tag{10.64}$$

In what follows, we present the acoustic analogue of Hawking radiation in Bose Einstein condensates. Because Hawking radiation has never been observed (either in astrophysical or in other physical scenarios), experimental results on its acoustic analogue in BEC would be of major importance.

10.4.2 Effective Metric in a Condensate

Let us consider the quantum hydrodynamical Eqs. (7.76) and (7.77) derived in section (7.6). In many relevant situations, the quantum Bohm potential can be neglected. The resulting equations then reduce to the continuity and momentum equations for a classical fluid. In this case, and making use of the irrotationality condition $\mathbf{v} = \hbar\nabla\varphi/M$, it is possible to derive the following equation for the condensate phase

$$\left(\frac{\partial}{\partial t} + \nabla\cdot\mathbf{v}\right)\frac{n_0}{c_s^2}\left(\frac{\partial}{\partial t} + \cdot\nabla\mathbf{v}\right)\varphi = \nabla\cdot(n_0\nabla\varphi). \tag{10.65}$$

This equation can be compared wit the relativistic wave equation of the scalar field φ

$$\Box\varphi = \frac{1}{\sqrt{-g}}\partial_\mu\left(\sqrt{-g}g^{\mu\nu}\partial_\nu\varphi\right) = 0, \tag{10.66}$$

where $g^{\mu\nu}$ represents the acoustic analogue of the *Painlevé-Lemaitre metric*

$$g^{\mu\nu} = \frac{1}{n_0 c_s}\begin{pmatrix} \mathbf{1} & \mathbf{v} \\ \mathbf{v} & \mathbf{v}\otimes\mathbf{v} - c_s^2\mathbf{1}\end{pmatrix}, \tag{10.67}$$

with $\mathbf{1}$ representing the 3×3 unit matrix. This analogy was first proposed by Unruh [13]. In this class of metrics, a singularity can occur, exactly where the flow velocity equals the local sound velocity c_s. Different number of configurations can be imagined in the laboratory, where such a singularity could be observed. To better understand the correspondence of this emergent geometry and a black hole, let us consider the case of a one dimensional flow $\mathbf{v} = v_0(x)\mathbf{e}_x$. In the $(1+1)-$dimensional plane, the metric (10.67) yields the following invariant interval

$$ds^2 = \frac{1}{n_0 c_s}\left[-\left(1 - \frac{v_0(x)^2}{c_s^2(x)}\right)c_s^2 d\tau^2 + \left(1 - \frac{v_0(x)^2}{c_s^2(x)}\right)^{-1} dx^2\right], \tag{10.68}$$

where we have rescaled the time variable τ as

$$\frac{d\tau}{dt} = \frac{v_0}{c_s^2 - v_0^2} dx. \tag{10.69}$$

The case of a spherically symmetric flow with a 'sink' at the origin can be described by a velocity field of the form $\mathbf{v} = -v_0(r)\mathbf{e}_r/r$, which is associated with the following metric [14]

$$ds^2 = -\left(c_s^2 - \frac{v_0}{r}\right) dt^2 - 2\frac{v_0}{r} dr dt + dr^2 + r^2 d\theta^2, \tag{10.70}$$

which is the (2 + 1)-dimensional analogue of the Schwarzchild metric in (10.61) and, therefore, may be interpreted as the acoustic equivalent of a black hole. This suggests that the region at which $v_0(x) = c_s(x)$ can be regarded as the *sonic horizon*.

10.4.3 Acoustic Hawking (Unruh) Radiation

Following the original argument put forward by Hawking to describe the radiation of black holes, Unruh derived the equivalent temperature at which phonon pair creation should occur at the acoustic horizon. This equivalence is based on the definition of the acoustic analogue of *surface gravity*, as defined below. Although the original derivation does not depend on the characteristics of the fluid, we will see that the BECs dynamics contain special features that may allow for an easier experimental observation. We restrict our discussion to the important case of static metrics (associated to steady flows). This follows from the fact that the acoustic surface gravity depends on the gradients of the fluid flow.

Let us consider the one-dimensional flow described by the metric in Eq. (10.67). The interesting question to ask at this point is exactly what happens if we quantize the phonon modes at the vicinity of the sonic horizon? To do that, we make use of the following decomposition for the outgoing modes

$$\varphi = \sum_{\omega} \left(\varphi_\omega a_\omega + \varphi_\omega^* a_\omega^\dagger ger\right). \tag{10.71}$$

Following Unruh's derivation, the idea is to consider that far from the horizon, an observer traveling with the fluid as it flows through the sonic horizon will perceive the field φ as being essentially in the vacuum state. At the horizon, this corresponds to black hole radiation at the temperature [13–15]

$$T_{\mathrm{H,eff}} = \frac{\hbar}{2\pi}\frac{1}{2}\frac{\partial\left[v(x)^2 - c(x)^2\right]}{\partial x} = \frac{\hbar\kappa}{2\pi}, \tag{10.72}$$

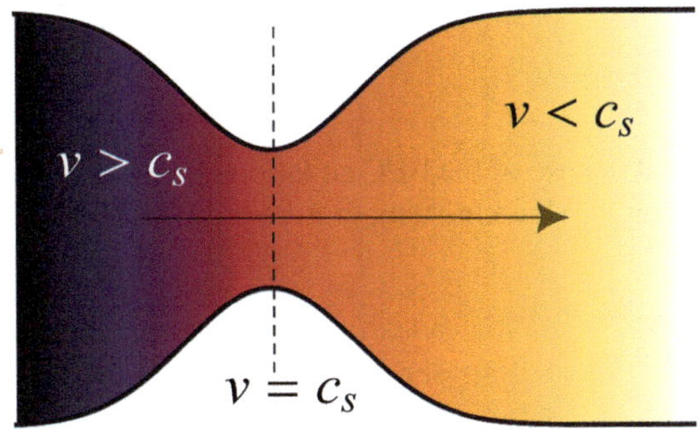

Fig. 10.3 de Laval nozzle configuration

where $\kappa = |\partial(v - c_s)/\partial x|$ represents the acoustic surface gravity. The derivation can easily be extended to different geometries, by redefining the gravity surface in terms of the normal $\hat{\mathbf{n}}$ of the horizon

$$\kappa = \nabla(v - c_s) \cdot \hat{\mathbf{n}}. \tag{10.73}$$

One of the most interesting proposals for experimental observation of acoustic horizons with condensates involves the *de Laval nozzle* configuration [10]. Consider a BEC flowing through a nozzle of transverse section $A(x)$ in the x direction as depicted in Fig. 10.3. Making use of the fluid equations for a stationary flow, it can be easily shown that the superfluid acceleration a along the nozzle can be expressed as

$$a = -\frac{v^2 c_s^2}{c_s^2 - v^2} \frac{A'}{A}, \tag{10.74}$$

where $A'(x)$ is the section derivative. At the horizon, both A' and $c_s^2 - v^2$ approach zero. Therefore, applying l'Hôpital's rule, we obtain the limit value

$$a = \left. \frac{-c^4 A''/A}{(c^2)' - 2a_H} \right|_H . \tag{10.75}$$

The derivative of the square of the local sound speed can be easily related with the event horizon as follows

$$(c_s^2)' \equiv \frac{1}{M} \frac{d^2 n}{dn^2} n' = -\frac{n}{M} \frac{d^2 p}{dn^2} \frac{(Av)'}{Av} = -\left. \frac{n_H}{M} \frac{d^2 p}{dn^2} \right|_H \left(\frac{A'_H}{A_H} + \frac{a_H}{c_{s,\mathrm{H}}^2} \right). \tag{10.76}$$

Further similar calculations allow to determine the acoustic gravity surface at the horizon, which simply reads

$$\kappa_H = \frac{c_{\rm s,H}^2}{\sqrt{2A}} \sqrt{1 + \frac{1}{M}\frac{n}{2c_s^2}\frac{d^2p}{dn^2}\bigg|_H} \sqrt{A_H''}. \tag{10.77}$$

Making use of the BEC equation of state, $c_s^2 = gn/M$, we can easily determine the acoustic Hawking temperature in Eq. (10.72), which yields

$$T_{\rm H,eff} = \frac{\hbar c_{\rm s,H}}{2\pi\sqrt{A_H}} \sqrt{\frac{3A_H''}{4}}. \tag{10.78}$$

This simple result can also be given in terms of the phonon wavelength at which the emission spectrum is peaked [11],

$$\lambda_{\rm peak} = 4\pi^2\sqrt{A_H}\sqrt{\frac{4}{3A_H''}}. \tag{10.79}$$

It is important to note, however, that A_H'' cannot be chosen as arbitrarily large, since this would violate the quasi-one-dimensional approximation used in this derivation. Without going into much detail (for which the reader is urged to check Ref. [11]), the present analysis shows that the best realistic situation is achieved for a spectrum peaked at $\lambda_{\rm peak} \approx \sqrt{A_H}$. In fact, this is exactly why the acoustic version of Hawking radiation is so difficult to observe, as the peak wavelength is of the order of the typical physical dimensions of the system. For realistic BEC experiments, the typical sound velocity is $c_S \sim 6\,$mm/s. If the nozzle diameter is then chosen to value $A_H \sim 1\,\mu$m, and the flare-rate to be $A_H'' \sim 1$, then acoustic Hawking radiation is expected to occur at

$$T_H \approx 7 \text{ nK}. \tag{10.80}$$

Because T_H is comparable (in magnitude) with the condensation temperature ($T_c \sim 90\,$nK), the de Laval nozzle configuration has gained much attention as a possible stage to observe the phenomenon. More likely is the scenario envisaged with the help of Feshbach resonances [16]. By increasing the scattering length up to 100 times [17], the sound speed could be enhanced by a factor of 10 (notice that $c_s \sim \sqrt{a_s}$, which would lead to $T_H \approx 70\,$nK).

References

1. T.D. Lee, C.N. Yang, Phys. Rev. **105**, 1119 (1957)
2. M. Ueda, *Fundamentals and new frontiers of Bose Einstein condensation* (World Scientific, Singapore, 2010)
3. A. Retzker, J.I. Cirac, M.B. Plenio, B. Reznik, Phys. Rev. Lett. **101**, 110402 (2008)
4. J.T. Mendonça, G. Brodin, M. Marklund, Phys. Lett. A **372**, 5621 (2008)
5. J.T. Mendonça, A. Guerreiro, Phys. Rev. A **72**, 063805 (2005)

6. V.V. Dodonov, Phys. Scr. **82**, 038105 (2010)
7. J.T. Mendonça, *Theory of Photon Acceleration* (Institute of Physics Publications, Bristol, 2001)
8. J.T. Mendonça, G. Brodin, M. Marklund, Phys. Lett. A **375**, 2665 (2011)
9. M. Novello, M. Visser, G. Volovik, *Artificial Black Holes* (World Scientific, Singapore, 2002)
10. C. Barceló, S. Liberati, M. Visser, Class. Quant. Gravit. **18**, 1137 (2001)
11. C. Barceló, A. Cano, L.J. Garay, G. Jannes, Phys. Rev. D **74**, 024008 (2006)
12. B.F. Schutz, *A First Course in General Relativity* (Cambridge University Press, Cambridge, 1985)
13. W.G. Unruh, Phys. Rev. Lett. **46**, 1351 (1981)
14. M. Visser, Class. Quant. Gravit. **15**, 1767 (1998)
15. S.W. Hawking, Nature **248**, 5443 (1974)
16. J.L. Roberts, N.R. Claussen, S.L. Cornish, C.E. Wieman, Phys. Rev. Lett. **85**, 728 (2000)
17. S.L. Cornish, N.R. Claussen, J.L. Roberts, E.A. Cornell, C.E. Wieman, Phys. Rev. Lett. **85**, 1795 (2000)

Chapter 11
Superfluidity

Superfluid is a fluid with an extremely low viscosity and high thermal conductivity. The current description of superfluidity, first proposed by Tisza [1] soon after the discovery of Helium-4 superfluidity, is based on the assumption that two different fluids coexist. One is the superfluid, which corresponds to the condensed gas. The other is the normal fluid, which in our current view is made of quasi-particle excitations, or phonons.

We have seen that the condensed gas can be described by a wave kinetic equation, which is equivalent to the Gross-Pitaevskii equation and is valid in the mean-field approximation. We will show here that the gas of phonons can also be described by a similar wave kinetic equation.

We then study the excitation of long wavelength perturbations in this mixture of two fluids. The corresponding dispersion relation will be established. We also show that these long wavelength perturbations can be Landau damped by phonons. The chapter is completed with a discussion of the rotons, in the context of condensates.

11.1 Phonon Kinetics

Let us start from the fluid equations for the condensate. We have shown that, by linearizing these equations with respect to density perturbations $\tilde{n} = n - n_0$, where n_0 is the equilibrium density, and $|\tilde{n}| \ll n_0$, we obtain a wave equation of the form

$$\left(\frac{\partial^2}{\partial t^2} - c_s^2 \nabla^2 + \frac{\hbar^2}{4M^2} \nabla^4 \right) \tilde{n} = 0. \tag{11.1}$$

We have also seen that, in a uniform gas and for perturbations of the form $\exp(i\mathbf{k}\cdot\mathbf{r} - i\omega t)$, we get the phonon dispersion relation $\omega^2 = c_s^2 k^2 + (\hbar^2 k^4 / 4M^2)$, where $c_s = \sqrt{g n_0 / M}$ is the Bogoliubov sound speed. We now consider a more

J.T. Mendonça and H. Terças, *Physics of Ultra-Cold Matter*, Springer Series on Atomic, Optical, and Plasma Physics 70, DOI 10.1007/978-1-4614-5413-7_11,

generic situation where the background density evolves in space and time as $n_0 \rightarrow n_0 + n'(\mathbf{r}, t)$ and the resulting sound speed becomes

$$c_s^2(\mathbf{r}, t) = (g/M)[n_0 + n'(\mathbf{r}, t)]. \tag{11.2}$$

Here, the quantity $n'(\mathbf{r}, t)$ is, for instance, associated with a long wavelength perturbation in the medium, as considered later. In order to study the solution of Eqs. (11.1) and (11.2), we use the generalized space-time Wigner-Moyal procedure already discussed in previous chapters, and derive the corresponding wave kinetic equation. This kinetic treatment of the phonon gas will revel a collective force that acts back on the BEC. In other words, this will give a different perspective of the interaction between the normal and superfluid fractions, thus representing an alternative to both the two-fluid theory of Landau [2] and Khalatnikov [3] and the kinetic ZNG theory [4].

In the same spirit of our previous derivations, we introduce the auto-correlation function, as

$$K_{12} = \tilde{n}_1 \tilde{n}_2^* , \quad \tilde{n}_j \equiv \tilde{n}(\mathbf{r}_j, t_j) \tag{11.3}$$

for $j = (1, 2)$, and use new space and time variables, $\mathbf{s}$ and τ, such that

$$\mathbf{r}_1 = \mathbf{r} + \mathbf{s} , \quad t_1 = t + \tau/2$$

$$\mathbf{r}_2 = \mathbf{r} - \mathbf{s} , \quad t_2 = t - \tau/2 \tag{11.4}$$

With the help of the wave equation (11.1), we can then derive an evolution equation for the auto-correlation function, which reads

$$\left[2\frac{\partial^2}{\partial \tau \partial t} - (c_1^2 - c_2^2)\nabla_s^2 - 2c_s^2(\nabla \cdot \nabla_2) + \frac{\hbar^2}{M}\nabla_s^2(\nabla \cdot \nabla_s) \right\} K_{12} = 0, \tag{11.5}$$

where $c_j = c_s(\mathbf{r}_j, t_j)$, and $\nabla_s \equiv \partial^2/\partial \mathbf{s}^2$. At this point, we introduce the double Fourier transformation of K_{12}, in order to get the Wigner function for the phonon field, as defined by

$$F(\mathbf{r}, t; \omega, \mathbf{k}) = \int d\mathbf{s} \int d\tau K_{12} \exp(-i\mathbf{k} \cdot \mathbf{s} - i\omega\tau). \tag{11.6}$$

Replacing this in Eq. (11.5), we can then obtain

$$i\left(\frac{\partial}{\partial t} + \mathbf{v}_k \cdot \nabla\right) F = \frac{k^2}{2\omega}\frac{g}{m} \int \frac{d\mathbf{q}}{(2\pi)^2} \int \frac{d\Omega}{2\pi} n'(\Omega, \mathbf{q})(\Delta F) \exp(-i\mathbf{q} \cdot \mathbf{r} - i\Omega t). \tag{11.7}$$

Here, we have used the Fourier decomposition of the fluctuating part of the background density, as given by

$$n'(\mathbf{r}, t) = \int \frac{d\mathbf{q}}{(2\pi)^2} \int \frac{d\Omega}{2\pi} n'(\Omega, \mathbf{q}) \exp(-i\mathbf{q} \cdot \mathbf{r} - i\Omega t), \tag{11.8}$$

where the wavevectors $\mathbf{q}$ and frequencies Ω describe the slow evolution of this background density. We have also used

$$(\Delta F) \equiv F(\omega - \Omega/2, \mathbf{k} - \mathbf{q}/2) - F(\omega + \Omega/2, \mathbf{k} + \mathbf{q}/2). \tag{11.9}$$

Finally, we have introduced the group velocity of the phonon modes, as defined by

$$\mathbf{v}_k \equiv \frac{\partial \omega}{\partial \mathbf{k}} = \left(c_s^2 + \frac{\hbar^2 k^2}{2M} \right) \frac{\mathbf{k}}{\omega}. \tag{11.10}$$

It can easily be recognized that this wave kinetic description is formally very similar to that used to study sound waves in the atomic clouds in magneto-optical traps, as well as to that used to describe the condensate field itself. The main difference is qualitative, due to the fact that we are using here a gas of quasi-particles, and not a gas of real atoms.

Is should also be noticed that c_s^2 is not a constant but a function of the background gas, showing that the velocity of the phonon quasi-particles can also vary in space and time. This results from assuming that the dispersion relation (11.2) is locally valid, and that ω is always univocally determined by the phonon wavevector $\mathbf{k}$. This allows us to introduce a reduced Wigner function $F_k(\mathbf{r}, t)$ in the spirit of the quasi-particle approximation, as determined by

$$F(\mathbf{r}, t; \omega, \mathbf{k}) = F_k(\mathbf{r}, t)\delta(\omega - \omega_k), \tag{11.11}$$

where ω_k is mode frequency as determined by the phonon dispersion relation. It is exactly this reduced Wigner function that we can identify with the phonon occupation number in the quasi-classical limit. In order to consider such a limit, we assume that $|\mathbf{q}| \ll |\mathbf{k}|$. This leads to

$$(\Delta F) \simeq -\mathbf{q} \cdot \frac{\partial F_k}{\partial \mathbf{k}}. \tag{11.12}$$

By replacing this in (11.10), we can then obtain a quasi-classical kinetic equation for phonons, of the Vlasov type, reading

$$\left(\frac{\partial}{\partial t} + \mathbf{v}_k \cdot \nabla + \mathbf{f}_k \cdot \frac{\partial}{\partial \mathbf{k}} \right) F_k = 0 \tag{11.13}$$

where we have defined the force acting on the phonons as

$$\mathbf{f}_k = -\frac{k^2}{2\omega_k} \frac{g}{M} \nabla n'(\mathbf{r}, t). \tag{11.14}$$

Notice that the phonon trajectories are determined by the characteristic equations of the kinetic equation (11.13), which are

$$\frac{d\mathbf{r}}{dt} = \mathbf{v}_k \;, \qquad \frac{d\mathbf{v}_k}{dt} = -\mathbf{f}_k . \tag{11.15}$$

So, in many respects, the phonon quasi-particles behave kinetic and dynamically as real particles, because they are described by formally identical equations. Therefore, the description of the superfluid medium as a mixture of two fluids can be made on very similar grounds for both the condensed fraction and the fluctuations above it. This will be discussed in what follows

11.2 Phonon Fluid Equations

A step further in the description of the quasi-particle excitations can be taken by deriving the corresponding fluid equations. These equations are derived from the kinetic equation (11.13), by computing the successive momenta of the group velocity $\mathbf{v}_k$. In what concerns the zeroth momentum, integration of the first term yields

$$\frac{\partial}{\partial t} \int F_k \frac{d\mathbf{k}}{(2\pi)^3} = \frac{\partial n_\nu}{dt} , \tag{11.16}$$

where n_ν is the mean density of the phonon gas, as defined by

$$n_\nu(\mathbf{r}, t) = \int F_k(\mathbf{r}, t) \frac{d\mathbf{k}}{(2\pi)^3} . \tag{11.17}$$

Integration of the second term in (11.13) leads to

$$\int \mathbf{v}_k \nabla F_k \frac{d\mathbf{k}}{(2\pi)^3} = \nabla \cdot (n_\nu \mathbf{u}) - \int F_k (\nabla \cdot \mathbf{v}_k) \frac{d\mathbf{k}}{(2\pi)^3} , \tag{11.18}$$

where we have defined the mean phonon velocity $\mathbf{u}$ as

$$\mathbf{u}(\mathbf{r}, t) = \frac{1}{n_\nu} \int \mathbf{v}_k F_k \frac{d\mathbf{k}}{(2\pi)^3} . \tag{11.19}$$

Finally, integration of the third term provides

$$\int \mathbf{f}_k \cdot \frac{\partial F_k}{\partial \mathbf{k}} \frac{d\mathbf{k}}{(2\pi)^3} = \int F_k (\nabla \cdot \mathbf{v}_k) \frac{d\mathbf{k}}{(2\pi)^3} . \tag{11.20}$$

The sum of these three terms results on the continuity equation for the phonon gas

$$\frac{\partial n_\nu}{\partial t} + \nabla \cdot (n_\nu \mathbf{u}) = 0. \tag{11.21}$$

Similarly, the (linear) momentum conservation equation for the gas of quasi-particles can be obtained by multiplying the kinetic equation (11.13) by $\mathbf{v}_k$, and integrating over the wavevector space. The first term leads to

$$\int \mathbf{v}_k \frac{\partial F_k}{\partial t} \frac{d\mathbf{k}}{(2\pi)^3} = \frac{\partial}{\partial t}(n_\nu \mathbf{u}) - \int \frac{\partial \mathbf{v}_k}{\partial t} F_k \frac{d\mathbf{k}}{(2\pi)^3}. \tag{11.22}$$

The contribution of the second term can be written as

$$\int \mathbf{v}_k \mathbf{v}_k \cdot \nabla F_k \frac{d\mathbf{k}}{(2\pi)^3} = \nabla \cdot (n_\nu \langle \mathbf{v}_k \mathbf{v}_k \rangle) - \int F_k \nabla \cdot (\mathbf{v}_k \mathbf{v}_k) \frac{d\mathbf{k}}{(2\pi)^3}, \tag{11.23}$$

where we have used the tensor

$$\langle \mathbf{v}_k \mathbf{v}_k \rangle = \frac{1}{n_\nu} \int (\mathbf{v}_k \mathbf{v}_k) F_k \frac{d\mathbf{k}}{(2\pi)^3}. \tag{11.24}$$

We now introduce the concept of *phonon pressure*, P_ν, such that

$$\langle \mathbf{v}_k \mathbf{v}_k \rangle = \mathbf{u}\mathbf{u} + \frac{P_\nu}{n_\nu} \mathbf{I}, \tag{11.25}$$

where $\mathbf{I}$ is the 3×3 identity matrix. Using (11.24), it can be easily shown that

$$P_\nu = \frac{1}{3} \int (\mathbf{v}_k - \mathbf{u})^2 F_k \frac{d\mathbf{k}}{(2\pi)^3} \equiv \frac{n_\nu}{3} u_{\text{th}}^2, \tag{11.26}$$

where u_{th} is an effective thermal velocity for the phonon gas. Finally, the contribution from the third term in the phonon kinetic equation takes the form

$$\int \mathbf{v}_k \left(\mathbf{f}_k \cdot \frac{\partial F_k}{\partial \mathbf{k}} \right) \frac{d\mathbf{k}}{(2\pi)^3}. \tag{11.27}$$

Adding these three contributions, we arrive at the momentum conservation for the phonon gas, as

$$\frac{\partial \mathbf{u}}{\partial t} + \mathbf{u} \cdot \nabla \mathbf{u} = -\frac{\nabla P_\nu}{n_\nu} + \mathbf{F}_\nu, \tag{11.28}$$

where $\mathbf{F}_\nu$ is the mean force acting on the the phonons, as determined by

$$\mathbf{F}_\nu = \frac{1}{n_\nu} \int F_k \left[\frac{\partial \mathbf{v}_k}{\partial t} + \nabla \cdot (\mathbf{v}_k \mathbf{v}_k) + \frac{\partial}{\partial \mathbf{k}} \cdot (\mathbf{f}_k \mathbf{v}_k) \right] \frac{d\mathbf{k}}{(2\pi)^3}. \tag{11.29}$$

Similarly, we could also derive the energy conservation of the phonon gas, by multiplying the kinetic equation (11.13) by $(\mathbf{v}_k \mathbf{v}_k)$ and integrating over the phonon spectrum, i.e., by computing its second momentum.

11.3 Slow Perturbations in the Superfluid

Let us now consider a perturbation in the phonon fluid, as described by the above fluid equations. Let us assume $n_\nu = n_{\nu 0} + \tilde{n}_\nu$, where $n_{\nu 0}$ is the unperturbed phonon density, and the perturbation evolves as $\tilde{n}_\nu \propto \exp(i\mathbf{q}\cdot\mathbf{r} - i\Omega t)$. Linearizing Eqs. (11.21) and (11.28) with respect to these perturbations, and neglecting for the moment the mean force $\mathbf{F}_\nu$, we get

$$\frac{\partial^2 \tilde{n}_\nu}{\partial t^2} - \frac{1}{3} u_{\text{th}}^2 \nabla^2 \tilde{n}_\nu = 0, \tag{11.30}$$

which leads to the dispersion relation

$$\Omega^2 = \frac{1}{3} u_{\text{th}}^2 q^2. \tag{11.31}$$

We can see from Eq. (11.26) that we usually have $u_{\text{th}}^2 \simeq c_s^2$. This shows that the phonon gas oscillations take the form of acoustic waves of a different kind, associated with pure oscillations of the quasi-particle mean density, with a sound speed which is $1/\sqrt{3}$ smaller than that of the condensed gas. This is usually called the *second sound*. It corresponds to oscillations of the phonon density only, where the interactions with the condensed gas are not considered. In general, however, we expect these oscillations in n_ν to be coupled with those of the condensed gas density n. Such a coupling is provided by the microscopic force $\mathbf{f}_k$ and by the corresponding average force $\mathbf{F}_\nu$, as defined by (11.29). This force can be roughly estimated by noting that $\omega_k \simeq c_s k$ and $v_k \simeq c_s$, if we account only the contributions in the long wavelength (purely acoustic) limit. This allows us to write

$$\frac{\partial}{\partial \mathbf{k}} \cdot \mathbf{f}_k \mathbf{v}_k \simeq -\frac{g}{2M} \nabla n, \tag{11.32}$$

where n is the density of the condensed gas. On the other hand, for isotropic phonon distributions F_k, the first term in (11.29) is zero and, neglecting nonlinear terms, we obtain for the average force the expression

$$\mathbf{F}_\nu = -\frac{1}{n_{\nu 0}} \frac{g}{2M} \nabla n. \tag{11.33}$$

Including this new contribution in the wave equation for the perturbations in the phonon gas, we get

$$\frac{\partial^2 \tilde{n}_\nu}{\partial t^2} - \frac{1}{3} u_{\text{th}}^2 \nabla^2 \tilde{n}_\nu = \frac{g}{2M} \nabla^2 n', \tag{11.34}$$

where n' is the perturbed condensed atom density. Assuming that $\tilde{n}_\nu$ and n' are related, and therefore n' also behaves in space and time as $\exp(i\mathbf{q}\cdot\mathbf{r} - i\Omega t)$, we obtain the following relation between the two density perturbations

$$\left(\Omega^2 - c_s^2 q^2\right)\tilde{n}_\nu = \frac{g}{2M} q^2 n'. \tag{11.35}$$

Let us now turn to the fluid equations of the condensed fraction. The coupling with the phonon gas can be described by including nonlinear corrections to the density wave equation (11.1). These nonlinearities are due to the slow ponderomotive effects associated with the presence of phonons. In the homogeneous case, $V_0 = 0$, the fluid equations lead to

$$\frac{\partial^2 n}{\partial t^2} - \frac{1}{M}\nabla \cdot [n\nabla(V_B + gn)] = \nabla \cdot [n(\mathbf{v}\cdot\nabla\mathbf{v}) + \nabla\cdot(n\mathbf{v})\mathbf{v}]. \tag{11.36}$$

Let us then consider a slow perturbation of the condensed gas $(n', \mathbf{v}')$ in the presence of an arbitrary phonon spectrum. We can then use

$$n = n_0 + n' + n_{\text{ph}}, \quad \mathbf{v} = \mathbf{v}' + \mathbf{v}_{\text{ph}}, \tag{11.37}$$

where

$$n_{\text{ph}}(\mathbf{r}, t) = \int n_k \exp(i\mathbf{k}\cdot\mathbf{r} - i\omega_k t)\frac{d\mathbf{k}}{(2\pi)^3} \tag{11.38}$$

and

$$\mathbf{v}_{\text{ph}}(\mathbf{r}, t) = \frac{1}{n_0}\int \frac{\omega_k \mathbf{k}}{k^2} n_k \exp(i\mathbf{k}\cdot\mathbf{r} - i\omega_k t)\frac{d\mathbf{k}}{(2\pi)^3}, \tag{11.39}$$

with ω_k given in terms of the following dispersion relation

$$\omega_k^2 = \frac{g}{M}(n_0 + n')k^2 + \frac{\hbar^2 k^4}{4M^2}. \tag{11.40}$$

Notice that the quantity n_{ph}, which represents the atom density perturbation associated with the phonon spectrum, should not be confuse with the quantity n_ν introduced above, which is the number density of phonon quasi-particles. An explicit relation between these two quantities will be establish below. Linearizing (11.36) with respect to the slow perturbations, we get

$$\left(\frac{\partial^2}{\partial t^2} - \frac{gn_0}{M}\nabla^2 + \frac{\hbar^2}{4M^2}\nabla^4\right) n' = n_0\nabla^2|\mathbf{v}_{\text{ph}}|^2 + \frac{g}{2M}\nabla^2|n_{\text{ph}}|^2. \tag{11.41}$$

Putting equations (11.38) and (11.39) together, we have

$$|n_{\text{ph}}|^2 = \int n_k n_{-k}\frac{d\mathbf{k}}{(2\pi)^3} = \int F_k \frac{d\mathbf{k}}{(2\pi)^3} = n_\nu \tag{11.42}$$

and

$$|\mathbf{v}_{\text{ph}}|^2 = \int \mathbf{v}_k \cdot \mathbf{v}_{-k} \frac{d\mathbf{k}}{(2\pi)^3} = \frac{1}{n_0} \int \frac{\omega_k^2}{k^2} F_k \frac{d\mathbf{k}}{(2\pi)^3} \equiv \frac{c_0^2}{n_0^2} n_\nu, \tag{11.43}$$

where c_0 is the average phase velocity of the phonon spectrum. Replacing this in Eq. (11.41), we obtain

$$\left(\frac{\partial^2}{\partial t^2} - \frac{g n_0}{M} \nabla^2 + \frac{\hbar^2}{4M^2} \nabla^4 \right) n' = \left(\frac{c_0^2}{n_0} + \frac{g}{2M} \right) \nabla^2 n_\nu. \tag{11.44}$$

We can see that the density perturbations of the condensate are again coupled with the phonon spectrum. Assuming as above that $n_\nu = n_{\nu 0} + \tilde{n}_\nu$, we get for perturbations evolving as $\exp(i\mathbf{q} \cdot \mathbf{r} - i\Omega t)$ a new relation between n' and $\tilde{n}_\nu$, of the form

$$\left(\Omega^2 - \Omega_q^2 \right) n' = \frac{c_1^2}{n_0} q^2 \tilde{n}_\nu \tag{11.45}$$

where we have used

$$\Omega_q^2 = c_s^2 q^2 + \frac{\hbar^2 q^4}{4M^2}, \quad c_s^2 = \frac{g n_0}{M} \tag{11.46}$$

and

$$c_1^2 = c_0^2 + \frac{1}{2} c_s^2. \tag{11.47}$$

Replacing this in Eq. (11.35), we finally arrive at the dispersion relation for the slow perturbations in the superfluid, as given by

$$\left(\Omega^2 - \Omega_q^2 \right) \left(\Omega^2 - c_s^2 q^2 \right) = \frac{g}{2M} \frac{c_1^2}{n_0} q^4. \tag{11.48}$$

This represents the combined oscillations of two different fluids, the condensed fluid and the thermal fluid, where the later has been described in terms of phonon quasi-particles. In the absence of the coupling provided by the quantity c_1^2, this dispersion relation would brake down into two distinct dispersion relations, one ($\Omega^2 = \Omega_q^2$) valid for the condensed gas alone, and the other ($\Omega^2 = u_{\text{th}}^2 q^2/3$) valid for the phonon gas alone, in the absence of condensed atom oscillations.

11.4 Superfluid Currents

Let us now consider the evolution of the total fluid mass density, as defined by

$$\rho = M(n_0 + n_x) \equiv Mn, \tag{11.49}$$

where n_0 is the density of the superfluid, and $n_x = n - n_0$, the density of the normal fluid. It is plausible to assume that n_x is proportional to the density of phonons, because these quasi-particles represent the normal fraction of the medium. This happens because we make no difference between quantum and thermal fluctuations in the present formulation, so all the fluctuations above the condensate are treated as phonon quasiparticles. We can then write, $n_x = \alpha n_\nu$, where n_ν is the phonon mean density and the constant of proportionality α has to be determined. Now, by using the two conservation equations for both the atoms and the phonons, we can easily construct the total mass conservation equation, as

$$\frac{\partial \rho}{\partial t} + \nabla \cdot \mathbf{J} = 0 \,, \quad \mathbf{J} = M(n_0 \mathbf{v} + \alpha n_\nu \mathbf{u}), \tag{11.50}$$

where $\mathbf{v}$ is the mean velocity of the condensed phase and $\mathbf{J}$ is the total mass current. Using the definition of the mean phonon velocity $\mathbf{u}$, we can write this current as

$$\mathbf{J} = M n_0 \mathbf{v} + M\alpha \int \mathbf{v}_k F_k \frac{d\mathbf{k}}{(2\pi)^3}. \tag{11.51}$$

It is now important to compare this quantity with the total momentum transported by the medium. By definition, we have

$$\mathbf{P} = M n_0 \mathbf{v} + M \int \hbar \mathbf{k} F_k \frac{d\mathbf{k}}{(2\pi)^3}, \tag{11.52}$$

where $\hbar \mathbf{k}$ is the single phonon momentum, and F_k the phonon distribution. On the other hand, we know that the phonon group velocity is determined by Eq. (11.10), allowing us to write

$$\mathbf{P} = M n_0 \mathbf{v} + \int M_k \mathbf{v}_k F_k \frac{d\mathbf{k}}{(2\pi)^3}, \tag{11.53}$$

where we have introduced the quantity

$$M_k = \frac{\hbar \omega_k}{c_s^2 + (\hbar^2 k^2 / 2M)}. \tag{11.54}$$

The latter can be regarded as the *phonon effective mass*, which depends on the momentum (mode) $\mathbf{k}$ of the phonon. Notice that phonons, just as photons in vacuum, have no finite rest mass. In other words, $M_k \to 0$ for $k \to 0$ is in agreement with the Goldstone theorem. Equating the expressions of $\mathbf{J}$ and $\mathbf{P}$, as given by Eqs. (11.51) and (11.52), we can determine the proportionality factor between thermal atoms and phonons, reading

$$\alpha \equiv \frac{n_x}{n_0} = \frac{\mathbf{u}}{n_\nu u^2} \cdot \int \frac{M_k}{M} \mathbf{v}_k F_k \frac{d\mathbf{k}}{(2\pi)^3}. \tag{11.55}$$

It is interesting to look at the limit of very small phonon wavenumbers, such that the quantum dispersion (or the free-particle like behavior) term can be neglected. In such limit, we have

$$M_k \simeq \frac{\hbar k}{c_s}, \quad \mathbf{v}_k \simeq c_s \frac{\mathbf{k}}{k}. \tag{11.56}$$

The current (11.51) reduces then to

$$\mathbf{J} \simeq M n_0 \mathbf{v} + M \alpha c_s \int \frac{\mathbf{k}}{k} F_k \frac{d\mathbf{k}}{(2\pi)^3} \tag{11.57}$$

and the proportionality factor becomes

$$\alpha \simeq \frac{\hbar \mathbf{u}}{M u^2} \cdot \left\langle \frac{\mathbf{v}_k}{c_s} \right\rangle. \tag{11.58}$$

Finally, we should notice that phonons can only be excited by atoms or by macroscopic objects moving with a velocity v_s larger than the phonon phase velocity ω_k / k. As a result, for

$$v_s \ < \ \frac{\omega_k}{k} \equiv \sqrt{c_s^2 + \frac{\hbar^2 k^2}{4M^2}}, \tag{11.59}$$

no phonons with wavenumber k are excited in the medium. Therefore, no phonons at all will exist in the condensate, as long as its velocity with respect to some atom or boundary or object is lower than the Bogoliubov sound speed, or

$$v_s \ < \ c_s = \sqrt{\frac{g n_0}{M}}. \tag{11.60}$$

This is the well known *Landau criterium* for superfluidity.

11.5 Phonon Landau Damping

We can now refine the above description of long wavelength excitations in the condensate, by using a kinetic description of the phonon gas. This can be done by going back to Eq. (11.35) and calculate the quantity n' with the help of the phonon kinetic equation (11.13). By using $F_k = F_0(\mathbf{k}) + F_k'$, such that $F_0(\mathbf{k})$ is the equilibrium phonon distribution, and $n' = \int F_k' d\mathbf{k}/(2\pi)^3$ is the phonon density perturbation, we have after linearization of the kinetic equation

$$F_k' = \frac{k^2}{2\omega_k} \frac{g n_\nu}{M} \frac{\Delta F_0(\mathbf{k})}{(\Omega - \mathbf{q} \cdot \mathbf{v}_k)}, \tag{11.61}$$

where $\Delta F_0(\mathbf{k}) = F_0(\mathbf{k} - \mathbf{q}/2) - F_0(\mathbf{k} + \mathbf{q}/2)$. Using this in Eq. (11.35), we can then obtain a kinetic dispersion relation for the excitations in the superfluid, as

$$\left(\Omega^2 - \Omega_q^2\right) = \frac{g}{2M}\frac{c_1^2}{n_0}q^2 \int \frac{k^2}{\omega_k} \frac{\Delta F_0(\mathbf{k})}{(\Omega - \mathbf{q}\cdot\mathbf{v}_k)} \frac{d\mathbf{k}}{(2\pi)^3}. \tag{11.62}$$

This kinetic dispersion relation can also be written as

$$\left(\Omega^2 - \Omega_q^2\right) = \frac{g}{2M}\frac{c_1^2}{n_0}q^2 \int F_0(\mathbf{k}) \left[\frac{k_+^2}{\omega_+}\frac{1}{(\Omega - \mathbf{q}\cdot\mathbf{v}_+)} - \frac{k_-^2}{\omega_-}\frac{1}{(\Omega - \mathbf{q}\cdot\mathbf{v}_-)}\right] \frac{d\mathbf{k}}{(2\pi)^3}, \tag{11.63}$$

where we have used $\mathbf{k}_\pm = \mathbf{k} \pm \mathbf{q}/2$, $\omega_\pm = \omega_{k_\pm}$ and $\mathbf{v}_\pm = \mathbf{v}(\mathbf{k}_\pm)$. It is useful to compare this kinetic result with Eq. (11.48). We can see the appearance of a resonance inside the integral, which leads to an imaginary part associated with *phonon Landau damping*. This implies the existence of wave damping proportional to the population difference in the phonon spectrum $\Delta F_0(\mathbf{k})$. It means that excitons can be damped (or excited) due to resonant interaction with phonons. These kinetic effects associated with resonant wave interactions with the phonon quasi-particles are similar to those found in the resonant interactions between phonons, described now as waves, interacting with atoms. On the other hand, atoms can also be described on a shorter scale by a wave function. This shows that both atoms and phonons can be described on a short length scale as waves and, on a much larger length scale as particles or quasi-particles. Atoms are seen as particles by phonons, and phonons are seen as particles by long wavelength excitations in the superfluid.

11.6 Roton Excitation

Rotons were first proposed by Landau [2], and stay as one of the main concepts in superfluid theory. Originally, they have been connected with the existence of strong interactions in the physical systems. Our recent view of rotons is that they are a particular case of phonon quasi-particles, associated to the existence of a minimum of the phonon dispersion curve, and their existence does not necessarily imply strong interactions. The existence of this minimum in superfluids has been demonstrated by molecular dynamic simulation models, and confirmed by experiments. Here we show that it can also be demonstrated by analytical methods.

It is known that the relation between Bose-Einstein condensates and superfluids is not (and has never been) obvious. This will be discussed in the next section. In recent years it has been recognized that rotons can be excited in condensates, in the presence of long range electric or magnetic dipole-dipole interactions between the atoms. This can be achieved for instance in quasi-2D pancake shaped condensates where the dipoles become aligned by an external static magnetic field [5]. The case of a cigar shaped quasi-1D in a static field was also examined [6]. Here we show that rotons can eventually be excited in condensates for the case dipole-dipole interactions. In order to establish this general analytical result, we use the wave kinetic approach.

11.6.1 Wave Kinetic Equation with Dipolar Interactions

Our starting point is the Gross-Pitaevskii equation in the presence of dipole-dipole interactions, which can be written as

$$i\hbar \frac{\partial \Psi}{\partial t} = -\frac{\hbar^2}{2M} \nabla^2 \Psi + (V_0 + V_{\mathrm{NL}})\Psi, \tag{11.64}$$

where $V_0 \equiv V_0(\mathbf{r})$ is the confining harmonic potential, and the nonlinear potential V_{NL} contains, apart from the usual interaction potential, an additional nonlocal term, as given by

$$V_{\mathrm{NL}}(\mathbf{r}, t) = g\, |\Psi(\mathbf{r}, t)|^2 + \int U_d(\mathbf{r} - \mathbf{r}') \left|\Psi(\mathbf{r}', t)\right|^2 d\mathbf{r}' \tag{11.65}$$

where $U_d(\mathbf{r} - \mathbf{r}')$ is the dipole-dipole interaction potential, to be specified later. We have seen that, by applying the Wigner-Moyal procedure, we can derive a wave kinetic equation equivalent to (11.64), of the generic form

$$i\hbar \left(\frac{\partial}{\partial t} + \mathbf{v}_q \cdot \nabla \right) W = \int V(\mathbf{k}, t) \Delta W \exp(i\mathbf{k} \cdot \mathbf{r}) \frac{d\mathbf{k}}{(2\pi)^3}, \tag{11.66}$$

where $\mathbf{v}_q = \hbar \mathbf{q}/M$ is the atom velocity, but where now the dipole interaction potential is included. $W(\mathbf{q}, \mathbf{r}, t)$ is the usual Wigner function corresponding to $\Psi(\mathbf{r}, t)$, or more explicitly

$$W(\mathbf{q}, \mathbf{r}, t) = \int \Psi^*(\mathbf{r} - \mathbf{s}/2, t) \Psi(\mathbf{r} + \mathbf{s}/2, t) e^{i\mathbf{q}\cdot\mathbf{s}} d\mathbf{s}. \tag{11.67}$$

In Eq. (11.66), we have also used the standard notation

$$\Delta W = W(\mathbf{q} - \mathbf{k}/2, \mathbf{r}, t) - W(\mathbf{q} + \mathbf{k}/2, \mathbf{r}, t). \tag{11.68}$$

We write these quantities explicitly, at the cost of being repetitive, because we want to state the difference between the wave kinetic equation used before in this chapter, and which concerns the phonon field, and the wave kinetic equation for the primary gas, the gas of condensate atoms. Finally, the quantity $V(\mathbf{k}, t)$ is the spatial Fourier transformation of the total potential, which includes the dipole interactions. In order to obtain an explicit expression for this quantity, we should notice that the atom probability density is

$$|\Psi(\mathbf{r}, t)|^2 = \int W(\mathbf{q}, \mathbf{r}, t) \frac{d\mathbf{q}}{(2\pi)^3}. \tag{11.69}$$

Replacing this in Eq. (11.65), we get

$$V_{\rm NL}(\mathbf{r},t) = g\int \frac{d\mathbf{k}}{(2\pi)^3}\left\{W(\mathbf{q},\mathbf{r},t) + \int d\mathbf{r}' U_d(\mathbf{r}-\mathbf{r}')W(\mathbf{q},\mathbf{r}',t)\right\}. \quad (11.70)$$

Now, using the convolution theorem, we have

$$\int U_d(\mathbf{r}-\mathbf{r}')W(\mathbf{q},\mathbf{r}',t)d\mathbf{r}' = \int U_d(\mathbf{k})W(\mathbf{q},\mathbf{k},t)e^{i\mathbf{k}\cdot\mathbf{r}}\frac{d\mathbf{k}}{(2\pi)^3}, \quad (11.71)$$

where $U_d(\mathbf{k})$ is the Fourier component of the dipole-dipole interaction potential $U_d(\mathbf{r})$. This means that the total potential can be written in the form

$$V(\mathbf{r},t) \equiv V_0(\mathbf{r}) + V_{NL}(\mathbf{r},t) = \int V(\mathbf{k},t)e^{i\mathbf{k}\cdot\mathbf{r}}\frac{d\mathbf{k}}{(2\pi)^3}, \quad (11.72)$$

where we have

$$V(\mathbf{k},t) = V_0(\mathbf{k}) + [g + U_d(\mathbf{k})]\, I(\mathbf{k},t)\,, \quad I(\mathbf{k},t) = \int W(\mathbf{q},\mathbf{k},t)\frac{d\mathbf{q}}{(2\pi)^3}. \quad (11.73)$$

Here, the quantity $I(\mathbf{k},t)$ represents the Fourier component of the probability density (11.69).

11.6.2 Dispersion Relation

Let us now consider the evolution of a perturbation $\tilde{W}$ with respect to some equilibrium quasi-distribution W_0. Linearizing the wave kinetic equation (11.66), we get

$$i\hbar\left(\frac{\partial}{\partial t} + \mathbf{v}_q\cdot\nabla\right)\tilde{W} = \int [g + U_d(\mathbf{k})]\, I(\mathbf{k},t)V(\mathbf{k},t)\Delta W_0 \exp(i\mathbf{k}\cdot\mathbf{r})\frac{d\mathbf{k}}{(2\pi)^3}, \quad (11.74)$$

where we have considered the homogeneous approximation, $V_0(\mathbf{r}) = 0$, which is valid for short wavelength perturbations, as it was already discussed. Let us now assume a sinusoidal perturbation, such that

$$\tilde{W}(\mathbf{q},\mathbf{k},t) = \tilde{W}(\mathbf{q},\mathbf{k})\exp(i\mathbf{k}\cdot\mathbf{r} - i\omega t). \quad (11.75)$$

We then have $\tilde{I}(\mathbf{k},t) = \tilde{I}(\mathbf{k})\exp(-i\omega t)$, and replacing in (11.74), we simply get

$$\tilde{W}(\mathbf{q},\mathbf{q}) = \frac{\tilde{I}(\mathbf{k})}{\hbar}[g + U_d(\mathbf{k})]\frac{\Delta W_0}{(\omega - \mathbf{k}\cdot\mathbf{v}_q)}. \quad (11.76)$$

Integrating over $\mathbf{q}$, one obtains the following dispersion relation

$$1 - \frac{1}{\hbar}\left[g + U_d(\mathbf{k})\right] \int \frac{\Delta W_0}{(\omega - \mathbf{k} \cdot \mathbf{v}_q)} \frac{d\mathbf{q}}{(2\pi)^3} = 0. \tag{11.77}$$

It is also useful to write it in a slightly different and more explicit way, as

$$1 - \frac{1}{\hbar}\left[g + U_d(\mathbf{k})\right] \int W_0(\mathbf{q}) \left[\frac{1}{(\omega_- - \mathbf{k} \cdot \mathbf{v}_q)} - \frac{1}{(\omega_+ - \mathbf{k} \cdot \mathbf{v}_q)}\right] \frac{d\mathbf{q}}{(2\pi)^3} = 0, \tag{11.78}$$

with

$$\omega_\pm = \omega \pm \frac{\hbar k^2}{2M}. \tag{11.79}$$

Let us consider the zero temperature limit, which can be described by the unperturbed distribution $W_0(\mathbf{q}) = (2\pi)^3 n_0 \delta(\mathbf{q} - \mathbf{q}_0)$, where n_0 is the unperturbed density of the condensate, which is allowed to move with a constant velocity $\mathbf{v}_0 = \hbar \mathbf{q}_0 / M$. Replacing this in (11.78), we get

$$1 - \frac{\hbar n_0}{M} \frac{\left[g + U_d(\mathbf{k})\right] k^2}{(\omega_- \mathbf{k} \cdot \mathbf{v}_q)^2 - (\hbar k^2/2M)^2} = 0, \tag{11.80}$$

or, more explicitly,

$$(\omega - \mathbf{k} \cdot \mathbf{v}_0)^2 = k^2 \left[c_s^2 + \frac{n_0}{M} U_d(\mathbf{k})\right] + \frac{\hbar^2 k^4}{4M^2}. \tag{11.81}$$

This generalizes our previous result for sound waves in a condensate, by including a new term due to atomic dipole-dipole interactions. It should be noticed that these interactions modify the usual sound speed c_s, as discussed next.

11.6.3 Roton Instability

Let us now consider possible forms of the interaction potential $U_d(\mathbf{k})$. In principle, significant dipole-dipole interactions can exist inside the condensate if we excite some of the atoms in high Rydberg states, as recently proposed by Henkel et al. [7]. An useful model is provided by a potential of the form

$$U_d(\mathbf{r}) = \frac{C_6}{r^6 + R_0^6}, \tag{11.82}$$

where C_6 is an effective van der Waals coefficient, and R_0 is the effective range of the interaction. For such a case, we get

$$U_d(\mathbf{k}) = C_6 \int \frac{\exp(i\mathbf{k} \cdot \mathbf{r})}{r^6 + R_0^6} d\mathbf{r} = 4\pi C_6 \int_0^\infty \frac{\cos(kr)}{r^6 + R_0^6} dr. \tag{11.83}$$

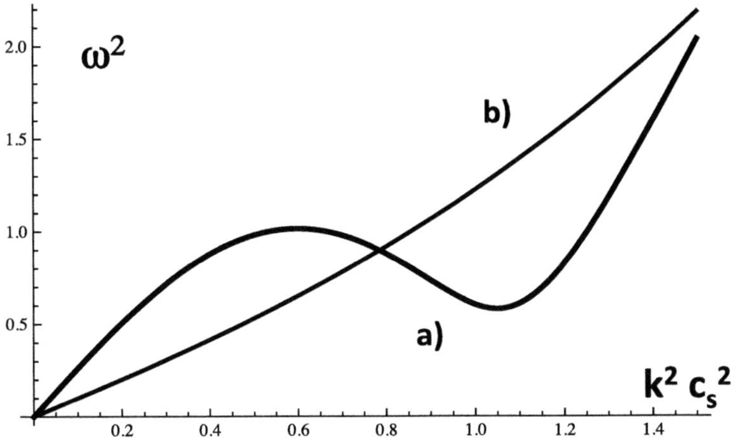

Fig. 11.1 Dispersion relation of Bogoliubov oscillations in a condensate, modified by the long dipole-dipole interactions, (*a*) U_d as described by a C_6 van der Waals potential; and (*b*) $U_d = 0$ for reference

It is now useful, for the sake of a qualitative and analytical discussion, to use the approximation

$$U_d(\mathbf{k}) \simeq 4\pi C_6 \int_0^{R_0} \cos(kr)dr = \frac{\Omega}{4\pi} k C_6 \sin(kR_0). \tag{11.84}$$

This shows a the existence of a maximum at $k = \pi/2R_0$, and a negative minimum at $k = 3\pi/2R_0$. This is in good qualitative agreement with the molecular dynamical simulations of [7], and can be used to illustrate the occurrence of a roton minimum. By replacing (11.84) in the dispersion relation (11.81), and considering a condensate at rest, $\mathbf{v}_0 = 0$, we obtain

$$\omega^2 = k^2 c_s^2 + \frac{\hbar^2 k^4}{4M^2} + kR_0 \alpha \sin(kR_0), \tag{11.85}$$

with

$$\alpha = \frac{n_0}{M} \frac{4\pi}{R_0} C_6. \tag{11.86}$$

In the particular case when $k = k_0 \equiv 3\pi/2R_0$, this takes the form

$$\omega_0^2 = k_0^2 c_s^2 + \frac{\hbar^2 k_0^4}{4M^2} - kR_0, \tag{11.87}$$

where $\omega_0 \equiv \omega(k_0)$. See Fig. 11.1 for an illustration.

We can see that a critical value of the parameter α exists such that $\omega_0 = 0$. This corresponds to

$$\alpha_c = \frac{3\pi}{2R_0^2}\left(c_s^2 + \frac{9\pi^2}{2^4}\frac{\hbar^2}{M^2R_0^2}\right), \tag{11.88}$$

For $\alpha > \alpha_c$, a *roton instability* at zero temperature can occur, such that $\Re(\omega) = 0$. We can also establish the same condition for a generic form of potential $U_d(\mathbf{k})$, and not just for the simplified version (11.84) of the van der Waals potential. Going back to the dispersion relation (11.81), and using $\mathbf{v}_0 = 0$, we can see that a roton instability corresponding to $\omega^2 \leq 0$ will occur for $U_d(\mathbf{k}) < 0$, and in addition

$$\frac{n_0}{M}|U_d(\mathbf{k})| \geq c_s^2 + \frac{\hbar^2k^2}{4M^2}. \tag{11.89}$$

When these two conditions are satisfied, roton excitation will become possible in BEC with dipolar interactions. The influence of a finite temperature, and the contribution of kinetic effects to the roton dispersion properties can also be discussed using the general dispersion relation (11.77) or (11.78).

References

1. L. Tisza, Nature **141**, 913 (1938)
2. L. Landau, The theory of superfluidity of Helium II. J. Phys. USSR **5**, 71–90 (1941)
3. I.M. Khalatnikov, *An Introduction to the Theory of Superfluidity* (Westview, New York, 2000)
4. E. Zaremba, A. Griffin, T. Nikuni, Phys. Rev. A **57**, 4695 (1998)
5. L. Santos, G.V. Shlyapnikov, M. Lewenstein, Phys. Rev. Lett. **90**, 250403 (2003)
6. S. Giovanazzi, D.H.J. O'Dell, Eur. Phys. J. D **31**, 439 (2004)
7. N. Henkel, R. Nath, T. Pohl, Phys. Rev. Lett. **104**, 195302 (2010)

Chapter 12
Rotating BECs

In this chapter we deal with the important vorticity problem in a condensate. We first explain how quantum vortices can occur in a superfluid, and what is their expected structure. These vortices can spontaneously form in a rotating condensate, and they have tendency to nucleated and to form regular arrays, taking typically triangular lattice shapes. These vortex lattices can oscillate, as first discussed by Tkatchenko.

After discussing the dispersion relation of the Tkatchenko modes, we consider a less conventional type of wave modes, the Rossby waves, which are formally analogous to those known in geophysics. This is done by deriving a generalized form of Charney equation, valid for rotating condensates. Large nonlinear vortical structures, are also shown to exist. We then propose a new type of dispersion relation, associated with Rossby-Tkatchenko modes, which tend to both the Tkatchenko and the Rossby modes if we take appropriate limits. The chapter concludes with a discuss of the coupling between a condensed gas and a photon beam carrying orbital angular momentum, which can be used as a method to induce rotation on a condensate.

12.1 Quantum Vortices

We have seen that the velocity field of the condensed gas can be defined as the gradient of a phase function, as $\mathbf{v} = (\hbar/M)\nabla\varphi$. This shows that the condensate is irrotational, $\nabla \times \mathbf{v} = 0$, except when the phase is singular. Such a singularity can occur at the centre of a quantum vortex, located at some position $\mathbf{r}_\nu$. If we integrate the fluid velocity $\mathbf{v}$ along a closed contour C surrounding the singularity, we have to satisfy

$$\int_C \mathbf{v} \cdot d\mathbf{r} = \frac{\hbar}{M} \int_C d\varphi = \kappa \nu \, , \quad \kappa = \frac{\hbar}{M} 2\pi, \tag{12.1}$$

J.T. Mendonça and H. Terças, *Physics of Ultra-Cold Matter*, Springer Series on Atomic, Optical, and Plasma Physics 70, DOI 10.1007/978-1-4614-5413-7_12,

where ν is an integer and κ is the quantum of circulation. Such a quantization rule was first considered by Onsager [24], and then by Feynman [15]. This becomes necessary to guarantee that the condensate wavefunction is a single valued function. We can use the Stoke's theorem to write the above circulation as $\int \nabla \times \mathbf{v} \cdot d\mathbf{S} = \nu\kappa$, which implies a singular vorticity locates at the vortex centre, as determined by

$$\nabla \times \mathbf{v} = \nu\kappa\delta^{(2)}(\mathbf{r} - \mathbf{r}_\nu). \tag{12.2}$$

Such an expression states the local break down of irrotationallity associated with a vortex. Let us study the local structure of the vortex, by writing the wavefunction in cylindrical coordinates $\mathbf{r} = (r, \theta, z)$, as

$$\Phi(\mathbf{r}, t) = \psi(r, z, t) \exp(i\nu\theta), \tag{12.3}$$

where the amplitude $\psi(r, z, t)$ is real. Replacing this wavefunction in the Gross-Pitaevskii equation, we obtain

$$i\hbar\frac{\partial\psi}{\partial t} = -\frac{\hbar^2}{2M}\left[\frac{1}{r}\frac{\partial}{\partial r}\left(r\frac{\partial\psi}{\partial r}\right) + \frac{\partial^2\psi}{\partial z^2} - \frac{1}{r^2}\nu^2\psi\right] + V_{\text{ext}}(\mathbf{r}) + g\psi^3. \tag{12.4}$$

We notice the appearance of the term proportional to l^2, which results from the rotational motion around the singularity. Obviously, this equation is valid for a condensate with axial symmetry. Therefore rotation takes place around the z-axis.

Let us ignore the influence of the confining potential, $V_{\text{ext}}(\mathbf{r}) = 0$. In this case, the wave function amplitude ψ is independent of z, and we can take $\partial/\partial z = 0$. In steady state, we can also use $i\hbar\partial\psi = \mu\psi$, where μ is the chemical potential. The above wave equation simplifies to

$$\mu\psi = -\frac{\hbar^2}{2M}\left[\frac{1}{r}\frac{\partial}{\partial r}\left(r\frac{\partial\psi}{\partial r}\right) - \frac{1}{r^2}\nu^2\psi\right] + g\psi^3. \tag{12.5}$$

It is now convenient to introduce dimensionless variables, ρ and χ, such that

$$\rho = \frac{r}{\xi} = \frac{r}{\hbar}\sqrt{2Mgn_\infty}\,, \quad \chi = \frac{\psi}{\psi_\infty}, \tag{12.6}$$

where $\xi = \hbar/\sqrt{2Mgn_\infty}$ is the *healing length* (or coherence length), and $n_\infty = \psi_\infty^2$ is the asymptotic value of the gas density, far away from the vortex centre. We should also take the chemical potential equal to $\mu = gn_\infty$, in order to satisfy the wave equation at infinity. This allows us to rewrite Eq. (12.5) as

$$\chi = -\frac{1}{\rho}\frac{\partial}{\partial\rho}\left(\rho\frac{\partial\chi}{\partial\rho}\right) + \frac{l^2\nu^2\chi}{\rho^2} + \chi^3. \tag{12.7}$$

This equation can be solved numerically. But, for $\nu = 1$, a good approximate solution is given by

$$\chi(\rho) = \frac{\rho}{\sqrt{2 + \rho^2}}, \tag{12.8}$$

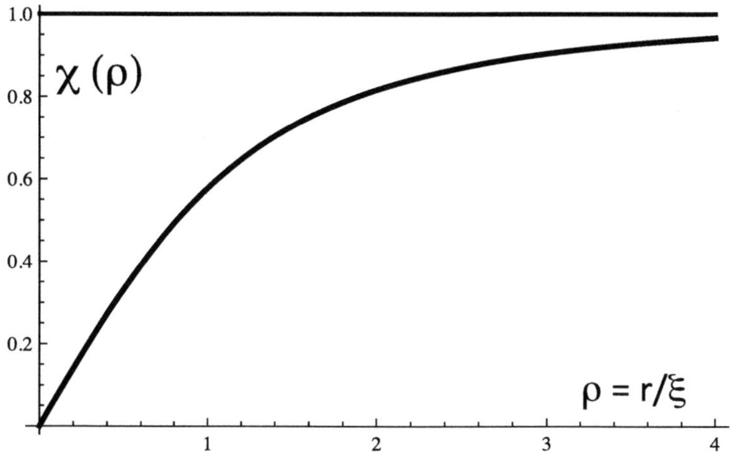

Fig. 12.1 The approximate vortex solution (12.8) is shown in normalized radial coordinates. The size of the vortex is determined by the healing length ξ

which is illustrated in Fig. 12.1. This shows that, for long radial distances $\rho \gg 1$, the condensate wavefunction tends monotonically to its asymptotic value $\psi_\infty = \sqrt{\mu/g}$. In contrast, for short distances $\rho \ll 1$, the centrifugal force dominates and $\psi(r) \simeq r$, as it would correspond to a free particle with angular momentum $l = 1$, turning around the z-axis. It then becomes clear that the vortex size is of the order of the healing length ξ.

Vortex structures have been observed in rotating condensates, in both simulations and experiments. The first experimental observation of vortices was reported in 1999 [21,22]. Depending on the value of the angular frequency Ω, the number of observed vortices can change, from a single vortex to an array of several vortices, resembling the Abrikosov vortex arrays in superconductors. Arrays of hundreds of vortices have already been reported in condensates [1, 11]. For a review of rotating BECs, the reader should refer to the reviews [6, 14]. The lattice array is typically triangular, with a surface density given by

$$n_v = \frac{M\Omega}{\pi\hbar} = \frac{1}{\pi a_v^2}, \tag{12.9}$$

where a_v is the distance between nearby vortices. The density n_v is defined as the number of vortices per unit area, in the plane perpendicular to the rotation axis. Noticing that the circulation of the velocity field around a single vortex with $\nu = 1$ is equal to $\kappa = 2\pi\hbar/M$, as shown by (12.1), Eq. (12.9) shows that the coarse-grained velocity field $\bar{\mathbf{v}}$, obtain by circulating over several vortices $\bar{\mathbf{v}} = \Omega \times \mathbf{r}$, corresponds to $\nabla \times \bar{\mathbf{v}} = 2\Omega$, which coincides with that of a rigid rotation at the angular velocity Ω. This means that, although the condensate is a superfluid, with zero vorticity everywhere except in the immediate vicinity of a vortex, it acquires a

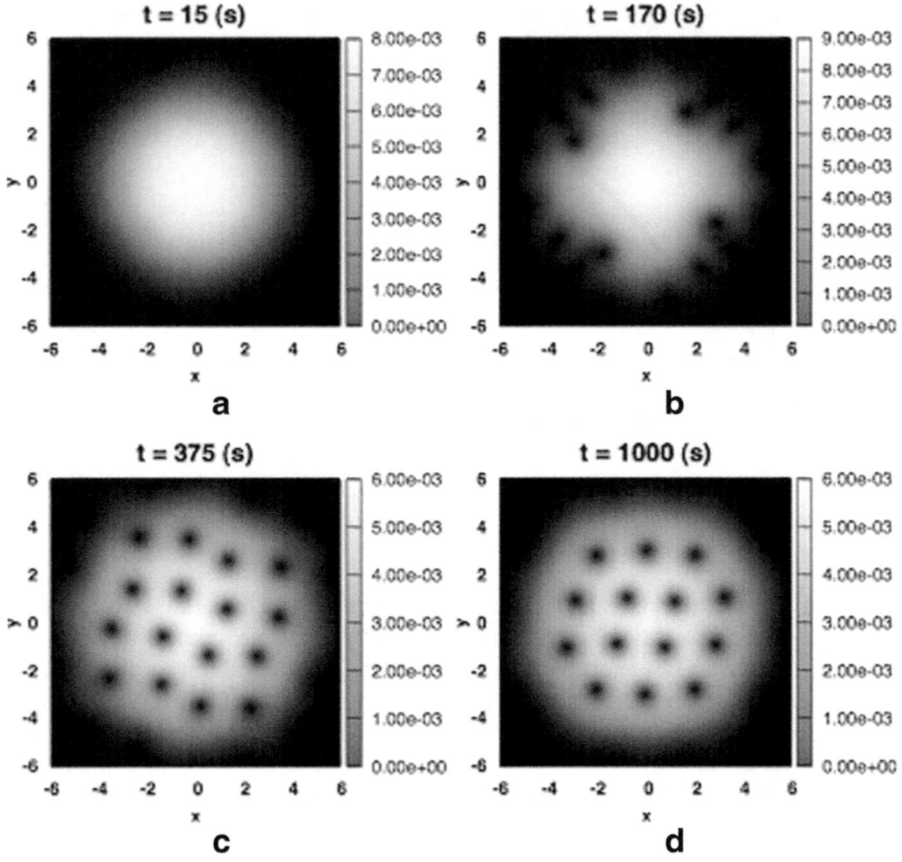

Fig. 12.2 Formation of a vortex array, as shown by solving numerically the Gross-Pitaevskii equation for a rotating BEC

diffused vorticity equal to that of a rigid body with the same angular velocity, where the total vorticity is accommodated by the vortex lattice. Formation of a vortex array can be seen by numerical integration of the GP equation, as shown in Fig. 12.2.

12.2 Vortex Nucleation

The quoted experiments in condensates confirmed the nucleation of quantized vortices, which is a clear manifestation of their superfluid properties. Since then, much effort has been made to understand the dynamics of the rotating BEC [19] and, in particular, the mechanisms of vortex nucleation [18, 28]. Particular interesting features of quantized vortices in Bose Einstein condensates of alkali atoms are related to the formation of vortex arrays, where singly quantized vortices typically arrange in highly regular triangular lattices. Such a configuration is only possible when a sufficient amount of angular momentum is effectively transferred to the system, corresponding to a situation of rapid rotation. The acquired angular velocity then tends to enlarge the rotating cloud and the centrifugal force is responsible for the flattening of the density profile towards a two-dimensional configuration. In the limit where the rotation frequency Ω approaches the transverse trapping frequency $\omega_\perp$, the quadratic centrifugal and the harmonic trapping potentials cancel out and the system is no longer bounded. The possibility of reaching high angular velocities is therefore provided by the addition of anharmonic terms to the trapping potential, making worthy the investigation of new equilibrium configurations with different vortex states and new collective modes [13].

12.3 Tkachenko Modes

Let us now focus on the oscillations that can occur in a vortex lattice. This was first discussed by Tkachenko [30], and then observed experimentally in liquid Helium-4 [3] and in condensates [11]. In order to describe such oscillations, we make use of extended fluid equations where the contributions from the motion of the vortices around their equilibrium positions are taken into account. These extended fluid equations in the rotating frame can take the form [9]

$$\frac{\partial n}{\partial t} + \nabla \cdot (n\mathbf{v}) = 0 \,, \quad \frac{\partial \mathbf{v}}{\partial t} = -\frac{g}{M}\nabla n - 2\Omega \times \mathbf{v} - \frac{\mathbf{F}_{\mathrm{el}}(\mathbf{s})}{Mn}, \tag{12.10}$$

where $\mathbf{F}_{\mathrm{el}}(\mathbf{s})$ is the so-called *Magnus force*, which is an elastic force due to the deformation of the vortex lattice. It therefore depends on the displacement field $\mathbf{s} \equiv \mathbf{s}(\mathbf{r})$, describing the deviations of the lattice with respect to the equilibrium. The components of this elastic force can be defined as the derivatives of a strain tensor, σ_{jk}, as given by $F_{\mathrm{el},j}(\mathbf{s}) = \partial\sigma_{jk}(\mathbf{s})/\partial x_k$. In the momentum equations, we have neglected the confining potential as well as the quantum Bohm term, but retained the relevant forces, which are the Coriolis force and the nonlinear force. The displacement field $\mathbf{s}$ can be related with the above fluid equations by using the equation of conservation of the number of vortices. This can be written in the form of a continuity equation

$$\frac{\partial n_v}{\partial t} + \nabla \cdot (n_v \dot{\mathbf{s}}) = 0. \tag{12.11}$$

Noting that the vortex density n_v is proportional to $2\Omega + \nabla \times \mathbf{v}$, we can transform this equation into

$$\frac{\partial \mathbf{v}}{\partial t} = -\frac{g}{M}\nabla n - 2\Omega \times \dot{\mathbf{s}}. \tag{12.12}$$

By taking a perturbation with respect to the equilibrium density $n = n_0$, and assuming that the perturbations in n, $\mathbf{v}$ and $\dot{\mathbf{s}}$ behave in space and time as $\exp(i\mathbf{k} \cdot \mathbf{r}_\perp - i\omega t)$, where $\mathbf{r}_\perp$ refers to the xy-plane perpendicular to Ω, we can then derive a dispersion relation which has two solutions for ω^2, of the form

$$\omega^2 = 4\Omega^2 + c_s^2 k^2 \,, \quad \omega^2 = \frac{c_T^2 c_s^2 k^4}{4\Omega^2 + c_s^2 k^2}, \tag{12.13}$$

where $c_s = \sqrt{g n_0 / M}$ is the Bogoliubov sound speed, and $c_T = \sqrt{\hbar\Omega/4M}$ is the *Tkachenko speed*, characterizing the lattice vibrations. The first solution is just that of the usual sound waves, $\omega = c_s k$, modified by the fluid rotation. The second solution, which depends on this new characteristic speed c_T corresponds to the *Tkachenko modes*. They can be observed as vibrations in the vortex lattice [12].

12.4 Rossby Waves

We now consider a different kind of fluid mode which can take place in a rotating condensate, similar to the *Rossby waves*, well known in geophysics. These waves, also called *planetary waves*, are recognized as the main pattern of long period variability of the upper tropospheric winds [26]. These are responsible for the well known cyclonic and anticyclonic systems that characterize the weather systems in mid latitudes, and can be observed both in the upper troposphere and in the oceans. These waves are due to the dependence of the Coriolis force with latitude, which acts as a restoring force for the particles in the atmosphere. In a rotating condensate, as we have seen, the Coriolis force parameter is given by twice the angular rotation frequency, Ω. We show below that such waves are dispersive, with a negative phase velocity, which means that they always propagate opposite to the condensate rotation.

As we have also seen, in the presence of a large number of vortices it is possible to define a coarse grained velocity, by averaging the velocity field over regions containing many vortex lines. This is known as the *diffused vorticity approach* [12] and corresponds to assuming a rigid-body rotation $\mathbf{v} = \Omega \times \mathbf{r}$, where the angular velocity is $\Omega = \Omega\hat{\mathbf{z}}$ with $\Omega = \pi\hbar n_v / M$, where $n_v(\mathbf{r})$ is the average vortex density Therefore, the irrotationality condition $\nabla \times \mathbf{v} = 0$ is replaced by $\nabla \times \mathbf{v} = 2\Omega$. In that case, the macroscopic dynamics of the rotating fluid is provided by the fluid equations in the rotating frame

$$\frac{\partial n}{\partial t} + \nabla \cdot (n\mathbf{v}) = 0 \tag{12.14}$$

$$\left(\frac{\partial}{\partial t} + \mathbf{v} \cdot \nabla\right)\mathbf{v} = -\frac{g\nabla n}{m} - \frac{\nabla V}{m} + \frac{\hbar^2}{2m^2}\nabla\left(\frac{\nabla^2\sqrt{n}}{\sqrt{n}}\right) - 2\Omega \times \mathbf{v}, \tag{12.15}$$

where $\mathbf{v}\cdot\nabla\mathbf{v} = \nabla(v^2)/2 - \mathbf{v}\times(\Omega\times\mathbf{v})$. The usual hydrodynamic calculations are based on the Thomas-Fermi approximation, which consists of neglecting the quantum pressure proportional to $\hbar^2$. Here, however, we include this quantum term, since we may be interested in Bogoliubov-like waves. Notice that this procedure does not contradict the diffused vorticity approximation, since the quantum features will be only included in the dynamics of the fluctuations and do not affect the equilibrium configuration of the system. This allows one to cast quantum features that may be relevant to describe the so-called quantum turbulence [20], where the healing length sets a lower scale for the Kolmogorov cascade. Here, $V(\mathbf{r},\Omega) = V_{trap}(\mathbf{r}) - m\Omega^2 r^2/2$, with $r = (x^2+y^2)^{1/2}$, stands for the effective trapping potential, which reads

$$V(\mathbf{r},\Omega) = \frac{\hbar\omega_\perp}{2}\left[\left(1-\frac{\Omega^2}{\omega_\perp^2}\right)\frac{r^2}{a_{ho}^2} + \beta\frac{r^4}{a_{ho}^4}\right], \tag{12.16}$$

with the characteristic harmonic oscillator length $a_{ho} = \sqrt{\hbar/m\omega_\perp}$, and the dimensionless anharmonicity parameter β. The term $2\Omega\times\mathbf{v}$ in Eq. (12.15) represents the Coriolis force, which acts as the restoring force for the oscillations considered here. We consider perturbations around the equilibrium configuration, such that $n = n_0 + \delta n$ and $\mathbf{v} = \delta\mathbf{v}$. In that case, the system can be described by the following set of perturbed equations

$$\frac{\partial}{\partial t}\delta n + \nabla\cdot(n_0\delta\mathbf{v}) = 0 \tag{12.17}$$

$$\left(\frac{\partial}{\partial t} + \delta\mathbf{v}\cdot\nabla\right)\delta\mathbf{v} = -gm\nabla\delta n - 2\Omega\times\delta\mathbf{v} + \frac{\hbar^2}{4m^2}\nabla\left(\frac{\nabla^2\delta n}{n_\infty}\right), \tag{12.18}$$

where n_∞ is the peak density. Here, we have considered the quantum pressure to be important only for the fluctuations,

$$\frac{\hbar^2}{4m^2}\nabla\left(\frac{\nabla^2(n_0+\delta n)}{n_0}\right) \approx \frac{\hbar^2}{4m^2}\nabla\left(\frac{\nabla^2\delta n}{n_\infty}\right), \tag{12.19}$$

where the local density n_0 is assumed to vary on a scale much longer than δn. The rotational velocity field can be split into to parts, $\delta\mathbf{v} \approx \delta\mathbf{v}_0 + \delta\mathbf{v}_p$, where

$$\delta\mathbf{v}_0 = \frac{1}{2\Omega}\hat{\mathbf{z}}\times\mathbf{S} \tag{12.20}$$

is the zeroth-order drift velocity, which results from taking $d/dt = \partial/\partial t + \delta\mathbf{v}\cdot\nabla = 0$ in Eq. (12.18) and the vector $\mathbf{S}$ is defined as

$$\mathbf{S} = -g\nabla\delta n + \frac{\hbar^2}{4m^2}\nabla\left(\frac{\nabla^2\delta n}{n_\infty}\right). \tag{12.21}$$

The polarization velocity $\delta \mathbf{v}_p$ is the first-order correction to the drift velocity (12.21) and satisfies the following equation

$$\left(\frac{\partial}{\partial t} + \delta \mathbf{v}_0 \cdot \nabla \right) \delta \mathbf{v}_0 = -2\Omega \times \delta \mathbf{v}_p, \tag{12.22}$$

which yields

$$\delta \mathbf{v}_p = -\frac{1}{4\Omega^2} \frac{\partial \mathbf{S}_\perp}{\partial t} - \frac{1}{8\Omega^3} (\hat{\mathbf{z}} \times \mathbf{S}) \cdot \nabla_\perp \mathbf{S}, \tag{12.23}$$

where $\mathbf{S}_\perp = (S_x, S_y)$ is the transverse component of $\mathbf{S}$. The continuity equation (12.17) can be written in the following fashion

$$\frac{d}{dt} \ln n + \nabla \cdot \boldsymbol{\delta} \mathbf{v}_p = 0, \tag{12.24}$$

where the material derivative can be approximated as $d/dt \approx \partial/\partial t + \delta \mathbf{v}_0 \cdot \boldsymbol{\nabla}$. Using the fact that $\ln n \approx \ln n_0 + \phi$, where $\phi = \delta n / n_\infty$, and putting equations (12.20), (12.23) and (12.24) together, one obtains

$$\left(1 - r_0^2 \nabla_\perp^2 + \frac{1}{2} r_0^2 \xi^2 \nabla_\perp^4 \right) \frac{\partial \phi}{\partial t} + 2\Omega \left\{ \psi, \phi - \nabla^2 \psi + \ln n_0 \right\} = 0. \tag{12.25}$$

This equation is formally similar and generalizes the *Charney equation* [10], also referred in the literature as Charney-Hasegawa-Mima (CHM) equation, which is widely used in the study of the dynamics of waves and turbulence in plasmas and in the atmosphere. Here, $r_0 = c_s / 2\Omega$ represents the generalized Rossby radius and

$$\psi = r_0^2 \phi - r_0^2 \xi^2 \nabla^2 \phi / 2. \tag{12.26}$$

The operator $\{a, b\} = r^{-1} (\partial_r a \partial_\theta b - \partial_r b \partial_\theta a)$ is simply the Poisson bracket in polar coordinates and θ represents the angular coordinate. Eq. (12.25) describes hydrodynamic drift waves in a rapidly rotating Bose Einstein condensate and includes new features relatively to the CHM equation. Namely, the terms proportional to ξ^2 cast the effects of the zero-point pressure, which are known to play no role in geophysics. According to typical experimental conditions in condensates, we estimate the sound speed to be $c_s \sim 1\,\mathrm{mm/s}$, $\Omega \sim \omega_\perp \approx 2\pi \times 65\,\mathrm{Hz}$ [7], a transverse harmonic oscillator radius of $a_{\mathrm{ho}} \sim 1.7\,\mu\mathrm{m}$ and a Rossby radius around $r_0 \sim 1.2\,\mu\mathrm{m}$. The *Rossby number*, Ro, defines the ratio of the inertial to Coriolis forces

$$\mathrm{Ro} = \frac{c_s}{Lf}, \tag{12.27}$$

where L is a typical length of the system and f is the Coriolis parameter [10]. We now show that a rotating BEC can sustain a new fluid mode corresponding to a drift-acoustic wave. For that purpose, we keep only the linear terms in Eq. (12.25), and look for perturbations of the form $\phi \sim e^{i(\mathbf{k}\cdot\mathbf{r}-\omega t)}$. The respective dispersion relation is then readily obtained and reads

$$\omega = -v_R k_\theta \frac{1+\xi^2 k^2/2}{1+r_0^2 k^2(1+\xi^2 k^2/2)}, \tag{12.28}$$

where $k_\theta = (\mathbf{k}\cdot\mathbf{e}_\theta)$ is the polar (or zonal) component of the wave vector $\mathbf{k} = (k_x, k_y)$. The term $v_R = -2\Omega r_0^2 \partial_r \ln n_0$ is the generalized *Rossby (drift) velocity*. Because the equilibrium profile is generally smooth, we expect v_R to be small (as compared with the Bogoliubov speed c_s), which suggests that these waves appear as low frequency oscillations (compared to both $\omega_\perp$ and Ω). The dispersion relation (12.28) is similar to the expression for barotropic Rossby waves in the atmosphere [26] and to the dispersion relation obtained for drift waves in a magnetized plasma [16]. For long wavelengths $r_0^2 k^2 \ll 1$ (and consequently $\xi^2 k^2 \ll 1$), Eq. (12.28) reduces to the *zonal flow* dispersion relation $\omega \approx -k_\theta v_R$. One of the remarkable features of the zonal, transverse acoustic, waves is that of having negative zonal phase and group velocities, $c_\theta^{(ph)} = c_\theta^{(g)} \approx -v_R$. It means that they propagate always "westward" with respect to the rotation of the condensate, which explains the negative values for the frequency in Eq. (12.28). For short wave lengths, one obtains the dispersion relation for the (actual) Rossby waves $\omega \approx -v_R k_\theta / \left(r_0^2 k^2\right)$, with phase and group velocities approximately given by

$$c_\theta^{(ph)} \approx -\frac{v_R}{r_0^2 k^2}$$

$$c_\theta^{(g)} \approx v_R \frac{2k_\theta/k - 1}{r_0^2 k^2}, \tag{12.29}$$

In Fig. 12.3, we plot the dispersion relation (12.28) for different values of the healing length ξ, using $\langle v_R \rangle = 0.1 c_s$, where $\langle v_R \rangle$ is the mean Rossby velocity inside the cloud. This procedure is similar to a local density approximation, which is valid for sufficiently smooth equilibrium profiles.

Although a single Rossby wave of arbitrary amplitude is a solution of Eq. (12.28), a superposition of waves, generally, is not. The nonlinear interaction between waves leads to a mechanism of spectral energy transfer. To study the wave interaction properties, one decomposes the solution into its Fourier series, $\phi_\mathbf{k} = \sum_\mathbf{k} \tilde{\phi}_\mathbf{k} \exp(i\mathbf{k}\cdot\mathbf{r})$ which, after plugging into Eq. (5.98), yields the following nonlinear equation

$$\frac{\partial \tilde{\phi}_\mathbf{k}}{\partial t} + i\omega_k \tilde{\phi}_\mathbf{k} = \sum_{\mathbf{k}_1,\mathbf{k}_2} \Lambda_{\mathbf{k}_1,\mathbf{k}_2}^{\mathbf{k}} \tilde{\phi}_{\mathbf{k}_1} \tilde{\phi}_{\mathbf{k}_2}, \tag{12.30}$$

where

$$\Lambda_{\mathbf{k}_1,\mathbf{k}_2}^{\mathbf{k}} = 2r_0^2 \frac{\left(1+\xi^2 k_1^2/2\right)\left(1+r_0^2 k_2^2 + r_0^2 \xi^2 k_2^4/2\right)}{1+r_0^2 k^2 + r_0^2 \xi^2 k^4/2} \delta\left(\mathbf{k}_1+\mathbf{k}_2-\mathbf{k}\right)\left(\mathbf{k}_2\times\mathbf{k}_1\right)\cdot\Omega \tag{12.31}$$

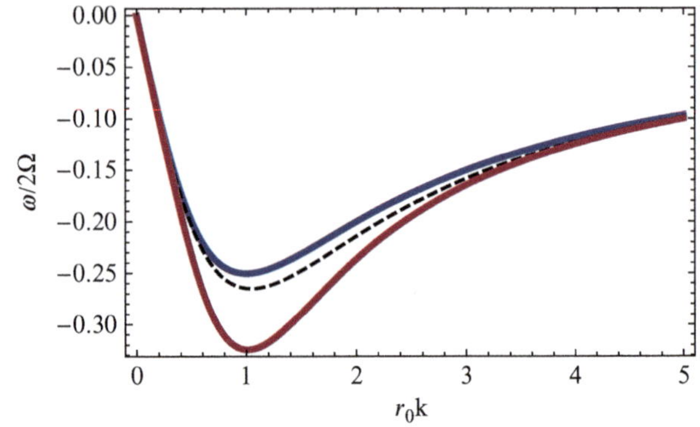

Fig. 12.3 Dispersion relation of the Rossby waves in a rapidly rotating BEC, for $v_R = 0.1c_s$. It is clearly shown that phase speed is generally negative. The *blue full line* corresponds to the Thomas-Fermi case, $\xi = 0$. The *black dashed* and *red full lines* respectively correspond to $\xi = 0.7r_0$ and $\xi = 1.3r_0$

is the nonlinear coupling operator, and ω_k is given by Eq. (12.28). Only waves that satisfy the condition $\mathbf{k}_1 + \mathbf{k}_2 = \mathbf{k}$ interact nonlinearly with each other. The set of the waves satisfying this condition is known in the literature as a *resonant triad*. This resonance mechanism is able to transfer energy between different length scales, being one of the sources of classical turbulence in plasmas and in the atmosphere [16, 25]. Here, due to the existence of additional terms that properly account for the quantum hydrodynamical features of the system, i.e., when large variations of the density profile are present, Eq. (12.30) may be used to describe turbulence in rotating Bose-Einstein condensates, opening a stage to explore the similarities and differences between classical and quantum turbulence.

Another interesting feature of the Rossby waves in Bose Einstein condensates is the possibility of finding localized structures, which may result, for example, from the saturation of the triadic resonance mechanism mentioned above. Such purely nonlinear solitary structures can be obtained from the stationary solutions of Eq. 5.98, which readily yields

$$\left(1 + \frac{\xi^2}{2}\nabla_\perp^2\right)\{\phi, \nabla_\perp^2\phi\} - \frac{\xi^2}{2}\{\phi, \nabla_\perp^4\phi\} = 0. \qquad (12.32)$$

In the Thomas-Fermi limit, the latter expression simply reduces to $\{\phi, \nabla_\perp^2\phi\} = 0$, which is satisfied for a family of functions $\nabla_\perp^2\phi = \mathcal{F}(\phi)$, where $\mathcal{F}(\phi)$ is an arbitrary function of its argument. The different choices for $\mathcal{F}$ will lead to different structures, which describe many physically relevant nonlinear stationary solutions. For example, for the choice $\mathcal{F}(\phi) \propto \exp(-2\phi)$, Stuart [29] showed that the so-called "cat-eye" solution describes a vortex chain in a magnetized plasma sheet, which has been observed experimentally in mixing layer experiments [8]. However, in the present case, there are physical limitations that impose specific constraints to the choice of the solutions. In particular, as discussed in [13], the equilibrium profile associated with the potential in Eq. (12.16), which is given by $n_0(r) = n_\infty(R_+^2 - r^2)(r^2 - R_-^2)$, where the peak density is $n_\infty = \beta\hbar\omega_\perp/2g$, must vanish at the Thomas-Fermi radii defined as follows

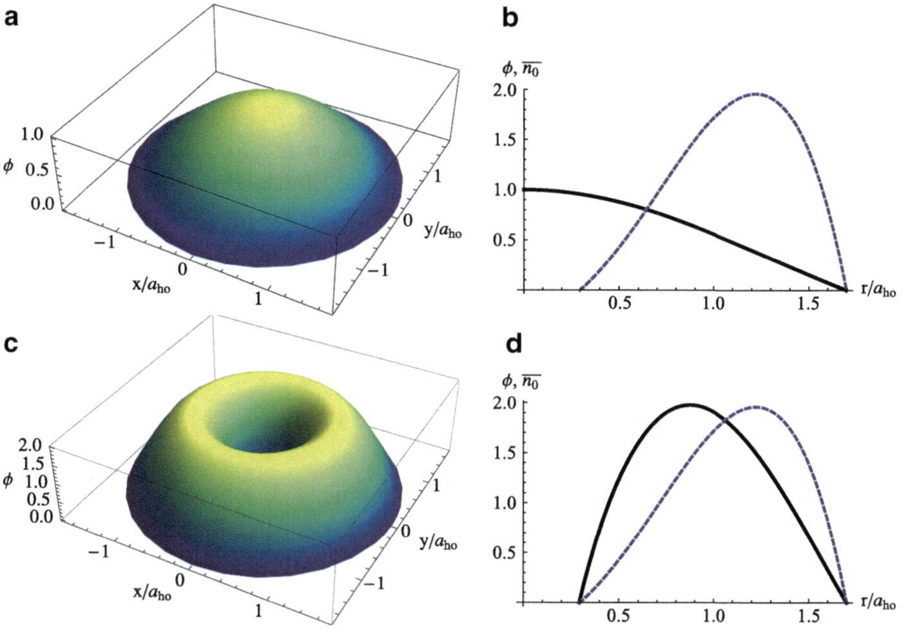

Fig. 12.4 Nonlinear stationary solitary wave resulting from the saturation of the triad resonant decay of Rossby waves, obtained for $\Omega = 2.4\omega_\perp$ and $\beta = 1.6$: (**a**) $\mu = 0.2\hbar\omega_\perp$ and (**c**) $\mu = -0.2\hbar\omega_\perp$. Plots (**b**) and (**d**) respectively compare the radial structures (*full lines*) of (**a**) and (**c**) with the corresponding Thomas-Fermi equilibria (*dashed lines*) discussed in the text, obtained for the same set of parameters

$$\frac{R_\pm^2}{a_{ho}^2} = \frac{\Omega^2 - \omega_\perp^2}{2\beta\omega_\perp^2} \pm \sqrt{\left(\frac{\Omega^2 - \omega_\perp^2}{2\beta\omega_\perp^2}\right)^2 + \frac{2\mu}{\beta\hbar\omega_\perp}}, \tag{12.33}$$

where μ represents the chemical potential. For $\mu > 0$, the radius R_- is purely imaginary and the density vanishes at $R = R_+$, while for $\mu < 0$ both R_- and R_+ are present. This reflects the transition occurring at $\mu = 0$, where a hole forms in the centre of the condensate and the equilibrium profile assumes an annular shape. The simplest nonlinear structure that verifies such constraints is obtained for $\mathcal{F}(\phi) = -\kappa\phi$, and the respective radial solution, for $\ell = 0$, yields $\phi(r) = AJ_0(\kappa r) + BY_0(\kappa r)$. The values of κ, A and B are such that the solution vanishes at the radii $R_\pm$. In Fig. 12.4, we plot two possible solitary structures in the overcritical rotation regime $\Omega > \omega_\perp$, obtained for both $\mu > 0$ and $\mu < 0$. It is interesting to observe that, even for the same set of parameters, the resulting solitary structures may differ from the usual Thomas-Fermi equilibrium profiles discussed before.

12.5 Rossby-Tkatchenko Modes

We have seen that a rotating condensate has at least two types of oscillating modes, the Tkachenko modes and the Rossby modes. It is therefore legitimate to consider the possible existence of a more general dispersion relation, which reduces to the two distinct modes when we take appropriate limits. The search for such a general result is not obvious and we propose here to solve the problem by using a simple regression method. As a starting point of this methods we can start from the dispersion relation for the Tkachenko modes, as determined by the second solution in (12.13). It can be easily realized, by doing a regression Fourier analysis, that this dispersion relation can be directly derived from the density wave equation

$$\left(4\Omega^2 - c_s^2\nabla_\perp^2\right)\frac{\partial^2 \tilde{n}}{\partial t^2} + c_s^2 c_T^2 \nabla_\perp^4 \tilde{n} = 0 \tag{12.34}$$

where $\tilde{n}$ is the perturbed fluid density. By assuming $\tilde{n} \propto \exp(i\mathbf{k}\cdot\mathbf{r}_\perp - i\omega t)$ we immediately recover (12.13). Similarly the linear dispersion relation for the Rossby waves (12.28) can be rewritten, in the long wavelength limit $\xi^2 k^2 \ll 1$, in the form

$$\omega = -\frac{2\Omega c_s^2 \beta k}{4\Omega^2 + c_s^2 k^2}, \tag{12.35}$$

where $\beta = \partial_r \ln n_0$. Again, a similar regression Fourier analysis will lead to the wave equation

$$\left(4\Omega^2 - c_s^2\nabla_\perp^2\right)\frac{\partial \tilde{n}}{\partial t} - 2\Omega c_s^2 \beta \frac{\partial \tilde{n}}{\partial \theta} = 0. \tag{12.36}$$

It is important to notice the striking similarities between the two wave equations (12.34) and (12.36). This allows us to combine them into a single equation, as

$$\left(4\Omega^2 - c_s^2\nabla_\perp^2\right)\frac{\partial^2 \tilde{n}}{\partial t^2} - 2\Omega c_s^2 \beta \frac{\partial^2 \tilde{n}}{\partial \theta \partial t} + c_s^2 c_T^2 \nabla_\perp^4 \tilde{n} = 0. \tag{12.37}$$

For a wave perturbation with frequency ω and wavenumber k, we obtain from here the *Rossby-Tkachenko dispersion relation*, as

$$\omega = \frac{-2\Omega c_s^2 \beta k_\theta + \sqrt{\Omega^2 c_s^4 \beta^2 k_\theta^2 + c_s^2 c_T^2 k^4 (4\Omega^2 + c_s^2 k^2)}}{4\Omega^2 + c_s^2 k^2}, \tag{12.38}$$

Notice that when we neglect the effects of the lattice, $c_T \to 0$, this reduces to the Rossby modes (12.36). On the other hand, in the limit of a uniform condensate, such that $\beta \to 0$, we recover the Tkachenko modes. This generalized dispersion relation

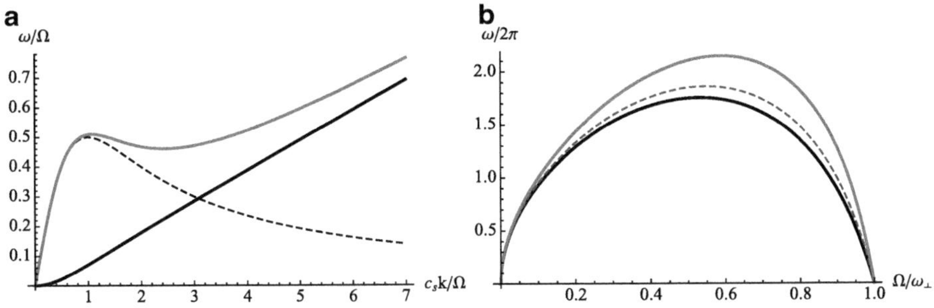

Fig. 12.5 Dispersion relation for the Rossby-Tkachenko modes in a rotating vortex, for $c_T = 0.2c_s$. (**a**) Homogeneous case. for comparison we reprent the Rossby dispersion curve (*dashed line*) and the Tkachenko dispersion curve (*full line*); (**b**) Trapped case: Eigen-frequencies obtained for the compatible wavenumber $k_0 = 5.45/T_\perp$. From *top* to *bottom*, $l = 5$, $l = 3$ and $l = 0$

is represented in Fig. 12.5a, for $k = k_\theta$, where a comparison is made with our two previous dispersion curves. We can see that the Tkachenko term dominates for large wavenumbers, and the Rossby terms for low wavenumbers.

Let us now discuss finite effects. As already mentioned, the inhomogeneity imposed by the trap plays an important role in the experiments, and may be the key to the detection of these Rossby-Tkachenko modes. We have seen that the equilibrium condition in the harmonic trap can be described by a parabolic profile, valid in the Thomas-Fermi approximation, as

$$n_0(r) = n(0)\left(1 - \frac{r^2}{R_\perp^2}\right), \quad R_\perp = \frac{\sqrt{2}c_s(0)}{\sqrt{\omega_\perp^2 - \Omega^2}}, \tag{12.39}$$

where $R_\perp$ is the size of the condensate under rotation, and $c_s(0)$ the sound velocity at the centre of the trap. The mode frequencies corresponding to such a profile can be obtained by replacing the parameter β in Eq. (12.38) by its mean value $\langle\beta\rangle$, taken over the entire profile of the condensed cloud, as defined by

$$\langle\beta\rangle = \frac{2}{R_\perp^2}\int_0^{R_\perp} \frac{1}{n_0(r)}\frac{dn_0(r)}{dr} r dr = -\frac{4(\pi - 1)}{R_\perp} \tag{12.40}$$

On the other hand, the poloidal wavenumber $k_\perp$ is replaced by lk_0, where l is the mode "winding" number. In order to find the compatible wavenumber k_0, we can now use the following argument [9]. For slow rotations, such that $\Omega \ll \omega_\perp$, the basic periodicity in the medium will be the vortex interspacing in the lattice, or

$$\lim_{\Omega \to 0} \omega(k_0) = \frac{8\pi}{\sqrt{3}}\frac{c_T}{R_\perp}, \tag{12.41}$$

which leads to the value $k_0 = 5.45/R_\perp$. Replacing this and (12.40) in the dispersion relation, we can estimate the vibration of the lattice as a function of the rotation frequency Ω. The result is represented in Fig. 12.5b. As shown in this figure, the difference between the original Tkachenko modes ($l = 1$) and the present Rossby-Tkachenko modes is very small for the rotation rates $\Omega/\omega_\perp \sim 0.85$–$0.99$, as used in the experiments [11]. A deviation from the Tkachenko modes will therefore become more evident for $l > 1$ and $\Omega/\omega_\perp \sim 0.6$–$0.8$.

12.6 Coupling with Photon OAM States

The excitation of rotating condensates can be made by using laser beams with orbital angular momentum. Let us first briefly discuss the properties of the orbital angular momentum for electromagnetic radiation and then consider the coupling with the condensate. In classical physics, the angular momentum $\mathbf{J}$ of the electromagnetic radiation is defined by [17]

$$\mathbf{J} = \epsilon_0 \int_{vol} [\mathbf{r} \times (\mathcal{E} \times \mathbf{B})] \, d\mathbf{r}. \tag{12.42}$$

This quantity can be split into two distinct terms,

$$\mathbf{J} = \int \mathbf{J}_\mathbf{k} \frac{d\mathbf{k}}{(2\pi)^3} , \quad \mathbf{J}_\mathbf{k} = \epsilon_0 \left[\sum_j \mathcal{E}^*_{j,\mathbf{k}} \left(\mathbf{k} \times \nabla_\mathbf{k}\right) A_{j,\mathbf{k}} + \left(\mathcal{E}^*_\mathbf{k} \times \mathbf{A}_\mathbf{k}\right) \right], \tag{12.43}$$

where $\mathbf{J}_\mathbf{k}$ is the angular momentum associated with each photon state of wavevector $\mathbf{k}$, and $j = x, y, z$ represent the three components of the electric field $\mathcal{E}$ and of the vector potential $\mathbf{A}$. The last term in the expression of $\mathbf{J}_\mathbf{k}$ represents the polarization or *photon spin*, $\mathbf{S}_\mathbf{k}$, while the first term is the usually ignored orbital angular momentum $\mathbf{L}_\mathbf{k}$. The corresponding quantum operators are

$$\mathbf{S}_\mathbf{k} = -i\hbar a^\dagger_\mathbf{k} a_\mathbf{k} \left(\mathbf{e}^*_\mathbf{k} \times \mathbf{e}_\mathbf{k}\right) , \quad \mathbf{L}_\mathbf{k} = -i\hbar \sum_j a^\dagger_{j,\mathbf{k}} \left(\mathbf{k} \times \nabla_\mathbf{k}\right) a_{j,\mathbf{k}}, \tag{12.44}$$

where $\mathbf{a}^\dagger_\mathbf{k}$ and $\mathbf{a}_\mathbf{k}$ are the creation and destruction operators of the photon state $\mathbf{k}$. For photon propagation along a given direction z, we can introduce the unit vector $\mathbf{e}_\pm = (\mathbf{e}_x \pm i\mathbf{e}_y)/\sqrt{2}$ and $\mathbf{e}_z$, allowing us to write the spin operator for the left and right circular polarization states as $\mathbf{S} = \pm\hbar a^\dagger a \mathbf{e}_z$, and for the longitudinal polarization as $\mathbf{S} = 0$. This last polarization state is absent in vacuum, or in a neutral medium, but it exists in a metal or in a plasma, where it corresponds to a longitudinal photon, or plasmon. We know that, in general, the eigenstates of the photon angular momentum can be written as [5]

$$J^2 \mathbf{Y}_{jm} = j(j+1)\mathbf{Y}_{jm} , \tag{12.45}$$

where the eigenfunctions $\mathbf{Y}_{jm}$ can be expressed in terms of spherical harmonics. But this is usually not very practical, because we usually think in terms of the plane wave presentation of the photon field. The situation simplifies when we deal with a laser beam, where the electromagnetic field can be represented in a basis of orthogonal Laguerre-Gauss (LG) functions. It as recently been demonstrated that each LG-mode corresponds to a well defined value of the axial orbital angular momentum $L_z = \hbar l$. This is the reason why the study of photon orbital angular momentum became so popular in recent years [2]. This concept has been studied both in theory and experiments in nonlinear and quantum optics. Its extension to scalar quasi particles, such as plasmons and phonons has also been considered [23]. In the paraxial approximation, a laser beam with frequency ω can be written as a superposition of LG-modes, in the form

$$\mathbf{E}(\mathbf{r}, t) = \sum_{pl} E_{pl}(\mathbf{r}) \exp\left(-i\omega t + i \int^z k(z')dz'\right), \tag{12.46}$$

where $\omega = kc$, and l and p are integers. In cylindrical variables $\mathbf{r} \equiv (r, \varphi, z)$, the modes $E_{pl}(\mathbf{r})$ can be described by

$$E_{pl}(r, \varphi, z) = C_{pl} X^{|l|} L_p^{|l|}(X) \exp(-X/2) e^{il\varphi}, \tag{12.47}$$

with C_{pl} standing for a normalization factor and $X = r^2/w^2(z)$. Here, w represents the laser beam waist, and $L_p^{|l|}(X)$ represent the modified Laguerre polynomials, as defined by

$$L_p^l(X) = \frac{\exp(X)}{p} \frac{d^p}{dX^p} \left[X^{l+p} \exp(-X)\right]. \tag{12.48}$$

By choosing the appropriate normalization factor, C_{pl} we obtain the orthogonality condition

$$\int_0^\infty r dr \int_0^{2\pi} d\varphi E_{pl}^* F_{p'l'} = \delta_{pp'} \delta_{ll'}. \tag{12.49}$$

Of particular interest is the field mode corresponding to $p = 0, l = 1$ mode, which can be simply written

$$E_{pl}(r, \varphi, z) = \frac{r}{\sqrt{\pi} w_0^2} \exp\left(-\frac{r^2}{2w_0^2}\right) e^{i\varphi}, \tag{12.50}$$

where the waist was assumed constant and equal to its focal value w_0. Coupling of this field with the atoms subsequent rotation of the condensate as been observed in recent years [1, 4, 27].

References

1. J.R. Abo-Shaeer, C. Raman, J.M. Vogels, W. Ketterle, Science **292**, 476 (2001)
2. L. Allen, M.W. Beijersbergen, R.J.C. Spreeuw, J.P. Woerdman, Phys. Rev. A **45**, 8185–8189 (1992)
3. C.D. Andereck, J. Chalups, W.I. Glaberson, Phys. Rev. Lett. **44**, 33 (1980)
4. M.F. Andersen, C. Ryu, P. Cladé, V. Natarajan, A. Vaziri, K. Helmerson, W.D. Philips, Phys. Rev. Lett. **97**, 170406 (2006)
5. V.B. Berestetskii, E.M. Lifshitz, L.P. Pitaevskii, *Quantum Electrodynamics* (Butterworth-Heinemann, Oxford, 1982)
6. I. Bloch, J. Dalibard, W. Zweger, Rev. Mod. Phys. **80**, 885 (2008)
7. V. Bretin, S. Stock, Y. Seurin, J. Dalibard, Phys. Rev. Lett. **80**, 885 (1998)
8. F.K. Browand, P.D. Wiedman, J. Fluid Mech. **76**, 127, (1976)
9. C. Byam, Phys. Rev. Lett. **91**, 110402 (2003)
10. J.G. Charney, A. Elliassen, Tellus **1**, 38 (1949)
11. I. Coddington, P. Engels, V. Schweikhard, E.A. Cornell, Phys. Rev. Lett. **91**, 100402 (2003)
12. M. Cozzini, S. Stringari, Phys. Rev. A **67**, 041602(R) (2003)
13. M. Cozzini, A.L. Fetter, B. Jackson, S. Stringari, Phys. Rev. Lett. **94**, 100402 (2005)
14. A. Fetter, Rotating trapped Bose-Einstein Condensates. Rev. Mod. Phys. **81**, 647 (2009)
15. R.P. Feynman, in *Progress in Low Temperature Physics 1*, ed. by C. J. Gorter (North-Holland, Amsterdam, 1955). Chap. 2
16. A. Hasegawa, K. Mima, Phys. Fluids **21**, 87 (1978)
17. J.D. Jackson, *Classical Electrodynamics*, 3rd edn. (Hamilton Printing Company, Castleton, New York 1999)
18. N.A. Jamaludin, N.G. Parker, A.M. Martin, Phys. Rev. A **77**, 051603(R) (2008)
19. S. Kling, A. Pelster, Phys. Cold Trapp. Atoms **19**, 1072 (2009)
20. M. Kobayashi, M. Tsubota, J. Low Temp. Phys. **150**, 587 (2007)
21. K.W. Madison, F. Chevy, W. Wohlleben, J. Dalibard, Phys. Rev. Lett. **84**, 806 (2000)
22. M.R. Matthews, B.P. Anderson, P.C. Haljan, D.S. Hall, C.E. Wieman, E.A. Cornell, Phys. Rev. Lett. **83**, 2498 (1999)
23. J.T. Mendonça, B. Thidé, H. Then, Phys. Rev. Lett. **102**, 185005 (2009)
24. L. Onsager, Nuovo Cimento **6**(Suppl. 2), 249 (1949)
25. J. Pedlosky, *Geophysical Fluid Dynamics*, 2nd edn. (Springer, New York, 1986)
26. C.-G. Rossby et al., J. Mar. Res. **1**, 38 (1939)
27. C. Ryu, M.F. Andersen, P. Cladé, V. Natarajan, K. Helmerson, W.D. Philips, Phys. Rev. Lett. **99**, 260401 (2007)
28. S. Sinha, Y. Castin, Phys. Rev. Lett. **87**, 190402 (2001)
29. J.T. Stuart, J. Fluid Mech. **29**, 417 (1967)
30. V.K. Tkachenko, Sov. Phys. JETP **22**, 1282 (1966)

Chapter 13
Quantum Coherence

In this chapter, we explore the topic of matter wave interferometry and of quantum coherence, which plays a central role in quantum theory and is also used for many experimental applications. Atom interferometers are briefly discussed, and decoherence processes are introduced. We then consider decoherence of atom interference fringes, associated with quantum fluctuations of gravitational space-time. In the frame of the quantum theory of gravitation, still under construction, a fluctuating space-time foam should exist at the Planck space-time scale. Based on recent theoretical models, we speculate on the possible observation of quantum gravitational fluctuations, by using matter wave interferometry. Finally, the interference and tunneling of two condensates, confined in nearby potential wells, is considered, and the condensate analogue of Josephson oscillations is described.

13.1 Atom Interferometry

Atom interferometers have been inspired in the optical interferometers, and in some sense are based on the same configurations, although we can also find conceptual differences such as those associated with the fact that we can more easily confine atoms in a volume, than photons. Matter wave interferometry has been first proposed for electrons [1] and neutrons [2]. Early proposals for atom interferometry date from the mid 1980s. Both the history and the concepts have been reviewed in detail [3]. The advantage of atom interferometers with respect to their optical counterpart is the much smaller value of the atom wavelength, typically 10^4 times smaller than the wavelength of visible photons. Atom interferometers can be also used with a very low momentum uncertainty, by using atom clouds in MOTs or Bose Einstein condensates. The momentum uncertainty can be estimated by $\Delta p \simeq \sqrt{T/2M}$, where T is the temperature (in energy units). For instance, for Na atoms at a temperature of $T \sim 100\,\mu\text{K}$, we have $\Delta p \sim 10\hbar\omega/c$, where ω is the frequency of the cooling laser. For a condensate, with $T \sim 10^{-8}$ K, we would have $\Delta p \sim 0.1$.

J.T. Mendonça and H. Terças, *Physics of Ultra-Cold Matter*, Springer Series on Atomic, Optical, and Plasma Physics 70, DOI 10.1007/978-1-4614-5413-7_13,

One of the basic elements of an atom interferometer is the *light grating*, which is made of two counter-propagating optical beams. The atoms interacting with such grating suffer a quantized diffraction. This possibility was first suggested by the famous work of Kapitza and Dirac [4], who considered electron wave functions. It was first demonstrated for atoms by Moskowitz et al. [5], and later applied to condensates [6]. For a two-level atom, interacting with an optical grating with detuning $\Delta = \omega - \omega_a$, we have an effective optical potential of

$$V(\mathbf{r}) = \frac{\hbar}{2} \frac{\Omega_R^2(\mathbf{r})}{(2\Delta + i\Gamma)}, \tag{13.1}$$

where $\Omega = \mathbf{p}_{ab} \cdot \mathcal{E}/\hbar$ is the Rabi frequency and Γ the atomic decay rate. This potential is proportional to the laser light intensity associated with the grating, $I(\mathbf{r})$. In this expression, the real part is associated with the ac Stark effect, and the imaginary part to spontaneous scattering. The simplest configuration that we can imagine for a light grating corresponds to that of two laser beams with the same frequency, forming a standing wave pattern with a periodicity given by the wave vector $\mathbf{k}_G = \mathbf{k}_1 + \mathbf{k}_2$. This is usually called an optical lattice.

Several different regimes of atom diffraction can be found, depending on the size of the light grating, and on the strength of the scattering potential $V(\mathbf{r})$. The relevant scale length of the optical grating is the grating period $d = \pi c/\omega$. This has to be compared with the de Broglie wavelength of the atom, λ_B. An optical lattice is considered thick if the laser beams are larger than d^2/λ_B. For a thin grating, we can observe *Kapitza-Dirac scattering*. In what concerns the potential strength, we should compare its maximum value, V_{max} with the typical energy scale of the grating, E_G, as given by

$$E_G = \frac{\hbar^2 k_G^2}{2M} = 4\hbar\omega_{\text{rec}}, \tag{13.2}$$

where $\hbar\omega_{\text{rec}}$ is the recoil kinetic energy associated with the emission or absorption of a photon from the standing wave lasers. For shallow optical lattices, such that $V_{\text{max}} \ll E_G$, we can observe a typical *Bragg diffraction*. In contrast, for deep optical lattices, we will have *atom channeling*. These different regimes are briefly described next.

First, we consider a thin grating, where the Kapitza-Dirac scattering is taking place. In this case, the atom interacts with light for a short time τ. As a result of such interaction, the atom is excited into different momentum states, which are multiples of $\hbar\mathbf{k}_G$. The probability for finding the atom in these diffracted states is given by

$$P_n = J_n^2(\Omega_R^2 \tau / 2\Delta), \tag{13.3}$$

where n is integer, and J_n are the first order Bessel functions. This is valid for normal incidence of the atom on the grating. This is also valid under the so-called *Raman-Nath approximation*, which states that the transverse motion of the atom stays small in comparison with its longitudinal motion. For a standing wave with a parabolic profile near the laser beam axis, this approximation is valid for small interaction times τ such that

$$\tau < \frac{1}{2}\sqrt{(\Omega_R^2 + \Delta^2)^{1/2} E_G/\hbar}. \tag{13.4}$$

If the interaction time is longer, the excitation of momentum states with large n is suppressed by phase mixing.

In the opposite case of thick gratings, we have Bragg diffraction. In this case, we need to consider the full matter wave propagation of the atom wave function inside the periodic optical potential. Diffraction will then occur at a specific angle, θ_B, defined by the well known Bragg's formula

$$n\lambda_B = \frac{2\pi}{k_G}\sin\theta_B. \tag{13.5}$$

The probability for light scattering through the Bragg angle is determined by

$$P(\tau) = \sin^2\left(\frac{\Omega_R^2}{4\Delta}\tau\right). \tag{13.6}$$

The temporal oscillations predicted by this expression have been experimentally observed. For higher order $n > 1$ Bragg diffraction, the probabilities are given by

$$P_n(\tau) = \sin^2\left\{\frac{[(n-1)!]^2}{2^{(4n-3)}}\Omega_R^{2n}\Delta^n\omega_{rec}^{(n-1)}\tau\right\}. \tag{13.7}$$

This expression is valid for large detuning, $\Delta \gg n^2\omega_{\text{rec}}$.

The Bragg diffraction regime described above is only valid for weak optical potentials. In the case of a strong potential, such that $V_{\text{max}} > E_G$, the atoms become trapped in the standing optical potential well, and can be guided through the optical crystal without being diffracted. This effect has been observed in both charged particle beams in material crystals and neutral atoms in optical lattices.

13.2 Time Interferometry

Many interesting effects arise in atomic interferometry when the system is modulated in time. Contrary to the case of light, where all wavelengths propagate at the same speed c, the vacuum itself constitute a dispersive medium for atoms when time modulation is applied. Therefore, two matter wave components interfering at (x, t) may have propagated from the same x but originated from there at different times t, if they have different velocity. Time-dependent boundary conditions can cause matter wave diffraction phenomena in time that are similar to spatial diffraction phenomena arising from spatially dependent boundary conditions.

This is in close connection to the concepts of time refraction [7] and time diffraction, which are associated to the temporal change of the optical properties of a dielectric medium. A first description of time diffraction was due to Moshinsky [8], who argued that temporal equivalent of Fresnel fringes can occur with matter waves after opening a shutter. Similarly, an opening and closing of a shutter results in a single slit diffraction in time; two successive openings makes a double slit; of course, a periodic change in the opening of the slit produces a diffraction pattern in time. As in the case of spatial diffraction, time diffraction can also exhibit both near-field and far-field regimes. Moreover, depending on the duration of the interaction (interference time), one may also observe Bragg and Raman-Nath regimes.

To illustrate some particular features of time interferometry with cold atoms, we discuss the case of an interferometer that uses optical standing waves as phase gratings operating in the time domain [9]. Let us consider that two electric pulses of equal amplitude E and frequency ω propagate inside the cold atomic gas, initially prepared with a velocity v_0. The time-dependent Hamiltonian of the laser-atom field is

$$H(x,t) = \frac{p^2}{2M} + \hat{V}_{\text{int}}(x,t), \tag{13.8}$$

where $V_{\text{int}}(x,t) = \hbar\Omega(t)\sin(kx)$ and

$$\Omega(t) = \frac{(\mu E)^2}{8\hbar^2\Delta} \tag{13.9}$$

is the Rabi frequency, μ is the off-diagonal dipole matrix element and $\Delta = \omega - \omega_a$ is the detuning with respect to the atomic transition. Immediately before the interaction, the atom can be described by a plane wave

$$\psi(x,0^-) = e^{ik_0x}. \tag{13.10}$$

For a short pulse of duration τ, the kinetic term $p^2/2M$ can be neglected provided that

$$\tau\omega_{\text{rec}} \sim \sqrt{\omega_{\text{rec}}\Omega(t)}\tau \ll 1, \tag{13.11}$$

which is equivalent to a temporal "thin grating" condition (or to the so called Raman-Nath approximation [10]). Under this approximation, the wave function takes the following form under the action of the Hamiltonian (13.8)

$$\psi(x,0^+) = e^{i\theta_1\sin(kx)+ik_0x}, \tag{13.12}$$

where the pulse integral is given by

$$\theta_1 = -\int_0^\tau dt\ \Omega(t). \tag{13.13}$$

In order to compute the temporal evolution of the wave function for $t > 0^+$, we Fourier expand the initial solution (13.12). Noting that each Fourier component is associated with the eigenvalue $\hbar(nq + k_0)^2/2m$, one obtains

$$\psi(x,t) = e^{i(k_0x-\omega_0t)} \sum_n J_n(\theta_1) e^{ink(x-v_0t)} e^{-in^2\omega_{\text{rec}}t}, \tag{13.14}$$

where $\omega_0 = \omega_{\text{rec}}(k_0) = \hbar k_0^2/2M$. After one period, i.e. for $t = T$, a second standing wave is applied, such that the resulting evolution is given by

$$\psi(x,t) = e^{i(k_0x-\omega_0t)} \sum_{n_1,n_2} J_{n_1}(\theta_1) J_{n_2}(\theta_2)\, e^{in_1kv_0T}\, e^{i(n_1+n_2)k[x-v_0(t-T)]} \times$$
$$\times\, e^{-in_1^2\omega_{\text{rec}}T}\, e^{-i(n_1+n_2)^2\omega_{\text{rec}}(t-T)}, \tag{13.15}$$

The corresponding atomic density $\rho(x,t)_{v_0} = \psi^*(x,t)\psi(x,t)$ with initial velocity v_0 is therefore given by

$$\rho(x,t) = \sum_{n_1,m_1,n_2,m_2} J_{n_1}(\theta_1) J_{m_1}(\theta_1) J_{n_2}(\theta_2) J_{n_2}(\theta_2) \times$$
$$\times\, e^{-i(n_1-m_1)kv_0T}\, e^{i[(n_1+n_2)-(m_1+m_2)]k[x-v_0(t-T)]} \times$$
$$\times\, e^{-i(n_1^2-m_1^2)\omega_{\text{rec}}T}\, e^{-i[(n_1+n_2)^2-(m_1+m_2)^2]\omega_{\text{rec}}(t-T)}. \tag{13.16}$$

Because we have implicitly assumed that the wave vectors of the standing waves, k_1 and k_2, have the same direction, we only retain the terms involving e^{ikx} such that $(n_1 + n_2) - (m_1 + m_2) = 1$. Moreover, averaging the atomic density matrix over an initial velocity distribution assuming that $k\langle u\rangle T \gg 1$, the only nontrivial contributions to Eq. (13.16) occur in the vicinity of echo times $t_N = (N + 1)T$. Therefore, one only needs to cast the terms at times $t = t_N + \Delta t$, where Δt is a delay accounting for the velocity broadening and fulfills the following relation

$$\Delta t = \frac{1}{k\langle u\rangle} \ll T. \tag{13.17}$$

Finally, the Fourier transform of the density matrix, or simply the interference signal, reads

$$\tilde{\rho}(k,\Delta t) = i(-1)^{N+1}\langle e^{-ikv_0\Delta t}\rangle J_N[2\theta_1 \sin(\omega_{\text{rec}}\Delta t)] \times$$
$$\times\, J_{N+1}[2\theta_2 \sin(N\omega_{\text{rec}}T + \omega_{\text{rec}}\Delta t)]. \tag{13.18}$$

Assuming that the atoms follow a Maxwellian velocity distribution

$$\langle e^{-ikv_0\Delta t}\rangle = e^{-(k\langle v\rangle\Delta t/2)^2}, \tag{13.19}$$

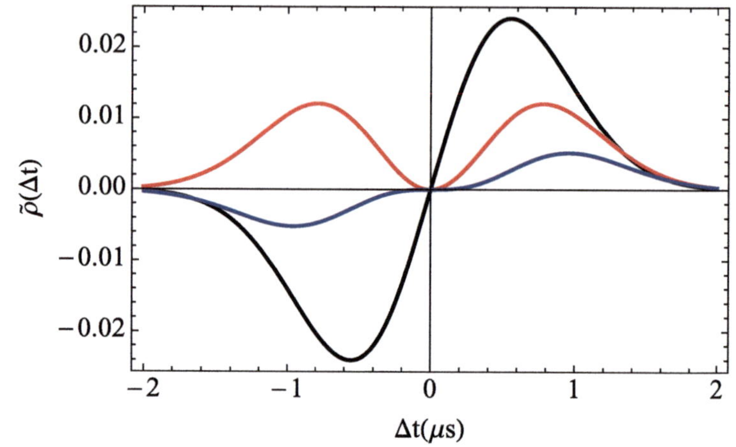

Fig. 13.1 Illustration of the interference signal (13.20) as function of the delay time for a time-interferometry setup with $^{8}5$Rb atoms cooled down to $\sim$150 μK, and a period of $T \sim 800\,\mu$s. *Black*, *red* and *blue curves* are respectively obtained for $N = 1, 2$ and 3 echos

and equal pulse durations, $\theta_1 = \theta_2$, together with the assumption of small recoils $\omega_{\rm rec} \ll k\langle v\rangle \sim 1/\Delta t$, we have

$$\tilde{\rho}(k,t) \simeq (\Delta t)^N e^{-(q\langle v\rangle \Delta t/2)^2} J_{N+1}\left[2\theta_2 \sin(N\omega_{\rm rec}T)\right]. \qquad (13.20)$$

In Fig. 13.1, we plot the expected signal for $N = 1, 2$ and 3. The parameters are taken from Ref. [9] and correspond to an experiment made with $^{8}5$Rb atoms.

13.3 Decoherence Processes

In be above discussion we have considered perfect interference processes, leading to ideal interference fringes. How this well defined wave phenomenon can be destroyed partial or totally by the environment is our concern here. Destruction of quantum coherence is an important ingredient of quantum mechanics, because without decoherence we could not even talk of single atom wave functions. It is therefore useful to approach the decoherence processes as intrinsically associated with any quantum measurement. Therefore, decoherence is a central piece of quantum theory (see for a review [11]).

Young's double-slit experiment provides the basis for quantum interferometry and helps to identify the different decoherence processes that can lead to the destruction of the interference patterns. The degree of decoherence depends on the interaction between the atom and the environment surrounding the double-slit system, therefore coupling the atom quantum states of motion (associated with

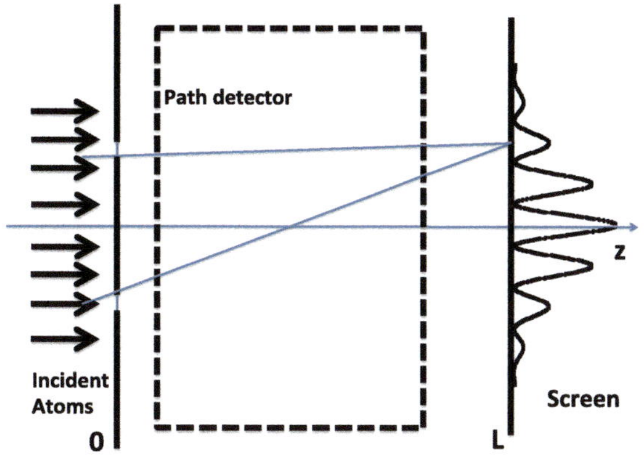

Fig. 13.2 Young's two-slit experiment for matter waves, with a path detector system

the different atom paths) with the quantum states of the environment. Here we follow the approach proposed by Tan and Walls, which identifies the environment with a "path detector" [12]. In order to discuss decoherence, we consider Young's experiment for atoms with mass M, to which we attach the path detector system, allowing to determine which path was actually followed by the atom (see Fig. 13.2). The atom propagates initially at $z < 0$, with momentum $\mathbf{p} = \hbar k \mathbf{e}_z$. We assume that the perturbations $\delta \mathbf{k}_\perp$ introduced by the slits are small in comparison with the initial momentum. The slits are located at positions $x_1 = d/2$ and $x_2 = -d/2$, according to the usual Young's configuration. We assume that the quantum state of the atom after passing the slits is a superposition of position eigenvalues, and can be written as

$$|\psi_0\rangle = \frac{1}{\sqrt{2}} \left(|\psi_1\rangle + |\psi_2\rangle \right) \tag{13.21}$$

where $|\psi_0\rangle \equiv |\psi(t = 0)\rangle$, and $t = 0$ is the instant of the interaction with the infinitesimally thin double-slit screen located at position $z = 0$. After moving into the region between the slitted and the detection screens, the atom attains the screen at $z = L$, where the detectors are located, with a final state vector determined by

$$|\psi(t)\rangle = U(t) \, |\psi_0\rangle \; , \quad U(t) = \exp\left(-\frac{i}{\hbar} \frac{p^2}{2M} t \right) . \tag{13.22}$$

This allows us to write

$$|\psi(t)\rangle = \frac{1}{\sqrt{2}} \int_{-\infty}^{\infty} \left[\psi_1(x) + \psi_2(x) \right] |x\rangle \, dx, \tag{13.23}$$

with

$$\psi_j(x) = \langle x| \, U(t) \, |\psi_j\rangle = C_j \exp\left[i \frac{M}{2\hbar t} (x - x_j)^2 \right], \tag{13.24}$$

where C_j are normalization factors. Here, $|\psi_j(x)|^2$ can be regarded as the probability for detection of the atom at position x on the final screen, if the atom passed through the slit $j = 1$ or 2. Noting that the time of flight between the slitted screen and the final screen is ML/p_z, we can write

$$\psi_j(x) = C_j \exp\left[i\frac{k}{2L}(x - x_j)^2\right]. \tag{13.25}$$

The total probability density to detect the atom at the position x is then given by the square of the projection of (13.22) on the eigenstate $|x\rangle$. In explicit terms, we have

$$|\langle x|\psi(t)\rangle|^2 = \frac{1}{2}\left\{|\psi_1(x)|^2 + |\psi_2(x)|^2 + 2\Re\left[\psi_1^*(x)\psi_2(x)\right]\right\}. \tag{13.26}$$

Using (13.25), we obtain

$$|\langle x|\psi(t)\rangle|^2 = 1 + \cos\left(\frac{kd}{L}x\right), \tag{13.27}$$

which describes the interference fringes, in the absence of decoherence. Let us now include the effect of the background medium, represented by our path detector, which is assumed initially in some pure quantum state $|D_0\rangle$. The passage of the atom will eventually excite it to the states $|D_j\rangle$, for $j = 1$ or 2, according to the slit used by the atom path. The atom and the detector (the medium) will become entangled and the total quantum probability for the total system is now given by

$$|\langle x|\psi(t)\rangle|^2 = \frac{1}{2}\left\{|\psi_1(x)|^2 + |\psi_2(x)|^2 + 2\Re\left[\psi_1^*(x)\psi_2(x)\langle D_1|D_2\rangle\right]\right\}. \tag{13.28}$$

We can see that, due to the presence of the path detector, the interference term is reduced by the factor $\langle D_1|D_2\rangle$, introduced by the presence of a background medium, the path detector system. In the limit of orthogonality, such that $\langle D_1|D_2\rangle = 0$, the fringes will simply disappear from the screen. This means that, in this case it will be possible to determine which slit was used by the atom path, because the atom traces its unequivocal mark in the detector system.

Let us now assume the the detector (or the environment) is described by a continuum of states, such that

$$|D\rangle = \int |\beta\rangle\, d\beta. \tag{13.29}$$

We then have, for each state $|\beta\rangle$, the probability to detect the atom at a position x on the screen given by

$$|\langle x|\psi(t)\rangle|_\beta^2 = \frac{1}{2}\left\{|\psi_1(x)\langle\beta|D_1\rangle|^2 + |\psi_2(x)\langle\beta|D_2\rangle|^2 + 2\Re\left[\psi_1^*(x)\psi_2(x)\langle D_1|\beta\rangle\langle\beta|D_2\rangle\right]\right\}. \tag{13.30}$$

By integrating this over all the possible states $|\beta\rangle$, we will recover Eq. (13.28) for the total probability to find the atom at the position x. On the other hand, the probability P_β, for the medium to be in a given state $|\beta\rangle$ after the passage of the atom, is given by

$$P_\beta \equiv \int |\langle x|\psi(t)\rangle|_\beta^2 \, dx = \frac{1}{2}\left(|\langle\beta|D_1\rangle|^2 + |\langle\beta|D_2\rangle|^2\right). \tag{13.31}$$

For each specific state of the path detector, we have a partial probability $P_\beta(x) = |\langle x|\psi(t)\rangle|_\beta^2$ to observe the atom at x. The visibility of the resulting fringes is associated with the amplitude of the interference term in Eq. (13.30). Dividing this amplitude by the probability to find the detector in this particular state, as given by (13.31), we can define the parameter as

$$\nu_\beta = \frac{2\,|\langle D_1|\beta\rangle\langle\beta|D_2\rangle|}{|\langle\beta\, D_1\rangle|^2 + |\langle\beta|D_2\rangle|^2}. \tag{13.32}$$

This definition is a consequence of the conditional probability, $P(x|\beta) = P_\beta(x)/P_\beta$. The quantity ν_β can be seen as the *decoherence factor*, associated with the existence of interactions between the atom and the background medium, that prevents the existence of ideal interference fringes as described by Eq. (13.27). Decoherence or dephasing can be evaluated for a series of different interaction processes between the atom and the environment, such as atom scattering from a background gas, or photon scattering by an atom moving in a background radiation spectrum. These two cases were analyzed by Uys et al. [13], where the decoherence factor was defined as

$$\nu(\mathbf{d}) = \left[\int P(\mathbf{q})d\mathbf{q}\right]^{-1} \int P(\mathbf{q}) \exp(-i\mathbf{q}\cdot\mathbf{d})d\mathbf{q}, \tag{13.33}$$

where $P(\mathbf{q})$ is the probability for a momentum change $\mathbf{q}$ of the atom, due to any of these scattering processes, and $\mathbf{d}$ is the separation vector between the two arms of the interferometer.

13.4 Gravitational Decoherence

We address here the fundamental question of quantum space-time fluctuations, and their possible contributions to quantum decoherence. We expect the granulation of the metric field to be observed on a time scale given by the *Planck time*, $T_P = \sqrt{\hbar G/c^5} \simeq 10^{-43}$ s, where G is Newton's gravitational constant, or at the Planck length $L_P = c\tau_P \simeq 10^{-35}$ m. This is obviously impossible to observe, for any conceivable experiment. However, we can use Einstein's paradigm of the *brownian motion*, and assume that microscopic fluctuations can be indirectly observed on a much larger macroscopic scale. In this way, the thermal fluctuations of the medium, associated with random collisions of the unobserved atoms on a large brownian particle, can change the character of its macroscopic motion, and therefore prove although on an indirect way the existence of the atomic background.

Similarly, for the gravitational space-time, we can imagine a situation where the brownian particle is replaced by a single atom, and the thermal fluctuations are replaced by the metric fluctuations. The atomic wave function, and in particular the measure of its coherence properties as observed by atom interferometers, can be modified due to the existence of gravitational quantum fluctuations. This idea was first advanced by Ellis et al. [14], who proposed the used of neutron interferometry. Models for atom interferometry were proposed afterwards [15, 16]. In the absence of a well established quantum theory of gravitation, we can base our estimations for the gravitational decoherence on the assumption that the Minkowski metric tensor $\eta_{ab} = \text{diag}(-1, 1, 1, 1)$, is modified by the presence of a fluctuating *scalar field* A, leading to a physical metric tensor of the form

$$g_{\mu\nu} = (1 + A)^2 \eta_{\mu\nu}. \tag{13.34}$$

This scalar field A, which can also be called a *conformal field*, is assumed to described a broad spectrum of fluctuations which are due to the quantum nature of space-time. The centre-of-mass motion of an atom with mass M in such a background can be described by the relativistic wave equation

$$g^{\mu\nu} \nabla_\mu \nabla_\nu \phi = \left(\frac{Mc}{\hbar} \right)^2 \phi, \tag{13.35}$$

where ϕ is the Klein-Gordon wave function of the atom, and ∇_μ is the covariant derivative associated with the metric (13.34). This equation can be explicitly written as

$$\left(\nabla^2 - \frac{1}{c^2} \frac{\partial^2}{\partial t^2} \right) \phi = (1 + A)^2 \left(\frac{Mc}{\hbar} \right)^2 \phi - 2\partial_\mu \left[\ln(1 + A) \right] \partial^\mu \phi. \tag{13.36}$$

The centre-of-mass motion of the atoms is usually non-relativistic, and we simplify this result by assuming that $p \ll cM$, where p is the expectation value for the atom momentum. Introducing a new wave function ψ such that

$$\phi = \psi \exp\left(-i \frac{Mc^2}{\hbar} t \right), \tag{13.37}$$

we can derive a modified Schrödingier equation for the atom motion in the background gravitational field, reading [17]

$$i\hbar \frac{\partial \psi}{\partial t} = \left[-\frac{\hbar^2}{2M} \nabla^2 + \frac{3}{2} A^2 \right] \psi. \tag{13.38}$$

We can see the appearance of a fluctuation potential term, which depends quadratically on the conformal field fluctuations, A. We therefore have to discuss the spectrum of these fluctuations. We can simply describe it as a superposition of stochastic space-time fluctuations, thus considering the following properties for the different spectral components

$$\langle A(\mathbf{k}, t)\rangle = 0, \quad \big\langle A(\mathbf{k}, t) A(\mathbf{k}', t + \tau)\big\rangle = \delta(\mathbf{k}, \mathbf{k}') C(\tau), \tag{13.39}$$

where $C(\tau)$ is the auto-correlation function, to be specified. It is related to the power spectral density $S(\omega)$ by the well known relation

$$C(\tau) = \frac{1}{(2\pi c)^3} \int \omega^2 S(\omega) \cos(\omega\tau) d\omega. \tag{13.40}$$

This is in fact a space-time generalization of the usual *Wiener-Khintchine theorem*. The atomic wavefunction decoherence will depend on the averaged quantity $\langle A^2\rangle$, as shown by the quadratic term in the Hamiltonian of the equation Schrödinger equation (13.38). This quantity will be determined by

$$\langle A^2\rangle = \int S(\mathbf{k}) \frac{d\mathbf{k}}{(2\pi)^3}, \tag{13.41}$$

where $\omega = kc$ is used. This means that, in order to get an explicit value for the atom decoherence we need to make a choice on the spectral function. A natural choice is provided by an argument first considered in random electrodynamics [18], namely that a function of the type $S(\omega) \sim 1/\omega$ is necessary to guarantee that the energy spectrum of a massless field is Lorentz invariant. In the case of a conformal field, we can use the expression

$$S(\mathbf{k}) = \frac{\hbar G}{2c^2} \frac{1}{\omega(k)}. \tag{13.42}$$

This form has the appropriate dimensions, and is compatible with the energy density resulting from the superposition of zero-point energy terms $\hbar\omega/2$. Note that to calculate the frequency integrals, we need to introduce a cut-off at $\omega_c = \omega_P/\lambda_R$, where $\omega_P = 2\pi/T_P$ is the Planck frequency. This is equivalent to assuming that below a given length scale $L_R = \lambda_R L_P$, the probing particle (the atom) is insensitive to the short wavelength fluctuations. This states the resolution of our brownian particle to detect fluctuations. The correlation time of the conformal field fluctuations can then be defined by the normalized integral over the frequency spectrum, as

$$\tau_R = \frac{\pi}{\Omega} \int_0^{\omega_c} \omega^4 S^2(\omega) d\omega \ , \quad \Omega = \left[\int_0^{\omega_c} \omega^2 S(\omega) d\omega\right]^2 . \tag{13.43}$$

With these considerations in mind, the atom decoherence can then be solved in terms of the density matrix of the atom wave packet $\rho(t)$. The result takes the form [16,17]

$$\left|\frac{\delta\rho}{\rho_0}\right| = \frac{1}{3}\left(\frac{L_P}{L_R}\right)^3\left(\frac{M}{M_P}\right)^2\left(\frac{T}{T_P}\right), \tag{13.44}$$

where $\rho_0 \equiv \rho_{\mathbf{rr}'}(0)$ is the initial value of the off-diagonal elements of the density matrix and $\delta\rho \equiv \rho_{\mathbf{rr}'}(T) - \rho_{\mathbf{rr}'}(0)$ is the difference observed after a dephasing time T, where this time was assumed large with respect to the effective correlation time $\tau_R \simeq \lambda_R T_P$. In order to see the eventual significance of this result, we consider a simple but quite extreme case of matter wave interferometry, as that associated with fullerene molecules, made of 70 carbon atoms, in order to increase the value of the quantum particle M, and a dephasing time of $T = 10^{-3}$ s, as those corresponding to actually performed experiments [19]. In this case we can write

$$T \simeq 10^{40} T_P\ , \qquad M \equiv M_{C_{70}} \simeq 10^{-17} M_P\ , \qquad \left|\frac{\delta\rho}{\rho_0}\right| \simeq \frac{10^6}{\lambda_R^3}. \tag{13.45}$$

This calculation could be used to estimate the scale length λ_R at which we can probe the conformal field fluctuations, if we could identify a residual decoherence not explained by any of the other possible decoherence mechanisms. For instance, an observed coherence of $\delta\rho/\rho_0 \simeq 0.01\,\%$ would correspond to an observed scale length of $\lambda_R \geq 10^3$.

Three different kinds of comments arise from this discussion. First, the possible influence of the fluctuations of space-time is surprisingly close to the present experimental conditions of matter wave interferometry; second, it will be very difficult to isolate the different decoherence mechanisms that can be present in a given experiment, although their dependence upon the different experimental parameters could eventually help to identify some of them; the third comment is related with the use of a conformal scalar field A to describe the space-time fluctuations. We have neglected the zero-point fluctuations of the tensor field itself $g_{\mu\nu}$, which would also contribute to decoherence. It can be shown, based on general arguments of conformal invariance [20], that the total gravitational Hamiltonian can be written as a sum of two terms

$$\mathcal{H} = \int \left(\mathcal{H}^{CF} + \mathcal{H}^{GW}\right) d^3x, \tag{13.46}$$

where the Hamiltonian density $\mathcal{H}^{CF}$ is associated with the conformal field, and the second term $\mathcal{H}^{GW}$ to the spin-2 gravitons, which are the elementary excitations of the tensor field g_{ab}. The energy $\mathcal{H}_0^{GW}$ due to the zero-point fluctuations of the graviton modes should exactly compensate the contribution of the conformal field fluctuations, leading to the constraint [16]

$$\mathcal{H}^{CF} = -\mathcal{H}_0^{GW}. \tag{13.47}$$

This means that, for a given frequency ω, a pair of gravitons with the same zero-point energy $\hbar\omega/2$, but in two orthogonal helicity states, should be canceled by a conformal field quantum $\hbar\omega$, thus maintaing the total zero state energy of gravity as equal to zero. This is a very simple and powerful argument which could also be used in other contexts. We can therefore conclude that the spin-2 fluctuations which were neglected in the above model, contribute a the same level as the conformal field fluctuations to the final decoherence in matter wave experiments.

A final remarks concerns the validity of the scalar-tensor model of gravity used here and the very existence of a scalar field. It should be noticed that the existence of scalar fields have not been confirmed by any direct astronomical observation. For instance, retardation radio signal experiments by the Solar gravity imposed an upper bound on the value of $|A| < 10^{-3}$ [21]. More recently, general relativity tests made with the Cassini spacecraft let to the result $|A| < 10^{-5}$ [22].

On the other hand, and in contrast with this negative observational evidence, many different theoretical arguments point to the possible existence of a scalar-tensor theory of gravity, from unification and string theories to the origin of mass and inflation. In a generic scalar-tensor metric theory, we can use a metric of the form [23]

$$\tilde{g}_{\mu\nu} = A^2(\phi) g_{\mu\nu}, \tag{13.48}$$

where $g_{\mu\nu}$ is the Einstein metric and the conformal factor $A(\phi)$ is a real function of the scalar field ϕ. Coupling between the scalar field and matter is determined by the quantity $\alpha(\phi)$, defined by

$$\alpha(\phi) \equiv \frac{\partial a(\phi)}{\partial \phi}, \quad a(\phi) = \ln A(\phi), \tag{13.49}$$

The choice of a particular conformal metric (13.34) is thus compatible with this general form. It can be shown that, if a significant scalar field existed in early stages of the universe, these general scalar-tensor fields tend quite fast in time to a purely relativistic metric [23], which is compatible with the absence of an observable scalar field. On the other hand, even in the absence of a global scalar field, local excitation of scalar fields and waves can become possible, due to non-stationary events such as supernova oscillations and explosions [24].

13.5 Josephson Tunneling of a Condensate

Let us now consider the interaction between two distinct Bose Einstein condensates and the associated coherence phenomena. It is useful to start with the case of a condensate confined in a double potential well, as that defined by $V(\mathbf{r}) = V(\mathbf{r}_\perp)V(z)$, where $V(\mathbf{r}_\perp)$ is the usual transverse harmonic potential, and the axial potential $V(z)$ can be generically described by

$$V(z) = -\frac{1}{2}\mu z^2 + \frac{1}{4!}\lambda z^4 - hz, \tag{13.50}$$

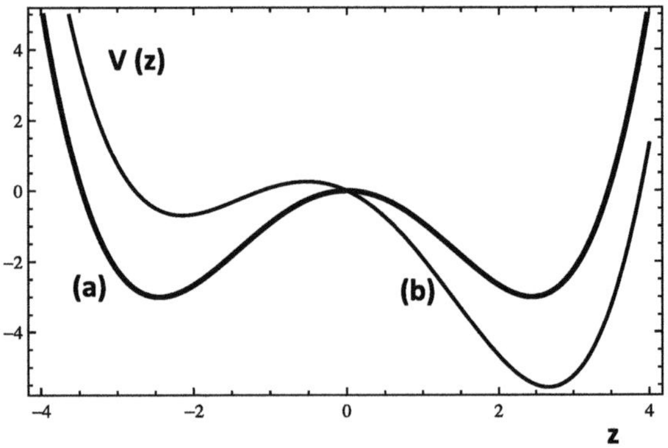

Fig. 13.3 Illustration of the double well axial potential, for $\mu = \lambda = 2$, (**a**) symmetric case, $h = 0$, and (**b**) asymmetric case, $h = 1$

where μ, λ and h are constants. See Fig. 13.3 for an illustration. Here we focus on the symmetric case, corresponding to $\mu, \lambda > 0$ and $h = 0$. Inside this double well, we have two possible atomic states $|\psi_1\rangle$ and $|\psi_2\rangle$, corresponding to atoms trapped in wells 1 and 2. Given the symmetry of the double well, these atomic states have the same energy E_0. We assume that the atoms can only occupy these ground states, which is a valid assumption at very low temperatures. Calling a_1 and a_2 the operators that destroy these quantum states, we can describe the Hamiltonian operator of the bosonic gas in the double well $\hat{H}$ as [25]

$$\hat{H} = E_0(\hat{N}_1 + \hat{N}_2) - \frac{\hbar}{2} J \left(a_1^\dagger a_2 + a_2^\dagger a_1 \right) + H_{\text{int}}, \tag{13.51}$$

where $\hat{N}_j = a_j^\dagger a_j$ are the number operators for the quantum states $j = 1, 2$, and the parameter J determines the strength of the tunneling effect between the two wells. We can see that the tunneling terms of $\hat{H}$ create an atom state in a given potential well and destroy an atom state in the other, thus satisfying the particle number conservation. The last term in (13.51), H_{int}, represents the interaction between the atoms inside each well, and can be written as

$$H_{\text{int}} = \frac{\hbar}{2} U \left[\hat{N}_1(\hat{N}_1 - 1) + \hat{N}_2(\hat{N}_2 - 1) \right], \tag{13.52}$$

where the interaction constant is given by

$$U = \frac{U_0}{\hbar} \int |\psi_1(\mathbf{r})|^4 \, d\mathbf{r} = \frac{U_0}{\hbar} \int |\psi_1(\mathbf{r})|^4 \, d\mathbf{r}. \tag{13.53}$$

In the absence of interactions, $H_{int} = 0$, the Hamiltonian (13.51) can simply be written as

$$\hat{H} = E_0(\hat{N}_+ + \hat{N}_-) - \frac{\hbar}{2} J(\hat{N}_+ - \hat{N}_-), \tag{13.54}$$

where we have used the new operators

$$a_{\pm} = \frac{1}{\sqrt{2}}(a_1 \pm a_2), \tag{13.55}$$

and $\hat{N}_{\pm} = a_{\pm}^{\dagger} a_{\pm}$ are the associated number operators. The eigenstates of this non-interacting gas of bosons in the double potential well are determined by symmetric and anti-symmetric state vectors

$$|\psi_{\pm}\rangle = \frac{1}{\sqrt{2}}\left(|\psi_1\rangle \pm |\psi_2\rangle\right), \tag{13.56}$$

with the corresponding eigenvalues

$$E_{\pm} = E_0 \mp \frac{\hbar}{2} J. \tag{13.57}$$

The ground state of the system is therefore given by the symmetric state $|\psi_+\rangle$, and the first excited state is $|\psi_-\rangle$. At very low temperatures such that $T \ll \hbar J$, the system of N identical boson atoms will condensate in a state

$$|\Psi\rangle = \frac{1}{\sqrt{N!}}\left(a_+^{\dagger}\right)^N |0\rangle, \tag{13.58}$$

which corresponds to N atoms in the same symmetric state $|\psi_+\rangle$.

In the presence of atomic interactions, $H_{\text{int}} \neq 0$, the generic situation is that of a different mean number of atoms in the two wells, $N_1 \neq N_2$, with $N = N_1 + N_2$. We can describe the temporal evolution of the system by using the Heisenberg representation, where the time-dependent operators a_j satisfy the equations

$$\frac{da_1}{dt} = \frac{1}{i\hbar}[a_1, H], \quad \frac{da_2}{dt} = \frac{1}{i\hbar}[a_2, H]. \tag{13.59}$$

Using Eq. (13.51) we can write the explicit form of these evolution equations as

$$\frac{da_1}{dt} = -i\omega_0 a_1 + \frac{i}{2} J a_2 - iU\hat{N}_1 a_1, \quad \frac{da_2}{dt} = -i\omega_0 a_2 + \frac{i}{2} J a_1 - iU\hat{N}_2 a_2, \tag{13.60}$$

with $\omega_0 = E_0/\hbar$. If initially, at time $t = 0$, the mean number of atoms is different $N_1 \neq N_2$, the state of the system will be determined by a linear combination of states $|\psi_1\rangle$ and $|\psi_2\rangle$, as given by

$$|N_1, N_2, \theta\rangle = \eta_1 |\psi_1\rangle + \eta_2 e^{i\theta} |\psi_2\rangle, \tag{13.61}$$

where θ is the relative phase, and η_j are normalization factors defined as $\eta_j = \sqrt{N_j/N}$, and satisfying $\eta_1^2 + \eta_2^2 = 1$. This allows us to define a *phase state* $|N, \theta\rangle$ for the N bosons, by using a new creation operator defined by

$$a^\dagger_{N_1,N_2,\theta} = \eta_1 a^\dagger_1 + \eta_2 e^{i\theta} a^\dagger_2. \tag{13.62}$$

We can then write

$$|N,\theta\rangle = \frac{1}{\sqrt{N!}} \left(a^\dagger_{N_1,N_2,\theta} \right)^N |0\rangle\,, \tag{13.63}$$

or, in a more explicit and alternative form

$$|N,\theta\rangle = \sum_{N_1=0,N_2=N-N_1}^{N} \sqrt{\frac{N!}{N_1!N_2!}} \eta_1^{N_1} \eta_2^{N_2} e^{iN_2\theta} |N_1,N_2\rangle\,. \tag{13.64}$$

This shows that the phase state $|N,\theta\rangle$ is a linear superposition of double Fock states $|N_1,N_2\rangle$, such that $N_1 + N_2 = N$. Such a phase state reveals the existence of quantum correlations between the two parts of the condensate existing in a double well confinement potential.

When the number of atoms in the condensed gas is large, $N \gg 1$, the operators in the evolution equations (13.60) can be replaced by classical quantities. This replacement can be formally justified by assuming that the state of the condensate is described in terms of two coherent states $|\alpha_1\rangle$ and $|\alpha_2\rangle$, such that $a_j \left|\alpha_j\right\rangle = \alpha_j \left|\alpha_j\right\rangle$ (see [25] for more detailed discussion). This means that we can replace the operators a_j in (13.60) by the complex numbers $\alpha_j = \sqrt{N_j}\exp(i\theta_j)$, for $j = 1,2$, where we can choose $\theta_2 = \theta_1 + \theta$. We then get

$$\frac{d\alpha_1}{dt} = -i\omega_0\alpha_1 + \frac{i}{2}J\alpha_2 - iUN_1\alpha_1\,, \quad \frac{d\alpha_2}{dt} = -i\omega_0\alpha_2 + \frac{i}{2}J\alpha_1 - iUN_2\alpha_2. \tag{13.65}$$

Multiplying the first of these equations by α_1^*, and the second one by α_2^*, and equating separately to zero their real and imaginary parts, we get from the imaginary part

$$\frac{dN_1}{dt} = -J\sqrt{N_1N_2}\sin\theta\,, \quad \frac{dN_2}{dt} = J\sqrt{N_1N_2}\sin\theta. \tag{13.66}$$

On the other hand, the real part leads to two additional equations

$$\frac{d\theta_1}{dt} = \frac{J}{2}\sqrt{\frac{N_2}{N_1}}\cos\theta - \omega_0 - UN_1 \tag{13.67}$$

$$\frac{d\theta_2}{dt} = \frac{J}{2}\sqrt{\frac{N_1}{N_2}}\cos\theta - \omega_0 - UN_2,$$

These equations show that, apart from the single atom potential frequency ω_0, the phases evolve in time due to both quantum tunneling and atom interactions.

Taking the difference between the two populations, $n = N_1 - N_2$ and the phase difference between the two condensate parts, $\theta = \theta_1 - \theta_2$, we obtain the following evolution equations

$$\frac{dn}{dt} = -J\sqrt{N^2 - n^2}\sin\theta, \tag{13.68}$$

and

$$\frac{d\theta}{dt} = \frac{Jn}{\sqrt{N^2 - n^2}}\cos\theta + Un. \tag{13.69}$$

It is important to notice that these two equations can be written in canonical form

$$\frac{dn}{dt} = -\frac{\partial H}{\partial \theta}, \quad \frac{d\theta}{dt} = \frac{\partial H}{\partial n}, \tag{13.70}$$

where $H \equiv H(n, \theta)$ is the classical Hamiltonian defined by

$$H(n, \theta) = -J\sqrt{N^2 - n^2}\cos\theta + \frac{1}{2}Un^2, \tag{13.71}$$

for the canonically conjugate pair of action-angle variables (n, θ). This is in fact the Hamiltonian of a non-rigid pendulum, with a variable length proportional to the quantity $\sqrt{N^2 - n^2}$, and angular momentum n.

Let us now discuss the physical meaning of these quasi-classical equations, describing the evolution of the atom number difference between the two sides of the double well, and the respective phase difference. First of all, we can define a current of atoms flowing from the potential well 1 to the well 2, as

$$I = -\frac{dN_1}{dt} = \frac{dN_2}{dt} = -J\sqrt{N_1 N_2}\sin\theta. \tag{13.72}$$

We can see that this current is proportional to the quantum tunneling factor J, as expected, but it also depends nonlinearly on the phase difference θ between the two condensed parts. Stationary states with no net current $I = 0$ correspond to the two phase values $\theta = 0$ and $\theta = \pi$, which in fact are associated with the two energy states $|\psi_1\rangle$ and $|\psi_2\rangle$ defined above. We can also see from (13.71) that the lowest energy state is attained for $n = 0$, or $N_1 = N_2$. It is then useful to consider how perturbations around these equilibria evolve in time. Assuming small deviations from the equilibrium state $n = 0$ and $\theta = 0$, and expanding the Hamiltonian (13.71) to the lowest order of these variables, we obtain the quadratic form

$$H(n, \theta) = \frac{1}{2}\left(U + \frac{J}{N}\right)n^2 + \frac{1}{2}JN\theta^2. \tag{13.73}$$

This local Hamiltonian describes a one-dimensional harmonic oscillator with frequency

$$\omega_J = \sqrt{J(NU + J)}. \tag{13.74}$$

Such oscillations are formally similar to the *Josephson oscillations* associated with the current between two superconductors separated by an insulator barrier. For a non-interacting gas, with $U = 0$, or in the weakly interacting limit $U \ll J/N$, we get $\omega_J \simeq J$, which coincides with the tunnel oscillating frequency of a single atom in a double well. In the opposite case of strong atom interactions $U \gg J/N$, the oscillating frequency tends to the so-called *plasma frequency*, $\omega_J \simeq \sqrt{UJN}$, which in the case of Josephson junctions is associated with the Coulomb oscillations between Cooper pairs. For negative interaction potential $U < 0$, corresponding to attractive interactions, the frequency decreases with U and can even become imaginary, leading to a collapse instability.

References

1. L. Marton, Phys. Rev. **85**, 1057 (1952)
2. H. Rauch, W. Treimer, U. Bonse, Phys. Lett. **47**, 369 (1974)
3. A.D. Cronin, J. Schmiedmayer, D.E. Pritchard, Optics and interferometry with atoms and molecules. Rev. Mod. Phys. **81**, 1051 (2009)
4. P.L. Kapitza, P.A.M. Dirac, Proc. Camb. Philos. Soc. **29**, 297 (1933)
5. P. Moskowitz, P. Gould, S. Atlas, D. Pritchard, Phys. Rev. Lett. **51**, 370 (1983)
6. Y.B. Ovchinnikov, J.H. Muller, M.R. Doery, E.J.D. Vredenbregt, K. Helmerson, S.L. Rolston, W.D. Phillips, Phys. Rev. Lett. **83**, 284 (1999)
7. J.T. Mendonça, A. Guerreiro, Phys. Rev. A **72**, 063805 (2005)
8. M. Moshinsky, Phys. Rev. **88**, 625 (1952)
9. S.B. Cahn, A. Kumarakrishnan, U. Shim, T. Sleator, Phys. Rev. Lett. **79**, 784 (1997)
10. H.J. Metcalf, P. van der Straten, *Laser Cooling and Trapping* (Springer, New York, 1999)
11. W.H. Zurek, Decoherence, einselection, and the quantum origins of the classic. Rev. Mod. Phys. **75**, 716 (2003)
12. S.M. Tan, D.F. Walls, Phys. Rev. A **47**, 4663 (1993)
13. H. Uys, J.D. Perreault, A.D. Cronin, Phys. Rev. Lett. **95**, 150403 (2005)
14. J. Ellis, J.S. Hagelin, D.V. Nanopolous, M. Srednicki, Nucl. Phys. B **241**, 381 (1984)
15. W.L. Power, I.C. Percival, Proc. R. Soc. Lond. A **456**, 955 (2000)
16. C.H.-T. Wang, R. Bingham, J.T. Mendonça, Class. Quantum Gravity **23**, L59 (2006)
17. P.M. Bonifacio, C. H.-T. Wang, J.T. Mendonça, R. Bingham, Class. Quantum Gravity **26**, 145013 (2009).
18. T.H. Boyer, Phys. Rev. D **11**, 790 (1975)
19. L. Hackermïler, K. Hornberger, B. Brezger, A. Zeilinger, M. Arndt, Nature **427**, 711 (2004)
20. C.H.-T. Wang, Phys. Rev. D **71**, 124026; **72**, 087501 (2005)
21. R.D. Reasenberg, I.I. Shapiro, P.E. MacNeil, R.B. Goldstein, J.C. Breitenthal, J.P. Brenkle, D.L. Cain, T.M. Kaufman, T.A. Komarak, A.I. Zygeilbaum, Astrophys. J. **234**, L219 (1979)
22. B. Bertotti, L. Iess, P. Tortora, Nature **425**, 374 (2003)
23. T. Damour, K. Nordtvedt, Phys. Rev. Lett. **70**, 2217 (1993)
24. C.H.-T. Wang, P.M. Bonifacio, R. Bingham, J.T. Mendonça, Phys. Lett. B **705**, 148 (2011)
25. C. Cohen-Tannoudji, D. Guéry-Odelin, *Advances in Atomic Physics* (World Scientific, Singapore, 2011)

Part III
The Physics of Ultracold Plasmas

Chapter 14
Ultra-cold Plasmas

Traditionally, the concept of plasma, is associated with a very hot gas. It is sometimes considered that the basic four *states of matter*, solid, liquid, gas and plasma, are the modern counterpart of the ancient *elements*: Earth, Water, Air and Fire. So, the plasma state corresponds to the possible highest internal energy of the medium, the star material being the most striking example. We can identify a plasma with a medium containing a large fraction of free charged particles, which interact between themselves by long range electromagnetic forces [1–3]. The Universe is dominated by plasma. The solar corona is a very hot plasma with temperatures around one million degrees Kelvin, from where a plasma flow called the solar wind is emitted and propagates to very large distances, interacting with the Earth magnetosphere. The Earth can be seen as a cold drop in the middle of a hot plasma environment. In the last 50 years, plasma physics has been mainly driven by the quest for a nuclear fusion reactor. It is known that fusion of light elements provides the source of energy of the stars, and is the basis for the Hydrogen bomb, but its use in a controlled and sustained way, for peaceful applications in electrical power plants is still in the far horizon for mankind.

In recent years, this traditional view of a plasma as a very hot medium dramatically changed when very cold plasmas were created by ionizing the ultra-cold gas contained in a trap [4, 5], or even by ionizing a Bose-Einstein condensate [6]. Electron temperatures in the range of 1–100 K, and ion temperatures of the order or below 1 K could be achieved. This new area of plasma physics was recently reviewed by Killian et al. [7]. In this first chapter devoted to ultra-cold plasmas, we first examine the basic properties of the plasma medium and stress some of the specific features of low temperature plasma creation and behavior. In subsequent chapters we will analyze the plasma quasi-equilibrium and subsequent expansion, and the basic properties of plasma waves.

J.T. Mendonça and H. Terças, *Physics of Ultra-Cold Matter*, Springer Series on Atomic, Optical, and Plasma Physics 70, DOI 10.1007/978-1-4614-5413-7_14,

14.1 Different Plasma Regimes

An ionized gas usually contains n_e electrons, n_i ions and n_a neutral atoms (or molecules) per unit volume. If the medium is electrically neutral, we should have $n_e = Zn_i$, where Ze is the charge of he ions. Such a medium can be characterized by its *degree of ionization* f_I, which is the fraction of ionized atoms, as defined by

$$f_I = \frac{n_i}{n_i + n_a}, \tag{14.1}$$

It can obviously vary from a minimum of $f_I = 0$ (neutral gas) to a maximum of $f_I = 1$ (fully ionized gas). Strictly speaking, the fully ionized gas is called a *plasma*, a name first introduced in physics by Langmuir in 1928. But this name is also frequently used for $f_I < 1$. The ultra-cold plasmas are called 'plasma' in this broad sense of the word.

In a weakly ionized plasma, the collision frequency between electrons and neutrals, ν_{ea}, is larger than the electron-electron and the electron-ion collision frequencies, ν_{ee} and ν_{ei}. These collision frequencies can generically be defined as

$$\nu_{ej} = n_j \langle \sigma_{ej} v_e \rangle , \quad j = e, i, a, \tag{14.2}$$

where v_e is the electron velocity, n_j the density of the various particle species j, and σ_{ej} the collision cross-sections, usually a function of the electron velocity v_e. The average is taken over the electron velocity distribution.

The inequality $\nu_{ea} > \nu_{ee}, \nu_{ei}$ occurs when the degree of ionization is lower than some critical value $f_{I,c}$. For instance, in the case of Helium at a temperature of $1\,\mathrm{eV} \simeq 10^4\,\mathrm{K}$, it is found $f_{I,c} \simeq 10^{-7}$ [8]. This counter-intuitive example shows that interactions between charged particles can dominate over the interactions with neutral atoms, even for very low ionization degrees, $f_I \ll 1$. This is due to the fact that long-range Coulomb interactions lead to much larger collision cross-sections than the short-range electron-neutral interactions.

Another important plasma parameter is the *coupling parameter* Γ, which is the ratio of potential to the kinetic energy of a given particle species $\alpha = e, i$ (for electrons or ions) in the plasma. It is defined by

$$\Gamma_\alpha = \frac{q_\alpha^2}{4\pi\epsilon_0 a_\alpha T_\alpha} , \quad a_\alpha = \left(\frac{3}{4\pi n_\alpha} \right)^{1/3} , \tag{14.3}$$

where $q_e = -e$ and $q_i = Ze$ are the electron and ion electric charges, and T_α stands for the temperature of the species (in units of energy). Here we have also introduced the *Wigner-Seitz radius*, a_α, which corresponds to the mean separation of two particles of the same species. For electron and ions, the coupling parameter simply reads

$$\Gamma_e = \frac{e^2}{3^{1/3}(4\pi)^{2/3}\epsilon_0} \frac{n_e^{1/3}}{T_e} , \quad \Gamma_i = Z^2 \frac{T_e}{T_i} \Gamma_e . \tag{14.4}$$

For thermal equilibrium and single-ionized ions, we have $\Gamma_e = \Gamma_i = \Gamma$. Usually, plasmas are produced with different ionic and electronic temperatures, and in the particular case of ultra-cold plasmas, we usually have $T_e \gg T_i$. This means that the electrons can be weakly coupled ($\Gamma_e \ll 1$), while the ions can be strongly coupled ($\Gamma_i \sim 1$). For large values of the ion coupling parameter $\Gamma_i > 1$, the plasma starts to behave as a liquid, and for larger values $\Gamma_i \gg 1$, even Coulomb crystalization can eventually occur.

At this point it is useful to introduce another useful plasma parameter, the *Landau distance*, r_α. It is defined as the distance at which the electrostatic energy interaction between two particles of the same species α become equal to their thermal energy. It simply relates with the coupling parameter as

$$r_\alpha = \frac{q_\alpha^2}{4\pi\epsilon_0 T_\alpha}, \quad \Gamma_\alpha = \frac{r_\alpha}{a_\alpha}. \tag{14.5}$$

If the Wigner-Seitz radius for a given particle species α, is smaller than the corresponding Landau length, $a_\alpha < r_\alpha$, the potential energy dominates over the kinetic thermal energy, and these particles become strongly coupled.

Moreover, the plasma behaves as a classical fluid as long as its temperatures T_α are larger than the corresponding Fermi temperatures $T_{F\alpha}$. Assuming, for simplicity, a plasma in thermal equilibrium ($T_e = T_i = T$), we can define a *degeneracy parameter*, χ, such that

$$\chi = \frac{T_F}{T}, \quad T_F = \frac{\hbar^2}{2m_e}\left(3\pi^2\right)^{2/3} n_e^{2/3}, \tag{14.6}$$

where T_F is the Fermi temperature of the electron gas (equal to the Fermi energy E_F in energy units). It is useful to write this quantity in terms of the electron thermal *de Broglie wavelength* λ_B, defined by

$$\lambda_B = \frac{\hbar}{m_e v_{\text{th,e}}}, \quad v_{\text{th,e}} = \sqrt{T/m_e}. \tag{14.7}$$

The degeneracy parameter can then be written as

$$\chi = \frac{1}{2}\left(3\pi^2\right)^{2/3}\left(n_e \lambda_B^3\right)^{2/3}. \tag{14.8}$$

Quantum effects will therefore become important for $\chi \geq 1$. This relation can also be written in terms of the electron Wigner-Seitz radius a_e,

$$\chi = \frac{1}{2}\left(\frac{3^2\pi}{4}\right)^{2/3}\left(\frac{\lambda_B}{a_e}\right)^2. \tag{14.9}$$

This shows that quantum effects become important when the electron thermal de Broglie wavelength becomes comparable to the inter-particle distance. In such case, the wavefunctions of nearby electrons overlap, and quantum degeneracy cannot be ignored.

In current experimental conditions for ultra-cold plasmas, we usually have $\chi \ll 1$, allowing plasma to be described in the frame of classical physics. However, in future experiments, density can eventually increase by a few orders of magnitude leading to quantum plasma regimes. The discussion about quantum effects in plasmas will be addressed in a later Chapter, for completeness.

14.2 Basic Plasma Properties

A local perturbation in the electron mean density will lead to oscillations in the plasma medium. This can simply be described by using the hydrodynamics equations, namely the electron continuity equation

$$\frac{\partial n_e}{\partial t} + \nabla \cdot (n_e \mathbf{v}_e) = 0 \tag{14.10}$$

and the momentum conservation equation

$$\left(\frac{\partial}{\partial t} + \mathbf{v}_e \cdot \nabla \right) \mathbf{v}_e = \frac{e}{m_e} \nabla V, \tag{14.11}$$

where the electrostatic potential V is determined by Poisson's equation

$$\nabla^2 V = -\frac{1}{\epsilon_0} \sum_\alpha n_\alpha q_\alpha = \frac{e}{\epsilon_0}(n_e - Z n_i). \tag{14.12}$$

In what follows, we assume that the ions are at rest due to their larger mass, $m_i \gg m_e$, and take the equilibrium ion density as $n_{i0} = n_{e0}/Z$, where n_{e0} is the electron equilibrium density, corresponding to a constant value of the potential. Taking the electron density perturbation $\tilde{n}_e = n_e - n_{e0}$, which is assumed to oscillate at a frequency ω, as $\tilde{n}_e \propto \exp(-i\omega t)$, we obtain the following linearized equations

$$\tilde{n}_e = \frac{n_{e0}}{i\omega} \nabla \cdot \mathbf{v}_e, \quad \mathbf{v}_e = -\frac{e}{i\omega m_e} \nabla V, \quad \nabla^2 V = \frac{e}{\epsilon_0} \tilde{n}_e, \tag{14.13}$$

which simply yield

$$\left(1 - \frac{\omega_{pe}^2}{\omega^2} \right) \nabla^2 V = 0, \quad \omega_{pe} = \sqrt{\frac{e^2 n_{e0}}{\epsilon_0 m_e}}. \tag{14.14}$$

The quantity ω_{pe} corresponds to the *electron plasma frequency*. Equation (14.14) simply shows that oscillations of the electron gas, such that $\nabla^2 V \neq 0$, can only exist at the plasma frequency $\omega = \omega_{pe}$. The electron plasma period $2\pi/\omega_{pe}$ provides the characteristic time scale for the electron response to plasma perturbations.

The above discussion is valid for a homogeneous plasma medium. If we want to consider center-of-mass oscillations of a finite plasma, for instance a cylinder or a sphere, we first define the centre-of-mass position as

$$\mathbf{R}(t) = \frac{1}{N_e} \int_{vol.} \mathbf{r}\, n_e(\mathbf{r}, t) d\mathbf{r}, \quad N_e = \int_{vol.} n_e(\mathbf{r}, t) d\mathbf{r}, \tag{14.15}$$

where N_e is the total number of electrons in the plasma. From the above electron fluid equations it is then possible to derive the centre of mass equation, of the form

$$\frac{d^2\mathbf{R}}{dt^2} - \omega_M^2 \mathbf{R} = 0, \quad \omega_M = A\omega_{pe}, \tag{14.16}$$

where the oscillating frequency is equal to ω_{pe} multiplied by a geometric factor A. For a cylindrical plasma we have $A = 1/\sqrt{2}$, and for a sphere we have $A = 1/\sqrt{3}$. This new oscillation frequency ω_M is usually called the *Mie frequency*. A similar concept can be used in the case of ultra-cold atomic cloud, as shown in previous Chapters.

Let us now discuss, in simple terms, the electron-ion correlations inside the plasma. The medium is electrically neutral, and in equilibrium, the local neutrality condition implies that $n_{e0} = Zn_{i0}$, yielding a vanishing net electric field. This is of course valid on a large scale, because locally the electrons and ions interact electrostatically, and local neutrality is lost at a microscopic scale. To illustrate this statement, let us consider the local potential perturbation created by an ion, located at the origin, $\mathbf{r} = 0$. In its immediate vicinity, the electrostatic potential is determined by Poisson's equation (14.12), which we can rewrite as[1]

$$\nabla^2 V = \frac{e}{\epsilon_0} [n_e(r) - Zn_{i0}]. \tag{14.17}$$

Here, we have assumed that the electron density is disturbed only in the vicinity of the ion, such that the average ion density remains constant. If the electron density population is in equilibrium in the electrostatic potential created by the ion, we can write

$$n_e(r) = n_{e0} \exp(eV/T_e) \simeq n_{e0} \left(1 + \frac{eV}{T_e}\right), \tag{14.18}$$

where we have assumed a weak potential $eV \ll T$, and taken $n_{e0} = Zn_{i0}$. Replacing this in Eq. (14.17), we get

---

[1]To be consistent with the usual notation in plasma physics, we consider the temperature in units of energy, by setting $k_B = 1$ from now on.

$$\nabla^2 V = \frac{V}{\lambda_D^2}, \quad \lambda_D = \sqrt{\frac{\epsilon_0 T_e}{e^2 n_{e0}}}, \tag{14.19}$$

where the quantity λ_D is the *electron Debye length*. Assuming spherical symmetry for the potential created around the ion, we are reduced to the radial equation

$$\frac{d}{dr}\left(r^2 \frac{dV}{dr}\right) = \frac{r^2 V}{\lambda_D^2}, \tag{14.20}$$

which possesses the following general solution

$$V(r) = \frac{1}{r}\left(A e^{-r/\lambda_D} + B e^{r/\lambda_D}\right). \tag{14.21}$$

Imposing the vanishing potential condition at infinity, $V \to 0$ as $r \to \infty$, and the unscreened Coulomb potential $V \to Ze/(4\pi\epsilon_0 r)$ in the immediate vicinity of the ion, $r \to 0$, we find $A = Ze/4\pi\epsilon_0$, and $B = 0$. From here, we simply get

$$V(r) = \frac{Ze}{4\pi\epsilon_0 r} \exp\left(-\frac{r}{\lambda_D}\right). \tag{14.22}$$

This takes the form of a *Yukawa potential*, with a finite interaction range determined by the electron Debye length λ_D. Replacing this in Eq. (14.18), we obtain the electron density profile around the ion

$$n_e(r) = n_{e0}\left[1 + Z\frac{r_e}{r} \exp\left(-\frac{r}{\lambda_D}\right)\right], \tag{14.23}$$

where r_e is the electron Landau distance, as defined by (14.5). This density profile represents the electron cloud which exists in the immediate vicinity of any given ion inside the plasma. For distances larger than the electron Debye length, $r > \lambda_D$, the cloud dissipates and the charge of the ion is completely neutralized, due to the electron screening. We conclude that the plasma is a neutral medium when observed on a scale larger than λ_D. This result also gives us information on the electron-ion correlation length. Correlations are important only at short distances, $r < \lambda_D$. On the other hand, the approximation used in (14.18) is only strictly valid for $r_e \ll r$.

We can also consider the ion-ion correlations, in a similar simple way by introducing an ion density profile $n_i(r)$, instead of considering a constant ion density n_{i0} in the potential equation (14.17). For a Boltzmann ion equilibrium around the ion located at $\mathbf{r} = 0$, we can write

$$n_i(r) = n_{i0} \exp(-ZeV/T_i) \simeq n_{i0}\left(1 - Z\frac{eV}{T_i}\right). \tag{14.24}$$

Let us restrict our discussion to a plasma in thermal equilibrium $T_i = T_e = T$. Replacing this expression in (14.17), we obtain after integration a new expression, which is formally identical to (14.22), but with λ_D replaced by a new screening distance, λ_s, as defined by

$$\lambda_s^2 = \frac{\epsilon_0 T}{e^2 n_{e0} + Z^2 e^2 n_{i0}} = \frac{\lambda_D^2}{1+Z}. \tag{14.25}$$

The screening distance is slightly reduced with respect to our previous result, due to ion screening. We notice that, for singly ionized ions, we get $\lambda_s = \lambda_D/\sqrt{2}$. Replacing this new solution in Eqs. (14.18) and (14.24) we obtain the electron and ion screening profiles

$$n_e(r) = n_{e0}\,[1 + g_{ei}(r)]\ , \quad n_i(r) = n_{i0}\,[1 + g_{ii}(r)] \tag{14.26}$$

where we have introduced the electron-ion and the ion-ion correlation functions, defined as

$$g_{ei}(r) = Z\frac{r_e}{r}\exp\left(-\frac{r}{\lambda_s}\right)\ , \quad g_{ii}(r) = -Z^2\frac{r_e}{r}\exp\left(-\frac{r}{\lambda_s}\right). \tag{14.27}$$

Using the above results, we can easily establish the mean value of the electron-ion interaction energy density, U_{int}, as

$$U_{\text{int}} = -\frac{1}{2}(n_{e0} + n_{i0})\frac{Ze^2}{4\pi\epsilon_0\lambda_s}. \tag{14.28}$$

This quantity can then be compared with the thermal energy density inside the plasma

$$U_{\text{th}} = \frac{3}{2}(n_{e0} + n_{i0})T. \tag{14.29}$$

The ratio between these two quantities provides

$$\frac{|U_{\text{int}}|}{U_{\text{th}}} = \frac{Z}{3}\frac{r_e}{\lambda_s} = \frac{1}{3^2 N_D}, \tag{14.30}$$

where $N_D = (4\pi/3)(n_{e0} + n_{i0})\lambda_s^3$ is the total number of particles inside the screening sphere, usually called the *Debye sphere*. When this number is large, $N_D \gg 1$, the interaction energy is negligible with respect to the thermal energy, and the plasma can be seen as a perfect gas. In the opposite case, $N_D \leq 1$, particle correlations become important, and the above simple fluid description of the electric screening is no longer valid.

14.3 Ionization Processes

The main process leading to the formation of ultra-cold plasma is photoionization. We first review this process, and then discuss other particular features of ultra-cold plasmas. Photoionization of ultra-cold plasma is usually due to two-photon transitions, where the first photon is provided by the laser cooling beam, exciting

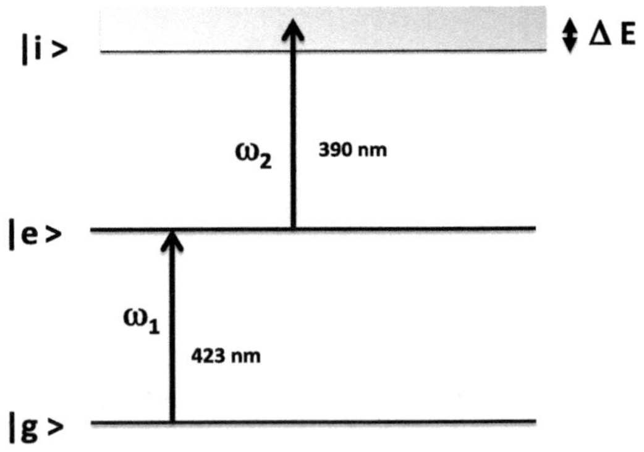

Fig. 14.1 Two-step photoionization process, using photons with energies $\hbar\omega_1$ and $\hbar\omega_2$. The second photon energy exceeds by a small amount ΔE the ionization energy

the atoms from the ground to an intermediate energy level, and the second photon is provided by a pulsed laser, exciting the atom up to the ionization continuum (see Fig. 14.1). As an example, Ca atoms can be cooled and trapped in a MOT using laser cooling beams at a wavelength of 423 nm, using transitions between the $^1S_0$ and the $^1P_1$ energy levels. A second transition, using photons at 390 nm, with an energy slightly exceeding the ionization level by an amount of $\Delta E \sim 1.5\ T_e$, creates Calcium ions Ca^+ [9]. A similar scheme is used for Strontium in Ref. [4].

In steady-state, the laser cooling process populates the upper level with a probability given by [10]

$$f_b = \frac{1}{2}\frac{s_0}{1 + s_0 + (2\Delta/\Gamma)^2}, \tag{14.31}$$

where $s_0 = I_0/I_{sat}$ is, as defined previously, the saturation parameter associated to the cooling transition, Δ the frequency detuning and $1/\Gamma$ the natural lifetime of the upper energy level $|e\rangle$. On the other hand, the ionization probability increases with the total intensity I_1 (integrated over time) of the second laser pulse, with frequency ω_1. The final fraction of ionized atoms can be written as

$$f_I = f_b\left(1 - e^{-\sigma I_1}\right) \tag{14.32}$$

where σ is the ionization cross-section divided by the photon energy $\hbar\omega_1$. Using Eqs. (14.31) and (14.32), we can determine degree of ionization f_I of an ultra-cold plasma. In typical experimental conditions, this quantity can vary between 20 and 50 %. But larger degrees of ionization can also be envisaged [11]. An important aspect of photoionization is that it transfers part of the photon energy to the resulting free electrons and ions. By using energy and momentum conservation for a single photon absorption, we can write

$$\hbar\mathbf{k}_1 = \mathbf{p}_e + \mathbf{p}_i, \quad \hbar\omega_1 = E_I + (\epsilon_e + \epsilon_i), \tag{14.33}$$

where $\epsilon_\alpha = p_\alpha^2/2m_j$ are the single electron and ion kinetic energies, for $\alpha = e, i$. Noting that the photon momentum $\hbar k_1$ is usually small in comparison with the free electron momentum p_e, we conclude that $\mathbf{p}_i = -\mathbf{p}_e$ and, due to the large mass ratio between electrons and ions, we get $\epsilon_i \ll \epsilon_e$. This allows us to write

$$\epsilon_e \simeq \Delta E = (\hbar\omega_1 - E_I) , \quad \epsilon_i \simeq \epsilon_e \left(\frac{m_e}{m_i} \right), \tag{14.34}$$

where ΔE is the excess of photon energy, as shown in Fig. 14.1. This quantity can vary between 1 and 10^3 K. This means that the ions will remain in the micro to the milli-Kelvin domain, even after thermal rearrangement. We can then conclude that the ultra-cold plasma created by the two-photon ionization process will be partly ionized ($f_I < 1$) and far from thermal equilibrium, $T_e \gg T_i$.

Spontaneous ionization can also occur, starting from a gas of highly excited Rydberg atoms. In this case, the photon energy of the second laser beam is red-tuned with respect to the ionization energy, $\Delta E < 0$ [5]. Ionization is provided by collisions between Rydberg atoms, or alternatively, by photoionization due to blackbody radiation photons, which provide the initial free electrons in the gas. These initial free electrons can further increase the number of ionized atoms, by inelastic scattering with the Rydberg population.

More traditional gaseous discharges can also be used to create ultra-cold plasmas. For instance, by using supersonic expansion of molecular beams, we can produce very dense plasmas with 10^{14} electrons per cm^{-3} [12]. With hot-filament discharges, assisted by liquid Helium cooling, it is also possible to obtain plasma temperatures in the range of 200 K [13]. Afterglow discharges also allow to obtain $T_e \sim 4$ K [14]. On the other hand, one component plasmas confined in Penning traps are known for very long time to create electrons with $T_e < 30$ K [15, 16], and ions with $T_i < 100$ mK [17].

14.4 Single Particle Motion

In order to understand the properties of a magnetized plasma it is useful to start with a discussion of single particle trajectories. Static and slowly varying magnetic fields can be created by external currents, like the Helmoltz coils in a magneto-optical trap, or by internal plasma currents. We first consider a static field $\mathbf{B}_0$. The motion of a particle with electric charge q and mass m in this field is described in the non-relativistic classical limit by the equation of motion

$$\frac{d\mathbf{v}}{dt} = s\omega_c(\mathbf{v} \times \mathbf{b}) , \quad \omega_c = \frac{|q|B_0}{m}, \tag{14.35}$$

where ω_c is the cyclotron frequency, s is the sign of the charge, and $\mathbf{b} = \mathbf{B}_0/B_0$. We can generically assume that the field is parallel the z-axis, $\mathbf{b} = \mathbf{e}_z$. This equation can be easily integrated, leading to

$$(v_x + i v_y) = v_\perp \exp(is\,\omega_c t) , \quad v_z = v_\parallel, \tag{14.36}$$

where $v_\parallel$ and $v_\perp$ are two constants. Further integration gives the particle trajectory, as

$$\mathbf{r}(t) = \mathbf{r}_0 + s r_c \left[\sin(s\omega_c t)\mathbf{e}_x - \cos(s\omega_c t)\mathbf{e}_y\right] + v_\parallel t \mathbf{e}_z \ , \tag{14.37}$$

where $\mathbf{r}_0$ is the initial particle position and $r_c = v_\perp/\omega_c$ is the cyclotron or *Larmor radius*. This is the well known helical trajectory, resulting from the superposition of an uniform motion along the magnetic field, and a circular motion, the *cyclotron motion*, around the magnetic field, with radius r_c. The parallel motion is usually called the *guiding centre* motion, because it corresponds to the motion of the centre of the cyclotron orbit. A more precise description of the guiding centre motion in a non-uniform field will be given below. Here it is useful to notice that the cyclotron motion can be seen as an elementary coil, which produces a small magnetic dipolar field. The magnetic moment of such a coil can be calculated as $\mu_m = \pi r_c^2 I$, where I is the elementary current. This leads to

$$\mu_m = -\frac{\epsilon_\perp}{B_0}, \quad \epsilon_\perp = \frac{1}{2} m v_\perp^2, \tag{14.38}$$

where $\epsilon_\perp$ is the energy of the cyclotron motion. We can see that the single particle magnetic moment is opposite to the static magnetic field. If we now calculate the cumulative result of all the elementary cyclotron motions in a plasma, we get the macroscopic magnetization, $\mathbf{M}$ defined as

$$\mathbf{M} = -M\mathbf{b}, \quad M = \frac{1}{B_0} \sum_{\alpha=e,i} \frac{1}{2} m_\alpha v_\perp^2. \tag{14.39}$$

If we assume that, in thermal equilibrium the average energy for each particle species is $\langle \epsilon_\perp \rangle_\alpha = T_\alpha$, we obtain an average plasma magnetization per unit volume equal to $M = (n_e T_e + n_i T_i)/B_0$, where n_e and n_i are the particle densities. For a plasma in thermal equilibrium, $T_e = T_i = T$ and $n_e = n_i = n$, we get $M = 2nT/B_0$. Including the resulting magnetization current $\mathbf{J}_M$ in the Ampere's law, we obtain

$$\nabla \times \mathbf{H} = \mathbf{J} + \mathbf{J}_M, \quad \mathbf{J}_M = \nabla \times \mathbf{M}, \tag{14.40}$$

where $\mathbf{H} = \mathbf{B}_0/\mu_0$. This means that the total magnetic field $\mathbf{B}$ inside the plasma can be written as

$$\mathbf{B} = \mathbf{B}_0 + \mu_0 \mathbf{M} = \frac{\mu_0 \mathbf{H}}{(1+\beta)} \ , \quad \beta = \frac{nT}{(B_0^2/2\mu_0)}. \tag{14.41}$$

This shows that the cyclotron motion of the charged particles reduces the magnetic field $\mathbf{B}$ inside the plasma. In other words, the plasma is a *diamagnetic medium*. The parameter β is the ratio between the kinetic pressure and the magnetic pressure, and plays an important role in plasma theory. Plasma diamagnetism is negligible as long as $\beta \ll 1$.

Let us now consider a more general equation of motion, replacing (14.35) by

$$m\frac{d\mathbf{v}}{dt} = \mathbf{F} + s\omega_c(\mathbf{v}\times\mathbf{b}) \ , \quad \mathbf{F} = q\boldsymbol{\mathcal{E}} + \mathbf{F}_g, \tag{14.42}$$

where $\mathbf{F}$ is a constant force, which includes the electric force associated with a static electric field $\boldsymbol{\mathcal{E}}$, and the gravitational force, $\mathbf{F}_g = m\mathbf{g}$. It is useful to split the parallel and the perpendicular motion with respect to the magnetic field, by using $\mathbf{F} = \mathbf{F}_\parallel + \mathbf{F}_\perp$ and $\mathbf{v} = \mathbf{v}_\parallel + \mathbf{v}_\perp$, which trivially reads

$$m\frac{d\mathbf{v}_\parallel}{dt} = \mathbf{F}_\parallel \ , \quad m\frac{d\mathbf{v}_\perp}{dt} = \mathbf{F}_\perp + s\omega_c(\mathbf{v}_\perp\times\mathbf{b}). \tag{14.43}$$

The first equation shows that the particle is uniformly accelerated along the magnetic field under the influence of the parallel force $\mathbf{F}_\parallel$. In what concerns the perpendicular equation, it can be reduced to the form of Eq. (14.35) by using the identity $(\mathbf{F}_\perp\times\mathbf{b})\times\mathbf{b} = -\mathbf{F}_\perp$, which leads to

$$\frac{d\mathbf{v}'}{dt} = s\,\omega_c(\mathbf{v}'\times\mathbf{b}) \ , \quad \mathbf{v}' = \mathbf{v}_\perp - \mathbf{v}_D, \tag{14.44}$$

where $\mathbf{v}_D$ is the particle *drift velocity* as defined by

$$\mathbf{v}_D = \frac{\mathbf{F}_\perp\times\mathbf{B}_0}{qB_0^2}. \tag{14.45}$$

If the static electric field is zero in the transverse plane, $\boldsymbol{\mathcal{E}}_\perp$, and $\mathbf{F}_\perp$ is the force of gravity, or a similar force not depending on the particle charge, than the drift velocity will depend on the sign of the charge. The electrons and ions will drift in opposite directions and create a plasma current. If, on the contrary, the transverse force is just the electrical force, $\mathbf{F}_\perp = q\boldsymbol{\mathcal{E}}_\perp$, the resulting drift velocity is independent of the charge, and both electrons and ions will move in the same direction. Instead of an electric current, the electric drift will generate plasma convection.

We can then say that the motion of a charged particle in constant electric, magnetic and gravitical forces is a superposition of three distinct kinds of motion. One is the uniformly accelerated motion along the magnetic field lines, due to the parallel force $\mathbf{F}_\parallel$. The second is a drift motion with velocity $\mathbf{v}_D$ in a direction perpendicular to both the magnetic and the transverse force $\mathbf{F}_\perp$. Finally, the third is the cyclotron motion in the transverse plane, as described by the equation of motion (14.44), with a rotating frequency equal to ω_c.

As we may observe from Eqs. (14.35) and (14.45), the cyclotron frequency is directly proportional to B_0, while the drift velocity is inversely proportional to B_0. This means that, for increasing magnetic fields, the rotation frequency increases and the Larmor radius decreases, while the drift motion slows down. We can call *guiding-centre motion* to the slow motion of the centre of rotation of the (fast) cyclotron orbits. It is therefore interesting to consider the guiding-centre motion in a more general class of electric and magnetic and gravitical fields, which are still

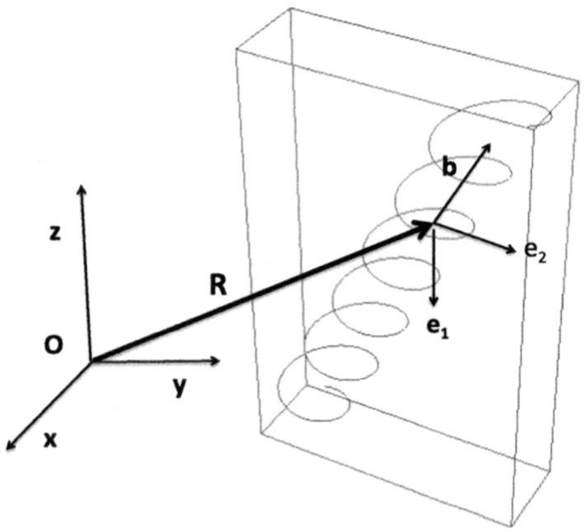

Fig. 14.2 Coordinate system for guiding centre motion of a charged particle in a static magnetic field $\mathbf{B}_0(\mathbf{r})$

constant in time but weakly varying in space. This means that, if L is the scale of variation of the guiding-centre motion, then we can define a small parameter, $\epsilon \equiv r_c/L \ll 1$.

Let us define the following coordinate system (see Fig. 14.2), where $\mathbf{r}$ is the particle position, and $\mathbf{R}$ the position of its guiding-centre. We can write

$$\mathbf{r} = \mathbf{R} + \mathbf{s} \,, \quad \mathbf{s} = r_c(\cos\theta\ \mathbf{e}_1 + \sin\theta\ \mathbf{e}_2) \tag{14.46}$$

where $\theta = \int \omega_c dt$, and $\mathbf{e}_j$ with $j = 1, 2$ are two orthogonal unit vectors in the plane locally perpendicular to the magnetic field lines. We have $\mathbf{e}_j \cdot \mathbf{b} = 0$. We can now develop the field $\mathbf{B}_0(\mathbf{r})$ and the force $\mathbf{F}(\mathbf{r}) = q\mathcal{E}(\mathbf{r}) + \mathbf{F}_g(\mathbf{r})$, around the position of the guiding-centre $\mathbf{R}$, assuming that the displacement $\mathbf{s}$ associated with the Larmor orbit is a perturbation. We can then use

$$\mathbf{B}_0(\mathbf{r}) = \mathbf{B}_0(\mathbf{R}) + \mathbf{s}\cdot\nabla\mathbf{B}_0(\mathbf{R}) \,, \quad \mathbf{F}(\mathbf{r}) = \mathbf{F}(\mathbf{R}) + \mathbf{s}\cdot\nabla\mathbf{F}(\mathbf{R}). \tag{14.47}$$

Replacing the latter in Eq. (14.42), we obtain

$$\frac{d\mathbf{v}}{dt} \equiv \frac{d^2\mathbf{R}}{dt^2} + \frac{d^2\mathbf{s}}{dt^2} \simeq \frac{1}{m}\left[\mathbf{F}(\mathbf{R}) + \mathbf{s}\cdot\nabla\mathbf{F}(\mathbf{R})\right] + \frac{q}{m}\left(\frac{d\mathbf{R}}{dt} + \frac{d\mathbf{s}}{dt}\right) \times \left[\mathbf{B}_0(\mathbf{R}) + (\mathbf{s}\cdot\nabla\mathbf{B}_0(\mathbf{R})\right]. \tag{14.48}$$

Notice that for slowly varying fields, we have

$$\frac{d^2\mathbf{s}}{dt^2} = \frac{q}{m}\left[\frac{d\mathbf{s}}{dt} \times \mathbf{B}_0(\mathbf{R})\right]. \tag{14.49}$$

We can establish an evolution equation for $\mathbf{R}$ by averaging (14.48) in time, over a period of the cyclotron motion. Keeping in mind that such an averaging process leads to

$$\langle\mathbf{s}\rangle \simeq 0 \,, \quad \left\langle\frac{d\mathbf{R}}{dt}\right\rangle \simeq \frac{d\mathbf{R}}{dt}, \tag{14.50}$$

Eq. (14.48) can then be reduced to

$$\frac{d^2\mathbf{R}}{dt^2} = \frac{\mathbf{F}(\mathbf{R})}{m} + \frac{q}{m}\left[\frac{d\mathbf{R}}{dt} \times \mathbf{B}_0(\mathbf{R}) + \left\langle \frac{d\mathbf{s}}{dt} \times \mathbf{s} \cdot \nabla \mathbf{B}_0(\mathbf{R}) \right\rangle\right]. \tag{14.51}$$

After an explicit calculation of the time averaging over the cyclotron period, we can rewrite the last term of Eq. (14.51) in a more appropriate form, leading to

$$\frac{d^2\mathbf{R}}{dt^2} = \frac{\mathbf{F}(\mathbf{R})}{m} + \frac{q}{m}\left[\frac{d\mathbf{R}}{dt} \times \mathbf{B}_0(\mathbf{R}) - \frac{\omega_c}{2} r_c^2 \nabla B_0(\mathbf{R})\right]. \tag{14.52}$$

This equation is sometimes written in terms of the elementary magnetic moment, $\mu_m = mv_\perp^2/2B_0$. Defining the guiding-centre velocity as $\mathbf{u} = d\mathbf{R}/dt$, we obtain

$$\frac{d\mathbf{u}}{dt} = \frac{\mathbf{F}(\mathbf{R})}{m} + \frac{q}{m}\mathbf{u} \times \mathbf{B}_0(\mathbf{R}) - \mu_m \nabla B_0(\mathbf{R}). \tag{14.53}$$

In order to understand the physical meaning of this equation, it is convenient to split the parallel and the perpendicular motion with respect to $\mathbf{B}_0$. Defining the parallel guiding-centre velocity as $u_\parallel = \mathbf{u} \cdot \mathbf{b}$, and using a more explicit expression for the parallel force as $F_\parallel = qE_\parallel + mg_\parallel$, we obtain

$$\frac{du_\parallel}{dt} = \frac{q}{m}E_\parallel + g_\parallel - \mu_m \frac{\partial B_0}{\partial l}, \tag{14.54}$$

where $\partial/\partial l = \mathbf{b} \cdot \nabla$ is the derivative taken along the magnetic field lines. The transverse component of the guiding-centre velocity can be determined by taking the external product of (14.53) with $\mathbf{b}$. Noting that $\mathbf{u}_\perp = -(\mathbf{u} \times \mathbf{b}) \times \mathbf{b}$, we obtain

$$\mathbf{u}_\perp = \frac{1}{qB_0}\left(\mathbf{F} - m\frac{d\mathbf{u}}{dt} - \mu_m \nabla B_0\right) \times \mathbf{b}. \tag{14.55}$$

This expression contains the different types of drift velocities. The first term contains both the electric and the gravitical drift. The second term is an inertial drift, and the third describes the magnetic drift due to the inhomogeneity of the magnetic field. These drift motions have an important influence on the plasma behavior and particle confinement in external fields. They also lead to different types of instabilities, as it will be illustrated later.

14.5 Adiabatic Invariants

An important aspect of particle trajectories in non-uniform magnetic fields, as those existing in magnetic traps, for example, is the existence of adiabatic invariants. These quantities can be defined in the following way. If the particle motion is described by some generic Hamitonian $H(q_i, p_i, t)$, and if a given coordinate q_k is a cyclical variable, then an adiabatic invariant exists, and can be defined as

$$J_k = \int_C p_k dq_k = \text{const.}, \tag{14.56}$$

where integration is taken over a close contour C. If the particle moves in a three-dimensional space, it is possible to define a maximum of three adiabatic invariants. We have seen before that the charged particle trajectories in a static magnetic field $\mathbf{B}_0$ depend on a cyclical variable, which is the angle θ describing the cyclotron motion in the perpendicular plane. This means that we can define an adiabatic invariant of the form

$$J_\theta = \int_C \mathbf{p}_\perp \cdot d\mathbf{l} = \int_C p_\theta dq_\theta, \tag{14.57}$$

where integration is performed over a Larmor orbit, with $q_\theta \equiv \theta$. Noting that the perpendicular particle momentum in a magnetic field is $\mathbf{p}_\perp = m\mathbf{v}_\perp + q\mathbf{A}$, where m and q are the particle mass and charge, and $\mathbf{A}$ is the vector potential associated with the static magnetic field, we can rewrite the invariant (14.57) as the sum of two distinct terms. More precisely, it is possible to related each of these terms to the magnetic flux through the surface limited by the cyclotron orbit. After integration over the cyclotron orbit, we obtain the following

$$m \int_C \mathbf{v}_\perp \cdot d\mathbf{l} = -2\pi q B_0 r_c^2 , \quad q \int_C \mathbf{A} \cdot d\mathbf{l} = q \int_S \mathbf{B}_0 \cdot d\mathbf{S} = \pi q B_0 r_L^2, \tag{14.58}$$

where $d\mathbf{S}$ is surface element subtended by the cyclotron orbit. Replacing this in Eq. (14.57), we concluded that the adiabatic invariant is given by

$$J_\theta = \int \mathbf{B}_0 \cdot d\mathbf{S} = \pi r_c^2 B_0 \tag{14.59}$$

This means that the total magnetic flux embraced by a cyclotron orbit is an adiabatic invariant. Introducing the magnetic moment μ_m defined above, we conclude that

$$\mu_m = \frac{q}{2}\omega_c r_c^2 = \frac{E_\perp}{B_0} \propto J_\theta, \tag{14.60}$$

where $E_\perp = mv_\perp^2/2$ is the perpendicular kinetic energy of the particle. In other words, the adiabatic invariant associated with the cyclotron motion is the magnetic moment of the particle. It can easily be recognized, by a direct calculation, that μ_m stays constant even for a varying magnetic field $\mathbf{B}_0$, as long as the later time variation is slow with respect to $1/\omega_c$. This slow variation can be characterized by a small parameter ϵ, such that

$$\epsilon = \frac{2\pi}{\omega_c}\frac{1}{B_0}\frac{\partial B_0}{\partial t} \ll 1. \tag{14.61}$$

However, if the magnetic varies in time, an electric field $\mathcal{E}$ is induced, which changes the transverse particle velocity. The resulting variation of the transverse energy of the particle over a cyclotron period is given by

$$\Delta E_\perp = q \int_0^{2\pi/\omega_c} \mathcal{E} \cdot \mathbf{v}_\perp dt = q \int_C \mathcal{E} \cdot d\mathbf{l}. \tag{14.62}$$

Using Stokes' theorem and the Faraday equation, we get

$$\Delta E_\perp = q \int_S (\nabla \times \mathcal{E}) \cdot \mathbf{S} = -q \int_S \frac{\partial \mathbf{B}_0}{\partial t} \cdot d\mathbf{S} \simeq \pi q r_c^2 \left| \frac{\partial \mathbf{B}_0}{\partial t} \right|. \tag{14.63}$$

On the other hand, the magnetic field variation ΔB_0, on a time scale of $2\pi/\omega_c$, is given by $\Delta B_0 = (\partial B_0/\partial t)(2\pi/\omega_c)$. We conclude that $\Delta E_\perp = \Delta B_0 \mu_m$. The variation of the magnetic moment on the same time scale is then given by

$$\Delta \mu_m = \frac{E_\perp + \Delta E_\perp}{B_0 + \Delta B_0} - \mu_m = \mathcal{O}(\epsilon^2). \tag{14.64}$$

We conclude that the magnetic moment is constant to the lowest order in ϵ. The existence of this adiabatic invariant was first notice by Alfvén in 1950, when he studied the motion of a charged particle in a dipole magnetic field, which is a simple model for the terrestrial magnetic field. This problem was first addressed by Stormer at the beginning of the last century [1].

Let us now consider a second adiabatic invariant, now associated with the parallel motion. The particle velocity along the magnetic field lines $v_\parallel$ is related with the total energy of the particle E through the expression

$$v_\parallel = \pm \sqrt{v^2 - \frac{2}{m} \mu_m B_0} = \pm \sqrt{\frac{2}{m}(E - qV - \mu_m B_0)}, \tag{14.65}$$

where V is the electrostatic potential. Let us assume that the amplitude of the magnetic field is not uniform and varies with the position along the magnetic field lines. We can then define a turning point, $B_0 = B_M \equiv m v^2 / 2\mu_m$, where the parallel velocity is zero, $v_\parallel = 0$. At this point the particle is reflected ($v_\parallel$ changes its sign), and the particle motion stays confined in the regions where $B_0 \leq B_M$. This is the confinement mechanism used in *magnetic mirrors*, which is a particular magnetic configuration based on two parallel coils creating two symmetric magnetic field maxima. On the other hand, when the particle approaches its reflection point and its parallel energy goes down to zero, the perpendicular energy increases, due to an increase of the Larmor radius, thus keeping the total energy constant.

The position where reflection takes place and $B_0 = B_M$ depends on the particle energy, as shown by the definition of B_M. Let us define an angle α, characterizing the direction of the particle velocity $\mathbf{v}$, such that $\sin\alpha = v_\perp / v$. As a consequence, the particle stays confined in a magnetic mirror if its velocity at the centre of the trap (where $B_0 = B_{\min}$ is a minimum) satisfies the following inequality

$$v^2 \leq \frac{2}{m} \mu_m B_M = \frac{v_\perp^2}{B_{\min}} B_M. \tag{14.66}$$

The latter can also be written in terms of the angle α, as

$$\sin^2 \alpha \geq \frac{B_{\min}}{B_M} = \frac{1}{\mathcal{R}}, \tag{14.67}$$

where $\mathcal{R} > 1$ is called the *mirror ratio*. There is then a *critical angle* α_c, such that $\mathcal{R} = 1/\sin^2 \alpha_c$, defining the trapping condition. Particles with an angle larger than critical, $\alpha > \alpha_c$ will stay confined. The others can escape. This mechanism defines a *loss cone* in phase space, corresponding to $\alpha \leq \alpha_c$ where particles are absent. The resulting distortion of the particle distribution function will then lead to plasma instabilities.

Furthermore, in a magnetic mirror it is possible to observe a parallel oscillating motion, between two symmetric turning points. The resulting periodic motion will imply the existence of a new adiabatic invariant, defined as

$$J_\parallel = \int_C p_\parallel ds = \text{const.}, \tag{14.68}$$

where ds is the element of length along the magnetic field lines. Using $p_\parallel = mv_\parallel + qA_\parallel$ and keeping in mind that no magnetic flux is involved in the periodic parallel motion, we can simply write

$$J_\parallel = m \int_C v_\parallel ds = \text{const.} \tag{14.69}$$

Let us assume that, for a given particle, the two symmetric turning points are situated a distance $L/2$ apart. We can then write this expression as $J_\parallel = mL \langle v_\parallel \rangle = \text{const.}$, where $\langle v_\parallel \rangle$ us the mean parallel velocity. If during the particle oscillation the position of the maximum B_M could change, due for instance to a displacement of one of the external coils, then the distance L would decrease and the mean velocity $\langle v_\parallel \rangle$ would increase, because the quantity $J_\parallel$ is conserved. We therefore get an acceleration of the charged particle, known as *Fermi acceleration* [18]. The existence of such an adiabatic invariant was assumed implicitly by Fermi, when he proposed this mechanism for the acceleration of cosmic rays, but it was only explicitly demonstrated by Northrop and Teller in 1960 [19].

Finally, let us consider a third adiabatic invariant related with the transverse motion. We have seen in the previous section that magnetic field gradients lead to a slow drift motion, such that the guiding centre of the particle changes its position in the perpendicular plane, and the particle is displaced to adjacent magnetic field lines while executing parallel oscillations. If the magnetic configuration is such that the slow drift motion is also periodic, we can then define a third adiabatic invariant associated with the perpendicular drift velocity $u_\perp$, such that

$$J_{\perp} = m \int_C \mathbf{u}_{\perp} \cdot d\mathbf{l} + q \int_S \mathbf{B}_0 \cdot d\mathbf{S} = \text{const.} \tag{14.70}$$

These two integrals can be approximately solved by assuming that the perpendicular motion defines a circle of radius R. We then have

$$\int_C \mathbf{u}_{\perp} \cdot d\mathbf{l} \simeq 2\pi R u_{\perp}\,, \qquad \int_S \mathbf{B}_0 \cdot d\mathbf{S} \simeq \pi R^2 B_0. \tag{14.71}$$

We notice that the second term is much larger than the first. We can write

$$\frac{q\pi R^2 B_0}{2\pi m R u_{\perp}} = \frac{\omega_c}{\omega_D} \gg 1, \tag{14.72}$$

where $\omega_D = u_{\perp}/R$ is the angular frequency associated with the slow drift motion. Neglecting the first term in (14.70), we can then say that the magnetic flux across the periodic drift trajectory is an adiabatic invariant,

$$J_{\perp} = q \int_S \mathbf{B}_0 \cdot d\mathbf{S} = \text{const.} \tag{14.73}$$

As a result of our discussion on adiabatic invariants, we can represent the motion of trapped particle orbits in many geometrical configurations as a superposition of three distinct periodic orbits. The corresponding frequencies are the cyclotron frequency $\omega_c = |q|B_0/m$, the parallel bouncing frequency $\omega_{\parallel} = \langle v_{\parallel} \rangle / L$, and the drift frequency $\omega_D = u_{\perp}/R$. It should be noticed that the adiabatic invariants can eventually be destroyed by resonances between two of these orthogonal oscillations. The nonlinear coupling between the different oscillation modes can lead to a chaotic motion where the adiabatic invariants loose their meaning. This problem was treated by Chirikov in 1960, when studying the nonlinear coupling between the cyclotron and the parallel oscillation in a magnetic mirror [20]. Starting from the particle equations of motion, he was able to derive a two-dimensional discrete map, called the *standard map*, which can also be used to describe many other dynamical processes, such as the motion of a particle in the field of an infinity of electrostatic wave modes [21].

14.6 Plasma Equations

The description of both equilibrium and dynamical features of a plasma must be established beyond the single-particle limit. In other words, to properly understand and describe the properties of a plasma, especially its phenomenology, one must inexorably consider the collective effects. There are two distinct ways of thoroughly doing so, either by the use of kinetic equations or via a set of fluid equations. In the former case, transport equations for the phase-space density $f(\mathbf{r}, \mathbf{v}, t)$ are established in order to retain the microscopic properties of the system. On the other hand, the hydrodynamic description is obtained by taking the momenta (in terms of the velocity field $\mathbf{v}$) of the kinetic (or transport) equations, and therefore can be regarded as a macroscopic description of the system, as the information in the velocity coordinate is integrated. Depending on the specificity of the problems, both descriptions can be employed. In what follows, we discuss the foundations of the kinetic description of a plasma and then explain in detail how the hydrodynamic or fluid model can be derived from it.

14.6.1 Klimontovitch Equation

Let us now focus on plasma kinetic description. Plasma is a globally neutral gas, made of free electrons and ions. The main difference with respect to a neutral gas is that charged particles can interact with each other over large distances. Therefore, the usual concept of a close collision between two neutral particles in a gas is replaced by collective interactions between all the charged particles (electrons and ions) in a plasma. In order to study the collective behavior of such a medium we can adapt the BBGKY approach or, in alternative, use a different but equivalent approach provided by the Klimontovitch equation [2, 22]. Such an alternative approach is considered here.

The dynamical state of each particle in a plasma is determined, at a given instant of time t, by the value of its vector position $\mathbf{R}(t)$ and velocity $\mathbf{V}(t)$. The density distribution of each particle in the six-dimensional single particle phase space $(\mathbf{r}, \mathbf{v})$ can be determined by Dirac delta functions. So, we can write the microscopic density f_{N_α} each particle species as

$$f_{N_\alpha}(\mathbf{r}, \mathbf{v}, t) = \sum_{j=1}^{N_\alpha} \delta[\mathbf{r} - \mathbf{R}_j(t)]\, \delta[\mathbf{v} - \mathbf{V}_j(t)]. \tag{14.74}$$

Given the well-known properties of the Dirac delta function, it is obvious that the integration of f_{N_α} in phase space $(\mathbf{r}, \mathbf{v})$ gives exactly N_α. If we could determine the exact trajectory of each and every particle in the plasma, by solving the individual equations of motions, then $f_{N_\alpha}(\mathbf{r}, \mathbf{v}, t)$ would contain all the possible information about the kinetic state of the medium. Formally, this can be done by using the equations of motion:

$$\frac{d\mathbf{R}_j}{dt} = \mathbf{V}_j\,, \quad m_\alpha \frac{d\mathbf{V}_j}{dt} = \mathbf{F}_m(\mathbf{R}_j, t) \tag{14.75}$$

with

$$\mathbf{F}_m(\mathbf{R}_j, t) = q_\alpha \left[\mathcal{E}_m(\mathbf{R}_j, t) + \mathbf{V}_j \times \mathbf{B}_m(\mathbf{R}_j, t)\right]. \tag{14.76}$$

Here, $\mathcal{E}_m(\mathbf{R}_j, t)$ and $\mathbf{B}_m(\mathbf{R}_j, t)$ are the microscopic electric and magnetic fields acting on the j-particle at time t. They are determined by Maxwell's equations,

$$\nabla \times \boldsymbol{\mathcal{E}}_m = -\frac{\partial \mathbf{B}_m}{\partial t}, \quad \nabla \cdot \boldsymbol{\mathcal{E}}_m = \frac{\rho_m}{\epsilon_0} \tag{14.77}$$

$$\nabla \times \mathbf{B}_m = \mu_0 \mathbf{J}_m + \frac{1}{c^2}\frac{\partial \boldsymbol{\mathcal{E}}_m}{\partial t}, \quad \nabla \cdot \mathbf{B}_m = 0, \tag{14.78}$$

where the charges and currents are consistently determined by the microscopic density, as

$$\rho_m = \sum_\alpha q_\alpha \int f_{N_\alpha}(\mathbf{r}, \mathbf{v}, t) d\mathbf{v}\,, \quad \mathbf{J}_m = \sum_\alpha q_\alpha \int \mathbf{v}\; f_{N_\alpha}(\mathbf{r}, \mathbf{v}, t) d\mathbf{v}. \tag{14.79}$$

In order to obtain an evolution equation for the plasma medium, we calculate the time derivative of the microscopic density f_{N_α}. Using the derivative of an implicit function

$$\frac{d}{dt} f\,[g(t)] = \frac{dg}{dt}\frac{df}{dg}, \tag{14.80}$$

we can write the time derivative of Eq. (14.74) in the form

$$\frac{\partial f_{N_\alpha}}{\partial t} = \sum_{j=1}^{n_\alpha} \left(\frac{d\mathbf{R}_j}{dt} \cdot \frac{\partial}{\partial \mathbf{R}_j} + \frac{d\mathbf{V}_j}{dt} \cdot \frac{\partial}{\partial \mathbf{V}_j} \right) \delta[\mathbf{r} - \mathbf{R}_j(t)]\; \delta[\mathbf{v} - \mathbf{V}_j(t)]. \tag{14.81}$$

Now, using the relation

$$\frac{\partial}{\partial x} f(x-y) = -\frac{\partial}{\partial y} f(x-y), \tag{14.82}$$

we can rewrite Eq. (14.81) in the new form

$$\frac{\partial f_{N_\alpha}}{\partial t} = -\sum_{j=1}^{n_\alpha} \left[\mathbf{V}_j \cdot \frac{\partial}{\partial \mathbf{r}} + \frac{\mathbf{F}_m(\mathbf{R}_j, t)}{m_\alpha} \cdot \frac{\partial}{\partial \mathbf{V}_j} \right] \delta[\mathbf{r} - \mathbf{R}_j(t)]\; \delta[\mathbf{v} - \mathbf{V}_j(t)]. \tag{14.83}$$

This expression we can be simplified by using the following property of the Dirac delta function $x\;\delta(x-y) = y\;\delta(x-y)$. This allows us to replace $\mathbf{R}_j$ by $\mathbf{r}$, and $\mathbf{V}_j$ by $\mathbf{v}$, inside the parenthesis of Eq. (14.83), and to move the differential operators outside the sum. Using (14.74), we finally get

$$\frac{\partial f_{N_\alpha}}{\partial t} + \mathbf{v} \cdot \frac{\partial f_{N_\alpha}}{\partial \mathbf{r}} + \frac{\mathbf{F}_m(\mathbf{r}, t)}{m_\alpha} \cdot \frac{\partial f_{N_\alpha}}{\partial \mathbf{v}} = 0. \tag{14.84}$$

This is the *Klimontovitch equation* for the plasma particle species α. It contains all the information concerning the kinetic state of the medium and exactly describes its evolution in space and time. In this sense, it is a complete equation, in the same way as the Liouville equation is complete but defined on a different phase space. The main difference is that it describes the physical system in the single-particle six dimensional phase space $(\mathbf{r}, \mathbf{v})$, and not in the $6N_\alpha$ dimensional phase space of the Liouville equation.

It can easily be recognized that Eq. (14.84) states the conservation of the microscopic distribution density in the phase space $(\mathbf{r}, \mathbf{v})$. The differential operator acting on $N_\alpha(\mathbf{r}, \mathbf{v}, t)$ is nothing but the total time derivative

$$\frac{d}{dt} f_{N_\alpha}(\mathbf{r}, \mathbf{v}, t) = 0 \,, \quad \frac{d}{dt} \equiv \frac{\partial}{\partial t} + \mathbf{v} \cdot \frac{\partial}{\partial \mathbf{r}} + \frac{d\mathbf{v}}{dt} \cdot \frac{\partial}{\partial \mathbf{v}}. \tag{14.85}$$

This shows that the Klimontovitch equation is formally identical to the Liouville equation, but uses a different microscopic distribution defined in a different phase space.

14.6.2 Vlasov Equation

The Klimontovitch equation is exact, but of little practical use, because it is in generality difficult, if not impossible, to solve. It is therefore useful to derive an approximate kinetic equation, which can give a coarse-grain or average description of the plasma behavior. Let us start by defining the mean value of the microscopic density, identifying it with the one-particle distribution of a given particle species

$$f_\alpha(\mathbf{r}, \mathbf{v}, t) = \langle f_{N_\alpha}(\mathbf{r}, \mathbf{v}, t) \rangle \,. \tag{14.86}$$

The averaging is made over a statistical ensemble. We can also define macroscopic electric and magnetic field as the ensemble averages of the microscopic fields $\mathcal{E}_m$ and $\mathbf{B}_m$. Designing by δf_{N_α}, $\delta \mathcal{E}$ and $\delta \mathbf{B}$ the fluctuations with respect to such ensemble averages, we can write

$$f_{N_\alpha}(\mathbf{r}, \mathbf{v}, t) = f_\alpha(\mathbf{r}, \mathbf{v}, t) + \delta f_{N_\alpha}(\mathbf{r}, \mathbf{v}, t), \tag{14.87}$$

and similarly for the fields, $\mathcal{E}_m = \mathcal{E} + \delta \mathcal{E}_m$, and $\mathbf{B}_m = \mathbf{B} + \delta \mathbf{B}$. Replacing this in the Klimontovitch equation (14.84), we can write an evolution equation for the one-particle distribution function f_α, in the form

$$\frac{\partial f_\alpha}{\partial t} + \mathbf{v} \cdot \frac{\partial f_\alpha}{\partial \mathbf{r}} + \frac{\mathbf{F}}{m_\alpha} \cdot \frac{\partial f_\alpha}{\partial \mathbf{v}} = -\frac{q_\alpha}{m_\alpha} \left\langle (\delta \mathcal{E} + \mathbf{v} \times \delta \mathbf{B}) \cdot \frac{\partial}{\partial \mathbf{v}} \delta f_{N_\alpha} \right\rangle, \tag{14.88}$$

where $\mathbf{F} = \langle \mathbf{F}_m \rangle$ is the average force associated with the macroscopic field $\mathcal{E}_m$ and $\mathbf{B}$. The left hand side of this equation is an average over products of functions varying rapidly in space and time, and describes the effects of field fluctuations associated with close particle collisions. It can be represented as $(\partial f_\alpha / \partial t)_{\text{coll}}$.

Many approximate expressions have been used for this collision term, some of them discussed later. The most drastic approximation is to neglect it completely, leading to the famous *Vlasov equation*

$$\frac{\partial f_\alpha}{\partial t} + \mathbf{v} \cdot \frac{\partial f_\alpha}{\partial \mathbf{r}} + \frac{\mathbf{F}}{m_\alpha} \cdot \frac{\partial f_\alpha}{\partial \mathbf{v}} = 0. \tag{14.89}$$

This equation is valid in the limit $\Lambda \gg 1$, where the parameter $\Lambda = 4\pi n \lambda_D^2$ is the number of particles in the Debye sphere. In this limit, the collective interactions between particles are much stronger than close binary collisions. The Vlasov equation is a simple but very powerful equation, which can be used to explains many important features of the kinetic plasma behavior. The collective particle interactions are hidden inside the force term $\mathbf{F}$, which is self-consistently determined by Maxwell's equations, with the macroscopic electric charge and current densities, ρ and $\mathbf{J}$ determined by the one-particle distribution functions f_α, according to

$$\rho = \sum_\alpha q_\alpha \int f_\alpha(\mathbf{r}, \mathbf{v}, t) d\mathbf{v}, \quad \mathbf{J} = \sum_\alpha q_\alpha \int \mathbf{v} f_\alpha(\mathbf{r}, \mathbf{v}, t) d\mathbf{v}. \tag{14.90}$$

This shows that the third term in the Vlasov equation (14.89) is a nonlinear term in f_α. It should also be noticed that the Vlasov equation conserves the particle distribution in the single particle phase space $(\mathbf{r}, \mathbf{v})$, and can be written in the form

$$\frac{d}{dt} f_\alpha(\mathbf{r}, \mathbf{v}, t) = 0, \tag{14.91}$$

where d/dt is the total time derivative. This is formally analogous to both the Klimontovitch and the Liouville equations, but with an obviously different meaning, as it describes the evolution of the average distribution function in phase-space. It should however be noticed that the Vlasov equation still conserves entropy. This can be shown by starting from the definition of entropy S of the plasma, as

$$S_\alpha = -\sum_\alpha \int f_\alpha \ln f_\alpha(\mathbf{r}, \mathbf{v}, t) d\mathbf{r} d\mathbf{v}. \tag{14.92}$$

The evolution of this quantity can then be written as

$$\frac{dS}{dt} = -\sum_\alpha \int (1 + \ln f_\alpha) \frac{\partial f_\alpha}{\partial t} d\mathbf{r} d\mathbf{v}. \tag{14.93}$$

Now, using the fact that f_α is governed by Eq. (14.89), we conclude that

$$\frac{dS}{dt} = 0. \tag{14.94}$$

This means that the plasma entropy is constant if the plasma is described by the Vlasov equation. This is equivalent to say that the Vlasov equation conserves information about the state of the system. This conservation property will be lost if we introduce particle collisions. In that case, the entropy will increase. It is also interesting to notice that Eq. (14.94) implies that the distribution functions are always a positive quantities, as it should be for probability densities. If we had $f_\alpha < 0$, the entropy S would become complex.

14.6.3 Kinetic Equations with Collisions

Binary collisions cannot always be ignored, and their effects can be included in the plasma kinetic equation. The simplest way is to include on the right hand side of Vlasov equation, a collision term of the form

$$\left(\frac{\partial f_\alpha}{\partial t}\right)_{\text{coll}} = -\nu_\alpha (f_\alpha - f_{0\alpha}), \tag{14.95}$$

where $f_{0\alpha}$ is the equilibrium distribution, and ν_α is the mean collision frequency. This contribution due to collisions can be understood in the following way. The term $-\nu_\alpha f_\alpha$ represents the number of particles of the species α lost at a given element of phase space volume, $d\mathbf{r}d\mathbf{v}$, centered at $(\mathbf{r}, \mathbf{v})$, due to binary collisions. The term $\nu_\alpha f_{0\alpha}$ determines the number of particles entering this elementary volume due to collisions taking place outside this volume. Equilibrium is attained when these two terms exactly compensate each other. We can generalize Eq. (14.95), by including all the possible collisions between particles of different species, as

$$\left(\frac{\partial f_\alpha}{\partial t}\right)_{\text{coll}} = -\sum_{\beta\alpha} \nu_{\alpha\beta} (f_\alpha - f_{0\alpha}), \tag{14.96}$$

where $\nu_{\alpha\beta}$ are the various mean collision frequencies between particle species α and β.

A more sophisticated version of the collision term can be obtained by transforming the plasma kinetic equation (14.88) into a Fokker-Planck equation,

$$\left(\frac{\partial f_\alpha}{\partial t}\right)_{\text{coll}} = \frac{\partial}{\partial v_i}\left[-A_i f_\alpha + \frac{1}{2}\frac{\partial}{\partial v_j} B_{ij} f_\alpha\right], \tag{14.97}$$

where the vector A_i and the tensor B_{ij}, with $i, j = (x, y, z)$, represent the friction and the diffusion coefficients in velocity space. This equation can be justified by noticing that most of the binary collisions have large impact parameters, and produce infinitesimal velocity deviations. This allows us to describe the collisional processes in differential form. A simple way to estimate the coefficients A_i and B_{ij} is to consider conservation of the total number of particles, their total momentum and their total energy. This means that

$$\int \left(\frac{\partial f_\alpha}{\partial t}\right)_{\text{coll}} a(\mathbf{v}) d\mathbf{v} = 0, \tag{14.98}$$

where $a(\mathbf{v})$ is equal to 1, $\mathbf{v}$ and v^2. The first integral, corresponding to $a(\mathbf{v}) = 1$ is automatically satisfied, if the distribution function f_α tends to zero at infinity. The other two integrals imply that

$$A_i = -\nu_\alpha (v_i - v_{0i}), \quad B_{ij} = 2\nu_\alpha \frac{T_\alpha}{m_\alpha} \delta_{ij}, \tag{14.99}$$

where $v_{\alpha i}$ are the components of the mean velocity $\mathbf{v}_\alpha$ of the population α. These two expressions are exact for a Maxwellian velocity distribution with temperature T_α, and can be used if f_α is not too far from equilibrium. Such a procedure is also known in the literature as the *relaxation-time approximation* and has been extensively employed to describe the dynamics of both Fermi and Bose gases nearly in equilibrium [23, 24]. Replacing these expressions in Eq. (14.97), we obtain the following plasma kinetic equation

$$\frac{df_\alpha}{dt} = \nu_\alpha \frac{\partial}{\partial \mathbf{v}} \cdot \left[(\mathbf{v} - \mathbf{v}_\alpha) f_\alpha + \frac{T_\alpha}{m_\alpha} \frac{\partial f_\alpha}{\partial \mathbf{v}} \right]. \tag{14.100}$$

More elaborate forms of the plasma kinetic equation can also be derived from the exact collision term defined in Eq. (14.88).

14.7 Fluid Equations

The plasma kinetic equation, even in the simplified Vlasov form, contains an excessive amount of information for a large number of physical problems. It is therefore useful to derive simpler evolution equations for the average plasma quantities, such as the mean particle densities, mean velocities, pressures and temperatures. These new equations are called hydrodynamic or fluid equations, as we have done for the cold atomic gas or for the condensates.

When we multiply the Vlasov equation by successive powers of $\mathbf{v}$ and integrate over the velocity space, we obtain an infinite series of equations. These equations are called the momenta of the Vlasov equation. By truncating this infinite series of equations, we obtain the simplified description of the plasma as a fluid medium, based on the continuity and the momentum conservation equations for each species.

Let us then go back to the Vlasov equation (14.89) and integrate it over the velocity space. Using the definition of the mean density of the particle species n_α, we see that the first term simply leads to

$$\int \frac{\partial}{\partial t} f_\alpha d\mathbf{v} = \frac{\partial}{\partial t} n_\alpha, \quad n_\alpha = \int f_\alpha d\mathbf{v}_\alpha. \tag{14.101}$$

The calculation of the second term implies the use of the mean velocity $\mathbf{v}_\alpha$, which allows us to write

$$\int \mathbf{v}\cdot\frac{\partial}{\partial \mathbf{r}} f_\alpha d\mathbf{v} = \nabla\cdot(n_\alpha \mathbf{v}_\alpha), \quad \mathbf{v}_\alpha = \frac{1}{n_\alpha}\int \mathbf{v} f_\alpha(\mathbf{r},\mathbf{v},t)d\mathbf{v}. \tag{14.102}$$

The third term, in its turn, can be integrate by parts and, noticing that f_α tends to zero at infinity, we get

$$\int \frac{\mathbf{F}}{m_\alpha}\cdot\frac{\partial}{\partial \mathbf{v}} f_\alpha d\mathbf{v} = -\frac{1}{m_\alpha}\int_\alpha \left(\frac{\partial}{\partial \mathbf{v}}\cdot\mathbf{F}\right) d\mathbf{v} = 0 \tag{14.103}$$

Therefore, the contribution of the third term of the Vlasov equation is equal to zero and, using Eqs. (14.101) and (14.102), we conclude that the first momentum of the Vlasov equation is the continuity equation

$$\frac{\partial n_\alpha}{\partial t} + \nabla\cdot(n_\alpha \mathbf{v}_\alpha) = 0. \tag{14.104}$$

Let us now consider the second momentum of the Vlasov equation, by multiplying Eq. (14.89) by $\mathbf{v}$ and integrating over velocity. From the first term, we get

$$\int \mathbf{v}\frac{\partial f_\alpha}{\partial t} d\mathbf{v} = \frac{\partial}{\partial t}\int \mathbf{v} f_\alpha d\mathbf{v} = \frac{\partial}{\partial t} n_\alpha \mathbf{v}_\alpha, \tag{14.105}$$

and, from the second term

$$\int \mathbf{v}\mathbf{v}\cdot\frac{\partial f_\alpha}{\partial \mathbf{r}} d\mathbf{v} = \nabla\cdot\int \mathbf{v}\mathbf{v} f_\alpha d\mathbf{v} = \nabla\cdot n_\alpha \langle\mathbf{v}\mathbf{v}\rangle \tag{14.106}$$

where the tensor $\langle\mathbf{v}\mathbf{v}\rangle$ is defined by

$$\langle\mathbf{v}\mathbf{v}\rangle = \frac{1}{n_\alpha}\int \mathbf{v}\mathbf{v} f_\alpha d\mathbf{v}. \tag{14.107}$$

By integrating the third term by parts, we simply obtain the momentum equation

$$\frac{\partial}{\partial t} n_\alpha \mathbf{v}_\alpha + \frac{\partial}{\partial \mathbf{r}}\cdot n_\alpha \langle\mathbf{v}\mathbf{v}\rangle = \frac{q_\alpha}{m_\alpha} n_\alpha [\boldsymbol{\mathcal{E}} + \mathbf{v}\times\mathbf{B}]. \tag{14.108}$$

In order to obtain a closed set of equations for the mean quantities, n_α and $\mathbf{v}_\alpha$, we write the tensor $\langle\mathbf{v}\mathbf{v}\rangle$ in terms of these quantities. This could in principle be done rigorously, by calculating the higher order momenta of the Vlasov equation. But such a process would lead to a third order tensor $\langle\mathbf{v}\mathbf{v}\mathbf{v}\rangle$, and so on. We would have an infinite sequence of coupled macroscopic equations. In order to avoid this infinite chain, we can use an estimate of the second order tensor $\langle\mathbf{v}\mathbf{v}\rangle$.

For such an estimate, let us start with a simple and relatively trivial case, corresponding to a cold plasma, $T_\alpha = 0$. In this extreme situation, the particle distribution functions will reduce to a Dirac delta function, stating that the all the particles of species α have the same velocity, $\mathbf{v}_\alpha$. In that case, we have

$$f_\alpha(\mathbf{r}, \mathbf{v}, t) = n_\alpha \delta(\mathbf{v} - \mathbf{v}_\alpha), \tag{14.109}$$

from which we thus get

$$\langle \mathbf{v}\mathbf{v} \rangle = \int \mathbf{v}\, \mathbf{v}\, \delta(\mathbf{v} - \mathbf{v}_\alpha) d\mathbf{v} = \mathbf{v}_\alpha \mathbf{v}_\alpha. \tag{14.110}$$

Let us now consider another simple case, corresponding to an isotropic distribution with zero mean velocity. In this case f_α only depends on the absolute value of the particle velocity, and not on its direction. The integrand in (14.107) is an odd function of v_i and v_j, because of the isotropy of f_α. When we integrate over v_i or v_j, we get zero off-diagonal tensor components $\langle v_i v_j \rangle = 0$, for $i \neq j$. In contrast, for the diagonal terms, we simply have $\langle v_i v_j \rangle = \langle v^2 \rangle$. When f_α is Maxwellian, with temperature T_α, this mean quadratic velocity is $3T_\alpha / m_\alpha$. We can then say that, for an isotropic particle velocity distribution $\langle \mathbf{v}\, \mathbf{v} \rangle = \langle v^2 \rangle \mathbf{1}$, where $\mathbf{1} \equiv \delta_{ij}$ is the unit tensor. The second term of the momentum equation (14.108) can then be written as

$$\nabla \cdot n_\alpha \langle \mathbf{v}\mathbf{v} \rangle = \nabla \cdot \mathbf{1} n_\alpha \langle v^2 \rangle = \nabla \frac{P_\alpha}{m_\alpha}, \tag{14.111}$$

where $P_\alpha = n_\alpha m_\alpha \langle v^2 \rangle$ is the pressure. Finally, let us consider a more general case, where the fluid has a finite mean velocity $\mathbf{v}_\alpha$, but the distribution stays isotropic in its own reference frame, where it is characterized by a temperature T_α. We can then write $\mathbf{v} = \mathbf{v}_\alpha + \mathbf{u}$, where $\mathbf{u}$ is isotropic. In the reference frame moving with the particles, we can use the results of the previous case and write

$$\langle \mathbf{u}\, \mathbf{u} \rangle = \frac{P_\alpha}{n_\alpha m_\alpha} \mathbf{1}. \tag{14.112}$$

We can then establish for the quadratic velocity tensor

$$\langle \mathbf{v}\, \mathbf{v} \rangle = \mathbf{v}_\alpha \mathbf{v}_\alpha + \frac{P_\alpha}{n_\alpha m_\alpha} \mathbf{1}. \tag{14.113}$$

Using the continuity equation (14.104), we may write

$$\frac{\partial}{\partial \mathbf{r}} \cdot n_\alpha \langle \mathbf{v}\, \mathbf{v} \rangle = \left(\frac{\partial n_\alpha}{\partial t} + n_\alpha \mathbf{v}_\alpha \cdot \nabla \right) \mathbf{v}_\alpha + \frac{1}{m_\alpha} \nabla P_\alpha \tag{14.114}$$

Replacing this in Eq. (14.108), we get

$$\left(\frac{\partial}{\partial t} + \mathbf{v}_\alpha \cdot \nabla \right) \mathbf{v}_\alpha + \frac{\nabla P_\alpha}{n_\alpha m_\alpha} = \frac{q_\alpha}{m_\alpha} (\mathcal{E} + \mathbf{v}_\alpha \times \mathbf{B}). \tag{14.115}$$

As a final step, we need to write the pressure P_α in terms of the quantities n_α and T_α, by using an equation of state. In quite general conditions, we can use

$$\nabla P_\alpha = \gamma_\alpha T_\alpha \nabla n_\alpha, \tag{14.116}$$

where γ_α is the adiabatic constant. For a gas of particles with no internal degree of freedom, it can be determined by the rule $\gamma_\alpha = 1 + (2/D)$, where D is the number of dimensions contributing to the temperature variations. We can then finally write the equation of momentum conservation in its most popular version

$$\left(\frac{\partial}{\partial t} + \mathbf{v}_\alpha \cdot \nabla\right)\mathbf{v}_\alpha + S_\alpha^2 \nabla \ln n_\alpha = \frac{q_\alpha}{m_\alpha}(\mathcal{E} + \mathbf{v}_\alpha \times \mathbf{B}), \tag{14.117}$$

where $S_\alpha = (\gamma_\alpha T_\alpha / m_\alpha)^{1/2}$ represents the thermal velocity. Equations (14.104) and (14.117) form, together with Maxwell's equations, a closed set of equations, which describe the evolution of the mean densities, velocities and fields in a plasma.

If, instead of using the Vlasov equation as a starting point, we had used a more complete kinetic equation with a collision term $(\partial f_\alpha/\partial t)_{\text{coll}}$, an additional term would appear in the momentum equation (14.117), This term would take the generic form

$$\int \left(\frac{\partial f_\alpha}{\partial t}\right)_{coll} \mathbf{v}_\alpha d\mathbf{v}_\alpha = -\nu_\alpha \mathbf{v}_\alpha, \tag{14.118}$$

where ν_α is the dominant collision frequency. Elastic collisions of particles of the species α with the other species $\beta \neq \alpha$ could be included in a similar way. On the other hand, inelastic collisions due to ionization and recombination processes would lead to a souce term in the continuity equation.

References

1. T.J.M. Boyd, J.J. Sanderson, *The Physics of Plasmas* (Cambridge University Press, Cambridge, UK, 2003)
2. D.R. Nicholson, *Introduction to Plasma Physics* (Wiley, New York, 1983)
3. N.A. Krall, A.W. Trivelpiece, *Principles of Plasma Physics* (McGraw-Hill, New York, 1973)
4. T.C. Killian, S. Kulin, S.D. Bergeson, L.A. Orozco, C. Orzel, S.L. Rolston, Phys. Rev. Lett. **83**, 4776 (1999)
5. M.P. Robinson, B.L. Tolra, M.W. Noel, T.F. Gallagher, P. Pillet, Phys. Rev. Lett. **85**, 4466 (2000)
6. D. Ciampini, M. Anderlini, J.H. Muller, F. Fuso, O. Morsch, J.W. Thomsen, E. Arimondo, Phys. Rev. A **66**, 043409 (2002)
7. T.C. Killian, T. Pattard, T. Pohl, J.M. Rost, Phys. Rep. **449**, 77 (2007)
8. J.-L. Delcroix, A. Bers, *Physique des Plasmas*, 2 volumes (CNRS Editions, Paris, 1994)
9. S.D. Bergeson, A. Denning, M. Lyon, F. Robicheaux, Phys. Rev. A **83**, 023409 (2011)
10. H.J. Metcalf, P. van der Straten, *Laser Cooling and Trapping* (Springer, New York, 2001)
11. E.A. Cummings, J.E. Daily, D.S. Durfee, S.D. Bergeson, Phys. Rev. Lett **95**, 235001 (2005)

12. J.P. Morrison, C.J. Rennick, J.S. Keller, E.R. Grant, Phys. Rev. Lett. **101**, 205005 (2008)
13. S. Dickson, S. Robertson, Phys. Plasmas **17**, 033508 (2010)
14. K. Minami, C. Kojima, T. Ohira, O. Ishihara, IEEE Trans. Plasma Sci. **31**, 429 (2003)
15. B.R. Beck, J. Fajans, J.H. Malmberg, Phys. Plasmas **3**, 1250 (1996)
16. F. Anderegg, D.H.E. Dubin, T.M. O'Neil, C.F. Driscoll, Phys. Rev. Lett. **102**, 185001 (2009)
17. J.J. Bollinger, D.J. Wineland, Phys. Rev. Lett. **53**, 348 (1984)
18. E. Fermi, *Astrophys. J.* **119**, 1 (1954)
19. T.G. Northrop, *The Adiabatic Motion of Charged Particles* (Interscience, New York, 1963)
20. B.V. Chirikov, A Universal instability of many dimensional oscillator systems. Phys. Rep. **52**, 265 (1979)
21. A.J. Lichtenberg, M.A. Lieberman, *Regular and Stochastic Motion* (Springer, New York, 1983)
22. Yu. L. Klimontovich, *Statistical Theory of Non-equilibrium Processes in a Plasma* (Pergamon, Oxford, 1967)
23. P. Pedri, D. Guéry-Odelin, S. Stringari, Phys. Rev. A **68**, 043608 (2003)
24. A. Griffin, T. Nikuni, E. Zaremba, *Bose-Condensed Gases at Finite Temperatures* (Cambridge University Press, Cambridge, 2009)

Chapter 15
Physics of Rydberg Plasmas

In this chapter, we deal with the expansion and the quasi-equilibrium states of an ultra-cold plasma. We have seen that ultra-cold plasmas can be produced by photoionizing a small cloud of laser-cooled atoms confined in a magneto-optical trap, and subsequently expands into the surrounding vacuum. Two different situations are usually considered in the literature: (i) the ultra-cold atoms are firstly laser excited into high Rydberg states and then the Rydberg gas spontaneously evolves into a plasma; (ii) the ultra-cold atoms are directly ionized by the laser and Rydberg atoms are formed by electron-ion recombination as the plasma expands. In this case, a spherical cloud of cold atoms is photoionized using a laser pulse. At first, some electrons promptly leave the plasma until sufficient space charge builds up to trap low energy electrons. Since the electrons are generated with a narrow band laser, they start with a non-thermal energy distribution, but they are rapidly thermalized by electron-electron collisions. During the thermalization process, some high energy electrons boil way, reducing the electron number and temperature. After the electrons thermalize, the plasma as a whole expands [26]. The plasma expansion results from the electron thermal pressure, $P_e = n_e T_e$, which imprints a radial velocity to the ions. This expansion is faster than what would be expected using a simple low temperature model, as a consequence of the electron-ion recombination. Since the latter scales as $T_e^{-9/2}$, the observed faster expansion results from the enhancement of electron-ion recombination at low temperatures.

Given the complexity of the physical processes underlying the plasma evolution, we discuss only the main theoretical issues without digressing on a complete and self-consistent description. We start by introducing the concept of ambipolar diffusion, which is central for the understanding of plasma discharges, and can be used to clarify the results of the non-collisional expansion models, which are discussed next. We then consider the properties of strongly coupled ions, which can exist in the ultra-cold plasma. We give special attention to the possible occurrence of phase transitions.

We also discuss the quasi-equilibrium configurations and the influence of particle correlations in the expansion process. We then consider the disorder induced

J.T. Mendonça and H. Terças, *Physics of Ultra-Cold Matter*, Springer Series on Atomic, Optical, and Plasma Physics 70, DOI 10.1007/978-1-4614-5413-7_15,

heating mechanism which is an essential ingredient for the establishment of a quasi-equilibrium ion temperature. Finally, we focus our attention on the main properties of Rydberg atoms, which play an important role in ultra-cold plasma physics. In particular, we discuss radiative cascades, Rydberg blockade and the three-body recombination process.

15.1 Plasma Expansion in the Collisional Regime

15.1.1 Free Diffusion

The ultra-cold plasma is described by two fluid equations, for electrons and ions ($\alpha = e, i$), which can be written as

$$\frac{\partial n_\alpha}{\partial t} + \nabla \cdot \mathbf{\Gamma}_\alpha = 0, \quad \frac{\partial \mathbf{v}_\alpha}{\partial t} + \mathbf{v} \cdot \nabla \mathbf{v} = \frac{q_\alpha}{m_\alpha} \mathcal{E} - \frac{\nabla P_\alpha}{m_\alpha n_\alpha} - \nu_\alpha \mathbf{v}_\alpha \tag{15.1}$$

with electric charges $q_e = -e$ and $q_i = Ze$. For Rydberg plasmas, the multiplicity of ionization is usually $Z = 1$. We have used the particle pressures P_α, the collision frequencies ν_α, and the particle fluxes $\mathbf{\Gamma}_\alpha = n_\alpha \mathbf{v}_\alpha$. In the present discussion, we have ignored any static magnetic field, and the electrostatic field $\mathcal{E}$ is determined by Poisson's equation

$$\nabla \cdot \mathcal{E} = \frac{1}{\epsilon_0} \sum_\alpha q_\alpha n_\alpha . \tag{15.2}$$

On a time scale much larger than the collision time, $\partial / \partial t \ll \nu_\alpha$, we can approximate the momentum equation in (15.1) by the steady-state relation

$$\mathbf{v}_\alpha = \frac{q_\alpha}{m_\alpha \nu_\alpha} \mathcal{E} - \frac{\nabla P_\alpha}{m_\alpha n_\alpha \nu_\alpha} . \tag{15.3}$$

For weak electric fields, the pressure force dominates over the electric force, and this relation reduces to

$$\mathbf{\Gamma}_\alpha = -D_\alpha \nabla n_\alpha \, , \quad D_\alpha = \frac{S_\alpha^2}{\nu_\alpha} , \tag{15.4}$$

where D_α is the diffusion coefficient for the particle species α. If we adopt an equation of state of the form $P_\alpha = T_\alpha n_\alpha^{\gamma_\alpha}$, where γ_α is the adiabatic constant, the particle thermal velocities will be determined by $S_\alpha = \sqrt{\gamma_\alpha T_\alpha / m_\alpha}$. The validity of this equation of state, and the choice of an adequate value for γ_α can only be elucidated by using a kinetic description. At this point, we should notice that

$$\frac{D_e}{D_i} = \frac{2}{Z} \frac{T_e}{T_i} \frac{m_i}{m_e} \frac{\nu_i}{\nu_e} . \tag{15.5}$$

This shows that, even for large electron temperatures $T_e \gg T_i$, we normally have a much stronger electron diffusion $D_e \gg D_i$, as long as $(\nu_i/\nu_e)(m_i/m_e) \gg (T_e/T_i)$. This is certainly true for ultra-cold plasmas. Using Eq. (15.4) in the continuity equation, we obtain a diffusion equation

$$\frac{\partial n_\alpha}{\partial t} - \nabla \cdot (D_\alpha \nabla n_\alpha) = 0. \tag{15.6}$$

Assuming spherical symmetry, and using the fact that the spatial scale at which D_α varies is much longer than the scale of variation of the density, this is reduced to

$$\frac{\partial n_\alpha}{\partial t} - \frac{D_\alpha}{r^2}\frac{\partial}{\partial r}\left(r^2 \frac{\partial n_\alpha}{\partial r}\right) = 0. \tag{15.7}$$

This spherical equation satisfies the well known Gaussian solution

$$n_\alpha(r,t) = \frac{N_\alpha}{(4\pi D_\alpha t)^{3/2}} \exp\left(-\frac{r^2}{4D_\alpha t}\right), \tag{15.8}$$

where N_α is the total number of particles in the expanding cloud, as defined by

$$N_\alpha = 4\pi \int_0^\infty n_\alpha(r,t) r^2 dr. \tag{15.9}$$

The size of the cloud of each particles species, is determined by the mean square distance

$$\langle r^2 \rangle_\alpha = \frac{4\pi}{N_\alpha} \int_0^\infty n_\alpha(r,t) r^4 dr = 6 D_\alpha t \tag{15.10}$$

Assuming that the plasma is instantaneously created at time $t = 0$, in a region with size a_0, with a Gaussian density profile, the free evolution of each particle species will then be determined by

$$n_\alpha(r,t) = \frac{N_\alpha}{[2\pi a_\alpha^2(t)]^{3/2}} \exp\left(-\frac{r^2}{2a_\alpha^2(t)}\right), \tag{15.11}$$

where the size of the cloud is determined by

$$a_\alpha(t) = \sqrt{\frac{1}{3}\langle r^2 \rangle_\alpha} = \sqrt{2D_\alpha(t + t_\alpha)} \tag{15.12}$$

with $a_0 = \sqrt{2D_\alpha t_\alpha}$. From this free expansion solution, we can conclude that, for a realistic situation where we have $D_e \gg D_i$, the electron cloud will duplicate its radius after a time $t \simeq 3t_e$,[1] while the ion cloud will take a much longer time $t \simeq 3(D_e/D_i)t_e$ to expand to a similar size. This means that, for $t \geq t_e$, we will have a situation where most of the electrons would have left the ion cloud. However, before that occurs, the assumption of a weak electric field, needed to justify the free expansion solution will breake down, and a radial electric field due to charge separation will take form, thus preventing the electrons to expand faster. These finite field effects will give rise to the ambipolar diffusion, to be discussed next.

15.1.2 Ambipolar Diffusion Regime

Let us then go back to Eq. (15.3), where the electric field is now retained, and which can be written as

$$\mathbf{\Gamma}_\alpha = n_\alpha \mu_\alpha \mathcal{E} - D_\alpha \nabla n_\alpha, \tag{15.13}$$

where $\mu_\alpha = q_\alpha / m_\alpha \nu_\alpha$ represent the species mobility. Using the above definition for the diffusion coefficient D_α, given in Eq. (15.4), we can state the famous *Einstein relation* between diffusion and mobility, as

$$\mu_\alpha = \frac{q_\alpha}{m_\alpha S_\alpha^2} D_\alpha. \tag{15.14}$$

Let us now assume the *ambipolar approximation*, which results from the assumption of absence of a net current in the plasmas, as stated by

$$\mathbf{J} = \sum_\alpha q_\alpha \mathbf{\Gamma}_\alpha = e(Z\Gamma_i - \Gamma_e) \simeq 0. \tag{15.15}$$

Simplifying the notation, we write $\Gamma \equiv \Gamma_e \simeq Z\Gamma_i$, using Eq. (15.13), and assuming quasi-neutrality $n \equiv n_e \simeq Zn_i$, we get the following approximate relations

$$\mathbf{\Gamma} = n\mu_e \mathcal{E} - D_e \nabla n \simeq n\mu_i \mathcal{E} - D_i, \nabla n \tag{15.16}$$

which allows us to write the electron flux as

$$\mathbf{\Gamma} = -D\nabla n, \quad D = \frac{D_i \mu_e - D_e \mu_i}{\mu_e - \mu_i}. \tag{15.17}$$

The new quantity D is called the *ambipolar diffusion coefficient*. On the other hand, the electric field is determined by

$$\mathcal{E} = \frac{D_i - D_e}{\mu_e - \mu_i} \frac{\nabla n}{n}, \tag{15.18}$$

[1] t_α is the inverse of the species collision frequency, $t_\alpha = \nu_e^{-1}$.

Coming back to the electron continuity equation (15.6), and replacing the diffusion coefficient by the ambipolar diffusion coefficient, we may write

$$\frac{\partial n}{\partial t} - \nabla \cdot (D \nabla n) = 0. \tag{15.19}$$

Given the assumption of quasi-neutrality, this equation describes the expansion of both the electron and ion populations, which are tied by the ambipolar electric field (15.18) and for that reason expand with the same diffusion coefficient D. Noticing that $D_e \gg D_i$ and $\mu_e \gg \mu_i$, we immediately observe, from the definition of D in Eq. (15.17), the following

$$D \simeq D_e \frac{\mu_i}{\mu_e} \sim D_e \frac{m_e}{m_i} \sim D_i. \tag{15.20}$$

This shows that, due to the electrostatic field, the electrons will diffuse much slower than in the free diffusion approximation, on a time scale determined by the ion diffusion coefficient. The electron expansion is therefore significantly retarded by the ambipolar field.

15.1.3 Recombination in Volume

Let us now assume that, during the diffusion process, significant recombination occurs, leading to a reduction of the number of expanding charged particles. This can be described by a quadratic loss term in the diffusion equation (15.19), which becomes

$$\frac{\partial n}{\partial t} = \nabla \cdot (D \nabla n) - \alpha_{\text{rec}} n^2, \tag{15.21}$$

where α_{rec} is the *recombination coefficient*. The existence of this new term can be justified by binary electron-ion reactions or by three body electron-electron-ion processes, both leading to the formation of neutral atoms. As discussed later, the three-body recombination can be the dominant process at low temperatures, which in its turn is responsible for the formation of Rydberg atom states. Such a process is very important for anti-matter research.

Equation (15.21) can be written in a more convenient form by noting that, for a given radial density profile, such as the Gaussian profiles considered above, we can use the relation

$$\nabla \cdot (D_\alpha \nabla n) = -\nu n \ , \quad \nu \equiv \frac{D}{\Lambda^2}, \tag{15.22}$$

where Λ is a length scale characterizing the density profile (say, its radius). We can then rewrite (15.21) in a much simpler form, as

$$\frac{\partial n}{\partial t} = -\nu n - \alpha_{\text{rec}} n^2. \tag{15.23}$$

This clearly states that the plasma density decreases during the expansion process, not only because ambipolar diffusion is taking place, but also due to recombination. Integrating, we obtain

$$f(t) = f(0)\exp(-\nu t), \quad f(t) = \frac{n(t)}{\nu + \alpha_{\rm rec} n(t)}. \tag{15.24}$$

It should be noticed that, although we have assumed a constant density profile characterized by the length scale Λ, recombination always tends to change the initial profile, because it is stronger in regions of high plasma density, as recombination scales as n^2.

15.2 Collisionless Plasma Expansion

Let us now consider the opposite regime where collisions are not relevant on the time scale of plasma expansion, $\partial/\partial t \gg \nu_\alpha$. This regime has been considered by several authors along the years, in particular those focusing on ultra-cold plasmas [22, 26]. Here again, we assume that the ions control the expansion process, as in the case of ambipolar diffusion, because the electrons remain tied to the ions by the electrostatic field. We can still use a variational Gaussian profile for the ions, similar to Eq. (15.11), but where collision dependent diffusion coefficients are not present, reading

$$n_i(r,t) = N_i \left[\frac{\beta(t)}{\pi}\right]^{3/2} \exp\left(-\beta(t) r^2\right), \tag{15.25}$$

where $\beta(t)$ is the variational parameter describing the time evolution ion cloud radiux, $a_i(t)$, according to

$$a_i(t) = \sqrt{\frac{1}{3}\langle r^2\rangle_i} = \frac{1}{\sqrt{2\beta(t)}}. \tag{15.26}$$

This would reduce to the previously assumed ambipolar density profile, for $\beta(t) = 1/4Dt$, where $D \sim D_i$. In order to determine $\beta(t)$ we have to introduce some simplifying assumptions concerning the role played by the electrons, and in particular, the electron density profile. For that purpose, let us go back to the electron fluid equations (15.1), with $\alpha = e$. Assuming a quasi-static equilibrium such that $d/dt \simeq 0$, and neglecting collisions, $\nu_e \simeq 0$, the electron momentum equation can be reduced to

$$S_e^2 \nabla n_e + \frac{e n_e}{m_e}\boldsymbol{\mathcal{E}} \simeq 0. \tag{15.27}$$

Introducing the electrostatic potential V, such that $\boldsymbol{\mathcal{E}} = -\nabla V$, we obtain

$$\frac{\nabla n_e}{n_e} \simeq \nabla\left(\frac{eV}{m_e S_e^2}\right). \tag{15.28}$$

After integration, and making use of an isothermal equation of state for the electron, $\gamma_e = 1$, the latter equation leads to the usual Boltzmann equilibrium. For a spherical cloud, such an equilibrium takes the form

$$n_e(r,t) = n_{e0}(t) \exp\left[\frac{eV(r,t)}{T_e(t)}\right], \tag{15.29}$$

where we allow the electron temperature to evolve during expansion. Here we choose $V(r = 0, t) = 0$ and $V(r > 0, t) < 0$, in order to describe an electron density profile of the form $n_e(r > 0) < n_{e0}$. If we insert this expression in Poisson's equation, we obtain a nonlinear equation for the potential. In order to avoid solving this equation, we can write from (15.29) an expression for the electrostatic potential, as

$$V(r,t) = \frac{T_e(t)}{e} \ln\left[\frac{n_e(r,t)}{n_{e0}(t)}\right]. \tag{15.30}$$

We can then assume the quasi-neutrality condition, $n_e(r,t) \simeq n_i(r,t)$, and use Eq. (15.25) for the ion density profile to obtain

$$V(r,t) = \frac{T_e(t)}{e} \beta(t) r^2 + C(t) \tag{15.31}$$

with

$$C(t) = \frac{T_e(t)}{e} \ln\left\{\frac{N_i}{n_{e0}(t)} \left[\frac{\beta(t)}{\pi}\right]^{3/2}\right\}. \tag{15.32}$$

The electrostatic field can then be determined by

$$E(r,t) = -\frac{\partial V}{\partial r} = 2\frac{T_e(t)}{e} \beta(t) r. \tag{15.33}$$

We can now go back to the ion momentum equation and, assuming a very cold ion population, such that $T_i \simeq 0$, we simply get

$$\frac{d\mathbf{v}_i}{dt} = \frac{Ze}{m_i} \boldsymbol{\mathcal{E}}, \tag{15.34}$$

where we have used the total time derivative $d/dt = \partial/\partial t + \mathbf{v}_i \cdot \nabla$. Using the field solution (15.33), we can see that the ion acceleration increases with the radius r, suggesting the use of a expression of the form

$$v_i(r,t) = \gamma(t) r, \tag{15.35}$$

where $\gamma(t)$ satisfies the equation

$$\frac{d\gamma}{dt} + \gamma^2(t) = 2\frac{ZT_e(t)}{m_i}\beta(t) \tag{15.36}$$

This establishes a relation between the variational functions $\gamma(t)$ and $\beta(t)$. A second relation can be obtained from the ion continuity equation, which in a spherical geometry can be written as

$$\frac{\partial n_i}{\partial t} + \frac{1}{r^2}\frac{\partial}{\partial r}\left(r^2 n_i v_i\right) = 0. \tag{15.37}$$

Using Eq. (15.25) for n_i, and (15.35) for v_i, we obtain

$$\frac{d\beta}{dt} = -2\gamma(t)\beta(t), \tag{15.38}$$

or, after integration

$$\beta(t) = \beta(0)\exp\left[-2\int_0^t \gamma(t')dt'\right]. \tag{15.39}$$

In order to provide the lacking relation between the electron temperature $T_e(t)$ with the variational functions $\gamma(t)$ and $\beta(t)$, one makes use of the energy conservation equation, which is stated as

$$\sum_{\alpha=e,i}\left(\frac{3}{2}T_\alpha(t) + \frac{1}{2}m_\alpha\left\langle v_\alpha^2\right\rangle\right) = \text{const.} \tag{15.40}$$

By neglecting both the ion temperature ($T_i \simeq 0$) and the electron kinetic energy, this reduces to

$$\frac{3}{2}T_e(0) = \frac{3}{2}T_e(t) + \frac{1}{2}m_i\left\langle v_i^2\right\rangle. \tag{15.41}$$

Using Eqs. (15.35) and (15.26), we realize that

$$\left\langle v_i^2\right\rangle = \gamma^2(t)\left\langle r^2\right\rangle_i = \frac{3}{2}\frac{\gamma^2(t)}{\beta(t)}, \tag{15.42}$$

which allows us to write

$$T_e(0) = T_e(t) + \frac{1}{2}m_i\frac{\gamma^2(t)}{\beta(t)}. \tag{15.43}$$

Taking the time derivative of the latter, we finally obtain

$$\frac{dT_e}{dt} = -\frac{1}{2}m_i\left[\frac{2\gamma(t)}{\beta(t)}\frac{d\gamma}{dt} - \frac{\gamma^2(t)}{\beta^2(t)}\frac{d\beta}{dt}\right] = -2Z\gamma(t)T_e(t), \tag{15.44}$$

which, for $Z = 1$ leads to $T_e(t)/\beta(t) = \text{const}$. A simple analytical solution for Eqs. (15.36), (15.38) and (15.44) is provided by

$$T_e(t) = \frac{T_e(0)}{\left(1 + t^2/\tau_{\exp}^2\right)}, \tag{15.45}$$

to which correspond

$$\beta(t) = \frac{\beta(0)}{\left(1 + t^2/\tau_{\exp}^2\right)}, \quad \gamma(t) = \frac{t/\tau_{\exp}^2}{\left(1 + t^2/\tau_{\exp}^2\right)}, \tag{15.46}$$

where the constant $\tau_{\exp}$ is the characteristic time for the ion expansion. This can be estimated as

$$\tau_{\exp} \simeq \frac{a_i(0)}{u_s(0)} = \sqrt{\frac{m_i}{T_e(0)\beta(0)}}, \tag{15.47}$$

where $u_s(t) = \sqrt{T_e(t)/m_i}$ is the ion acoustic velocity, and $a_i(t) \simeq 1/\sqrt{\beta(t)}$ is the radius of ion cloud. As it can be seen from Eq. (15.43), u_s is the characteristic velocity of the plasma expansion. In what concerns the ion temperature, which was disregarded in the above discussion, we can assume that it follows the same expansion law, allowing us to write

$$T_i(t) = \frac{T_i(0)}{\left(1 + t^2/\tau_{\exp}^2\right)}. \tag{15.48}$$

The above results are typical of a plasma expansion into vacuum. An acceleration linear in r, implies a velocity which is also linear in r, therefore preserving the Gaussian spatial profile along the expansion process. Finally, we should notice that this expansion will preserve the characteristic expansion velocity $u_s = \sqrt{T_e/m_i}$, also valid for the ambipolar diffusion. The collisionless plasma predicts the main properties of an expanding ultra-cold plasma, and is particularly adequate for experimental verification.

15.3 Strongly Coupled Ions

15.3.1 Ion-Neutral Coupling

The above discussion is strictly valid for a fully ionized plasma. When neutral atoms are presents, the ions tend to be strongly correlated with the neutrals, due to recombination and ionization processes. Such coupling can be described by using a kinetic approach, where the Vlasov equation for the ions is complemented by a ion-neutral coupling source term, as

$$\left(\frac{\partial}{\partial t} + \mathbf{v} \cdot \nabla + \frac{e\mathcal{E}}{m_i} \cdot \frac{\partial}{\partial \mathbf{v}}\right) f_i = \mathcal{S}_i, \tag{15.49}$$

where $\mathcal{E} = -\nabla V$ is the electrostatic field seen by the ions, $f_i \equiv f_i(\mathbf{r}, \mathbf{v}, t)$ is the ion distribution function, and $\mathcal{S}_i$ is the source term, which can be defined as [6]

$$\mathcal{S}_i = \sum_n \left[K_{\text{ion}}(n) f_a(n) - K_{\text{tbr}}(n) f_i\right], \tag{15.50}$$

where $K_{\text{ion}}(n)$ and $K_{\text{tbr}}(n)$ are the electron impact ionization and the three-body recombination rate coefficients, which generally depend not only on the principal quantum numbers n of the neutral atoms, with a distribution function $f_a(n) \equiv f_a(n, \mathbf{r}, \mathbf{v}_a, t)$, but also on the electron density and temperature, n_e and T_e. The evolution of the atom distribution function $f_a(n)$ is, on the other hand, determined by a similar kinetic equation, of the form

$$\left(\frac{\partial}{\partial t} + \mathbf{v}_a \cdot \nabla\right) f_a(n) = \mathcal{S}_a(n), \tag{15.51}$$

where the neutral atom source term $\mathcal{S}_a(n)$ is determined by

$$\mathcal{S}_a(n) = \sum_{n'} \left[K_{bb}(n', n) f_a(n') - K_{bb}(n, n') f_a(n)\right] + K_{\text{tbr}}(n) f_i - K_{\text{ion}} f_a(n). \tag{15.52}$$

Here, $K_{bb}(n, n')$ are the rate coefficients for the bound-bound transitions between neutral atomic states n and n', due to electron impact collisions. Obviously, these coefficients also depend on n_e and T_e. Assuming a Gaussian profile for the spherical atomic cloud, we can use the following expression for the ion distribution function

$$f_i(\mathbf{r}, \mathbf{v}, t) = C_i \left[\frac{\beta(t)}{\pi}\right]^{3/2} \exp\left(-\beta(t) r^2\right) \exp\left[-\frac{m_i}{2T_i(t)} \left(\mathbf{v} - \gamma(t) r \mathbf{e}_r\right)^2\right], \tag{15.53}$$

where C_i is an appropriate normalization constant. Integration over the ion velocities $\mathbf{v}_i$ leads to the density profile (15.25). A similar expression is assumed to be valid for the atom distribution functions $f_a(n)$. This allows us to calculate the mean square deviation and the mean square velocity, as

$$\langle r^2 \rangle_i = \frac{1}{N_i} \int d\mathbf{r} \int d\mathbf{v} r^2 f_i(\mathbf{r}, \mathbf{v}, t) = \frac{3}{2\beta(t)} \tag{15.54}$$

and

$$\langle v^2 \rangle_i = \frac{1}{N_i} \int d\mathbf{r} \int d\mathbf{v} v^2 f_i(\mathbf{r}, \mathbf{v}, t) = 3\left(\frac{T_i}{m_i} + \frac{\gamma^2(t)}{2\beta(t)}\right). \tag{15.55}$$

Putting the latter equations together, we obtain

$$\frac{d\gamma}{dt} = \frac{2N_i\beta(t)}{N_i + N_a}\frac{T_e + T_i}{m_i} - \gamma^2(t)\,, \quad \frac{dT_i}{dt} = -2\gamma(t)T_i(t), \tag{15.56}$$

which generalizes Eqs. (15.36) and (15.44). On the other hand, integration over the velocity space of the kinetic equation (15.51) leads to

$$\frac{\partial}{\partial t}N_a(n) = \sum_{n'\neq n}\left[R_{bb}(n',n)N_a(n') - R_{bb}(n,n')N_a(n)\right] + R_{\text{tbr}}(n)N_i - K_{\text{ion}}N_a(n), \tag{15.57}$$

where $N_a(n)$ are the populations of the different neutral atom states n, satisfying $\sum_n N_a(n) = N_a$, and the quantities R_{bb}, R_{ion} and R_{tbr} are the density averaged collision rates resulting from the integration of the quantities K_{bb}, K_{ion} and K_{tbr} over the ion density profile, as defined by the generic expression

$$R = \frac{1}{N_i}\int_0^\infty n_i(r)K dr. \tag{15.58}$$

Equation (15.57) determines the temporal evolution of the different neutral atom states. Finally, the energy conservation equation (15.40) in the presence of neutrals is replaced by a more complete expression taking into account the ion-atom coupling, as

$$\frac{3}{2}N_i\left\{[T_e(t) + T_i(t)] + \frac{m_i\gamma^2(t)}{2\beta(t)}\right\} - \sum_n N_a(n)\frac{\text{R}_\text{y}}{n^2} = \text{const.}, \tag{15.59}$$

where R_y is the Rydberg constant. Equations (15.56) and (15.59) show that inelastic collisions not only change the neutral atom populations $N_a(n)$, and consequently the ion populations N_i (or, in other words, the degree of ionization of the plasma cloud), but they also influence the electron temperature T_e. This is due to the fact that in the three-body recombination process the free electron caries the excess of energy resulting from it. Therefore, recombination tends to increase T_e. The same occurs for the Rydberg cascading due to electron-atom collisions. In contrast, T_e will decrease due to electron impact ionization and to the inverse Rydberg cascading.

Nevertheless, we should keep in mind that three-body recombination is extremely sensitive to the electron temperature, scaling as $T_e^{-9/3}$. This means that the influence of inelastic collisions will sharply decrease for increasing electron temperatures, where the models of the previous section (where ion-atom correlations were ignored) becomes more accurate.

15.3.2 Ion-Ion Coupling

A different type of strong coupling occurs for high densities, in the absence of neutral atoms, when the ion coupling parameter becomes non negligible, $\Gamma_i \geq 1$, and ion-ion correlations have to be taken into account. In this case the Vlasov equation for the ions is replaced by

$$\left(\frac{\partial}{\partial t} + \mathbf{v} \cdot \nabla + \frac{e\mathcal{E}}{m_i} \cdot \frac{\partial}{\partial \mathbf{v}}\right) f_i = G_{ii}, \tag{15.60}$$

where the ion-ion coupling term is defined by

$$G_{ii} = \frac{\partial}{\partial \mathbf{v}} \int d\mathbf{r}' \int d\mathbf{v}' \left(\frac{\partial V_{ii}}{\partial \mathbf{r}}\right) g_{ii}(\mathbf{r}, \mathbf{v}, \mathbf{r}', \mathbf{v}', t). \tag{15.61}$$

Here, V_{ii} stands for the two-particle interaction potential and g_{ii} represents the two-particle correlation function. Following the strategy of the previous sections, Eq. (15.56) easily leads to

$$\frac{d\gamma}{dt} = \frac{2\beta(t)}{m_i}\left[T_e(t) + T_i(t) + \frac{1}{3}U_{ii}\right] - \gamma^2(t) \tag{15.62}$$

and

$$\frac{dT_i}{dt} = -2\gamma(t)\left[T_i(t) + \frac{1}{3}U_{ii}\right] - \frac{2}{3}\frac{\partial U_{ii}}{\partial t}, \tag{15.63}$$

where we have included the total correlation energy, as defined by

$$U_{ii} = \frac{1}{N_i}\int n_i(\mathbf{r}, t) u_{ii}(\mathbf{r}) d\mathbf{r}, \tag{15.64}$$

with the single correlation energy given by

$$u_{ii}(\mathbf{r}) = \frac{e^2}{8\pi\epsilon_0} n_i(\mathbf{r}) \int \frac{g_{ii}(\mathbf{y})}{y} d\mathbf{y}, \quad \mathbf{y} = (\mathbf{r}' - \mathbf{r}). \tag{15.65}$$

It is plausible to assume that the total correlation energy tends to some quasi-equilibrium value U_{ii}^{eq} on a time scale of the order to $1/\omega_{pi}$, which is the typical time scale for the ion population [3]. This equilibrium quantity can be estimated as

$$U_{ii}^{\mathrm{eq}} = T_i \Gamma_i \left(\tilde{u} + \frac{1}{2}\frac{a}{\lambda_D}\right), \tag{15.66}$$

where quantity $\tilde{u}$ can be obtained from molecular dynamic simulations [7].

15.3.3 Phase Transitions

In ultra-cold plasmas, the ions can be strongly coupled in the sense that their average Coulomb interaction energy is much larger than the average kinetic energy, or in other words, the ion coupling parameter can be quite large, $\Gamma_i \gg 1$. This leads to a significant change in the plasma behavior and in particular to phase transitions. We now discuss the possible criteria for such phase transitions. We consider here a classical approach, based on the *one-component plasma model*, which was pioneered by Brush in [4] and reviewed by Ichimaru [13] (see also [7]).

We consider an ion gas immersed in a sea of neutralizing electrons. As a basic element of such a medium, we can identify the *ion sphere*, which can be defined as a sphere with radius equal to the Wigner-Seitz radius for the ions, a_i and containing the neutralizing electric charge that would exactly cancel a single ion charge Ze located inside this volume, with charge density ρ_e, as given by (see the Fig. 15.1a)

$$\rho_e = -\frac{3Ze}{4\pi a_i^3}, \quad a_i = \left(\frac{3}{4\pi n_i}\right)^{1/3}. \tag{15.67}$$

The total electrostatic energy contained in the ion sphere is $U_{is} = U_i + U_e$, where U_i is the interaction energy between the ion and the negative charge centered at a distance r from the ion, and U_e is the electrostatic energy of the negative charge itself, as given by

$$U_i = \frac{q_i^2}{a_i}\left[-\frac{3}{2} + \frac{1}{2}\left(\frac{r}{a_i}\right)^2\right], \quad U_e = \frac{3}{5}\frac{q_i^2}{a_i}. \tag{15.68}$$

where we use $q_i^2 = (Ze)^2/4\pi\epsilon_0$. Using the definition of the ion coupling parameter $\Gamma_i = q_i^2/a_i T_i$, where T_i is written in energy units, we obtain for U_{is} the expression

$$\frac{U_{is}}{T_i} = \Gamma_i\left[\left(\frac{3}{5} - \frac{3}{2}\right) + \frac{1}{2}\left(\frac{r}{a_i}\right)^2\right] \equiv -0.9\Gamma_i + 0.5\Gamma_i\left(\frac{r}{a_i}\right)^2. \tag{15.69}$$

The first term in these expressions gives the lowest possible energy value of the ion sphere, corresponding to the equilibrium position $r = 0$. The second term, proportional to r^2, can be seen as the potential energy of an harmonic oscillator centered at the ion position. Noting that, for thermal equilibrium, the mean energy of the harmonic oscillator is equal to $T_i/2$ per degree of freedom, we conclude that the mean value of the total electrostatic energy of the ion sphere will be given by

$$\frac{\langle U_{is}\rangle}{T_i} = -0.9\Gamma_i + \frac{3}{2}. \tag{15.70}$$

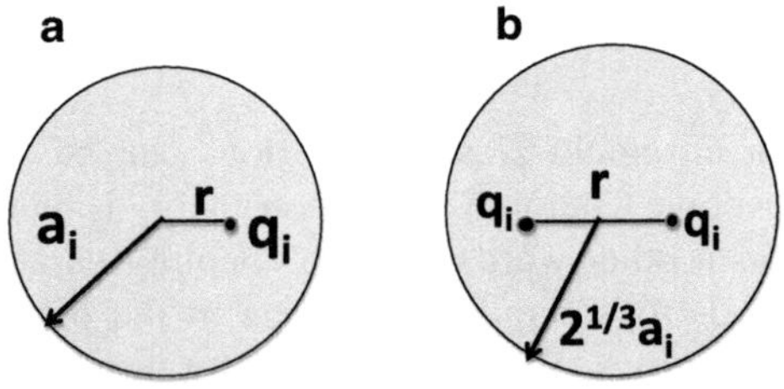

Fig. 15.1 (**a**) Single ion sphere, with radius a_i; (**b**) two-ions sphere, with radius $2^{1/3}a_i$

Let us now take a step forward and consider the interaction between nearby ions. For that purpose, we introduce the correlation function between two ions, $g(\mathbf{r}_1, \mathbf{r}_2)$, located at positions $\mathbf{r}_1$ and $\mathbf{r}_2$. Formally speaking, it can be defined as a relation between the single-particle distribution function $f_1(\mathbf{r})$ and the two-particle distribution function $f_{12}(\mathbf{r}_1, \mathbf{r}_2)$, as

$$f_{12}(\mathbf{r}_1, \mathbf{r}_2) = f_1(\mathbf{r}_1) f_1(\mathbf{r}_2) \left[1 + g(\mathbf{r}_1, \mathbf{r}_2)\right]. \tag{15.71}$$

If the medium is homogeneous, it will only depend on the vector uniting the two particles $\mathbf{r} = \mathbf{r}_1 - \mathbf{r}_2$, and if the medium is also isotropic, it will only depend of its distance $r = |\mathbf{r}_1 - \mathbf{r}_2|$. In the limit of weakly correlated ions, such that $\Gamma_i \ll 1$, the correlation function $g(r)$ can be determined by

$$g(r) = -\frac{q_i^2}{T} \frac{e^{-r/\lambda_D}}{r}, \tag{15.72}$$

in accordance with the elementary calculations of Sect. 14.2. However, for strongly correlated ions, $\Gamma_i \geq 1$, the correlation function exhibits a weaker dependence on the temperature, becoming mainly dependent of the plasma density. Such functional dependence can be modeled by Ichimaru [13]

$$g(r) = \exp\left[-\frac{w(r)}{T_i}\right] - 1\,, \quad w(r) = \frac{q_i^2}{r} - h(r), \tag{15.73}$$

where $w(r)$ is the mean potential energy, and $h(r)$ represents the deviations with respect to the bare Coulomb interaction q_i^2/r. For this reason, we can call $h(r)$ the *screening potential*. An estimate of the screening effect can easily be made for the particular case of very short distances $r \ll a_i$, in the following way.

Let us define a two-ion sphere, with a total electron charge of $-2Ze$, and a radius given by $2^{1/3}a_i$ (see Fig. 15.1b). The corresponding screening potential is then given as the energy difference between the two-ion sphere and the potential of two separate single ion spheres. Using (15.68), we can easily obtain

$$h(r) = \frac{q_i^2}{a_i}\left[0.9\,(2^{5/3} - 2) + \left(\frac{r}{2a_i}\right)^2\right] \equiv \frac{q_i^2}{a_i}\left[1.057 - \frac{1}{4}\left(\frac{r}{a_i}\right)^2\right]. \tag{15.74}$$

On the other hand, for longer distances, $r \simeq a_i$, we can evaluate $h(r)$ by numerical methods, leading to the following result

$$h(r) = \frac{q_i^2}{a_i} \left[c_0 - c_1 \left(\frac{r}{a_i} \right) \right]. \tag{15.75}$$

This is valid for moderate coupling coefficients, such that $4 < \Gamma_i < 160$, and in the range of distances $0.4 < (r/a_i) < 1.8$, with $c_0 = 1.25$ and $c_1 = (c_0/2)^2 = 0.39$. Based on these results for the ion-ion interaction potential, we can devise a body centered cubic lattice model (bcc lattice), with cubic cells of half diagonal distance d, such that, at $r = d$, the potential interaction vanishes and attains a minimum, as given by the conditions

$$w(r = d) = 0, \quad \left(\frac{dw}{dr} \right)_{r=d} = 0. \tag{15.76}$$

Replacing (15.75) in the expression of $w(r)$, and using these assumptions, we get

$$c_0 = 2\frac{a_i}{d}, \quad c_1 = \left(\frac{c_0}{2} \right), \tag{15.77}$$

which are confirmed by the above mentioned numerical results.

Knowing the expression for $g(r)$, we can now define the excess internal energy of the system as

$$\frac{U_{\text{exc}}}{T_i} = \frac{n_i q_i^2}{2T_i} \int \frac{g(r)}{r} d\mathbf{r}. \tag{15.78}$$

This is in fact the average potential energy associated with the Coulomb interactions, or the correlation energy. For the weakly correlated case, $\Gamma_1 \ll 1$, we can use Eq. (15.72) and obtain

$$\frac{U_{\text{exc}}}{T_i} = -\frac{\sqrt{3}}{2} \Gamma_i^{3/2}. \tag{15.79}$$

For the strongly correlated case, this quantity has been evaluated by particle Monte-Carlo calculations. In the liquid phase, for moderate values of the coupling parameter $\Gamma_1 \geq 1$, we have [7, 29]

$$\frac{U_{\text{exc}}}{N_i T_i} = a\Gamma_i + b\Gamma_i^{1/3} + c, \tag{15.80}$$

with the parameters estimated as $a = -0.90$, $b = 0.59$, $c = -0.27$, valid in the range of $1 \leq \Gamma_i \leq 160$. One can recognize in the first term of this expression the contribution of the single ion sphere, as estimated in (15.70).

A different expressions for U_{exc} can be found if, instead of assuming that the ions form a liquid, we assume an bcc lattice. An expansion of U_{exc} in power series of $1/\Gamma_i$, of the form

$$\frac{U_{\text{exc}}}{N_i T_i} = a_{-1}\Gamma_i + a_0 + \frac{a_1}{\Gamma_i} + \frac{a_2}{\Gamma_i^2}, \tag{15.81}$$

with $a_{-1} = -0.896$, $a_0 = 1.5$ and $a_1 = 2.5 \times 10^3$, and $a_2 = 10.84$, valid in the range $160 < \Gamma_i < 300$. In the first two terms of this expression we can recognize the above results of Eq. (15.70), obtained with the simple ion sphere model.

We can determine the value of Γ_i for which phase transition from liquid to solid takes place, by computing the *Helmoltz free energy* F for these two phases. In a normalized form, this can be defined as

$$f(\Gamma_i) \equiv \frac{F(\Gamma_i)}{N_i T_i} = \int_0^{\Gamma_i} \frac{U_{\mathrm{exc}}}{N_i T_i} \frac{d\Gamma}{\Gamma}. \tag{15.82}$$

Using (15.80), we get for the liquid phase

$$f(\Gamma_i) = a\Gamma_i + 3b\Gamma_i^{1/3} + (3 + c)\ln \Gamma_i - [1.15 + a + 3b], \tag{15.83}$$

while for the solid phase, using (15.81), we obtain

$$f(\Gamma_i) = a_{-1}\Gamma_i + \frac{9}{2}\ln \Gamma_i - 1.89 - \frac{1{,}490}{\Gamma_i^2}. \tag{15.84}$$

The intersection between these two curves, (15.83) and (15.84), determines a critical point at $\Gamma_m \simeq 170$, where a liquid-solid phase transition due to the occurrence of a *Wigner crystallization* will occur. Other similar values have been quoted in the recent literature [7]. Let us finally discuss the total internal energy of the system, as given by

$$U = \frac{3}{2} N_i T_i + U_{\mathrm{exc}} \tag{15.85}$$

and the corresponding ion pressure, as determined by

$$P_i V = N_i T_i + \frac{1}{3} U_{\mathrm{exc}}, \tag{15.86}$$

where the first term corresponds to the ideal gas and the second accounts for the excess of internal energy associated with the particle interactions.

In contrast with the liquid-solid phase, which occurs for a well defined value of Γ_i, the gas-liquid transition is usually considered to be a smooth transition, over a quite broad band around $\Gamma_i \sim 1$, where the correlation function slowly changes its shape. A proposal for a more precise response was recently advanced [28]. The starting point is the expression for the total internal plasma energy, including the electron thermal energy and the electrostatic energy associated with deviations from the total charge neutrality. This takes the form [11]

$$U = \frac{3}{2}(N_i T_i + N_e T_e) + \frac{1}{2}\int_V \rho\psi(\mathbf{r}) d\mathbf{r} - U_{\mathrm{self}}, \tag{15.87}$$

where N_e and T_e are the total number and temperature of the electrons, ρ and $\psi(\mathbf{r})$ are the charge density and electrostatic potential as determined by

$$\nabla^2 \psi = -\frac{\rho}{\epsilon_0}, \quad \rho = (Ze) \sum_{j=1}^{N_i} \delta(\mathbf{r} - \mathbf{r}_j) - e n_e. \tag{15.88}$$

In Eq. (15.87), the term U_{self} represents the self-energy of the discrete set of ions, which has to be subtracted from the electrostatic energy term. From here, an equation of state for the ion pressure can be derived, which can be written as [28]

$$P_i = \frac{1}{6} n^{7/3} \left(4 + \frac{1}{n^{1/3}}\right) \exp\left(-\frac{1}{n^{1/3}}\right) - \frac{3}{2} n^2 + nT \tag{15.89}$$

where n and T are normalized ion density and temperature, as defined by

$$n = \frac{4\pi}{3} n_i \lambda_D^3, \quad T = T_i \frac{\lambda_D}{(Ze)^2}. \tag{15.90}$$

A critical point is attained if the following conditions are satisfied [5]

$$\left.\frac{dP}{d(1/n)}\right|_T = 0, \quad \left.\frac{d^2 P}{d(1/n)^2}\right|_T = 0, \quad \left.\frac{d^3 P}{d(1/n)^3}\right|_T < 0. \tag{15.91}$$

This can be satisfied for the critical values of $n = n_c \equiv 6.66$ and $T = T_c \equiv 6.65$, which correspond to the following values of the critical parameter: $\Gamma_i = \Gamma_c \equiv 0.3$ and $\kappa = \kappa_c \equiv 0.53$. We can then say that, for super-critical ion temperatures $T > T_c$, the ion fluid will behave as a gas and, for sub-critical temperatures $T < T_c$ as a liquid. This range of parameters is attained in current experiments on ultra-cold plasmas.

15.4 Disorder Induced Heating

As we have seen, the ultra-cold plasma can be created in a short time scale by photoionization of the atomic cloud. Initially, the ions are created with a very low temperature, of the order of the temperature of the laser cooled atoms, $T_i(0) \simeq T_a$. This corresponds to a very high coupling parameter, typically $\Gamma_i \geq 100$. It means that the potential interaction energy between the ions is much larger than their average kinetic energy. As a consequence, the ions will rapidly start to move, in a disordered way, leading to the increase of the ion temperature on a time scale of the order to the inverse of the ion plasma frequency ω_{pi}. Such a disordered motion corresponds to transferring the excess of potential energy into kinetic energy, leading to an increase of the effective ion temperature. This is the so called *disorder induced heating* [24]. On a latter stage of the plasma evolution, such that $t \gg 1/\omega_{pi}$, the ions are characterized by a much smaller value of the ion coupling parameter, of the order of unity $\Gamma_i \simeq 1$ [15]. The final temperature can be estimated by assuming equipartition at the asymptotic equilibrium state. A simple model for such an equilibrium state will be described in the next Section. Assuming than that the final kinetic and potential ion energies are nearly equal, we get for the final ion temperature T_f the value of

$$T_f \simeq \frac{2}{3} \frac{(Ze)^2}{4\pi\epsilon_0 a_i} \equiv \frac{2}{3} \frac{q_i^2}{a_i}. \tag{15.92}$$

Of course, this disordered heating process would not occur if the ions were initially created in a stable configuration, where every ion would be seated on a local potential minimum.

In order to study this heating process, a molecular dynamic approach can be used, where the equations of motion of the ions are solved numerically and the average kinetic energy is used to determine the effective ion temperature. The ion equations of motion can be established by using the total Hamiltonian of the system. Before ionization, for $t < 0$, the centre-of-mass motion of the neutral atoms in the ultra-cold cloud can be determined by the classical Hamiltonian

$$H_a = \sum_{j=1}^{N_a} \frac{p_j^2}{2M}, \tag{15.93}$$

where $\mathbf{p}_j$ are the momenta of the neutral atoms with mass M. After ionization, for $t > 0$, and assuming that most of the atoms are ionized $N_i \simeq N_a$, the dynamical state of the ion cloud is determined by the new Hamiltonian

$$H_i = \sum_{j=1}^{N_i} \frac{p_j^2}{2m_i} + \sum_{j<k}^{N_i} \frac{q_i^2}{r_{jk}} \exp\left(-\frac{r_{jk}^2}{\lambda_D^2}\right) \tag{15.94}$$

where $m_i \simeq M$, and the Yukawa interaction between ions is assumed, with a screening length given by λ_D. The ion equations of motion will therefore be

$$\frac{d\mathbf{r}_j}{dt} = m_i \mathbf{v}_j \ , \quad m_i \frac{d\mathbf{v}_j}{dt} = -\frac{\partial H_i}{\partial \mathbf{r}_j} \equiv \mathbf{F}_j, \tag{15.95}$$

which can be formally solved as

$$\mathbf{v}_j(t) = \frac{1}{m_i} \int_0^t \mathbf{F}_j(t')dt'. \tag{15.96}$$

The effective ion temperature can then be defined as

$$T_i(t) = \frac{m_i}{3N_i} \sum_{j=1}^{N_i} \left\langle v_j^2(t) \right\rangle. \tag{15.97}$$

In the early stages of the heating process, the ion acceleration is nearly constant, and the ion temperature can be determined by a temporal expansion

$$T_i(t) \simeq T_i(0) + \left(\frac{t}{\tau_2}\right)^2 + \ldots, \tag{15.98}$$

were the constant τ_2 can be determined by

$$\tau_2 = \frac{3}{\sqrt{\langle F^2 \rangle}}, \quad \langle F^2 \rangle = \sum_{j=1}^{N_i} \int_0^t dt' \int_0^t dt'' \langle \mathbf{F}_j(t') \cdot \mathbf{F}_j(t'') \rangle \tag{15.99}$$

Numerical simulations [24] estimate this constant as a function of the parameter $\kappa = a_i / \lambda_D$, as

$$\tau_2 = \frac{3}{(33 - 4\kappa + 0.1\kappa^2)^{1/2}}. \tag{15.100}$$

Using appropriate units, and assuming that the initial ion temperature is negligible $T_i(0) \simeq 0$, we can describe the early stages of the disorder induced heating by the expression

$$T_i(t) \simeq \left(\frac{q_i^2}{a_i}\right) \left(\frac{\omega_{pi} t}{\tau_2}\right)^2. \tag{15.101}$$

We can see that the initial acceleration decreases for a stronger screening parameter κ, as should be expected, and is confirmed by experiments. For longer times, we can extrapolate these results and write

$$T_i(t) = \frac{2}{3} \frac{q_i^2}{a_i} \left[1 - \exp\left(-\frac{3\omega_{pi}^2 t^2}{2\tau_2^2}\right)\right]. \tag{15.102}$$

This expression reduces to (15.101) for short times $t \ll \tau_2$, and tends asymptotically to the final temperature on a time scale of the order of $1/\omega_{pi}$. This is qualitatively in agreement with the observations. In addition, a strongly damped oscillation on the same time scale is also observed in the experiments. A consistent theoretical model of the disorder induced heating is still missing. In particular, kinetic and fluid plasma description could be envisaged, in alternative to the molecular dynamics approach, and eventually lead to a more explicit analytical description of the ion heating process.

15.5 Quasi-equilibrium States

Let us now discus the possible form of electrostatic equilibrium that can be established in the ultra-cold plasma cloud, assuming for simplicity that we stay in the weakly coupling regime, $\Gamma_i \ll 1$, and neglecting the contributions coming from the ion-ion coupling. This is marginally valid for the quasi-equilibrium conditions which follow the fast process of disorder-induced heating. The main aim here is to show that significant electron trapping can occur, leading to a plasma equilibrium profile which is very similar to the well-known Thomas-Fermi profile for heavy atoms [16]. When the trapped electron population dominates over the "free" electrons, the Rydberg plasma qualitatively looks like a giant atom, where the ions play the role of a nucleus and the electrons compose the electronic cloud.

In the early stages of the plasma formation by photoionization, the ions are approximately described by a gaussian profile, which, taking $Z = 1$, reads

$$\nabla^2 V = \frac{e}{n_0}(n_e - n_i) \,, \quad n_i(r) = n_0 e^{-(r/a)^2} \,, \tag{15.103}$$

where a is the size of the ion cloud. This ion profile will create a positive electrostatic potential $V > 0$, such that an electron with velocity $\mathbf{v}$, will have an energy given by

$$E_e(\mathbf{r}, \mathbf{v}) = \frac{1}{2} m_e v^2 - eV(\mathbf{r}). \tag{15.104}$$

We immediately notice that this energy can be negative, leading to electron trapping in the collective ion potential. Such trapping occurs for the slowest (or coldest) electrons of the distribution, with velocities satisfying the condition $v < v_t$, where the trapping velocity is defined as

$$v_t = \sqrt{\frac{2e}{m_e}|V|}. \tag{15.105}$$

Putting Eqs. (15.103)–(15.105) together, and defining $\phi = e|V|/T_e$, we can write

$$\nabla^2 \phi = \frac{1}{\lambda_D^2}\left[\frac{n_e}{n_0} - e^{-(r/a)^2}\right] , \tag{15.106}$$

where $\lambda_D = v_{\text{the}}/\omega_{pe}$, and $\omega_p = (e^2 n_0/\epsilon_0 m_e)^{1/2}$ is the electron plasma frequency and $v_{\text{the}} = \sqrt{T/m_e}$ is the electron thermal velocity. The free electrons (or, more precisely, the untrapped electrons with $v_e > v_t$) follow the Boltzmann distribution associated with the energy (15.104). In contrast, the trapped electrons approximately follow an uniform distribution, since we assume that they cannot leave the trapping radius $R \sim a$ (at least during the early stages of the plasma evolution and before a significant expansion takes place). As a consequence of these two different electron populations, the electron density is now be determined as follows [18]

$$\frac{n_e}{n_0} = \frac{4}{\sqrt{\pi}}\left[\int_0^{u_t} u^2 du + \int_{u_t}^{\infty} e^{-(u^2-\phi)} u^2 du\right] , \tag{15.107}$$

where we have used the dimensionless velocity variable $u = v/v_{\text{th,e}}$. Using Eqs. (15.106) and (15.107), we can obtain a general expression for the electrostatic potential that casts the effects of the electron trapping, as

$$\nabla^2\phi = \frac{1}{\lambda_D^2}\left[\frac{4}{3\sqrt{\pi}}\phi^{3/2} - f(\phi) - e^{-(r/a)^2}\right], \tag{15.108}$$

where the function $f(\phi)$ is defined as

$$f(\phi) = e^{\phi}\left(1 - \frac{4}{\sqrt{\pi}}\int_0^{\sqrt{\phi}} e^{-u^2}u^2 du\right). \tag{15.109}$$

Equation (15.108) is very difficult to solve in general. Fortunately, approximate expressions can be provided in some limiting cases. First, for a weak trapping potential. $\phi \ll 1$, we can use the approximation

$$f(\phi) \approx 1 - \frac{4}{3\sqrt{\pi}}\phi^{3/2} + \phi - \frac{8}{15\sqrt{\pi}}\phi^{5/2} \tag{15.110}$$

and the potential equation becomes

$$\nabla^2\phi = \frac{1}{\lambda_D^2}\left[1 + \phi - \frac{8}{15\sqrt{\pi}}\phi^{5/2} - e^{-(r/a)^2}\right]. \tag{15.111}$$

More interesting for ultra-cold plasmas is the strong confinement case, $\phi \gg 1$, where we have $f(\phi) \approx 0$. The potential equation reduces to

$$\nabla^2\phi = \frac{1}{\lambda_D^2}\left[\frac{4}{3\sqrt{\pi}}\phi^{3/2} - e^{-(r/a)^2}\right]. \tag{15.112}$$

It should be noticed that, in this strong confinement regime, the potential equation is very similar to that of the Thomas-Fermi potential obtained for heavy atoms. The only difference is rooted in the fact that here the ions are not homogeneously distributed, and therefore we have included the ion inhomogeneity.

In order to solve Eq. (15.112) numerically, we define the dimensionless variable $\rho = r/\lambda_D$ and the reduced potential $\psi = \rho\phi$. Notice that ψ represents the potential relative to the Coulomb potential $\phi_{\text{Coul}} \sim 1/\rho$, such that $\psi = 1$ for an unscreened plasma. Here, we restrict the discussion to a spherically symmetric plasma, such that the Laplacian operator contains only the radial derivatives. Due to trapping, the resulting potential significantly differs from the Coulomb case, as it can be observed in Fig. 15.2. The corresponding electron profile is not Gaussian. Indeed, the numerical simulations reveal that n_e decays very quickly as a function of r, and it turns out that most of the electrons are trapped inside the radius $R \sim a$ defined by the width of the ion profile n_i. Such results suggest that the ions can efficiently trap a significant fraction of the electron population inside their cloud.

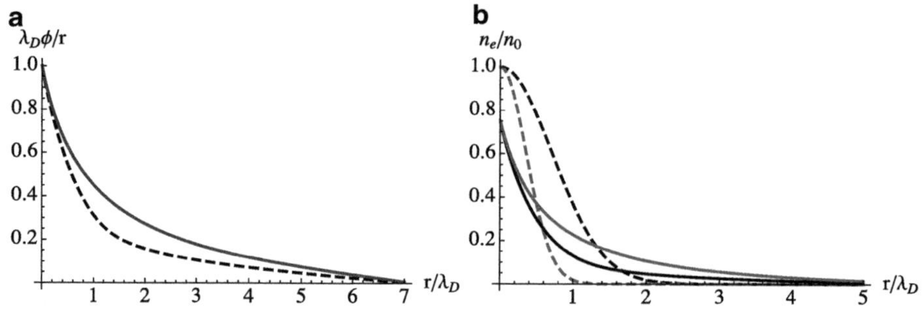

Fig. 15.2 (**a**) Thomas-Fermi potential obtained in the strong trap regime $\phi \gg 1$ for two different ion gaussian profiles: $a = \lambda_D$, (*dashed line*) and $a = 0.5\lambda_D$ (*full line*). (**b**) Ion (*dashed lines*) and electron (*full lines*) profiles in the early stages of a Rydberg plasma: $a = \lambda_D$ (*black lines*) and $a = 0.5\lambda_D$ (*gray lines*)

Because a fraction of the electron population is sufficiently cold to remain trapped inside the ion cloud, the Boltzmann statistics does not hold generally. Casting the effects of the electron trapping, we derived a nonlinear equation to describe the potential which reduces to the case of a Thomas-Fermi potential obtained in a different context to describe heavy atomic systems in the semi-classical regime. This discussion shows that a more realistic description of the plasma expansion process should include electron trapping effects, and use these Thomas-Fermi density and potential profiles, instead of the Gaussian profiles considered in the usual expansion models.

15.6 Rydberg Atoms

15.6.1 Basic Properties

Rydberg atoms are atoms in highly excited states, with one electron with a very large principal quantum number, $n \gg 1$, very close to ionization. In such a configuration, the core electrons provide an electronic shielding of the nucleus, in such a way that the excited electron perceives an effective nuclear charge of $+e$, like the electron in the Hydrogen atom. The energy quantum levels of a Rydberg atom can therefore be determined by an expression similar to that of Hydrogen, of the form [9]

$$E_n = -\frac{\mathrm{R_y}}{(n-\delta)^2}, \quad \mathrm{R_y} = 13.6 \quad \mathrm{eV}, \tag{15.113}$$

where $\mathrm{R_y}$ is the Rydberg constant, and the quantity δ is the *quantum defect.* The latter quantifies the energy correction associated with the fact that the orbits of the outer electron are not circular. As a result, even for $n \gg 1$, this electron will pass through the inner regions of the atom, which is occupied by the shielding electrons.

This quantum defect depends on the angular momentum state of atom, and is larger for strong deviations from a circular orbit. The extreme case corresponds to the s-states, which have zero angular momentum, and for which we have $\delta \sim 5 - 7$. For instance, for Caesium, we have $\delta \simeq 5.1$ for s-state, and $\delta \simeq 2.5$ for the d-state.

The size of the Rydberg atoms is also very large with respect with atoms in their ground level, or low n excited states. For instance, at $n = 40$, it can attain 200 nm, more than 10^3 times larger than a ground state atom. Due to this large size, the outer electron orbit becomes quasi-classical, and the old *Bohr model* of the atom can efficiently be used. For a circular orbit of radius r, the shielded nucleus with charge $+e$ will create an electrostatic force, which is balanced by the centrifugal kinetic force of the outer electron motion, allowing us to write

$$\frac{e^2}{4\pi\epsilon_0}\frac{1}{r^2} = \frac{m_e v^2}{r}. \tag{15.114}$$

Using the Bohr quantization for the electron orbital motion, we can write $n\hbar = m_e v r$, where n is an integer. This allows us to calculate the value of the quantized radial states $r = r_n$, as a function of the principal quantum number n, as

$$r_n = \frac{4\pi\epsilon_0}{e^2}\frac{n^2\hbar^2}{m_e}. \tag{15.115}$$

This shows that the orbit size scales as n^2, and the geometric cross-section of the atom as n^4. For $n = 1$, we recover the value of the Bohr radius $a_0 = 4\pi\epsilon_0\hbar^4/e^2 m_e \simeq 0.529$ A.

As an interesting example of application of this quasi-classical description, let us consider the classical *radiative cascade* of Rydberg states [8]. This occurs when a Rydberg atom is formed and then radiates down to lower excited states. We consider a Rydberg atom initially in a state $n \gg 1$, in nearly circular orbit, to which corresponds a high angular momentum, such that $l \simeq n$. From a classical point of view, the instantaneous power radiated by an accelerated particle with charge e, and velocity v, is given by the *Larmor formula*

$$P_{\rm rad} = \frac{e^2}{6\pi\epsilon_0 c^3}\left|\frac{dv}{dt}\right|^2. \tag{15.116}$$

Using this formula, we can determine the energy lost by an electron moving in elliptic orbit around the nucleus, as given by [17]

$$P_{\rm rad} = \frac{(2|E|)^{3/2} m_e^{1/2} e^5}{c^3 L^5}\left(1 - \frac{2|E|L^2}{e^2 m_e}\right), \tag{15.117}$$

where E can be identified with the energy of the Rydberg atom, as given by Eq. (15.113), and $\mathbf{L} = m_e \mathbf{v} \times \mathbf{r}$ is the angular momentum. By noting that in a quantized orbit, $L^2 = \hbar^2 l(l+1)$, and assuming $l \gg 1$, we can use $L \simeq \hbar l$. Neglecting the quantum defect for simplicity, we can rewrite the above equation in terms of the atomic quantum numbers (n, l), as

$$P_{nl} = \frac{(2\mathrm{R_y})^{3/2} m_e^{1/2} e^5}{n^3 c^3 \hbar^5 l^5} \left(1 - \frac{2\mathrm{R_y} \hbar^2 l^2}{n^2 e^2 m_e}\right). \tag{15.118}$$

This energy radiation process will make the atom decay into a lower energy state, which can be determined by the relation

$$P_{nl} = -\frac{d|E|}{dt} = \frac{\mathrm{R_y}}{n^3} \frac{dn}{dt}. \tag{15.119}$$

The temporal evolution of the quantum number of the atom is then determined by the expression

$$\frac{dn}{dt} = -\left(1 - \frac{l^2}{3n^2}\right) \frac{1}{\tau_{\mathrm{rad}} l^5}, \quad \tau_{\mathrm{rad}} = \frac{\hbar^5 c^3}{(2\mathrm{R_y})^{3/2} m_e^{1/2} e^5}, \tag{15.120}$$

where τ_{rad} is the characteristic time for radiative decay. Going back to the classical radiation theory, we know that the rate of change of the angular momentum of the radiating charge is given by

$$\frac{d\mathbf{L}}{dt} = -\frac{2e^2}{3c^3} \left(\mathbf{v} \times \frac{d\mathbf{v}}{dt}\right). \tag{15.121}$$

From this we get, for the elliptic electron orbit around the nucleus a decay rate [17]

$$\frac{dL}{dt} = -\frac{2e^4 (2|E|)^{3/2} m_e^{1/2}}{3c^3 L^2}. \tag{15.122}$$

Introducing Bohr quantization, we can then obtain the rate of change for the angular momentum state, as

$$\frac{dl}{dt} = -\frac{2}{3} \frac{1}{\tau_{\mathrm{rad}} l^2 n^3}. \tag{15.123}$$

Comparing with Eq. (15.120), and taking the ratio between dn/dt and dl/dt, we get

$$\frac{dn}{dl} = \frac{3}{2} \frac{n^3}{l^3} \left(1 - \frac{l^2}{3n^2}\right). \tag{15.124}$$

Solving the latter for $n(l)$, one finds

$$n^2 = \frac{l^2}{1 - C_0 l^3}, \quad C_0 = \frac{1}{l_0^3} \left(1 - \frac{l_0^2}{n_0^2}\right), \tag{15.125}$$

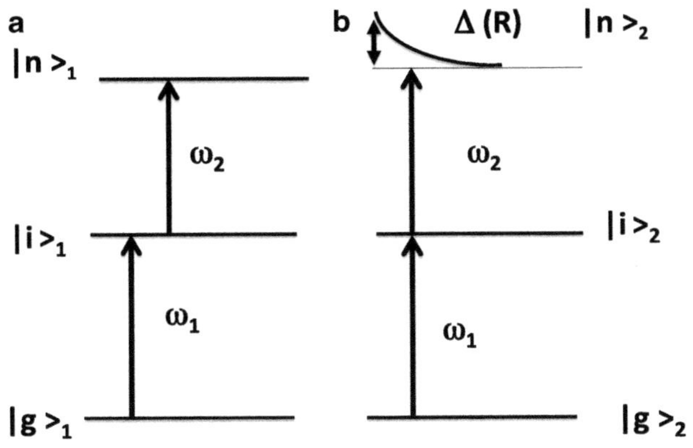

Fig. 15.3 Energy levels of two nearby atoms: if atom 1 is excited by a two-photon transition, from the ground level $|g\rangle_1$ to the Rydberg state $|n\rangle_1$, atom 2 located at a distance R will not be excited into the same state by the same photon transitions, due to the energy shift $\Delta(R)$

where the constant C_0 is determined by the initial conditions, $(n = n_0, l = l_0)$. This constant clearly depends on the initial angular momentum, and can vary between $C_0 = 0$, for $l_0 = n_0$, which would correspond to a circular electron orbit, to the maximum value of $C_0 = 1$, characterizing very elongated electron orbits, with $l_0 = 1 \ll n_0$. It is clear from this result that initial circular orbits decay along the diagonal line $n^2 = l^2$ in the (l, n) phase space, while orbits with $l_0 < n_0$, or $C_0 > 0$, decay along the line defined by (15.125), which therefore characterized the quantum cascading of highly excited Rydberg atoms. This quasi-classical theory of radiative decay, based on the simple Bohr quantization method, was shown to be accurate for atomic quantum numbers as low as $n \geq 10$ and $(l/n) \geq 0.1$ [8].

15.6.2 Rydberg Blockade

Rydberg atoms are characterized by having very strong interactions due to their large dipolar electric moment. The interaction between two Rydberg atoms is dominated by van der Waals forces at short distances, scaling as $1/R^6$, where R is the interatomic distance. For long distances, dipole-dipole interactions are dominant, and scale as $1/R^3$. For instance, the transition distance R_c between the van der Waals short-range and the dipole-dipole long-range interactions is of the order of $10\,\mu$m, for a Rubidium atom excited up to the $n = 100$ Rydberg state. At this critical distance, the interaction energy between such states is 10^{12} times larger than that of a magnetic dipole-dipole interaction between two ground state atoms [27]. As a consequence, energy shifts of the atomic levels induced by two Rydberg atom interactions will be significant, as illustrated in the Fig. 15.3. This means that, if one atom is excited into such a Rydberg state by some resonant photon transition, a second nearby atom cannot be excited by the same photon transitions, because its Rydberg energy states were shifted due to the interaction with the first atom. Experimental observation of such a blockade process was demonstrated up to distances of 10 μm for Rb atoms excited to the $79d_{5/2}$ level [31].

In order to understand this blockade mechanism, let us consider the dipole-dipole interaction between two atoms, 1 and 2, located at positions $\mathbf{r}_1$ and $\mathbf{r}_2$. Assuming that the interatomic distance $R = |\mathbf{r}_1 - \mathbf{r}_2|$ is much larger than the atomic size, or $R \gg n^2 a_0$, where n is the principal quantum number of the Rydberg state and a_0 is the Bohr radius, we can write the interaction potential as

$$V_{dd} = \frac{e^2}{R^3}\left(\mathbf{r}_1 \cdot \mathbf{r}_2 - 3(\mathbf{r}_1 \cdot \mathbf{R})(\mathbf{r}_2 \cdot \mathbf{R})\right). \tag{15.126}$$

The atoms are assumed excited at the same quantum level, represented by the quantum numbers (nlj), and therefore the asymptotic expression for the two-atom state vector $|\psi_{12}\rangle$ at long distances $R \to \infty$ is simply given by

$$|\psi_{12}\rangle = |\psi_1 \psi_2\rangle = \left|\psi_{nlj}\psi_{nlj}\right\rangle. \tag{15.127}$$

In the absence of external fields, this state is $(2j + 1)^2$ degenerate, where j is the total electronic angular momentum. The dipole-dipole interaction induces transitions to other two-atom states, where the angular momentum of each electron is determined by the usual dipole selection rules $(l_1, l_2 = l \pm 1)$ and $(j_1, j_2 = j \pm 0, 1)$. Although there is an infinite number of such states, the dipole-dipole interaction is dominated by a small number of the closest two-atom states. For instance, for Rubidium at zero temperature, the state $\left|60_{p1/2}\, 60_{p3/2}\right\rangle$ is mainly coupled to the state $\left|60_{s1/2}\, 61_{s1/2}\right\rangle$. This means that we can consider the long-range interactions between two Rydberg atoms to be established between the channels (nlj, nlj) and $(n_1 l_1 j_1, n_2 l_2 j_2)$, with an energy defect given by (in units of $\mathrm{R_y}$)

$$\delta = E(n_1 l_1 j_1) + E(n_2 l_2 j_2) - 2E(nlj) \quad (\mathrm{R_y}). \tag{15.128}$$

If we represent the (nlj, nlj) state by the state vector $|\varphi\rangle$, and the $(n_1 l_1 j_1, n_2 l_2 j_2)$ state by $|\chi\rangle$, the time-independent Schrödinger equation for the interaction can be written as

$$\begin{bmatrix} \delta & V_{dd} \\ V_{dd}^{\dagger} & 0 \end{bmatrix} \begin{bmatrix} |\chi\rangle \\ |\varphi\rangle \end{bmatrix} = \Delta \begin{bmatrix} |\chi\rangle \\ |\varphi\rangle \end{bmatrix}, \tag{15.129}$$

where Δ is the expected energy shift. This leads to the two following expressions

$$V_{dd}^{\dagger}\,|\chi\rangle = \Delta\,|\varphi\rangle\;, \quad V_{dd}\,|\varphi\rangle = (\Delta - \delta)\,|\chi\rangle\,, \tag{15.130}$$

from which the eigenvalue equation for $|\varphi\rangle$ can be extracted as

$$H_{dd}\left|\varphi\right\rangle = \Delta\left|\varphi\right\rangle, \quad H_{dd} = \frac{V_{dd}^{\dagger} V_{dd}}{(\Delta - \delta)}. \tag{15.131}$$

As we can see from Eq. (15.126), the matrix elements of $V_{dd}^{\dagger} V_{dd}$ vary with distance as $1/R^6$, corresponding to a van der Waals type of interaction, and we can expect the corresponding eigenvalues to behave with distance as C_3^2/R^6, where C_3 is the appropriate coefficient. Solving the eigenvalue equation (15.131), we then get for the $\Delta \equiv \Delta(R)$, the expression

$$\Delta(R) = \frac{\delta}{2}\left[1 - s_\delta\sqrt{1 + \frac{4C_3^2}{\delta^2 R^6}}\right], \tag{15.132}$$

s_δ is the sign of δ. This determines the frequency shifts due to dipole-dipole interactions as a function of the distance between the two atoms, such that for large enough distances ($R \to \infty$) the two-atom eigenstates will reduce to the (nlj, nlj) state. It is useful to define a critical distance $R_c = \sqrt{C_3}(2/\delta)^{1/3}$ such that the fraction inside the square root in this equation becomes equal to one, i.e. the distance for which the van der Walls potential equals the energy defect. At large inter-atomic distances such that $R \gg R_c$, the energy shift will behave as $1/R^6$, characteristic of the van der Waals regime, while for short distances $R \ll R_c$ the dipole-dipole $1/R^3$ dependence will become dominant, a shown by the asymptotic expressions

$$\Delta(R \gg R_c) \simeq \frac{C_3^2}{\delta R^6}, \quad \Delta(R \ll R_c) \simeq -s_\delta \frac{C_3}{R^3}. \tag{15.133}$$

The $1/R^3$ term will therefore represent the largest possible interaction energy between two non-overlapping Rydberg atoms. This scales typically as n^{11} [2, 33].

Let us now consider a cloud of N atoms with ground state $|g\rangle$ and Rydberg states $|\gamma\rangle$, where g and γ represent here the set of internal quantum numbers. And let us assume that the previously considered blockade mechanism is so efficient that only one atom, or at most two atoms, in the entire cloud can be excited in a Rydberg state. This means that the atomic cloud can only be found among the following three possible states: (i) the state $|\psi(g)\rangle$, corresponding to all the atoms in the ground state, (ii) the state $|\psi_k(\gamma)\rangle$, where only the k-th atom is in the Rydberg state γ, and (iii) the state $|\psi_{kl}(\gamma)\rangle$ where both the k-th and the l-th atoms are in the Rydberg state. In the latter, the two Rydberg atoms will interact according to the eigenvalue equation (15.131).

The interaction of such an atomic cloud with photons, assuming that they are resonant with a given atomic transition onto the considered Rydberg energy state $|\gamma\rangle$ can be described in the following way. The photon coupling with the kth-atom is described by the single atom Rabi frequency

$$\Omega_k = \frac{2}{\hbar}\left\langle\psi_k(\gamma)\right| H_k \left|\psi(g)\right\rangle, \tag{15.134}$$

where H_k is the Hamiltonian for electric dipole interactions. We can also define a *collective Rabi frequency* associated with the cloud of N atoms as

$$\Omega_N = \sqrt{\sum_k |\Omega_k|^2} = \Omega\sqrt{N}, \tag{15.135}$$

where Ω is the averaged single-atom Rabi frequency. The state vector of the cloud, $|\psi\rangle$, can now be represented as a superposition of the three allowed states, in the form of

$$|\psi\rangle = c_g\,|\psi(g)\rangle + c_s\,|\psi_s\rangle + \sum_{k<l} c_{kl}\,|\psi_{kl}(\gamma)\rangle\,, \tag{15.136}$$

where c_g, c_s and c_{kl} are appropriated coefficients, and the state vector $|\psi_s\rangle$ is the normalized singly excited state, as defined by

$$|\psi_s\rangle = \frac{1}{\Omega}\sum_k \Omega_k\,|\psi_k(\gamma)\rangle\,. \tag{15.137}$$

Now, inserting this in the time-dependent Schrödinger equation, we can derive the evolution equations for the atom populations, as determined by the following generalized Bloch equations

$$\begin{aligned} i\frac{dc_g}{dt} &= \frac{1}{2}\Omega_N^* c_s \\ i\frac{dc_s}{dt} &= \frac{1}{2}\Omega_N c_g + \frac{1}{N}\Omega_N^*\sum_{k<l}\kappa_{kl}^* c_{kl} \\ i\frac{dc_{kl}}{dt} &= \Delta_{kl}c_{kl} + \frac{1}{N}\Omega_N\kappa_{kl}c_s, \end{aligned} \tag{15.138}$$

where we have used the matrix elements

$$\kappa_{kl} = \frac{4}{\hbar^2\Omega^2}\,\langle\psi_{kl}(\gamma)|\,H_k H_l\,|\psi(g)\rangle\,. \tag{15.139}$$

Let us then assume that we start with all the atoms in the ground state, $|\psi\rangle = |\psi(g)\rangle$, and apply a pulse of photons with a duration such that $\Omega_N t = \pi$ (also known as π-pulse). If the blockade effect is active, the doubly excited amplitudes will always stay very small, and we can take the approximately constant value $c_s \simeq 1$. From here, we can easily integrate the last of the above Bloch equations, allowing us to conclude that

$$c_{kl} = -\frac{\Omega_N\kappa_{kl}}{N\Delta_{kl}}c_s. \tag{15.140}$$

The resulting probability for double Rydberg atom excitation inside the cloud is then given by

$$P_2 = \sum_{k<l} |c_{kl}|^2 = \frac{|\Omega_N|^2}{N^2} \sum_{k<l} \frac{|\kappa_{kl}|^2}{\Delta_{kl}^2}. \tag{15.141}$$

This can be written in a more suggestive way by introducing the definition of a *mean blockade shift* B, such that

$$\frac{1}{B^2} = \frac{2}{N(N-1)} \sum_{k<l} \frac{|\kappa_{kl}|^2}{\Delta_{kl}^2}, \tag{15.142}$$

in terms of which the probability for double Rydberg excitation simply follows

$$P_2 = \frac{N-1}{N} \frac{|\Omega_N|^2}{2B^2}. \tag{15.143}$$

We then arrive at the conclusion that, for a given collective Rabi frequency Ω_N, this probability is nearly independent of the number of atoms in the cloud, and only depends on the mean blockade shift B. But, of course, the blockade mechanism can only work if the interatomic distance R is not too large, and it will vanish at infinity. This means that there will be a domain of effective blockade inside the cloud. This leads to the concept of a *blockade sphere* [30].

We can assume that, due to the blockade mechanism, in a large volume of atoms with size R_{db} only one Rydberg atom exists, prevents the other atoms in the ground state to be excited into the same Rydberg state. Noting that the N_d atoms inside this volume are identical and indiscernible, they can be seen as forming a single gigantic atom, or a *superatom* [12, 32], which interacts with the photon radiation through an effective Rabi frequency, which is enhanced by a factor of $\sqrt{N_b}$ with respect to the Rabi frequency of a single atom, as shown by Eq. (15.135). The blockade radius can be estimated by comparing the collective Rabi frequency with the energy shift associated with the dipole-dipole interaction [19], as

$$\sqrt{n_a R_{\text{db}}^3}\Omega \simeq V(R_b), \tag{15.144}$$

where n_a is the atom density. For a van der Waals interaction as considered above, the maximum possible density of Rydberg atoms that can be formed inside the atomic cloud is given by

$$n_{\text{db}} \sim \frac{1}{R_{\text{db}}^3} \propto \left(\frac{n_a \Omega}{C_3^2} \right)^{1/5}. \tag{15.145}$$

This means that the fraction of excited Rydberg states will decrease with the atom density n_a and increase with the radiation intensity I as

$$\frac{n_{\text{db}}}{n_a} \propto n_a^{-4/5} I^{1/5}. \tag{15.146}$$

Such a dependence can be used as an experimental signature of the formation of superatoms due to the blockade mechanism.

15.7 Three-Body Recombination

Three-body collisions is the dominant recombination process for Rydberg atom formation in the ultra-cold expanding plasma. Progress in this area has been mainly driven by research on anti-matter. Several collaborations have been established at CERN [15], aiming at the creation and confinement of anti-Hydrogen. In the existing anti-matter devices, strong magnetic field are used in a *Penning trap*, to confine the ultra-cold anti-matter plasma. The Penning trap is composed of an axial magnetic field $\mathbf{B}_0 = B_0\mathbf{e}_z$, and an electrostatic quadrupole field, $\boldsymbol{\mathcal{E}}_0 = -\nabla V_0(\mathbf{r})$, as described the potential energy

$$U_0(\mathbf{r}) \equiv qV_0(\mathbf{r}) = \frac{1}{2}m\omega^2\left(z^2 - \frac{r_\perp^2}{2}\right), \tag{15.147}$$

where ω is the trap frequency. The charged particles with charge q and mass m are confined inside the trap by following trajectories that can be described as a superposition of three different oscillations: (i) the fast cyclotron motion, slightly perturbed by the electrostatic field, with the characteristic frequency

$$\omega_c' = \frac{\omega_c}{2}\left(1 + \sqrt{1 - 2\frac{\omega^2}{\omega_c^2}}\right) \simeq \omega_c\left(1 - \frac{\omega^2}{2\omega_c^2}\right), \tag{15.148}$$

with $\omega_c = |q|B_0/m$; (ii) an oscillation along the z-axis, due to the electric quadrupole axial confinement, at a frequency $\omega \ll \omega_c$; and (iii) a very slow *magnetron rotation* in the perpendicular plane, associated with the electric drift velocity $\mathbf{v}_E = \boldsymbol{\mathcal{E}}_0 \times \mathbf{B}_0/B_0^2$, with the characteristic angular frequency

$$\omega_{E\times B} = \frac{\omega_c}{2}\left(1 - \sqrt{1 - 2\frac{\omega^2}{\omega_c^2}}\right) \simeq \frac{\omega^2}{2\omega_c}. \tag{15.149}$$

We therefore have the angular frequency hierarchy, $\omega_c \gg \omega \gg \omega_{E\times B}$. See [20] for more details. Cold Rydberg atoms created and confined in such an environment have quite unique properties, which have been reviewed by Pohl et al. [25]. However, they retain the characteristic temperature dependence $T_e^{-9/2}$ for the three-body recombination rate in the ultra-cold plasma, as will be explained next. The reason for such a temperature scaling can easily be understood on the basis of the following qualitative arguments. First, the typical electron separation distance b, for a given kinetic energy T_e (in energy units), is of the order of the electron Landau distance, $b \sim r_e$ which scales as $1/T_e$ as shown before. Second, the electron-ion collision frequency for an impact parameter of the order of b is $\nu_{ei} \simeq n_e v_{\text{th,e}} b^2$. Third, the probability for a second electron to be close to the colliding electron and to receive an energy of T_e is proportional to be number of atoms in the collision volume, or $n_e b^3$. The three body recombination rate is therefore approximately given by the product

$$R_{\text{tbr}} \sim \left(n_e v_{\text{the}} b^2\right)\left(n_e b^3\right) \propto \frac{T_e^{1/2}}{T_e^5} = T_e^{-9/2}. \tag{15.150}$$

This strong dependence on the electron temperature is the most striking feature of recombination in the ultra-cold plasma and determines the creation rate of the Rydberg atom population.

The ultra-cold plasma can be assumed as strongly magnetized, because the electron Larmor radius $r_{ce} = v_{\text{the}}/\omega_{ce}$ is much smaller than the electron Landau distance, $b \sim e^2/T_e$, which is the mean inter-particle distance. This means that the cyclotron frequency is very large as compared with the other characteristic frequencies of the electron motion, allowing us to average over the cyclotron motion in order to look at the electron guiding-centre motion, as discussed in the previous chapter. For the Rydberg energy range of the atoms, which is slightly lower than the atom ionization energy E_I, but much larger than the electron thermal energy $T_e \ll E_I$, the magnetized Rydberg atom can be seen as a *guiding-centre atom*, where the electron guiding-centre moves along the magnetic field lines in the Coulomb field of the ion, and also moves slowly in the perpendicular plane around the ion due to the $\mathcal{E} \times \mathbf{B}_0$ drift. This means that the guiding-centre atom is a kind of microscopic Penning trap, where the axial confinement os imposed by the local ion Coulomb field $\mathcal{E}$. The corresponding to a confinement frequency is $\omega \sim v_{\text{the}}/b \sim \sqrt{e^2/m_e b^3}$, and the $\mathcal{E} \times \mathbf{B}_0$ drift motion leads to a magnetron frequency of the order of $\omega_{E\times B} \sim e/B_0 b^3$. The inequality $r_{Le} \ll b$ implies the frequency scaling $\omega_{ce} \gg \omega \gg \omega_{E\times B}$, similar to that of the Penning trap, but valid at a microscopic scale.

The three body recombination rate is controlled by a *kinetic bottleneck* [21], which occurs at a binding energy of a few times T_e below the ionization energy E_I. In order to describe the physics in the proximity of this bottleneck, we can use a variational approach [14], which can be formulated in the following way. We assume that the two-electron distribution function $f_2(1,2) \equiv f_2(\mathbf{r}_1, \mathbf{v}_1, \mathbf{r}_2, \mathbf{v}_2)$ is in thermal equilibrium if electron 1 has a bound state with binding energy E_1 above some limit E, and is zero for $E_1 < E$. When $f_2(1,2) = 0$ a Rydberg state is formed. On the other hand, electron 2 is always a free electron which is colliding with the bound electron 1. This means that the interaction between passing (2) and bound (1) electrons will produce a flux of electrons into the Rydberg states for which the boundary $E_1 < E$ is crossed. The one-way rate for the Rydberg state formation, $R(E)$, is determined by an integral over the electron collisions at the boundary E, as given in dimensionless form by

$$R(\epsilon) = \left(n_e b^3\right)^2 \int d\mathbf{r}_1 \int d\mathbf{r}_2 \int d\mathbf{v}_2 \left|\frac{\partial \phi_{12}}{\partial z_1}\right| f_2(1,2), \tag{15.151}$$

where $\epsilon = E/T_e$ and $\phi_{12} = 1/|\mathbf{r}_1 - \mathbf{r}_2|$, and where we use

$$f_2(1,2) = \frac{1}{2\pi} \exp(\epsilon_1 + \epsilon_2 - \phi_{12}) . \tag{15.152}$$

The integration of Eq. (15.151) is performed in a domain where

$$\epsilon_1 + \epsilon_2 - \phi_{12} < \epsilon, \tag{15.153}$$

which implies that the electron 1 can move from $\epsilon_1 = \epsilon$ down to a deeper binding energy state, while electron 2 can escape to infinity. Such a restriction limits the domain of integration over $\mathbf{v}_2$ to the domain $v_2^2/2 + \phi_{12} + \phi_{20} > 0$, where $\phi_{20} = -1/|r_2|$ represents the interaction between electron 2 and the ion, supposedly located at $\mathbf{r}_0 = 0$. It is also necessary to consider an additional restriction to the integration in (15.151), in order to avoid the diverges associated with Coulomb collisions. We therefore introduce a cut-off at $|\mathbf{r}_1 - \mathbf{r}_2| \geq C/\epsilon$, where C is a constant. The result is [10]

$$R(\epsilon) \leq \frac{2}{3} C (2\pi)^{3/2} \left(n_e b^3\right)^2 \frac{\exp(\epsilon)}{\epsilon^4}. \tag{15.154}$$

This expression has a strong minimum at $\epsilon = 4$ with the value $R_{\min} \simeq C(n_e b^3)^2$, which defines the energy bottleneck and results from the competition between the Boltzmann factor $\exp(\epsilon)$ and the phase-space factor ϵ^{-4}. The variational theory identifies this particle flux minimum with the recombination rate, or in other words, with the rate of formation of Rydberg atoms. A Monte-Carlo numerical calculation [10] shows that this recombination rate is quite insensitive the value of the constant C, and leads to the value (in dimensional units) of

$$R_{\text{tbr}} \simeq 0.07 n_e^2 b^5 v_{\text{th,e}}. \tag{15.155}$$

This value was obtained in the limit of an infinite magnetic field $B_0 \to \infty$, or equivalently, for a negligible Larmor radius. This is an order of magnitude smaller that the recombination rate in the absence of magnetic confinement, $B_0 \to 0$, as derived by [1] and [23].

References

1. D.R. Bates, I. Mendas, J. Phys. B **8**, 1770 (1971)
2. C. Boisseau, I. Simbotin, R. Coté, Phys. Rev. Lett. **88**, 133004 (2002)
3. M. Bonitz, Phys. Lett. A **221**, 85 (1996)
4. S.G. Brush, H.L. Sahlin, E. Teller, J. Chem. Phys. **45**, 2102 (1966)
5. P.M. Chiakin, T.C. Lubensky, *Principles of Condensed Matter Physics* (Cambridge University Press, Cambridge, UK, 1998)
6. J.-L. Delcroix, A. Bers, *Physique des Plasmas*, 2 volumes (CNRS Editions, Paris, 1994)

7. D.H.E. Dubin, T.M. O'Neil, Trapped nonneutral plasmas, liquids, and crystals (the thermal equilibrium states). Rev. Mod. Phys. **71**, 87 (1999)
8. M.R. Flannely, D. Vrinceanu, Phys. Rev. A **68**, 030502 (R) (2003)
9. T.F. Gallagher, *Rydberg Atoms* (Cambridge University Press, Cambridge, UK, 1994)
10. M.E. Glinski, T. O'Neil, Phys. Fluids B **3**, 1279 (1991)
11. S. Hamaguchi, R.T. Farouki, J. Chem. Phys. **101**, 9876 (1994)
12. R. Heidemenn, U. Raitzsch, V. Bendkowsky, B. Butscher, R. Löw, T. Pfau, Phys. Rev. Lett. **100**, 033601 (2008)
13. S. Ichimaru, Strongly coupled plasmas: high-density classical plasmas and degenerate electron liquids. Rev. Mod. Phys. **54**, 1017 (1982)
14. J.C. Keck, Adv. Chem. Phys. **13**, 85 (1967)
15. T.C. Killian, S.L. Rolston, Phys. Today **63**, 46 (2011)
16. L.D. Landau, E.M. Lifshitz, *Quantum Mechanics* (Butterworth-Heinemann, Oxford, 1981)
17. L.D. Landau, E.M. Lifshitz, *The Classical Theory of Fields* (Butterworth-Heinemann, Oxford, 1981)
18. E.M. Lifshitz, L.P. Pitaevskii, *Physical Kinetics* (Butterworth-Heinemann, Oxford, 1981)
19. R. Löw, H. Weimar, U. Raitzsch, R. Heidemann, V. Bendkowsky, B. Butscher, H.P. Büchler, T. Pfau, Phys. Rev. A **80**, 033422 (2009)
20. F.G. Major, V.N. Gheorghe, G. Werth, *Charged Particle Traps* (Springer, Berlin, 2005)
21. B. Makin, J.C. Keck, Phys. Rev. Lett. **11**, 281 (1962)
22. G. Manfredi, A. Mola, M.R. Feix, Phys. Fluids B **5**, 388 (1993)
23. P. Mansbach, J.C. Keck, Phys. Rev. **181**, 275 (1969)
24. M.S. Murillo, Phys. Rev. Lett. **87**, 115003 (2001)
25. T. Pohl, H.R. Sadeghpour, P. Schmelcher, Cold and ultracold Rydberg atoms in strong magnetic fields. Phys. Rep **484**, 181 (2009)
26. F. Robicheaux, J.D. Hanson, Phys. Plasmas **10**, 2217 (2003)
27. M. Saffman, T.G. Walker, K. Mølmer, Quantum information with Rydberg atoms. Rev. Mod. Phys. **82**, 2313 (2010)
28. P.K. Shukla, K. Avinash, Phys. Rev. Lett. **107**, 135002 (2011)
29. W.L. Slattery, G.D. Doolen, H.E. DeWitt, Phys. Rev. A **21**, 2087 (1980)
30. D. Tong et al., Phys. Rev. Lett. **93**, 063001 (2004)
31. E. Urban, T.A. Johnson, T. Henage, L. Isenhower, D.D. Yavuz, T.G. Walker, M. Saffman, Nat. Phys. **5**, 110 (2009)
32. V. Vuletic, Nat. Phys. **2**, 801 (2006)
33. T.G. Walker, M. Saffman, Phys. Rev. A **77**, 032723 (2008)

Chapter 16
Waves in Rydberg Plasmas

In this chapter we give an overview of the wave modes that can be excited in a plasma, by focusing our attention on problems which can be directly relevant to the area of ultra-cold Rydberg plasmas. This means that our presentation will be biased, deviating from the usual presentation of the subject in standard plasma physics books [1–3]. In particular, we give relevance to low frequency waves and to electrostatic oscillations, which can be more relevant to the ultra-cold plasma.

We start by discussing the electrostatic waves in isotropic plasmas, in order to compare them with the previously considered elementary excitations in the other two phases of ultra-cold matter. In contrast with the previous cases, two distinct modes can be considered here: electron plasma waves and ion acoustic waves. We will see that the ion acoustic waves are formally similar, but physically distinct, to both the hybrid sound of the non-condensed gas, and the Bogoliubov oscillations of the condensates. We also study the ion acoustic waves in a background expanding plasma, which have been recently observed experimentally. The electron plasma waves, which are the high frequency counterpart of ion acoustic waves, will be discussed first. Similarities with the hybrid sound will also be pointed out.

In order to complete our overview of wave theory, we also briefly discuss the electromagnetic waves in isotropic plasmas. This is important for the understanding of photon dynamics in such media. An element of originality with respect to the usual plasmas can be found in Rydberg plasmas, due to the existence of a large number of neutral atoms in highly excited states. The internal atomic states can contribute to photon dispersion, leading to the occurrence of polariton modes, with interesting dispersive properties. In particular, these modes can have very low group velocities, which is a characteristic feature of *slow light*. The polariton modes are only considered in the unmagnetized plasma case, but it is important to keep in mind that strong enough electromagnetic pulses (giving place to ponderomotive forces associated with these polariton modes) can eventually lead to the excitation of quasi-static magnetic fields.

In this chapter, we also focus our attention to wave phenomena in inhomogeneous plasmas. Recent experiments revealed the existence of electric drift instabilities [4],

J.T. Mendonça and H. Terças, *Physics of Ultra-Cold Matter*, Springer Series on Atomic, Optical, and Plasma Physics 70, DOI 10.1007/978-1-4614-5413-7_16,

which can be described with the help of a simple fluid model. We also study a different kind of wave modes that are related with the existence of density gradients. Noting that such gradients are always present in plasma experiments, this is an universal type of wave which can easily become unstable and lead to an increase of particle losses.

The chapter is then completed by a general formulation of wave dispersion in cold magnetized plasmas. Several types of waves are discussed, paying particular attention to the modes that can propagate along and across the static magnetic field. For parallel propagation, the transverse wave modes have right and left hand circular polarization, and display resonances associated with single particle cyclotron orbits. For perpendicular propagation, the modes are linearly or elliptically polarized, and display resonances associated with the collective particle response.

16.1 Isotropic Plasmas

We use fluid equations to describe the various types of waves that can propagate in a plasma. For simplicity we first focus on the case of isotropic and infinite plasmas. For high frequency electrostatic waves, the ions can be assumed at rest, by taking the limit of an infinite ion mass $m_i \to \infty$. The plasma motion is then simply described by the electron fluid equations

$$\frac{\partial n_e}{\partial t} + \nabla \cdot (n_e \mathbf{v}_e) = 0, \quad \frac{\partial}{\partial t}\mathbf{v}_e + \mathbf{v}_e \cdot \nabla \mathbf{v}_e + S_e^2 \nabla \ln n_e = -\frac{e}{m_e}\,\boldsymbol{\mathcal{E}}. \tag{16.1}$$

The electric field $\boldsymbol{\mathcal{E}}$ is determined by Poisson's equation

$$\nabla \cdot \boldsymbol{\mathcal{E}} = \frac{e}{\epsilon_0}(n_i - n_e), \tag{16.2}$$

where the ion density is assumed constant and equal to the equilibrium plasma density, $n_i = n_0$. At equilibrium, we have $\mathbf{v}_e = 0$ and $\boldsymbol{\mathcal{E}} = 0$. We then assume an infinitesimal perturbation, such that

$$n_e = n_0 + \tilde{n} \exp(i\mathbf{k} \cdot \mathbf{r} - i\omega t), \quad \mathbf{v}_e = \tilde{\mathbf{v}} \exp(i\mathbf{k} \cdot \mathbf{r} - i\omega t). \tag{16.3}$$

Linearizing the electron fluid equations (16.1), we get

$$\tilde{n} = n_0 \frac{\mathbf{k} \cdot \mathbf{v}}{\omega}, \quad \tilde{\mathbf{v}} = -\frac{ie\boldsymbol{\mathcal{E}}}{m_e \omega (1 - S_e^2 k^2/\omega^2)}. \tag{16.4}$$

Inserting this in Eq. (16.2), we obtain

$$i\mathbf{k} \cdot \left(1 - \frac{\omega_{pe}^2}{\omega^2 - S_e^2 k^2}\right) \boldsymbol{\mathcal{E}} = 0, \tag{16.5}$$

where we have used the electron plasma frequency, $\omega_{pe} = \sqrt{e^2 n_0/\epsilon_0 m}$. This can obviously be written in the form $i\mathbf{k} \cdot \mathbf{D} = 0$, where $\mathbf{D} = \epsilon_0 \epsilon(\omega, \mathbf{k})\ \boldsymbol{\mathcal{E}}$ is the displacement field vector, and $\epsilon(\omega, k)$ is the plasma dielectric function, as determined by

$$\epsilon(\omega, \mathbf{k}) = 1 - \frac{\omega_{pe}^2}{\omega^2 - S_e^2 k^2}. \tag{16.6}$$

Poisson's equation is only satisfied for $\epsilon(\omega, k) = 0$. The dispersion relation for the electron plasma waves is then given by

$$\omega^2 = \omega_{pe}^2 + 3k^2 v_{\text{the}}^2. \tag{16.7}$$

Here, we have used $S_e^2 = 3v_{\text{the}}^2$, where $v_{\text{the}} = \sqrt{T_e/m_e}$. This factor of 3 can be justified by the kinetic approach, to be discussed later. Equation (16.7) describe dispersive waves, where the phase and group velocities v_ϕ and v_g depend on the frequency and are related by

$$v_\phi v_g = 3v_{\text{the}}^2, \tag{16.8}$$

where the following definitions were used

$$v_\phi = \frac{\omega}{k}, \quad \mathbf{v}_g = \frac{\partial \omega}{\partial \mathbf{k}}. \tag{16.9}$$

We can see that the dispersion properties of electron plasma waves are formally analogous to those found in the first part of this book for the hybrid sound waves in a laser-cooled atomic gas. The differences are in the definition of the cut-off frequency ω_{pe}, and of the characteristic velocity v_{the}. In confined plasmas, the character of the electron plasma waves also change in a similar way, originating Tonks-Dattner modes similar to those already discussed in the first part of this book. In ultra-cold plasmas, electron plasma waves have been first observed by [5], and Tonks-Dattner modes were identified in [6].

Let us now retain the ion motion. The ion density is not a constant any more. Using an approach similar to the previous one, but where the ion fluid equations are also involved, we get for the perturbed quantities of electrons and ions ($\alpha = e, i$)

$$\tilde{n}_\alpha = \frac{n_0 \mathbf{k} \cdot \mathbf{v}_\alpha}{\omega}, \quad \mathbf{v}_\alpha = \frac{i q_\alpha \boldsymbol{\mathcal{E}}}{\omega m_\alpha (1 - S_\alpha^2 k^2/\omega^2)}, \tag{16.10}$$

and arrive at two equations involving the electron and ion density perturbations $\tilde{n}_e = n_e - n_0$ and $\tilde{n}_i = n_i - n_0$, as

$$(\omega^2 - S_e^2 k^2)\tilde{n}_e = \frac{e n_0}{m_e} i\mathbf{k} \cdot \boldsymbol{\mathcal{E}}, \quad (\omega^2 - S_i^2 k^2)\tilde{n}_i = -\frac{e n_0}{m_i} i\mathbf{k} \cdot \boldsymbol{\mathcal{E}}. \tag{16.11}$$

Using Poisson's equation (16.2), we establish two coupled equations for $\tilde{n}_e$ and $\tilde{n}_i$, in the form

$$\left(\omega^2 - S_e^2 k^2 + \omega_{pe}^2\right)\tilde{n}_e = \omega_{pe}^2 \tilde{n}_i, \quad \text{and} \quad \left(\omega^2 - S_i^2 k^2 + \omega_{pi}^2\right)\tilde{n}_i = \omega_{pi}^2 \tilde{n}_e, \tag{16.12}$$

which then yield the following dispersion relation

$$\frac{\omega_{pe}^2}{(\omega^2 - S_e^2 k^2 + \omega_{pe}^2)} = \frac{(\omega^2 - S_i^2 k^2 + \omega_{pi}^2)}{\omega_{pi}^2}. \tag{16.13}$$

This equation has two solutions, one corresponding to the electron plasma waves discussed above, and the other corresponding to much lower frequency waves, called ion acoustic waves. Noting that $S_i^2 \ll S_e^2$, due to the large mass difference between electrons and ions, and considering the limit $k^2 S_e^2 \ll \omega_{pe}^2$, we obtain for the ion acoustic waves

$$\omega \simeq \frac{k v_{ac}}{\sqrt{1 + k^2 \lambda_D^2}}, \quad v_{ac} = \sqrt{S_i^2 + S_e^2 (m_e/m_i)} = \sqrt{\frac{T_e + 3T_i}{m_i}}. \tag{16.14}$$

The factors 1 and 3 multiplying the electron and ion temperatures can be again justified within the kinetic plasma theory. We see that the ion acoustic waves are non dispersive in the limit of long wavelengths, $k^2 \lambda_D^2 \ll 1$, with phase and group velocities nearly equal to the ion sound velocity v_{ac}.

To complete our elementary discussion of plasma waves in isotropic media, let us now consider the case of high-frequency electromagnetic waves. Here again, we neglect the ion motion, and instead of Poisson's equation we use the full set of Maxwell's equations. From these equations, we can derive the propagation equation for the electric field, in the form

$$\left(\nabla^2 - \frac{1}{c^2}\frac{\partial^2}{\partial t^2}\right)\boldsymbol{\mathcal{E}} - \nabla(\nabla \cdot \boldsymbol{\mathcal{E}}) = \mu_0 \frac{\partial \mathbf{J}}{\partial t}, \tag{16.15}$$

where $\mathbf{J} = -e n_0 \mathbf{v}_e$ is the electron current. For perturbations evolving in space and time as $\exp(i\mathbf{k}\cdot\mathbf{r} - i\omega t)$, and assuming transverse waves such that $\mathbf{k}\cdot\boldsymbol{\mathcal{E}} = 0$, this can be transformed into

$$\left(k^2 - \frac{\omega^2}{c^2}\right)\boldsymbol{\mathcal{E}} = i\omega\mu_0 e n_0 \mathbf{v}. \tag{16.16}$$

Noticing that $n = 0$ and $\mathbf{v} = -ie\boldsymbol{\mathcal{E}}/m_e\omega$, we then derive the dispersion relation

$$\omega^2 = \omega_{pe}^2 + k^2 c^2. \tag{16.17}$$

This shows that the transverse electromagnetic waves can only propagate in a plasma with frequencies above the electron plasma frequency, which acts as a cut-off frequency. These are dispersive waves, and their phase and group velocities are related by

$$v_\phi \, v_g = c^2. \tag{16.18}$$

We should notice the similarity between these results and those for the electron plasma wave, as shown in Eqs. (16.7) and (16.8). We can see that, apart from the formal analogy, the electron thermal velocity is now replaced by the speed of light.

16.2 Polaritons and Slow Light

Let us consider electromagnetic wave propagation in a partially ionized plasma, where the neutral atoms can eventually be excited in Rydberg states, very close to ionization. We can start with the wave equation (16.15)

$$\left(\nabla^2 - \frac{1}{c^2} \frac{\partial^2}{\partial t^2} \right) \mathcal{E} = \mu_0 \frac{\partial^2 \mathbf{P}}{\partial t^2} + \mu_0 \frac{\partial \mathbf{J}}{\partial t}, \tag{16.19}$$

where we have now included the effect of the polarization $\mathbf{P}$ due to the presence of neutral atoms. We know that, for transverse waves, the amplitude of density perturbations is equal to zero. We then consider infinitesimal wave perturbations with frequency ω and wavevector $\mathbf{k}$, of the form

$$(\mathbf{v}_e, \mathcal{E}, \mathbf{P}) = (\mathbf{v}, \mathcal{E}_\omega, \mathbf{P}_\omega) \exp\left(i\mathbf{k} \cdot \mathbf{r} - i\omega t\right). \tag{16.20}$$

From the electron equations of motion, we get

$$\mathbf{v} = -i \frac{e}{m_e} \frac{\mathcal{E}_\omega}{(\omega + i\nu_e)}. \tag{16.21}$$

We retain the electron collision frequency ν_e because the case of very low frequency electromagnetic waves can eventually be considered. We can also use the constitutive relation

$$\mathbf{P}_\omega = n_a \chi_a(\omega) \mathcal{E}_\omega, \tag{16.22}$$

where n_a is the density of the neutral atoms, and $\chi_a(\omega)$ is the atomic susceptibility. From the above equations we can derive the dispersion relation

$$\frac{k^2 c^2}{\omega^2} = \epsilon(\omega) \equiv 1 + \chi_e(\omega) + N_a \chi_a(\omega), \tag{16.23}$$

where $\epsilon(\omega)$ is the dielectric function of the medium. The electron susceptibility of the plasma is simply determined by

$$\chi_e(\omega) = \chi_e'(\omega) + i\chi_e''(\omega), \tag{16.24}$$

where the real and the imaginary parts are given by

$$\chi_e'(\omega) = -\frac{\omega_{pe}^2}{(\omega^2 + \nu_e^2)}, \quad \chi_e''(\omega) = -\frac{\nu_e}{\omega}\chi_e'(\omega). \tag{16.25}$$

In order to establish the atomic susceptibility $\chi_a(\omega)$, we first notice that most of the Rydberg states of the neutral atoms inside the ultra-cold partially ionized gas are significantly populated. Neglecting the quantum defects, the atom energy levels can be written as (see Eq. (15.113))

$$E_\nu = -\frac{R_y}{\nu^2}, \quad \nu \gg 1, \tag{16.26}$$

where ν represents the principal quantum number.[1] The Rydberg energy spectrum provides an infinite set of nearly resonant transitions which contribute to wave dispersion. The dominant contribution to $\chi_a(\omega)$ will come from the closest radiative transition, such that $\hbar\omega \simeq E_{\nu+1} - E_\nu$. This can be approximately determined by $\omega \simeq (2R_y/\hbar\nu^3)$. Assuming a radiative transition between two quantum states $|\nu+1\rangle$ and $|\nu\rangle$, and defining $\hbar\omega_a \equiv E_{\nu+1} - E_\nu$, we can write

$$\chi_a(\omega) = \chi_a'(\omega) + i\chi_a''(\omega), \tag{16.27}$$

where the real and the imaginary parts are

$$\chi_a'(\omega) = -\frac{f_a}{n_0}\frac{\omega_{pe}^2\Delta}{(\Delta^2 + \Gamma^2)}D, \quad \chi_a''(\omega) = \frac{\Gamma}{\omega}\chi_a'(\omega) \tag{16.28}$$

with $D = [n_a^{(\nu+1)} - n_a^{(\nu)}]$ being the unperturbed population difference between the two states, Γ is the natural linewidth of the transition, and $\Delta = (\omega_a - \omega)$ is the frequency detuning. We have also defined the oscillator strength $f_a = (m_e/\hbar)|\langle\nu+1|\mathbf{r}|\nu\rangle|^2$. This is smaller by a factor of m_e/M_a, with respect to the usual definition. It should be noticed that the population difference D is nearly equal to zero over a considerable part of the Rydberg energy spectrum, which tends to become homogeneously populated by electron collisions or by the radiative cascade discussed in the previous chapter.

We first neglect the imaginary part of both susceptibilities, $\chi_e(\omega)$ and $\chi_a(\omega)$, which is valid in the limit $\nu_e \ll \omega$ and $\Gamma \ll \Delta$. Their influence will be discussed

[1] we conserve the symbol n to represent the particle density in this chapter.

later. The dispersion relation for the electromagnetic waves in a Rydberg plasma is then reduced to the simple expression [7]

$$\omega^2 = k^2c^2 + \omega_{pe}^2 \left[1 + \tau_a \frac{\omega^2}{(\omega_a - \omega)} \right], \quad \tau_a = f_a \frac{N_a}{n_0} D \tag{16.29}$$

where the quantity τ_a has the dimensions of time, and can be positive or negative, according to the sign of the population difference D. We can see that the existence of neutral Rydberg atoms introduces an additional term in the wave dispersion, containing an atomic resonance at $\omega = \omega_a$. This qualitatively changes the properties of the electromagnetic wave, which is now coupled with oscillations of the internal atomic energy states. In that sense we can call it a *polariton*. In order to understand the effect of the neutral atom contributions to the dispersion relation, let us consider $\omega = \omega' + i\omega''$, and assume the double resonance condition

$$\omega' = \omega_a = (\omega_{pe}^2 + k^2c^2)^{1/2}. \tag{16.30}$$

Replacing this in (16.29), and assuming that the imaginary part of the frequency is very small, $|\omega''| \ll \omega'$, we get

$$\omega'' = \pm \omega_{pe} \sqrt{\tau_a \omega'/2}. \tag{16.31}$$

This shows that ω'' can have real and positive values for $\tau_a > 0$, which corresponds to an inversion of population, or $D > 0$. The resonant electromagnetic wave with frequency (16.29) can become unstable. Such an instability results from the transfer of energy from the Rydberg atoms to the transverse coherent oscillations of the plasma medium.

Another interesting aspect of the dispersion relation (16.29) is the modification of the cut-off frequency. For $\tau_a = 0$ such a cut-off is simply determined by ω_{pe}, which is the usual plasma result for a fully ionized plasma. However, for $\tau_a \neq 0$, the cut-off is determined by the condition

$$\left(X^2 - 1\right)(1 - Y) = \alpha Y^2, \quad \alpha = \tau_b \omega_a, \tag{16.32}$$

where we have used the dimensionless quantities

$$X = \frac{\omega}{\omega_{pe}}, \quad Y = \frac{\omega}{\omega_a}. \tag{16.33}$$

We can see that, when $Y \to 0$, we get $X = 1$, which is the usual plasma cut-off. And, for $Y \to 1$, we have a resonance $X \to \infty$. Noting that $X = Y(\omega_a/\omega_{pe})$, we conclude that such a resonance can only be attained for $\omega_a \gg \omega_{pe}$. Furthermore, there is no real cut-off for $Y > 1$, or $\omega > \omega_a$. The properties of the solution $X = X(Y)$ are illustrated in Fig. 16.1, for both positive and negative values of α.

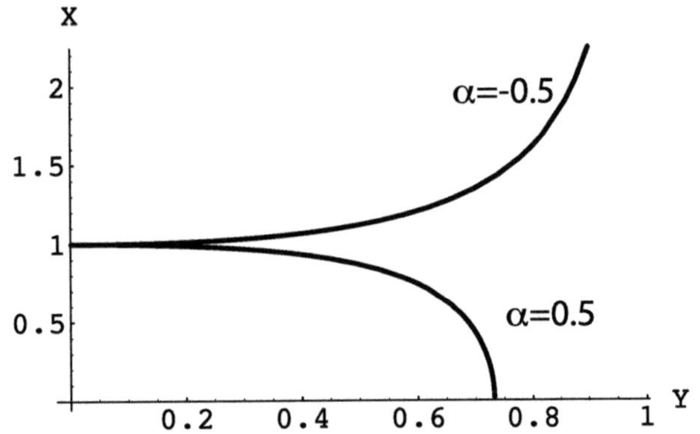

Fig. 16.1 Cut-off frequency for electromagnetic waves propagating in a Rydberg plasma, for $\alpha = \pm 0.5$ or $\tau_a = \pm 1/2\omega_{pe}$. Dimensionless parameters $X = \omega/\omega_{pe}$ and $Y = \omega/\omega_a$ are used

We can see that wave propagation below the plasma cut-off $X = 1$ becomes possible, due to the existence of Rydberg atoms.

Let us now retain the influence of the quantities ν_e and Γ. Going back to the dispersion relation (16.23), we can write

$$\frac{kc}{\omega} \equiv N(\omega) = \sqrt{1 + \chi_e(\omega) + N_a \chi_a(\omega)}. \tag{16.34}$$

The refractive index is obviously complex $N(\omega) = N'(\omega) + iN''(\omega)$, where the real part is $N'(\omega) \simeq \sqrt{1 + \chi'(\omega)}$, for $|\chi'(\omega)| \gg |\chi''(\omega)|$, and the sign of the imaginary part $N''(\omega)$ is equal to the sign of $\chi''(\omega)$. It is therefore useful to represent the quantities $\epsilon'(\omega)$ and $\chi''(\omega)$, which determine the wave dispersion and the wave damping, respectively, and to compare them with the case of a purely neutral medium. Here it is convenient to introduce the following variables $z = \Delta/\Gamma$, $a = \omega_a/\omega_{pe}$, and

$$b = \frac{f_a}{n_0} \frac{\omega_{pe}^2 D}{\gamma}, \quad \eta = \frac{\nu_e}{\omega_{pe}}, \quad g = \frac{\Gamma}{\omega_{pe}}. \tag{16.35}$$

The real part of the susceptibility can now be written in terms of these dimensionless quantities as

$$\chi'(z) = -\frac{bz}{z^2 + 1} - \frac{1}{(a - zg)^2 + g^2}, \tag{16.36}$$

where the first term represents the contribution from the neutral atoms, and the second term that of the free plasma electrons. Similarly, we have for the imaginary part

$$\chi''(z) = -\frac{b}{z^2 + 1} + \frac{\eta}{(a - zg)^3 + (a - zg)\eta^2}. \tag{16.37}$$

The quantity $\epsilon'(z) = 1 + \chi'(z)$ is represented in Fig. 16.2, for negative and positive inversion of populations, $b \pm 1$, respectively. Notice that wave propagation is

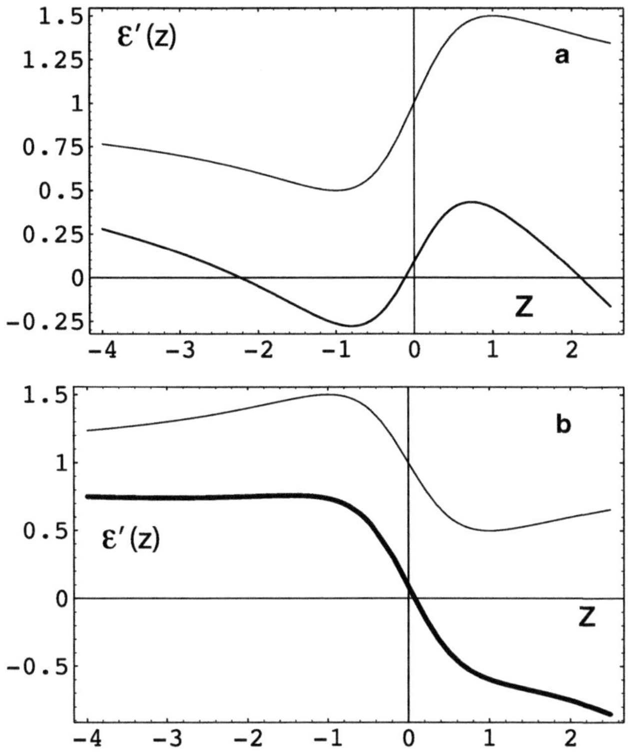

Fig. 16.2 Dielectric function $\epsilon'(z) = 1 + \chi'(z)$ are represented (in *bold*) as a function of the wave frequency detuning $z = \Delta/\gamma$, for $a = 1$, $g = \eta = 0.1$, and (**a**) $b = -1$ and (**b**) $b = 1$. For comparison, the curves of pure atomic dispersion corresponding to a non-ionized gas, are also shown

forbidden when this quantity is negative. These curves show a significant deviation with respect to the purely neutral gas. We can see the appearance of a plasma cut-off for both cases, with a significant qualitative difference for the case of $b = -1$ (no inversion of population).

Until now, we have assumed an infinitesimal wave amplitude, which allowed us to neglect saturation effects. Let us now take a further step and consider the case of a finite wave amplitude, such that a finite Rabi frequency Ω can be considered. In this case, the polariton dispersion relation can be written as

$$\omega^2 = k^2c^2 + \omega_p^2 \left[\frac{\omega^2}{\omega^2 + \nu_e^2} + \tau_a \frac{\omega^2(\omega - \omega_a)}{\Gamma^2 + \Omega^2 + (\omega - \omega_a)^2} \right]. \tag{16.38}$$

Under typical experimental conditions, the electron-atom collision frequency is in the range of $\nu_e/2\pi \sim 100\,\text{MHz}$. The typical plasma frequencies are $\omega_p/2\pi \sim 1\,\text{GHz}$ (for an electron density of $n_0 \sim 10^9\,\text{cm}^{-3}$) and the atomic frequency $\omega/2\pi \sim$

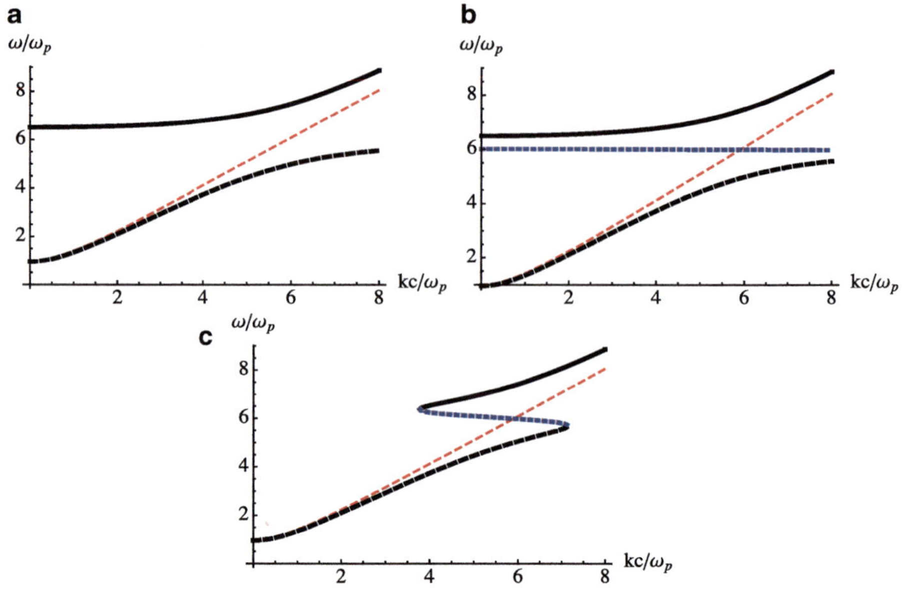

Fig. 16.3 Polariton dispersion relation in a Rydberg plasma, obtained for $\omega_a = 6\omega_p$ and a inverted-population parameter $\tau_a = 0.5\omega_p^{-1}$. (**a**) $\Omega = 0$, (**b**) $\Omega = 0.1\omega_p$ and (**c**) $\Omega = 0.5\omega_p$. In the above plots, the *thin dashed line* represents the usual electromagnetic wave dispersion relation $\omega^2 = \omega_p^2 + k^2c^2$

10 GHz (for ν ranging from 40–50) both belong to the microwave range, which allow us to safely neglect ν_e in the dispersion relation. We can also assume that Γ is small compared to both ω_a and ω_p, finally writing the lossless light-electron-atom polariton dispersion relation, accounting for quantum saturation in a simpler form

$$\omega^2 \approx k^2c^2 + \omega_p^2 \left[1 + \tau_a \frac{\omega^2(\omega - \omega_a)}{\Omega^2 + (\omega - \omega_a)^2} \right]. \tag{16.39}$$

The main qualitative features of this new polaritonic dispersion are summarized in Fig. 16.3. First, we can see the presence of new branches in the wave dispersion. Second, by allowing the atoms to be sensitive to the wave intensity (i.e., by setting $\Omega \neq 0$), we observe the emergence of a third dispersion branch, with frequencies ranging between the values of the two main branches and eventually connecting them. The interesting aspect of this new branch is that it describes waves with very low group velocities, as compared with the speed of light, $\partial\omega/\partial k \ll c$. This suggests that in Rydberg plasmas new slow-light phenomena can eventually occur. Moreover, we observe that an increasing value of the Rabi frequency Ω tends to reduce the range of slow-light dispersion and eventually suppress it. Indeed, for higher values of Ω, the dispersion relation Eq. (5.98) approaches to $\omega^2 \approx k^2c^2 + \omega_p^2[1 + \tau_a\omega^2(\omega - \omega_a)]$, which shows suppression of the slow-light band around $\omega \sim \omega_a$.

We can then say that the electromagnetic coupling between light and the atoms in a Rydberg plasma qualitatively changes the properties of wave dispersion. The resulting polaritonic dispersion exhibits new branches and, in particular, suggests that Rydberg plasmas can provide a stage to observe slow-light phenomena.

16.3 Ponderomotive Force

Let us now consider nonlinear corrections to the electromagnetic wave dispersion relation associated with ponderomotive force effects. In order to understand these effects, let us consider the electron momentum equation

$$\frac{d\mathbf{v}}{dt} = -\frac{e}{m_e}[\mathcal{E} + \mathbf{v} \times \mathbf{B}]. \tag{16.40}$$

To the lowest order of the wave field amplitude, we can neglect the Lorentz force, because it is of order v/c with respect to the electric force. Assuming that the wave field evolves in time as $\exp(-i\omega t)$, we then have $\mathbf{v} = -i(e/m_e\omega)\mathcal{E}$. This allows us to use for the Lorentz force the approximate expression

$$(\mathbf{v} \times \mathbf{B}) = -\frac{e}{m_e\omega^2}\left[\mathcal{E} \times (\nabla \times \mathcal{E})\right]. \tag{16.41}$$

If we now use the vector identity $\mathbf{a} \times (\nabla \times \mathbf{a}) = (1/2)\nabla a^2$, and average over the wave period in order to eliminate the terms oscillating at twice the wave frequency, we get the ponderomotive force, as defined by

$$\langle \mathbf{F} \rangle = -\frac{1}{2}\left(\frac{e}{m_e\omega}\right)^2 \nabla|\mathcal{E}|^2. \tag{16.42}$$

This is the slowly varying part of the Lorentz force acting on a single plasma electron. We can write, for the total force per unit volume, $\mathbf{F}_p = n_0 \langle \mathbf{F} \rangle$ acting on the plasma electron fluid, the following result

$$\mathbf{F}_p = -\frac{\epsilon_0}{2}\frac{\omega_{pe}^2}{\omega^2}\nabla|\mathcal{E}|^2. \tag{16.43}$$

This shows that the ponderomotice force is proportional to the gradient of the electromagnetic energy density. A more detailed analysis of the nonlinear plasma response would reveal that the ponderomotive force actually contains two distinct terms, $\mathbf{F} = \mathbf{F}_{ps} + \mathbf{F}_{pt}$, where the first one is the stationary force described in (16.43) and the second one is a non-stationary (temporal) force due to slow-time dependence of the wave field envelope. The expressions for these two force terms can be written as [8]

$$\mathbf{F}_{ps} = (N-1)\frac{\epsilon_0}{2}\nabla|\mathcal{E}_0|^2, \quad \mathbf{F}_{pt} = \frac{\epsilon_0}{2}\frac{\mathbf{k}}{\omega^2}\frac{\partial[\omega^2(N-1)]}{\partial\omega}\frac{\partial|\mathcal{E}_0|^2}{\partial t}. \tag{16.44}$$

where $N \equiv N(\omega)$ is the plasma refractive index. Notice that, in a fully ionized plasma ($\tau_a = 0$), this non-stationary ponderomotive term is absent, $\mathbf{F}_{pt} = 0$, and the stationary term will be reduced to the more familiar expression (16.43). The existence of the ponderomotive force leads to the occurrence of new plasma phenomena, for example the excitation of quasi-static magnetic field. This effect will be illustrated here. First of all, we notice that the low-frequency ponderomotive force pushes the electrons locally, from out of the regions with a larger wave energy density. This creates a slowly varying space charge electric field, as determined by

$$\boldsymbol{\mathcal{E}}_s = \frac{1}{n_0 e}\mathbf{F}_{pf} \equiv \frac{\epsilon_0 (N_1)}{2n_0 e}\nabla|\mathcal{E}_0|^2 + \frac{\epsilon_0}{2n_0 e}\frac{\mathbf{k}}{\omega^2}\frac{\partial[\omega^2(N-1)]}{\partial\omega}\frac{\partial|\mathcal{E}_0|^2}{\partial t}. \tag{16.45}$$

The induced slowly varying magnetic field $\mathbf{B}_s$ is then determined from Faraday's law, $\partial \mathbf{B}_s/\partial t = -\nabla \times \boldsymbol{\mathcal{E}}_s$, which, together with (16.45), leads to

$$\mathbf{B}_s = -\frac{\epsilon_0}{2n_0 e\omega^2}\frac{\partial[\omega^2(N-1)]}{\partial\omega}\nabla \times \left(\mathbf{k}|E_0|^2\right). \tag{16.46}$$

The magnitude of the magnetic field can then be written more explicitly as [9]

$$|\mathbf{B}_s| = \frac{ek\tau_a(2\omega_a - \omega)|\mathcal{E}_0|^2}{4m_e L\omega(\omega - \omega_a)^2}, \tag{16.47}$$

where L is scale length of the envelope amplitude variation $|E_0|^2$. This expression reveals that the magnetic field strength is proportional to τ_a, which (as we have seen earlier) depends on the atomic processes in Rydberg plasmas. Furthermore, there is a resonant enhancement of the magnetic field when $\omega \sim \omega_a$. The resulting electron cyclotron frequency ω_{ce} is

$$\omega_{ce} = \frac{k\tau_a\omega(2\omega_a - \omega)v_0^2}{4L(\omega - \omega_a)^2}, \tag{16.48}$$

where $v_0 = e|\mathcal{E}_0|/m_e\omega$ is the electron quiver velocity in the electromagnetic field. The resonant character of this effect will be limited by the existence of a spontaneous life-time of the transition radiation at the frequency ω_a. By assuming a finite value of the spontaneous bandwidth Γ, we can transform equation (16.47) into

$$|\mathbf{B}_s| = \frac{eck\beta(2\omega_a - \omega)(\omega - \omega_a)|E_0|^2}{4m_e L\omega[(\omega - \omega_a)^2 + \Gamma^2]}, \tag{16.49}$$

where the field singularity disappears, allowing us to estimate the maximum value of the quasi-static magnetic field. This is illustrated in Fig. 16.4, where the dimensionless quantity $b = 4m_e L\omega|\mathbf{B}_s|/ek\tau_a|\mathcal{E}_0|^2$ is represented as a function of the relative frequency. This shows that magnetic fields can be generated spontaneously in a Rydberg plasma by the ponderomotive force. This force, resulting from the high-frequency electromagnetic wave, pushes the electrons locally away and creates a slowly varying space charge electric field, and a quasi-static magnetic field. This can eventually influence the expansion of the ultra-cold plasma bubbles.

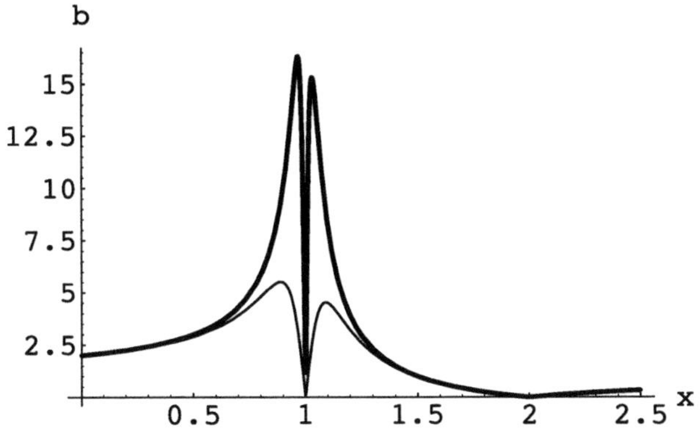

Fig. 16.4 Dimensionless magnetic field b, as a function of the relative wave frequency $x = \omega/\omega_a$, for two different values of the natural bandwidth parameter $g = \gamma^2/\omega_a^2$, for $g = 0.01$ and $g = 0.001$ (*bold*)

16.4 Electron Drift Instability

Until now we have only considered wave phenomena in isotropic plasmas. The variety of waves and instabilities in magnetized plasma is however much broader. Here, we consider the case of a plasma with crossed static electric and magnetic fields $\boldsymbol{\mathcal{E}}_0$ and $\mathbf{B}_0$. As it has been discussed in the recent literature [10], this is important as electron drift instabilities have been experimentally identified in the context of ultra-cold plasma studies.

We have seen in a previous chapter that, in such a case, the plasma particles acquire an electric drift velocity, which is independent of the sign of their charge, $\mathbf{v}_E = (\boldsymbol{\mathcal{E}}_0 \times \mathbf{B}_0)/B_0^2$. The resulting plasma convection can become unstable, even in the absence of any density gradient. The influence of theses gradients will be discussed in the next section, but we first consider, for the sake of clarity, that the plasma is uniform, with an equilibrium density n_0. In order to discuss such an instability, we assume perturbed electron density and velocity, as $n_e = n_0 + \tilde{n}_e$, and $\mathbf{v}_e = \mathbf{v}_E + \tilde{\mathbf{v}}_e$. On the other hand, the ions can be assumed unmagnetized, which is valid if the ion Larmor radius is larger than the characteristic scale of the plasma medium. Let us then consider the ion fluid equations in the ultra-cold limit $T_i \to 0$. We can therefore use

$$\frac{\partial n_i}{\partial t} + \nabla \cdot (n_i \mathbf{v}_i) = 0, \quad \left(\frac{\partial}{\partial t} + \mathbf{v}_i \cdot \nabla \right) \mathbf{v}_i = -\frac{e \nabla V}{m_i}, \tag{16.50}$$

where V is the electrostatic potential. Linearizing and solving for a perturbation of the form $\exp(i\mathbf{k} \cdot \mathbf{r} - i\omega t)$, we get

$$\tilde{n}_i = \frac{n_0 e \tilde{V}}{m_i \omega^2} k^2. \tag{16.51}$$

Replacing this in the potential equation (16.2), we obtain

$$\nabla^2 \phi = \frac{e\tilde{n}_e}{\epsilon_0} - \frac{\omega_{pi}^2}{\omega^2} k^2 \phi. \tag{16.52}$$

We can now determine the electron density perturbation $\tilde{n}_e$, by using the linearized electron fluid equations. From the continuity equation, we get

$$\frac{\partial n_e}{\partial t} + n_0 \nabla \cdot \mathbf{v}_e + \mathbf{v}_E \cdot \nabla \tilde{n}_e = 0, \tag{16.53}$$

and, from the momentum equation

$$\left(\frac{\partial}{\partial t} + \mathbf{v}_E \cdot \nabla \right) \tilde{\mathbf{v}}_e = \frac{e}{m_e} \nabla \phi - \omega_{ce} (\tilde{\mathbf{v}}_e \times \mathbf{b}) - \frac{\nabla \tilde{P}_e}{n_0 m_e}, \tag{16.54}$$

where we have introduced the electron cyclotron frequency $\omega_{ce} = eB_0/m_e$, and the unitary vector $\mathbf{b} = \mathbf{B}_0/B_0$. Here, $\tilde{P}_e = \tilde{n}_e T_e$ represents the perturbed electron pressure. For the assumed wave perturbation we can then easily get

$$\tilde{n}_e = n_0 \frac{\mathbf{k} \cdot \mathbf{v}_e}{(\omega - \mathbf{k} \cdot \mathbf{v}_E)} \tag{16.55}$$

and

$$(\omega - \mathbf{k} \cdot \mathbf{v}_E) \tilde{\mathbf{v}}_e = -\frac{ek}{m_e} \phi - i \omega_{ce} (\tilde{\mathbf{v}}_e \times \mathbf{b}) - \frac{T_e}{n_0 m_e} \mathbf{k} \tilde{n}_e. \tag{16.56}$$

In order to be more specific, we assume that the static magnetic field is defined along the z-direction, and the static electric field in the x-direction, such that

$$\boldsymbol{\mathcal{E}}_0 = \mathcal{E}_0 \mathbf{e}_x, \quad \mathbf{B}_0 = B_0 \mathbf{e}_z, \quad \mathbf{v}_E = -\frac{\mathcal{E}_0}{B_0} \mathbf{e}_y. \tag{16.57}$$

We also restrict our discussion to the case where the electrostatic perturbation propagates along the electron drift motion. We therefore take $\mathbf{k} = k\mathbf{e}_y$. In such conditions, the above equations can be solved for the two electron velocity components, as

$$\tilde{v}_x = -i \omega_{ce} \frac{\tilde{v}_y}{(\omega - k v_E)}, \tag{16.58}$$

and

$$\left[1 - \frac{(\omega_{ce}^2 - v_{\text{th,e}}^2 k^2)}{(\omega - k v_E)^2} \right] \tilde{v}_y = \frac{ek}{m_e} \frac{\phi}{(\omega - k v_E)}. \tag{16.59}$$

Using Eqs. (16.55) and (16.52), we can derive the dispersion relation for the electrostatic perturbations, as

$$\left(1 - \frac{\omega_{pi}^2}{\omega^2} \right) \left[(\omega - k v_E)^2 - (\omega_{ce}^2 - k^2 v_{\text{th,e}}^2) \right] = \omega_{pe}^2. \tag{16.60}$$

It is useful to consider this result for some particular cases, relevant to current experiments in ultra-cold plasmas. We assume the cold plasma limit, where $k^2 v_{\text{the}}^2 \ll \omega_{ce}^2$. The dispersion relation now reduces to

$$\left(1 - \frac{\omega_{pi}^2}{\omega^2}\right) \simeq \frac{\omega_{pe}^2}{(\omega - k v_E)^2 - \omega_{ce}^2}. \tag{16.61}$$

In the strong-drift limit, $\omega^2 \ll (k v_E)^2$, the dispersion relation simpler reads

$$\frac{\omega^2}{\omega_{pi}^2} = \frac{(k v_E)^2 - \omega_{ce}^2}{\omega_{uh}^2 - (k v_E)^2}, \tag{16.62}$$

where we have introduced the definition of the *upper hybrid frequency*, as $\omega_{uh} = (\omega_{pe}^2 + \omega_{ce}^2)^{1/2}$. This expression shows that a purely growing mode exists, with $\omega^2 < 0$, in the following conditions

$$\omega_{ce}^2 < (k v_E)^2 < \omega_{uh}^2. \tag{16.63}$$

The instability therefore occurs for wavenumbers of the order of $k \sim \omega_{ce}/v_E$. This effect can eventually explain recent experimental results in ultra-cold plasmas, where a radial electric field can provide the static field $\mathcal{E}_0$ during the plasma expansion phase [10]. Similar effects have also been considered in other contexts, such as the Hall thruster for space propulsion [11, 12].

16.5 Drift Waves at Plasma Gradients

In density gradient regions of a magnetized plasma, particle drift motion can take place. This motion is usually associated with low frequency oscillations. These are the so-called drift waves, which propagate across the magnetic field lines and can be the source of important particle and energy de-confinement. In order to describe such waves, we should also assume that the ions are cold, which allows us to neglect their temperature ($T_i \simeq 0$). Using the electron fluid equations, we can see that an equilibrium electron velocity $\mathbf{v}_0$ exists, resulting from the hydrostatic condition

$$e(\mathbf{v}_0 \times \mathbf{B}_0) + \frac{\nabla P_e}{n_e} = 0, \tag{16.64}$$

where $\mathbf{B}_0$ is the static magnetic field, and P_e is the electron pressure. By introducing the electron cyclotron frequency $\omega_{ce} = eB_0/m_e$, and using the simple equation of state $P_e = n_e T_e$ (with T_e written in energy units), we can rewrite this expression as

$$(\mathbf{v}_0 \times \mathbf{b}) = -\frac{T_e \nabla n_e}{n_e m_e \omega_{ce}}. \tag{16.65}$$

Notice that the temperature is assumed constant and uniform, but a temperature gradient term could also be included here. From Eq. (16.65) follows that the equilibrium drift velocity is perpendicular to both the static magnetic field and the density (or more generally pressure) gradient. For definiteness, we take $\mathbf{b} = \mathbf{e}_z$ and assume that n_e only depends on the x-coordinate, therefore getting

$$\mathbf{v}_0 = -\frac{T_e}{n_e m_e \omega_{ce}} \frac{dn_e}{dx} \mathbf{e}_y. \tag{16.66}$$

We now assume perturbations around this equilibrium. They satisfy the linearized fluid equations for the two particle species $\alpha = (e, i)$, for electrons and ions, where we use $n_\alpha = n_0 + \tilde{n}_\alpha$. For slow waves, we can assume that the electrons are thermalized and attain the Boltzmann equilibrium in the wave potential V. This means that we can write

$$n_e = n_0 \exp\left(\frac{e\tilde{V}}{T_e}\right) \simeq n_0 \left(1 + \frac{e\tilde{V}}{T_e}\right). \tag{16.67}$$

Inserting the latter in Poisson's equation (16.2), we obtain

$$\nabla^2 \tilde{V} = \frac{e^2 \tilde{V}}{\epsilon_0 T_e} - \frac{e}{\epsilon_0} \tilde{n}_i. \tag{16.68}$$

This has to be coupled with the ion fluid equations, which take the form

$$\frac{\partial \tilde{n}_i}{\partial t} + n_0 \nabla \cdot \tilde{\mathbf{v}}_i + (\tilde{\mathbf{v}}_i \cdot \mathbf{e}_x) \frac{dn_0}{dx} = 0, \tag{16.69}$$

$$\frac{\partial \tilde{\mathbf{v}}_i}{\partial t} = -\frac{e}{m_i} (\nabla \tilde{V} - \tilde{\mathbf{v}}_i \times \mathbf{B}_0).$$

Assuming a perturbation that evolves in the yz-plane, which is the plane that contains both the static magnetic field and the drift velocity vectors, as $\exp(i\mathbf{k} \cdot \mathbf{r} - i\omega t)$, with $\mathbf{k} = k_y \mathbf{e}_y + k_z \mathbf{e}_z$, we can derive then the following relation between the perturbed ion density and the electrostatic potential

$$\tilde{n}_i = \left(-\frac{ek_z}{m_i \omega_{ci} \omega} \frac{dn_0}{dx} + \frac{en_0 k_z^2}{m_i \omega^2}\right) \tilde{V}. \tag{16.70}$$

Introducing the drift velocity $\mathbf{v}_0$ defined in (16.66), and using the ion acoustic velocity $v_{ac} = \sqrt{T_e/m_i}$, this expression can be written as

$$\tilde{n}_i = n_0 \left(-\frac{v_0 k_z}{\omega} + \frac{v_{ac}^2 k_z^2}{\omega^2}\right) \frac{e\tilde{V}}{T_e}. \tag{16.71}$$

Now, replacing this in the potential equation (16.68), we get

$$\epsilon(\omega, \mathbf{k})\, k^2 \tilde{V} = 0, \tag{16.72}$$

where $k^2 = k_y^2 + k_z^2$. The condition $\epsilon(\omega, \mathbf{k}) = 0$ is necessary to guarantee the existence of an electrostatic potential perturbation $\tilde{\phi} \neq 0$. This condition can be explicitly written as

$$\omega^2 = \frac{\omega_{pi}^2}{v_{ac}^2 k^2} \left[\omega^2 - \omega(\mathbf{k} \cdot \mathbf{v}_0) - k_z^2 v_{ac}^2 \right]. \tag{16.73}$$

This is the dispersion relation for the low-frequency drift waves. In the long wavelength limit, where $\omega_{pi}^2 / v_{ac}^2 \gg k^2$, this can be simplified to give

$$\omega^2 = \omega(\mathbf{k} \cdot \mathbf{v}_0) + v_{ac}^2 k_z^2. \tag{16.74}$$

This looks very much like an ion acoustic wave modified by the electron drift motion. When the waves are propagating along the magnetic field lines, and $k_y = 0$ or $(\mathbf{k} \cdot \mathbf{v}_0) = 0$, we get $\omega = v_{ac} k_z$, the ion acoustic mode. In contrast, when the waves propagate across the magnetic field and $k_z = 0$, we simply have $\omega = k_y v_0$. This means that the transverse phase velocity coincides with the electron drift velocity.

Drift waves were first considered by Rudakov and Sagdeev [13]. Their stability and dispersion properties have been discussed for very long time in the plasma physics literature, due to their relevance to both space plasmas and fusion research. As a consequence of their universal character, they will probably also be relevant to the new area of ultra-cold plasmas.

16.6 Waves in Magnetized Cold Plasmas

16.6.1 General Dispersion Relation

We now give a general overview of waves that can propagate in magnetized cold plasmas. Here we mainly follow the approach of Ref. [3]. We restrict our discussion to the case of homogeneous and cold plasmas, where particle collisions are neglected. We start from Maxwell's equations with the plasma charge and current densities, ρ and $\mathbf{J}$ as defined by

$$\rho = \sum_\alpha q_\alpha n_\alpha, \quad \mathbf{J} = \sum_\alpha q_\alpha n_\alpha \mathbf{v}_\alpha, \tag{16.75}$$

where the mean densities and velocities n_α and $\mathbf{v}_\alpha$ are determined by the fluid equations of the different plasma species $\alpha = (e, i)$, where thermal effects are neglected. We assume propagation of an infinitesimal wave perturbation in uniform plasma, with equilibrium density n_0 and confining magnetic field $\mathbf{B}_0 = B_0 \mathbf{e}_z$. We can write $n_\alpha = n_0 + \tilde{n}_\alpha$, and assume that the particle and field perturbations $\tilde{n}_\alpha, \mathbf{v}_\alpha, \boldsymbol{\mathcal{E}}$ and $\mathbf{B}$ all evolve in time as $\exp(-i\omega t)$. The linearized fluid equations can then be written as

$$\tilde{n}_\alpha = -i \frac{n_0}{\omega} \nabla \cdot \mathbf{v}_\alpha, \quad \mathbf{v}_\alpha = -i \frac{q_\alpha}{\omega m_\alpha} (\boldsymbol{\mathcal{E}} + \mathbf{v}_\alpha \times \mathbf{B}_0) . \tag{16.76}$$

This last equation can be solve as $\mathbf{v}_\alpha = \bar{\bar{\mu}}_\alpha \cdot \boldsymbol{\mathcal{E}}$, where $\bar{\bar{\mu}}_\alpha$ is the *mobility tensor*. The total current density will then take the form

$$\mathbf{J} = \bar{\bar{\sigma}} \cdot \boldsymbol{\mathcal{E}}, \quad \bar{\bar{\sigma}} = n_0 \sum_\alpha q_\alpha \bar{\bar{\mu}}_\alpha \tag{16.77}$$

where $\bar{\bar{\sigma}}$ is the plasma *conductivity tensor*. From Maxwell's equations we then get

$$\nabla \times (\nabla \times \boldsymbol{\mathcal{E}}) = \frac{\omega^2}{c^2} \bar{\bar{\epsilon}} \cdot \boldsymbol{\mathcal{E}}, \quad \bar{\bar{\epsilon}} = \bar{\bar{1}} + \frac{i}{\omega \epsilon_0} \bar{\bar{\sigma}}, \tag{16.78}$$

where $\bar{\bar{\epsilon}}$ is the plasma dielectric tensor, and $\bar{\bar{1}}$ is the unit matrix. This can be written in explicit terms as

$$\bar{\bar{\epsilon}} = \begin{bmatrix} \epsilon_1 & i\epsilon_2 & 0 \\ -i\epsilon_2 & \epsilon_1 & 0 \\ 0 & 0 & \epsilon_3 \end{bmatrix}, \tag{16.79}$$

where the different matrix elements are defined as follows

$$\epsilon_1 = 1 - \sum_\alpha \frac{\omega_{p\alpha}^2}{(\omega^2 - \omega_{c\alpha}^2)}, \quad \epsilon_2 = \sum_\alpha s_\alpha \frac{\omega_{c\alpha}}{\omega} \frac{\omega_{p\alpha}^2}{(\omega^2 - \omega_{c\alpha}^2)}, \quad \epsilon_3 = 1 - \sum_\alpha \frac{\omega_{p\alpha}^2}{\omega^2}, \tag{16.80}$$

with s_α being the sign of the particle charge q_α, and the cyclotron frequencies $\omega_{c\alpha} = |q_\alpha| B_0 / m_\alpha$ defined as positive quantities. For very high frequencies, such that $\omega^2 \gg \omega_{ci}, \omega_{pi}$, we can neglect the ion terms in the dielectric tensor. If the wave perturbation evolves as $\exp(i\mathbf{k} \cdot \mathbf{r})$, the wave equation (16.78) reduces to an expression of the form $\bar{\bar{D}} \cdot \boldsymbol{\mathcal{E}} = 0$, where

$$\bar{\bar{D}} = \bar{\bar{1}} - \frac{\mathbf{k}\mathbf{k}}{k^2} - \frac{\omega^2}{k^2 c^2} \bar{\bar{\epsilon}}. \tag{16.81}$$

Nontrivial solutions for the wave electric field $\boldsymbol{\mathcal{E}} \neq 0$ can exist if the determinant of $\bar{\bar{D}}$ is equal to zero. If we assume a coordinate system where the wavevector $\mathbf{k}$ is in the Oxz plane, making an angle θ with the static magnetic field, $\mathbf{k} \cdot \mathbf{B}_0 = k B_0 \cos\theta$, this determinant equation can be written in the form

$$\mathcal{A}N^4 - \mathcal{B}N^2 + \mathcal{C} = 0, \tag{16.82}$$

where $N = kc/\omega$ is the refractive index, and

$$\begin{aligned}
\mathcal{A} &= \epsilon_1 \sin^2\theta + \epsilon_3 \cos^2\theta, \\
\mathcal{B} &= \left(\epsilon_1^2 - \epsilon_2^2\right)\sin^2\theta + \epsilon_1\epsilon_3\left(1 + \cos^2\theta\right), \\
\mathcal{C} &= \epsilon_3\left(\epsilon_1^2 - \epsilon_2^2\right).
\end{aligned} \tag{16.83}$$

The solution to Eq. (16.82) is then given by

$$N^2 = \frac{1}{2\mathcal{A}}(\mathcal{B} \pm \mathcal{D}), \tag{16.84}$$

with

$$\mathcal{D} = \left(\epsilon_1^2 - \epsilon_2^2 - \epsilon_1\epsilon_3\right)^2 \sin^4\theta + 4\epsilon_2^2\epsilon_3^2\cos^2\theta. \tag{16.85}$$

This shows that we have two distinct solutions for N^2, or two distinct modes of wave propagation. An alternative and eventually more suggestive way of writing $\det \bar{\bar{D}} = 0$ is

$$\tan^2\theta = -\frac{\epsilon_3[(N^2 - \epsilon_1)^2 - \epsilon_2^2]}{(\epsilon_1 N^2 - \epsilon_1^2 + \epsilon_2^2)(N^2 - \epsilon_3)}. \tag{16.86}$$

Of particular relevance are the conditions for which the wave refractive index becomes zero or infinite. We have a *cut-off* when $N = 0$, and a *resonance* when $N \to \infty$. Equation (16.82) shows that a cut-off occurs when $\mathcal{C} = 0$, which corresponds to one of the two independent conditions

$$\epsilon_3 = 0, \quad \epsilon_1 \pm \epsilon_2 = 0. \tag{16.87}$$

These conditions are independent of the angle of propagation θ. On the other hand, Eq. (16.84) shows that a resonance exists when $\mathcal{A} = 0$. This, however, depends on the propagation angle, as shown by the definition of $\mathcal{A}$ in Eq. (16.84), and takes place for

$$\tan^2\theta = -\frac{\epsilon_3}{\epsilon_1}. \tag{16.88}$$

Going back to Maxwell's equations (16.78), we can easily realize that a cut-off corresponds to the condition $\mathbf{k} \times \mathbf{H} = 0$, which means that the displacement current exactly compensates the convection current. On the other hand, at resonance ($N \to \infty$), the transverse electric field vanishes, $\boldsymbol{\mathcal{E}}_\perp = 0$, in order to guarantee that $\mathbf{k} \times \boldsymbol{\mathcal{E}}$ remains finite. This means that the electric field of the wave is purely longitudinal at resonance.

Let us now comment on the wave polarization. Given the condition $\bar{\bar{D}} \cdot \boldsymbol{\mathcal{E}} = 0$, the electric field components have to be proportional to the corresponding minor determinants. Developing the determinant $\det \bar{\bar{D}} = 0$ with respect to the first line, we obtain

$$\frac{E_y}{E_x} = \frac{i\epsilon_2}{N^2 - \epsilon_1}, \quad \frac{E_z}{E_x} = \frac{N^2 \sin^2\theta - \epsilon_3}{N^2 \sin\theta \cos\theta}. \tag{16.89}$$

In order to illustrate these generic results, we will discuss next the particular but important cases of parallel and perpendicular wave propagation with respect to the magnetic field lines.

16.6.2 Parallel Propagation

Let us first examine the case where the wave propagates along the magnetic field lines, and the wavevector $\mathbf{k}$ is oriented along $\mathbf{B}_0$. This corresponds to $\theta = 0$. As can be seen from Eq. (16.86), this can occur in three different situations. First, we can have $\epsilon_3 = 0$, defining the electrostatic oscillations. They can in fact be identified with the long wavelength limit of electron plasma waves. This also implies that $\mathcal{D} = 0$, as shown by (16.85). On the other hand, for $\epsilon_3 \neq 0$, the condition of parallel propagation is satisfied for two different values for N^2, as shown by both Eqs. (16.82) and (16.86), corresponding to $N_\pm^2 = \epsilon_1 \pm \epsilon_2$. This can be explicitly written as

$$N_\pm^2 = 1 - \frac{\omega_p^2}{(\omega \mp \omega_{ce})(\omega \pm \omega_{ci})}, \tag{16.90}$$

where we have used $\omega_p^2 = \omega_{pe}^2 + \omega_{pi}^2$. These are two transverse electromagnetic waves with $\mathbf{k}_\pm \cdot \boldsymbol{\mathcal{E}} = 0$, polarized in the plane perpendicular to $\mathbf{B}_0$. It can seen from Eq. (16.89) that

$$\frac{E_y}{E_x} = \frac{i\epsilon_2}{N_\pm^2 - \epsilon_1} = \pm i. \tag{16.91}$$

The plus sign, $+i$, corresponds to the right-hand circular polarization, or R *wave*, where the electric field turns in the same direction as the rotation of the electron cyclotron orbits around the magnetic field lines. Similarly, the minus sign, $-i$, describes the L *wave* with left-hand circular polarization, where the electric field turns with the positively charged ions. It is therefore not surprising to notice that the R and L waves are resonant at the electron and ion cyclotron frequencies respectively, with $N_\pm \to \infty$ for $\omega \to \omega_{ce}, \omega_{ci}$. Around such resonances the transfer of energy from the waves to the particles can be significant, a process used for plasma heating in magnetic fusion devices.

In what concerns the cut-off frequencies, such that $N = 0$, and $k = 0$, we can see directly from (16.90), or from the above general condition (16.87) that they are attained for

$$\omega^2 \mp \omega(\omega_{ce} - \omega_{ci}) - (\omega_p^2 + \omega_{ce}\omega_{ci}) = 0. \tag{16.92}$$

Noting that $\omega_{ci}/\omega_{ce} = m_e/m_i \ll 1$, we can define the two cut-off frequencies for the R and the L mode (corresponding to the plus and minus signs) as

$$\omega_R = \frac{1}{2}\left(\sqrt{\omega_{ce}^2 + 4\omega_p^2} + \omega_{ce}\right), \quad \omega_L = \omega_R - \omega_{ce}. \tag{16.93}$$

It is interesting to consider the limit of a weak magnetic field, such that $\omega_p^2 \gg \omega_{ce}^2$. These two cut-off frequencies reduce to $\omega_R \simeq \omega_p + \omega_{ce}/2$ and $\omega_L \simeq \omega_p - \omega_{ce}/2$. Comparing this with the cut-off for the unmagnetized plasma, which is $\omega = \omega_p$, we can see that the cut-off frequencies show a kind of *Zeeman splitting*, similar to that of the atom energy levels in a magnetic field.

These R and L modes are sometimes known under different names, in different regions of the frequency spectrum, because they have been explored by different scientific communities. As an illustration, let us first consider the high frequency regions, where $\omega^2 \gg \omega_{pi}^2, \omega_{ci}^2$. In this case, the refractive index (16.90) reduces to

$$N_\pm^2 \simeq 1 - \frac{\omega_p^2}{\omega(\omega \mp \omega_{ce})}. \tag{16.94}$$

An interesting intermediate region corresponds to $\omega_{ce} \gg \omega \gg \omega_{ci}$. In this case, we can write the dispersion relation for the R mode, corresponding to $N_+^2 \gg 1$ in the form

$$\omega \simeq k^2 c^2 \frac{\omega_{ce}}{\omega_p^2}. \tag{16.95}$$

This is the well known dispersion relation for the *whistler modes*, which can be excited by lightening flares in the *ionosphere*, and are observed in the audio-frequency region of the spectrum. The name is due to the fact that higher frequencies have higher phase and group velocities, due to the dependence $\omega \propto k^2$, and they arrive first at an observer located at low altitudes, when they propagate down to the Earth along the terrestrial magnetic field lines.

Let us now consider the low frequency limit, such that $\omega^2 \ll \omega_{pi}^2, \omega_{ci}^2$. In this case, the two modes, R and L, tend to a single mode with dispersion relation

$$N^2 \simeq 1 + \frac{\omega_p^2}{\omega_{ce}\omega_{ci}} = 1 + \frac{c^2}{v_A^2}, \quad v_A = \frac{B_0}{\sqrt{\mu_0 n_0 m_i}}, \tag{16.96}$$

where v_A is the *Alfvén speed*. This dispersion relation can also be written as

$$\omega^2 = \frac{k^2 v_A^2}{1 + (v_A/c)^2}. \tag{16.97}$$

These low frequency waves are called the *Alfvén waves*. For low magnetic fields and high plasma densities, we can have $v_A \ll c$. If we retain the lowest order corrections due to the ion cyclotron motion in (16.90), two distinct modes reappear, with dispersion relations

$$\omega^2 = k_\pm^2 v_A^2 \left(1 \pm \frac{\omega}{\omega_{ci}}\right). \tag{16.98}$$

In that case, the R wave is called the *compressional Alfvén wave*, with fast phase velocity, $\omega/k_+ > v_A$, and the L wave is often associated to the *shear Alfvén wave*, with slow phase velocity, $\omega/k_- < v_A$.

16.6.3 Perpendicular Propagation

Let us now consider propagation across the static magnetic field, such that $\mathbf{k}\cdot\mathbf{B}_0 = 0$, or $\theta = \pi/2$. Going back to the general dispersion relation (16.82) and (16.86), we obtain two new modes called the *ordinary mode*, $N^2 = N_O^2$, and the *extraordinary mode*, $N^2 = N_X^2$, with

$$N_O^2 = \epsilon_3 = 1 - \frac{\omega_p^2}{\omega^2}, \quad N_X^2 = \frac{\epsilon_1^2 - \epsilon_2^2}{\epsilon_1}. \tag{16.99}$$

The ordinary mode (or O mode) is linearly polarized in the direction of the static magnetic field, $\boldsymbol{\mathcal{E}} = \mathcal{E}\mathbf{b}$, as can be seen from (16.89). The currents associated with this electric field are not affected by the existence of static field $\mathbf{B}_0$, because the particles can move freely along the magnetic field lines. This mode is a transverse mode identical to that of a non-magnetized plasma, and has no electron density perturbations $\tilde{n} = 0$.

In contrast, the extraordinary mode (or X mode) has polarization in the perpendicular xy-plane. This means that it has both transverse and longitudinal components, as determined by

$$\frac{E_x}{E_y} = i\frac{\epsilon_2}{\epsilon_1}, \tag{16.100}$$

where, in general, we have $\epsilon_2 \neq \epsilon_1$. We should notice that, at resonance, $N_X \to \infty$, we have $\epsilon_1 \to 0$, and the mode becomes linearly polarized along the propagation direction, or, in other words, it behaves as an electrostatic mode. Near the cut-off, $N_X^2 = 0$, defined by $\epsilon_1 = \pm\epsilon_2$ the X mode becomes circularly polarized. The resonances for the X mode are defined by the condition $\epsilon_1 = 0$. This determines two resonance frequencies called the upper and lower *hybrid resonances*, as defined by

$$\omega_{uh} = \sqrt{\omega_p^2 + \omega_{ce}^2}, \quad \omega_{lh} = \left[\omega_{ce}\omega_{ci}\frac{(\omega_{pi}^2 + \omega_{ci}^2)}{\omega_{pi}^2 + \omega_{pe}\omega_{ce}}\right]^{1/2}. \tag{16.101}$$

In the limit of low plasma density, or strong magnetic field, such that $\omega_{pe}^2 \ll \omega_{ce}\omega_{ci}$, this reduces to $\omega_{uh} \simeq \omega_{ce}$ and $\omega_{lh} \simeq \omega_{ci}$. In the opposite limit of high density and weak confining field, $\omega_{pe}^2 \gg \omega_{ce}^2$, Eq. (16.101) become $\omega_{uh} \simeq \omega_p$ and $\omega_{lh} \simeq$

$\sqrt{\omega_{ce}\omega_{ci}}$. The explicit expression for the refractive index of the X mode can be written as

$$N_X^2 = \frac{(\omega^2 - \omega_p^2 - \omega_{ce}\omega_{ci})^2 - (\omega_{ce} - \omega_{ci})^2\omega^2}{(\omega^2 - \omega_{ce}^2)(\omega^2 - \omega_{ci}^2) - \omega_p^2(\omega^2 - \omega_{ce}\omega_{ci})} \tag{16.102}$$

In the very high frequency limit, $\omega^2 \gg \omega_{ce}^2, \omega_p^2$, we have $N_X^2 \simeq N_O^2$, and the two modes can hardly differentiate. For weak magnetic fields and high frequencies such that $\omega^2 > \omega_p^2 \gg \omega_{ci}^2$, we can write

$$N_X^2 \simeq N_O^2 - \frac{\omega_p^2\omega_{ce}^2}{\omega^4 N_O^2}. \tag{16.103}$$

This shows that, in the region mentioned above, we always have $N_X < N_O$. Finally, for very low frequencies such that $\omega^2 \ll \omega_{ci}^2, \omega_{pi}^2$, we obtain

$$N_X^2 \simeq 1 + \frac{\omega_p^2}{\omega_{ce}\omega_{ci}} = \frac{c^2}{v_A^2}. \tag{16.104}$$

In this limit, the X mode becomes an Alvén wave.

16.7 Waves in Expanding Plasmas

We conclude this chapter by discussing two different aspects of ion acoustic wave propagation in ultra-cold plasmas, which considerably differs from the usual plasma wave phenomena. First we discuss the case of a non-stationary plasma. In ultra-cold plasmas, the time scale for plasma bubble expansion is large compared with a typical ion acoustic wave period, but short as compared with the time scale for acoustic wave propagation inside the medium. As a result, the temporal evolution of the background plasma parameters can influence the wave oscillations that are excited in the medium. This finite size effect has been observed in recent experiments by [14].

The excitation of ion acoustic waves in a non-stationary plasma, can be described by the ion fluid equations for the ion number density n_i and the ion fluid velocity $\mathbf{v}$. We have the ion continuity and momentum equations, respectively,

$$\frac{\partial n_i}{\partial t} + \nabla \cdot (n_i \mathbf{v}_i) = 0, \quad \frac{\partial \mathbf{v}}{\partial t} + (\mathbf{v} \cdot \nabla)\mathbf{v} = -\frac{Ze}{m_i}\nabla V - \frac{v_{\text{thi}}^2}{n}\nabla n, \tag{16.105}$$

where the ion thermal speed is $v_{\text{thi}} = \sqrt{T_i/m_i}$, with the ion mass m_i. As we have previously seen, for low frequency waves, the electron density n_e can be assumed at Boltzmann equilibrium in the wave potential V, at the temperature $T_e \neq T_i$. This Boltzmann equilibrium remains valid as long as the phase velocity of the ion acoustic waves is much smaller than the electron thermal velocity. Finally, the electrostatic potential V is determined by Poisson's equation

$$\nabla^2 V = \frac{e}{\epsilon_0}(n_e - Zn), \quad n_e = n_{eq}(t)\exp(eV/T_e), \tag{16.106}$$

where $n_{eq}(t)$ is the equilibrium density. The plasma equilibrium, with $n = n_0/Z$ and $\mathbf{v} = 0$, is established on a very short timescale as compared with the long ion-time scale. This allows the equilibrium ion plasma density and temperature to vary on such a long time scale. This means that we can use a time-dependent equilibrium density $n_0(t) = n_0 f(t)$, with $|f(t)| \simeq 1$. For simplicity, the temperatures are assumed constant, but this assumption can easily be relaxed. We then consider $n_i = n_0(t)/Z + \tilde{n}$. Linearizing equations (16.105), we obtain the evolution equation for the ion density perturbation

$$\left\{\frac{\partial^2}{\partial t^2} + \omega_{pi}^2(t)\left[1 - \frac{n_0 f(t)}{Z}\frac{eV}{T_e}\right] - v_{\text{th,i}}^2 \nabla^2\right\}\tilde{n} = \nu(t)\left[\frac{\partial \tilde{n}}{\partial t}\right], \tag{16.107}$$

where the quantity $\nu(t)$ and the ion plasma frequency $\omega_{pi}(t)$ are defined by

$$\nu(t) = d\ln f(t)/dt, \quad \omega_{pi}^2(t) = \frac{Ze^2 n_0 f(t)}{\epsilon_0 m_i}. \tag{16.108}$$

Equation (16.107) can be coupled with the linearized Poisson's equation (viz. $eV/T_e \ll 1$), which can be written as

$$\nabla^2 V = \omega_{pi}^2(t)\frac{m_i V}{ZT_e} - Z\frac{e\tilde{n}}{\epsilon_0}. \tag{16.109}$$

For a slow temporal evolution of the plasma background, such that the source term on the right-hand side of Eq. (16.107) is negligible, we can seek solutions of the form

$$\tilde{n}(\mathbf{r}, t) = N_k e^{\pm i\mathbf{r}\cdot\mathbf{r}} \exp\left[-i\int^t \omega_k(t')dt'\right], \tag{16.110}$$

with a similar solution for the electrostatic potential V. Replacing $\tilde{n}$ in Eqs. (16.107) and (16.109), we can derive the dispersion relation for the ion acoustic waves in the non-stationary plasma, as [15]

$$\omega_k(t)^2 = k^2 v_{\text{th,i}}^2 \left[1 + \frac{ZT_e}{T_i}\frac{1}{1 + k^2\lambda_D^2(t)}\right], \tag{16.111}$$

where the time dependent electron Debye length is defined, $\lambda_D(t) = \left[ZT_e/m_i\omega_{pi}^2(t)\right]^{1/2}$. This shows that the wave frequency changes as a result of the temporal changes in the background medium. Similar effects have been first identified in sound waves propagating in neutral plasma [16, 17]. This is actually a more general characteristic feature of waves propagating in time-varying media, and was mainly studied in the context of electromagnetic radiation. Such a characteristic feature, already identified for phonons in condensates, is sometimes called *time-refraction* [18]. These frequency shifts are, however, less pronounced in the case of ion acoustic waves, and can even be negligible for very long wavelengths ($k^2\lambda_D^2 \ll 1$), as long as the electron and ion temperature variations are ignored. This situation strongly contrasts with the cases of the photons and plasmons, where the frequency shift is directly associated with time variations of the plasma density [19].

Let us now consider the case where the quantity $\nu(t)$ is not negligible. This corresponds to a plasma bubble, expanding in the absence of any ionization process. An exact solution of Eqs. (16.107)–(16.109) can be found, involving now two distinct solutions $\tilde{n}_k$ and $\tilde{n}_{-k}$, similar to (16.110), with the same axial wave vector $\mathbf{k}$ but propagating in opposite directions. These two ion acoustic wave modes are coupled in the time-varying plasma. It can be shown that the wave solution (16.110) is replaced by a more adequate solution of the form

$$\tilde{n}(\mathbf{r}, t) = N_k(t)\exp[i\mathbf{k}\cdot\mathbf{r} - i\varphi(t)] + N_{-k}(t)\exp[i\mathbf{k}\cdot\mathbf{r} + i\varphi(t)] + \text{c.c.}, \quad (16.112)$$

with the phase given by $\varphi(t) = \int^t \omega_k(t')dt'$ and $\omega_k(t)$ is defined by the dispersion relation (16.111). Replacing this in the wave equation, and retaining only the terms with the same spatial dependence in $\exp(\pm i\mathbf{k}\cdot\mathbf{r})$, we obtain two coupled equations for the counter-propagating wave amplitudes N_k and N_{-k}, as given by

$$\frac{dN_k}{dt} = \eta(t)N_{-k}\exp[2i\varphi(t)], \qquad \frac{dN_{-k}}{dt} = \eta(t)N_k\exp[-2i\varphi(t)], \qquad (16.113)$$

with

$$\eta(t) = \frac{1}{2}\left[\frac{1}{\omega_k}\frac{d\omega_k}{dt} - \nu(t)\right] = \frac{1}{2}\frac{d}{dt}\ln\left(\frac{\omega_k}{f}\right). \qquad (16.114)$$

Here we have assumed slowly varying amplitudes, such that $|dN_k/dt| \ll |\omega_k N_k|$, and $|dN_{-k}/dt| \ll |\omega_k N_{-k}|$. These coupled equations can be easily solved, when the reflected wave amplitude is small with respect to a given wave initially excited in the medium, $|N_{-k}| \ll |N_k|$, and $N_k(t) \simeq$ const. In such limit, we may have

$$N_{-k}(t) = R(t)N_k(t), \quad R(t) \simeq \int^t \eta(t')\exp[-2i\varphi(t')]dt', \qquad (16.115)$$

where $R(t)$ is a temporal reflection coefficient. This is formally very similar to what can be derived for the reflection coefficient $R(z)$ associated with the usual (space) refraction [2, 18], with the space variable z replaced by the temporal one t. However, the physical meaning is different, because there are no physical discontinuities from where the initial wave could be reflected. Such a discussion runs in parallel with that for the Casimir effect, as considered before in the frame of the quantum theory of Bose Einstein condensates. An explicit expression for $\eta(t)$ is given by

$$\eta(t) = \frac{\nu(t)}{2}\left[\frac{1}{2}\frac{k^2\lambda_D(t)^2}{1+k^2\lambda_D(t)^2}\left(1-\frac{k^2 v_{\mathrm{th,i}}^2}{\omega_k^2}\right)-1\right]. \tag{16.116}$$

We therefore conclude that any temporal change in the background plasma density (or temperature) will lead to wave reflection, even for an infinite and homogeneous plasma. This somewhat counter-intuitive result is impose by total momentum conservation.

16.8 Waves in Strongly Coupled Plasmas

We have seen that, in ultra-cold plasmas, the thermal energy of the charged particles (specially ions) can be smaller than the Coulomb interaction energy between nearest neighbors, making them strongly coupled systems. It is then useful to examine the properties of ion acoustic waves propagating in ultra-cold plasmas, when the ions are strongly coupled. For our purposes, we can still use the Boltzmann law for the electron density perturbation. In what concerns the ions, they can be described by a modified ion momentum equation, which can be established by including a phenomenological viscoelastic term, as [20]

$$\frac{d\mathbf{v}_i}{dt} = -\frac{\nabla P_i}{m_i n_i} + \frac{Ze}{m_i}\mathcal{E} + \int_{-\infty}^{t} dt' \int d\mathbf{r}' \eta_i(\mathbf{r}-\mathbf{r}', t-t')\mathbf{v}_i(\mathbf{r}', t'), \tag{16.117}$$

where P_i is the ion pressure and η_i is a non-local viscoelastic operator, describing memory effects and short range correlations between the ions. This term will obviously increase with the ion coupling parameter Γ_i. If we introduce an exponential memory decay described by a relaxation time τ_m, and all the other features of a viscosity operator, we can write the space-time Fourier components of this operator as

$$\eta_i(\omega, \mathbf{k}) = \frac{\eta k^2 + \eta^* \mathbf{k}(\mathbf{k}\cdot\ )}{m_i n_i (1 - i\omega\tau_m)}, \tag{16.118}$$

where $\eta^* = (\eta/3 + \zeta)$. The bulk viscosity coefficient η, the shear viscosity ζ and τ_m are introduced phenomenologically. Doing a regression Fourier analysis and combining this with Eq. (16.117), we can get the modified ion momentum equation, for one-dimensional propagation [21]

$$\left(1 + \tau_m \frac{\partial}{\partial t}\right)\left[\frac{\partial \tilde{V}}{\partial t} + \frac{Ze}{m_i}(1-R)\frac{\partial V}{\partial x} + \frac{\gamma_i T_i}{\rho_{i0}}\frac{\partial \tilde{n}}{\partial x}\right] = \frac{\eta_*}{\rho_{i0}}\frac{\partial^2 v}{\partial x^2}, \tag{16.119}$$

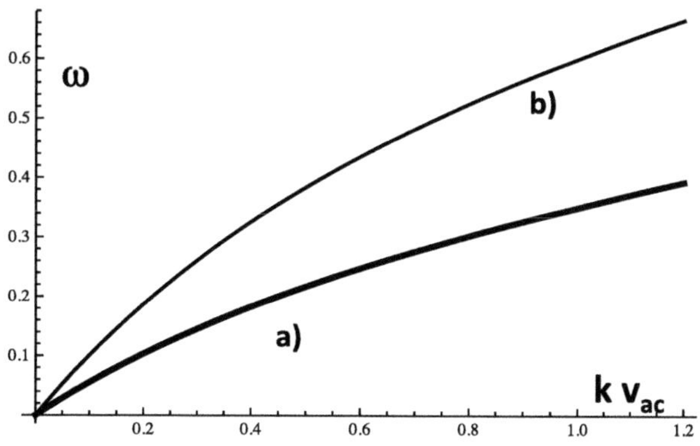

Fig. 16.5 (**a**) Dispersion relation of ion acoustic waves in a strongly coupled plasma, and (**b**) in the absence of ion coupling

where for practical purposes we can choose

$$\tau_m = \frac{\eta_i}{n_{i0}T_i}\left[1 - \mu_i + \frac{4}{15}U(\Gamma_i)\right]^{-1}, \quad \mu_i = \frac{1}{T_i}\frac{\partial P_i}{\partial n_{i0}} \simeq 1 + 0.33U(\Gamma_i) + \frac{\Gamma_i}{9}\frac{\partial U(\Gamma_i)}{\partial \Gamma_i}, \tag{16.120}$$

with $\rho_i = m_i n_{i0}$ being the ion mass density, $U(\Gamma_i)$ the measure of the excess internal ion energy, $\tilde{n}$ and $\tilde{v}$ the ion (number) density and ion velocity perturbations, and γ_i the ion adiabatic index. We have also introduced $R = e^2/4k_B T_e \lambda_D \equiv (16 n_{e0}\lambda_D^3)^{-1} \leq 1$. The term proportional to $(1 - R)$ in Eq. (16.119) accounts for the polarization force arising from interactions between the thermal electrons and strongly coupled ions in the ultra-cold plasma. Combining this with Poisson's equation (16.2) for the electrostatic potential, and using the electron Boltzmann law such that $\tilde{n}_e \simeq n_{e0}e\tilde{V}/T_e$, we obtain the propagation equation along the arbitrary x-direction

$$\left(1 + \tau_m \frac{\partial}{\partial t}\right)\left[\frac{\partial^2 \tilde{n}}{\partial t^2} - v_{\text{th,i}}^2 \frac{\partial^2 \tilde{n}}{\partial x^2} - \frac{Z_i n_{i0} e}{m_i}(1 - R)\frac{\partial^2 \tilde{V}}{\partial x^2}\right] - \frac{\eta_*}{\rho_i}\frac{\partial^3 n_{i1}}{\partial x^2 \partial t} = 0. \tag{16.121}$$

Assuming that $\tilde{n}_i$ and $\tilde{V}$ are proportional to $\exp(-i\omega t + ikx)$, where ω and k are the frequency and the wave number, we obtain the dispersion relation for the modified ion acoustic waves [21]

$$1 + \frac{1}{k^2\lambda_D^2} - \frac{(1 - R)\omega_{pi}^2}{\left[\omega^2 - k^2 V_{Ti}^2 + i\omega k^2 \eta_*/\rho_i(1 - i\omega\tau_m)\right]} = 0. \tag{16.122}$$

where $\omega_{pi} = (n_{i0}Z_i^2 e^2/\epsilon_0 m_i)^{1/2}$ is the ion plasma frequency. Equation (16.122) reveals that the electrostatic force is weakened by the polarization force, leading to a significant reduction of the ion plasma frequency in ultra-cold plasmas. In the limit of $|\omega| \gg \tau_m^{-1}$, Eq. (16.122) reduces to

$$\omega^2 = \frac{k^2\eta_*}{\rho_i \tau_m} + k^2 V_{Ti}^2 + \frac{(1-R)k^2 v_{ac}^2}{(1+k^2\lambda_D^2)}, \tag{16.123}$$

where $v_{ac} = \omega_{pi}\lambda_D \equiv (Z_i T_e/m_i)^{1/2}$ is the ion sound speed. In the opposite limit, such that $|\omega| \ll \tau_m^{-1}$, the modified ion acoustic wave is weakly damped due to the viscoelastic effect in ultra-cold plasmas, at it should be expected. The wave damping rate is determined from

$$\omega = -i\frac{k^2\eta_*}{2\rho_i} \pm \left[-\frac{k^4\eta_*^2}{4\rho_i^2} + k^2 V_{thi}^2 + \frac{(1-R)k^2 v_{ac}^2}{(1+k^2\lambda_D^2)}\right]^{1/2}. \tag{16.124}$$

This result shows that the ion-ion coupling can strongly modify the dispersion properties of ion acoustic waves, as illustrated in Fig. 16.5. In particular, the wave frequency is substantially reduced due to the polarization force, since R is typically of order 1 in ultra-cold plasmas.

References

1. T.W. Stix, *Theory of Plasma Waves* (McGraw-Hill, New York, 1962)
2. V.L. Ginzburg, *Propagation of Electromagnetic Waves in Plasma* (Gordon and Breach, New York, 1961)
3. D. Quemada, *Ondes dans les Plasmas* (Hermann, Paris, 1968)
4. J-H. Choi, B. Knuffman, X.H. Zhang, A. P. Povilus, G. Raithel, Phys. Rev. Lett. **89**, 233401 (2002)
5. S. Kulin, T.C. Killian, S.D. Bergeson, S.L. Rolston, Phys. Rev. Lett. **85**, 318 (2000)
6. R.S. Fletcher, X.L. Zhang, S.L. Rolston, Phys. Rev. Lett. **96**, 105003 (2006)
7. J.T. Mendonça, J. Loureiro, H. Terças, J. Plasma Phys. **51**, 024007 (2009)
8. V.I. Karpman, H. Washimi, Sov. Phys. JETP **44**, 528 (1976)
9. J.T. Mendonça, N. Shukla, P.K. Shukla, J. Plasma Phys. **76**, 19–23 (2010)
10. X.L. Zhang, R.S. Fletcher, S.L. Rolston, Phys. Rev. Lett. **101**, 195002 (2008)
11. A. Smirnov, Y. Raitses, N.J. Fisch, Phys. Plasmas **14**, 057106 (2007)
12. Y. Raitses, N.J. Fisch, Phys. Plasmas **8**, 2579 (2001)
13. L.I. Rudakov, R.Z. Sagdeev, Sov. Phys. Dokl. **6**, 415 (1961)
14. J. Castro, P. McQuillen, T.C. Killian, Phys. Rev. Lett. **105**, 065004 (2010)
15. J.T. Mendonça, P.K. Shukla, Phys. Plasmas **18**, 042101 (2011)
16. J. Castro, P. McQuillen, T.C. Killian, Phys. Rev. Lett. **105**, 065004 (2010); T.C. Killian, P. McQuillen, T.M. O'Neil, J. Castro, Phys. Plasmas **19**, 055701 (2012)
17. T. C. Killian, Science **316**, 705 (2007).
18. J.T. Mendonça, *Theory of Photon Acceleration* (Institute of Physics, Bristol, 2001)
19. J.T. Mendonça, New J. Phys. **11**, 013029 (2009)
20. P.K. Kaw, A. Sen, Phys. Plasmas **5**, 3552 (1998)
21. P.K. Shukla, Phys. Lett. A **374**, 3656 (2010)

Chapter 17
Kinetic Theory of Waves

We complete the wave description by extending it into three different directions. They include wave kinetic phenomena, waves in quantum plasmas and waves in turbulent plasmas. The discussion will be restricted to electrostatic waves, namely, electron plasma waves and ion acoustic waves. Our main aim will be to stress the analogy between the kinetic plasma effects, in particular, kinetic susceptibilities and the electron and ion Landau damping, with those pertaining to the ultra-cold atoms discussed in the first part of this book. Of particular interest is the propagation window of undamped ion acoustic waves, which depends on both the electron and ion Landau damping contributions, and depart from the single particle Landau damping considered in Chap. 5.

We then extend the wave description to the case of quantum plasmas. Although current plasma experiments remain distant from quantum plasma conditions, this extension is useful for two kinds of reasons. One is that we cannot exclude the possible experimental exploration of quantum plasma conditions in the long term. This could be achieved by a considerable increase of the plasma density and reduction of the electron temperature. Second, the quantum plasma regime is conceptually very interesting, because it introduces quantum dispersion terms which are formally identical to those already identified for waves propagating in the ultra-cold neutral gas and in condensates.

The chapter ends with a classical but non-trivial description of ion acoustic waves, where the plasma is prepared in a turbulent state, and not in the usual steady state. Plasma turbulence is described as a superposition of plasmon modes, where these modes are seen as quasi-particles, in the same way that we did for the sound waves in a superfluid. In this sense the medium will be composed by two different kinds of particles, the real plasma particles (electrons and ions) and the plasmon quasi-particles. The model is then conceptually analogous to the model of two fluids used to describe superfluidity, as shown in Chap. 11. The difference here is that the long wavelength excitations are replaced by ion acoustic waves, and the background (Bogoliubov) phonons are replaced by plasmons. The result of this turbulent background is the modification of the wave dispersion properties and the appearance of an effective sound velocity.

J.T. Mendonça and H. Terças, *Physics of Ultra-Cold Matter*, Springer Series on Atomic, Optical, and Plasma Physics 70, DOI 10.1007/978-1-4614-5413-7_17,

17.1 Kinetic Dispersion Relation

We first follow the standard description of kinetic wave theory, which can be found in many textbooks [1, 2]. Let us consider a collisionless plasma, described by the Vlasov equation. And let us study the evolution of infinitesimal oscillations, in a spatially uniform and isotropic plasma. We consider electrostatic oscillations, not only for simplicity but also because the kinetic effects are not important for transverse electromagnetic waves. The phase velocity of the electromagnetic waves in unmagnetized plasmas is always larger than the velocity of light c, implying that wave-particle interactions are not relevant. In the absence of magnetic fields, the Vlasov equation for each particle species $\alpha = e, i$ (for electrons or ions) can be written as

$$\frac{\partial f_\alpha}{\partial t} + \mathbf{v} \cdot \frac{\partial f_\alpha}{\partial \mathbf{r}} + \frac{q_\alpha}{m_\alpha} \boldsymbol{\mathcal{E}} \cdot \frac{\partial f_\alpha}{\partial \mathbf{v}} = 0. \tag{17.1}$$

As it is known, the electric field in this equation is self-consistently determined by Maxwell's equations or, for purely electrostatic waves, by Poisson's equation

$$\nabla \cdot \boldsymbol{\mathcal{E}} = \frac{1}{\epsilon_0} \sum_\alpha q_\alpha n_{0\alpha} \int f_\alpha(\mathbf{r}, \mathbf{v}, t) d\mathbf{v}. \tag{17.2}$$

We can see that the third term in the Vlasov equation is nonlinear, because the electric field $\boldsymbol{\mathcal{E}}$ depends on the distribution function f_α. However, in the limit of infinitesimal perturbations, we can linearize this term around the equilibrium, $f_\alpha = f_{0\alpha}$, and $\boldsymbol{\mathcal{E}} = 0$. This can be done by considering a small deviation with respect to equilibrium, $f_\alpha = f_{0\alpha} + \tilde{f}_\alpha$. Let us now consider a Fourier transformation in space coordinates, as defined by

$$\boldsymbol{\mathcal{E}}(\mathbf{r}, t) = \int \boldsymbol{\mathcal{E}}(\mathbf{k}, t) e^{i\mathbf{k}\cdot\mathbf{r}} \frac{d\mathbf{k}}{(2\pi)^3}, \quad \tilde{f}_\alpha(\mathbf{r}, \mathbf{v}t) = \int \tilde{f}_\alpha(\mathbf{k}, \mathbf{v}, t) e^{i\mathbf{k}\cdot\mathbf{r}} \frac{d\mathbf{k}}{(2\pi)^3}. \tag{17.3}$$

For each Fourier component, the linearized Vlasov equation takes the form

$$\frac{\partial \tilde{f}_\alpha(\mathbf{k})}{\partial t} + i\mathbf{k} \cdot \mathbf{v}\, \tilde{f}_\alpha(\mathbf{k}) + \frac{q_\alpha}{m_\alpha} \boldsymbol{\mathcal{E}}(\mathbf{k}) \cdot \frac{\partial f_{0\alpha}}{\partial \mathbf{v}} = 0. \tag{17.4}$$

On the other hand, we can also replace the operator ∇ by $i\mathbf{k}$ in Eq. (17.2). Let us now turn to the temporal evolution and assume that, at $t = 0$ we have a given perturbation described by

$$\tilde{f}_\alpha(\mathbf{k}, \mathbf{v}, t = 0) = g_\alpha(\mathbf{k}, \mathbf{v}). \tag{17.5}$$

We want to determine the evolution of such a perturbation. In order to do that, we add a source term representing the initial conditions to Eq. (17.4), and obtain

$$\left(\frac{\partial}{\partial t} + i\mathbf{k} \cdot \mathbf{v}\right) \tilde{f}_\alpha(\mathbf{k}) + \frac{q_\alpha}{m_\alpha} \boldsymbol{\mathcal{E}}(\mathbf{k}) \cdot \frac{\partial f_{0\alpha}}{\partial \mathbf{v}} = g_\alpha(\mathbf{k}) \delta(t). \tag{17.6}$$

In order to integrate this equation, we can use the temporal Laplace transformation, as defined by

$$\tilde{f}(\omega) = \int_0^\infty \tilde{f}(t) e^{i\omega t} dt, \tag{17.7}$$

where ω is a complex quantity with a positive imaginary part, in order to guarantee that this integral exists. Multiplying equation (17.6) by $\exp(i\omega t)$, and integrating in time, we obtain

$$\tilde{f_\alpha}(\omega) = \frac{i g_\alpha(\mathbf{v})}{(\omega - \mathbf{k}\cdot\mathbf{v})} - \frac{q_\alpha}{m_\alpha} \frac{i\mathcal{E}(\omega)\cdot\partial f_{0\alpha}/\partial\mathbf{v}}{(\omega - \mathbf{k}\cdot\mathbf{v})}. \tag{17.8}$$

We immediately observe that the distribution perturbation has two distinct terms. The first one depends on the initial conditions, and can be called the ballistic term, and the second one depends on the equilibrium distribution function $f_{0\alpha}$. Replacing this in the Laplace transformation of Poisson's equation, we obtain

$$\mathcal{E}(\omega, \mathbf{k}) = \frac{S(\omega, \mathbf{k})}{\epsilon(\omega, \mathbf{k})}, \quad \epsilon(\omega, \mathbf{k}) = 1 + \sum_\alpha \chi_\alpha(\omega, \mathbf{k}), \tag{17.9}$$

where $\epsilon(\omega, \mathbf{k})$ is the plasma dielectric function, with the particle susceptibilities defined by

$$\chi_\alpha(\omega, \mathbf{k}) = \frac{\omega_{p\alpha}^2}{k^2} \int \frac{(\mathbf{k}\cdot\partial f_{0\alpha}/\partial\mathbf{v})}{(\omega - \mathbf{k}\cdot\mathbf{v})} d\mathbf{v}. \tag{17.10}$$

The source term $S(\omega, \mathbf{k})$ is

$$S(\omega, \mathbf{k}) = \frac{1}{\epsilon_0} \sum_\alpha q_\alpha n_{0\alpha} \int \frac{g_\alpha(\mathbf{v}) d\mathbf{v}}{(\omega - \mathbf{k}\cdot\mathbf{v})}. \tag{17.11}$$

If we then use the inverse Fourier and Laplace transformations, we arrive at the final expression for the electric field

$$\mathcal{E}(\mathbf{r}, t) = \int \frac{d\mathbf{k}}{(2\pi)^3} \frac{\mathbf{k}}{k} e^{i\mathbf{k}\cdot\mathbf{r}} \int_{i\sigma-\infty}^{i\sigma+\infty} \frac{d\omega}{2\pi} \frac{S(\omega, \mathbf{k})}{\epsilon(\omega, \mathbf{k})} e^{i\omega t}. \tag{17.12}$$

This is the formal solution for the evolution of an electrostatic perturbation in a plasma, starting from given initial conditions. In order to understand its physical meaning, we need to write the quantities $\epsilon(\omega, \mathbf{k})$ and $S(\omega, \mathbf{k})$ more explicitly. We start by noting that the integration over the velocity component perpendicular to the wavevector $\mathbf{k}$ is trivial, and is not contributing to the singularity $(\omega - \mathbf{k}\cdot\mathbf{v}) = 0$. We are then left with an integration over the parallel component, $u = \mathbf{v}\cdot\mathbf{k}/k$. From Eqs. (17.10) and (17.11), we can then write the particle susceptibilities and the source term as

$$\chi_\alpha(\omega, \mathbf{k}) = -\frac{\omega_{p\alpha}^2}{k^2} \int \frac{(dF_{0\alpha}(u)/du)}{(u - \omega/k)} du, \quad S(\omega, \mathbf{k}) = -\frac{1}{\epsilon_0 k} \sum_\alpha q_\alpha n_{0\alpha} \int \frac{G_\alpha(u) du}{(u - \omega/k)}, \tag{17.13}$$

where $F_{0\alpha}(u)$ and $G_\alpha(u)$ are one-dimensional distribution functions, defined by

$$F_{0\alpha}(u) = \int f_{0\alpha}(\mathbf{v})\delta(u - \mathbf{k}\cdot\mathbf{v}/k)d\mathbf{v}\,, \quad G_\alpha(u) = \int g_\alpha(\mathbf{v})\delta(u - \mathbf{k}\cdot\mathbf{v}/k)d\mathbf{v}. \quad (17.14)$$

The integrals appearing in Eq. (17.13) have the generic form

$$I(u) = \int_{-\infty}^{\infty} \frac{h(u)}{(u - z)} du, \quad (17.15)$$

where $z = \omega/k$ with $z = z' + iz''$ and $z' \gg |z''|$, as a consequence of ω being a complex variable. In the limit of an infinitesimal and negative imaginary part, representing an eventual wave damping, we can write it as

$$I(u) = \lim_{\epsilon \to 0} \int_{-\infty}^{\infty} \frac{h(u)}{(u - a + i\epsilon)} du. \quad (17.16)$$

This integration is performed along the real axis in the complex u plane. By analytically extending the domain of $h(u)$ to the complex plane, we see that there is a pole at $u = a - i\epsilon$. When $\epsilon \to 0$, the pole goes into the integration path, which has to be slightly distorted in the vicinity of the singularity, in order to avoid it. Such a distorted path then leads to the following result

$$I(u) = \mathcal{P} \int_{-\infty}^{\infty} \frac{h(u)}{(u - a)} du + i\pi h(a)\,, \quad \mathcal{P} \int_{-\infty}^{\infty} = \lim_{\epsilon \to 0} \left[\int_{-\infty}^{a-\epsilon} + \int_{a+\epsilon}^{\infty} \right], \quad (17.17)$$

where the second term in $I(u)$ is the contribution of the pole, and $\mathcal{P}$ represents the principal part of the integral. We can formally extend the above result to the case of $\epsilon < 0$ and write the following expression, known as the *Plemelj formula*

$$\lim_{\epsilon \to 0} \frac{1}{u - a \pm i\epsilon} = \mathcal{P} \frac{1}{u - a} \mp i\pi\delta(u - a). \quad (17.18)$$

Such an expression has been previously used in this book. Applying this formula to the expression for the particle susceptibilities (17.13), we obtain

$$\chi_\alpha(\omega, \mathbf{k}) = -\frac{\omega_{p\alpha}^2}{k^2} \mathcal{P} \int \frac{dF_{0\alpha}(u)/du}{(u - \omega/k)} du - i\pi \frac{\omega_{p\alpha}^2}{k^2} \left(\frac{dF_{0\alpha}}{du} \right)_{u=\omega/k}. \quad (17.19)$$

We simply observe that the plasma dielectric function (17.9) is a complex quantity

$$\epsilon(\omega, \mathbf{k}) = \epsilon_r(\omega, \mathbf{k}) + i\epsilon_i(\omega, \mathbf{k}). \quad (17.20)$$

Given a real wavevector $\mathbf{k}$, the dispersion relation $\epsilon(\omega, \mathbf{k}) = 0$ will then lead to a complex frequency $\omega = \omega_r + i\gamma$. We assume that $|\gamma| \ll \omega_r$, for the above analysis to stay valid. This allows us to develop the dielectric function $\epsilon(\omega, \mathbf{k})$ around the real value $\omega = \omega_r$, which leads to the following dispersion relation

$$\epsilon(\omega_r, \mathbf{k}) + i\gamma \left(\frac{\partial \epsilon}{\partial \omega}\right)_{\omega=\omega_r} = 0. \tag{17.21}$$

Equating separately to zero the real and the imaginary parts, we obtain the dispersion relation which determines the wave frequency ω_r and the wave damping coefficient γ, for a given wavevector $\mathbf{k}$, as given by

$$\epsilon_r(\omega_r, \mathbf{k}) = 0, \quad \gamma = -\frac{\epsilon_i(\mathbf{k}, \omega_r)}{(\partial \epsilon_r / \partial \omega)_{\omega_r}} = 0. \tag{17.22}$$

17.2 Electron Plasma Waves

Let us first apply these results to electron plasma waves. Because of their high frequency with respect to the ion plasma frequency, we can assume that the ions have an infinite mass, $m_i \to \infty$, and ignore the ion motion. In this case, Eq. (17.22) leads to

$$1 - \frac{\omega_{pe}^2}{k^2} \mathcal{P} \int \frac{(dF_{0e}(u)/du)}{(u - \omega/k)} du = 0. \tag{17.23}$$

A small damping coefficient, $\gamma \ll \omega_r$, implies a large phase velocity ω/k, much larger than the electron thermal velocity, as will become apparent in the following. The dominant contribution to this integral is coming from the regions of low parallel velocity $u < \omega/k$. We can therefore develop the denominator as

$$\frac{1}{u - \omega/k} \simeq -\frac{k}{\omega}\left[1 + \frac{uk}{\omega} - \left(\frac{uk}{\omega}\right)^2 + \dots\right]. \tag{17.24}$$

By inserting the latter expansion in Eq. (17.23), and integrating by parts, we obtain

$$\int \frac{(dF_{0e}(u)/du)}{(u - \omega/k)} du \simeq \frac{k^2}{\omega^2} \int F_{0e}(u) \left[1 + 2u + 3\frac{u^2 k^2}{\omega^2}\right]. \tag{17.25}$$

If the distribution F_{0e} is normalized, and is an even function of the parallel velocity u, such that

$$\int_{-\infty}^{\infty} F_{0e}(u) du = 1, \quad \int_{-\infty}^{\infty} F_{0e}(u) u \, du = 0, \tag{17.26}$$

we can reduce the dispersion relation (17.23) to

$$1 - \frac{\omega_{pe}^2}{\omega^2}\left[1 + 3\frac{k^2}{\omega^2} \int u^2 F_{0e}(u) du\right] = 0. \tag{17.27}$$

Let us consider the specific case of a Maxwellian plasma with a given electron temperature T_e, as determined by the equilibrium distribution

$$F_{0e}(u) = \frac{1}{\sqrt{2\pi} v_{\text{the}}} \exp\left(-\frac{u^2}{2v_{\text{the}}^2}\right), \tag{17.28}$$

with $v_{\text{the}}^2 = T_e/m_e$. The integral in the above Eq. (17.27) will be equal to v_{the}^2, and the dispersion relation for electron plasma waves will take the form

$$1 - \frac{\omega_{pe}^2}{\omega^2}\left[1 + 3\frac{k^2 v_{\text{the}}^2}{\omega^2}\right] = 0. \tag{17.29}$$

Solving for ω^2, and assuming that $k^2 v_{\text{the}}^2 \ll \omega_{pe}^2$, we can immediately recover the result in Eq. (16.7)

$$\omega^2 = \omega_{pe}^2 + 3k^2 v_{\text{the}}^2. \tag{17.30}$$

Let us now find an explicit expression for the wave damping γ. Noting that

$$\left(\frac{\partial \epsilon_r}{\partial \omega}\right)_{\omega_r} = \frac{2\omega_{pe}^2}{\omega_r^3} \simeq \frac{2}{\omega_{pe}} \tag{17.31}$$

and replacing it in the above expression for γ, we get

$$\gamma = \frac{\pi \omega_{pe}^3}{2k^2}\left(\frac{dF_{0e}}{du}\right)_{u=\omega/k}. \tag{17.32}$$

Using the Maxwellian distribution, and introducing the electron Debye length $\lambda_D = v_{\text{the}}/\omega_{pe}$, we finally obtain

$$\gamma = -\sqrt{\frac{\pi}{8}} \frac{\omega_{pe}}{(k\lambda_D)^3} \exp\left(-\frac{1}{2k^2\lambda_D^2}\right). \tag{17.33}$$

This is the expression for the electron Landau damping for a plasma in thermal equilibrium. The damping coefficient tends to zero in the limit of very large wavelengths, or $k \to 0$. Wave damping increases rapidly with k, and becomes very large for $k\lambda_D \sim 1$. For $k\lambda_D = 0.5$, we get $\gamma = 0.93\omega_{pe}$, and the electron plasma waves are strongly damped. These waves can only propagate for wavelengths much larger than the electron Debye length, or $k\lambda_D \ll 1$. We can then conclude that the frequency of electron plasma waves are always nearly equal to ω_{pe}, otherwise they will be strongly attenuated by electron Landau damping.

It has to be noticed that Landau damping is not associated with any dissipating mechanism. It is a purely kinetic effect due to phase mixing: the electrons initially oscillate in phase with the wave, but start to oscillate out of phase due to their different velocities. As a consequence, the wave oscillations loose coherence and amplitude, even if the total amount of energy transported by the electrostatic field and by the oscillating particles stays constant. This is consistent with the fact that the Vlasov equation conserves entropy, as we have seen earlier in this book. Equation (17.33) also shows that γ can become positive, for parallel distribution functions $F_{0e}(u)$ with a positive derivative. This occurs for instance when an electron beam interacts with a plasma. In this case, instead of damping, we have a kinetic instability where electron plasma waves will grow spontaneously out of noise.

17.3 Ion Acoustic Waves

A similar kinetic treatment can also be employed for ion acoustic waves. In this case, both electron and ion Landau damping mechanisms are present, and the ion acoustic waves can only exist for phase velocities between the electron and the ion thermal velocities, $v_{\text{the}} \gg (\omega/k) \gg v_{\text{thi}}$. In this region, low frequency waves can exist with negligible damping. Let us consider first the electron distribution, by assuming that $v_{\text{the}} \gg (\omega/k)$, and retaining just its real part. This can be written, for an equilibrium distribution, as

$$\chi_e(\omega,\mathbf{k}) = -\frac{\omega_{pe}^2}{k^2}\mathcal{P}\frac{\partial F_{0e}/\partial u}{(u-\omega/k)}du \simeq -2\frac{\omega_{pe}^2}{k^2}\int\frac{\partial F_{0e}}{\partial u^2}du = \frac{1}{k^2\lambda_D^2}. \tag{17.34}$$

On the other hand, for the ion susceptibility, we assume that $(\omega/k) \gg v_{\text{thi}}$, and for an equilibrium Maxwellian distribution for the ions, we get

$$\chi_i(\omega,\mathbf{k}) = -\frac{\omega_{pi}^2}{k^2}\mathcal{P}\frac{\partial F_{0i}/\partial u}{(u-\omega/k)}du \simeq -\frac{\omega_{pi}^2}{k^2}. \tag{17.35}$$

This allows us to write the real part of the dielectric function for low frequency waves approximately as

$$\epsilon_r(\omega,\mathbf{k}) = 1 + \frac{1}{k^2\lambda_D^2} - \frac{\omega_{pi}^2}{\omega^2}, \tag{17.36}$$

and to determine the corresponding dispersion relation

$$\omega^2 = \frac{v_{ac}^2}{1+k^2\lambda_D^2}. \tag{17.37}$$

Here, we have defined the ion acoustic velocity $v_{ac} = \sqrt{ZT_e/m_i}$. On the other hand, the imaginary part of the dielectric function is determined by

$$\epsilon_i(\omega,\mathbf{k}) \simeq \sqrt{\frac{\pi}{2}}\sum_\alpha \frac{\omega_{p\alpha}^2}{k^2}\frac{\omega}{kv_{\text{th,ff}}^3}\exp\left(-\frac{\omega^2}{2k^2v_{\text{th,ff}}^2}\right), \tag{17.38}$$

from where we get the damping rate γ valid in the range $v_{\text{the}} \gg (\omega/k) \gg v_{\text{thi}}$, as

$$\gamma = -\sqrt{\frac{\pi}{8}}\frac{\omega}{(1+k^2\lambda_D^2)^{3/2}}\left\{\left(\frac{Zm_e}{m_i}\right)^{1/2} + \left(\frac{ZT_e}{T_i}\right)^{3/2}\exp\left[-\frac{ZT_e}{2T_i(1+k^2\lambda_D^2)}\right]\right\} \tag{17.39}$$

These waves, characterized by a real frequency ω and a damping rate γ, as given by the above Eqs. (17.37) and (17.39), are the ion acoustic waves, and as we have seen before, they are non-dispersive in the limit of $k^2\lambda_D^2 \ll 1$, where their phase velocity nearly coincides with a ion acoustic velocity v_{ac}. In contrast, the electron plasma waves exist in the same range of wavenumbers $k^2\lambda_D^2 \ll 1$ with different phase velocities but nearly the same frequency ω_{pe}. In the opposite limit of large wavenumbers, $k^2\lambda_D^2 \gg 1$ the ion acoustic waves will reduce to an ion oscillation with $\omega \simeq \omega_{pi}$.

Moreover, the electron Landau damping in Eq. (17.39) is always a small effect, of the order of $\omega(m_e/m_i)^{1/2}$, due to the fact that the slope of the electron distribution function at the Landau resonance $\omega = ku$ is very small, even if a large population of electrons exists in that region. In contrast, the ion Landau damping is only negligible if we have $T_e \gg T_i$, such that the ion acoustic velocity becomes much larger than the ion thermal velocity, $v_{ac} \gg v_{\text{thi}}$. The number of resonant ions at $\omega = ku$ is then negligibly small. We conclude that the ion acoustic waves can only be propagate as undamped modes if the electron temperature is much larger than the ion temperature, which is the case for most ultra-cold plasmas.

17.4 Waves in Quantum Plasmas

Plasmas are usually considered in the classical regime, but quantum degeneracy starts to play a role when the de Broglie wavelength of the thermal electrons, λ_B, becomes comparable with the inter-particle distance. Therefore, we have quantum plasma effects when

$$n_e\lambda_B^3 \geq 1\,, \quad \lambda_B = \frac{\hbar}{m_e v_{\text{the}}}, \tag{17.40}$$

where n_e is the electron plasma density. On the other hand, quantum statistics become important for electron temperatures below the Fermi temperature, $T_e \leq T_F$, where in energy units this coincides with the Fermi energy. When the quantum parameter is larger than one, $\chi = T_F/T_e \geq 1$, the equilibrium electron distribution is Fermi-Dirac, instead of Maxwell-Boltzmann. This means that the usual thermal velocity v_{the} has to be replaced by the Fermi velocity v_F, such that

$$v_F = \left(\frac{2T_F}{m_e}\right)^{1/2} = \frac{\hbar}{m_e}\left(3\pi^2 n\right)^{1/3}. \tag{17.41}$$

The plasma scaling length is now determined by $\lambda_F = v_F/\omega_p$, instead of the Debye length λ_D. Finally, in what concerns particle correlations, we can introduce a quantum version of the coupling parameter, still defined as the ratio between the interaction potential energy and the kinetic energy of the particles, given now by

$$\Gamma_Q = \frac{2}{(3\pi^2)^{2/3}} \frac{e^2 m_e}{\epsilon_0 \hbar^2 n_e^{1/3}} \simeq \left(\frac{1}{n_e \lambda_F^3} \right)^{2/3}. \tag{17.42}$$

This is similar to the classical definition, with λ_D replaced by λ_F. We can also write it as

$$\Gamma_Q \simeq \left(\frac{\hbar \omega_p}{E_F} \right)^2, \tag{17.43}$$

which directly compares the energy of a single plasmon $\hbar\omega_p$ with the Fermi energy. A similar discussion can be established to the case of ions.

A vast literature already exists for waves in quantum plasmas in [3–6]. Here we use a simple model, based on a coupled Schrödinger-Poisson system of equations. These equations describe the evolution of the collective electron wavefunction $\psi(\mathbf{r}, t)$ in a self-consistent electrostatic potential $V(\mathbf{r}, t)$, and can be written in the form

$$i\hbar \frac{\partial \psi}{\partial t} = -\frac{\hbar^2}{2m_e} \nabla^2 \psi + \left[U_{\text{eff}}(|\psi|^2) - eV \right] \psi, \tag{17.44}$$

and

$$\nabla^2 V = \frac{e}{\epsilon_0} \left(|\psi|^2 - n_0 \right), \tag{17.45}$$

where $|\psi|^2 = n_e$ is the electron density, and n_0 is the equilibrium plasma density. The ions are assumed immobile, and provide the neutralizing charge background. As for the effective potential $U_{\text{eff}}(|\psi|^2) \equiv U_{\text{eff}}(n_e)$, we can relate it with the quantum pressure $P(n)$, through the expression

$$U_{\text{eff}}(|\psi|^2) = \int^n \frac{dP(n')}{dn'} \frac{dn'}{n'}. \tag{17.46}$$

By choosing a convenient equation of state, we can determine this effective potential. For instance, if we consider electrons to behave as an ideal Fermi gas at the zero temperature limit [3, 4], we can use

$$P(n) = \frac{1}{5} \frac{m_e v_F^2}{n} = \frac{\hbar^2}{5m_e} \left(3\pi^2 \right)^{2/3} n^{2/3}. \tag{17.47}$$

We can now introduce the Wigner function associated with the electron wavefunction $\psi(\mathbf{r}, t)$, as defined by

$$F_e(\mathbf{r}, \mathbf{v}, t) = \int \psi^*(\mathbf{r} + \mathbf{s}/2, t) \psi(\mathbf{r} - \mathbf{s}/2, t)\, e^{i m_e \mathbf{v} \cdot \mathbf{s}/\hbar} d\mathbf{s}. \tag{17.48}$$

By following the Wigner-Moyal procedure, we can then arrive from (17.44) at the wave kinetic equation

$$\left(\frac{\partial}{\partial t} + \mathbf{v} \cdot \nabla\right) F_e = -\frac{i}{\hbar} \int V_{\text{tot}}(\mathbf{k}, t) \Delta F_e e^{i\mathbf{k}\cdot\mathbf{r}} \frac{d\mathbf{k}}{(2\pi)^3}, \tag{17.49}$$

where

$$\Delta F_e = F_e(\mathbf{v} - \mathbf{k}/2m_e) - F_e(\mathbf{v} + \mathbf{k}/2m_e), \tag{17.50}$$

and $V_{\text{tot}}(\mathbf{k}, t)$ are the space Fourier components of the total potential $V_{\text{tot}} = U_{\text{eff}}(n) - eV$. In terms of the Wigner function (17.48), Poisson's equation (17.45) can be written as

$$\nabla^2 V = -\frac{e}{\epsilon_0} \left(n_0 - \int F_e(\mathbf{v}) d\mathbf{v} \right). \tag{17.51}$$

Now, assuming perturbations of the form $\exp(i\mathbf{k} \cdot \mathbf{r} - i\omega t)$, and using a linear perturbation analysis, we arrive at a dispersion relation for the electron plasma waves of the form

$$\epsilon_Q(\omega, \mathbf{k}) \equiv 1 + \chi_e(\omega, \mathbf{k}) = 0, \tag{17.52}$$

where the electron susceptibility of the quantum plasma is now given by

$$\chi_e(\omega, \mathbf{k}) = -\frac{m_e \omega_{pe}^2}{k^2} \int \frac{F_{0e}(\mathbf{v} - \mathbf{k}/2m_e) - F_{0e}(\mathbf{v} + \mathbf{k}/2m_e)}{(\omega - \mathbf{k} \cdot \mathbf{v})} d\mathbf{v}. \tag{17.53}$$

The latter can be rewritten as

$$\chi_e(\omega, \mathbf{k}) = -\frac{m_e \omega_{pe}^2}{k^2} \int F_{0e}(\mathbf{v}) \left[\frac{1}{(\omega_+ - \mathbf{k} \cdot \mathbf{v})} - \frac{1}{(\omega_- - \mathbf{k} \cdot \mathbf{v})} \right] d\mathbf{v}, \tag{17.54}$$

where we have used $\omega_\pm = \omega \pm \hbar k^2/2m_e$. Rearranging terms we can finally write the electron susceptibility as

$$\chi_e(\omega, \mathbf{k}) = -\omega_{pe}^2 \int \frac{F_{0e}(\mathbf{v})}{(\omega - \mathbf{k} \cdot \mathbf{v})^2 - \hbar^2 k^4/m_e^2} d\mathbf{v}. \tag{17.55}$$

Using a Fermi-Dirac equilibrium distribution, we can then obtain from here the dispersion relation [4]

$$\omega^2 = \omega_{pe}^2 + k^2 v_F^2 + \frac{\hbar^2 k^4}{4m_e^2} \left(1 + \frac{4}{3} k^2 \lambda_F^2 \right). \tag{17.56}$$

This generalizes our previous results for the electron plasma waves onto the quantum regime. For weakly damped, long wavelength waves such that $k^2 \lambda_F^2 \ll 1$, we can see from this new dispersion relation that it becomes formally identical to the dispersion relation of hybrid phonons in a non-condensed laser-cooled neutral gas. It also displays the quantum dispersion term in k^4, common to the Bogoliubov oscillations in a condensate.

This quantum analysis can be generalized to take into account, not only low frequency ion acoustic waves, but also an eventual degeneracy of the plasma ions [7]. In this case, we can use collective wavefunctions, $\psi_\alpha(\mathbf{r}, t)$, for both electrons and ions, $\alpha = e, i$. In a normalized form, they can be determined by nonlinear Schrödinger equations

$$i\partial_t \psi_\alpha + A_\alpha \nabla^2 \psi_\alpha + (s_\alpha V - |\psi_\alpha|^\beta)\psi_\alpha = 0, \tag{17.57}$$

where $A_\alpha = m_\alpha e^2/2\epsilon_0 \hbar^2 \sqrt{n_0}$, and the electrostatic potential V between the particle species is determined from the Poisson equation

$$\nabla^2 V = |\psi_e|^2 - |\psi_i|^2. \tag{17.58}$$

We have also used the charge sign s_α, defined in such way that $s_e = +1$ and $s_i = -1$, and assumed singly charged ions. The wave functions ψ_α are normalized by $\sqrt{n_0}$, the potential V by T_F/e, time t is normalized by $\hbar/T_F$, and space $\mathbf{r}$ by the Debye-Fermi radius $\lambda_D = \sqrt{\epsilon_0 T_F/e^2 n_0}$. In Eq. (17.57), we have taken a particular choice for the effective potentials, of the form $|\psi_\alpha|^\beta$ where the exponent $\beta = 4/3$, can be used [3, 5]. A calculation similar to that used above for the electron plasma waves, then leads to the following expressions for the amplitude perturbation of the particle Wigner function F_α as

$$\tilde{F}_\alpha = -\frac{s_\alpha}{k^2}\frac{(F_{0\alpha-} - F_{0\alpha-})}{(\omega - \mathbf{v}\cdot\mathbf{k})}(\tilde{n}_i - \tilde{n}_e). \tag{17.59}$$

After integration over the particle velocities $\mathbf{v}$, we obtain a closed relation between the electron and ion density perturbations, $\tilde{n}_\alpha$, as

$$k^2 \tilde{n}_\alpha = -m_\alpha s_\alpha (\tilde{n}_i - \tilde{n}_e) I_\alpha(\omega, \mathbf{k}) \ , \quad I_\alpha(\omega, \mathbf{k}) = \int \frac{(F_{0\alpha-} - F_{0\alpha+})}{(\omega - \mathbf{v}\cdot\mathbf{k})} d\mathbf{v}. \tag{17.60}$$

From these equations we can derive the dispersion relation for electrostatic oscillations in a quantum plasma, which includes both the electron plasma waves and the ion acoustic waves. For that purpose, we first rewrite equations (17.60) in matrix form as

$$\begin{bmatrix} (k^2 - I_e) & I_e \\ I_i & (k^2 - I_i) \end{bmatrix} \begin{bmatrix} \tilde{n}_e \\ \tilde{n}_i \end{bmatrix} = 0. \tag{17.61}$$

The existence of a non-trivial solution for the perturbations $\tilde{n}_\alpha$ implies that the determinant of their coefficients have to be zero, which means that

$$(k^2 - I_e)(k^2 - I_i) - I_e I_i = 0. \tag{17.62}$$

This is the dispersion relation for electrostatic waves in a quantum unmagnetized plasma, where kinetic effects have been included. In particular, in the limit of immobile ions, $m_i \to \infty$, and we recover the above results for electron plasma waves in a quantum plasma.

17.5 Sound Waves in a Turbulent Plasma

We conclude this chapter on wave kinetic theory with a discussion on wave propagation in a classical turbulent plasma. More specifically, we study the evolution of an ion acoustic wave, when it propagates in a plasma which is in a turbulent state, due to the existence of an arbitrary plasmon spectrum. This means that the ion acoustic wave will propagate in a medium with two different kinds of fluids. One is the plasma itself, as a collection of free electrons and ions. The other is the fluid of quasi-particles, associated with short wavelength turbulence, which is a gas of plasmons.

The analogy with the theory of two fluids for superfluidity is striking. Such similarities with the superfluid theory can become even more evident if we use the Wigner-Moyal formulation, as illustrated next. We first start by deriving the wave kinetic equation for the plasmon field, which is physically equivalent to the wave equation for electron plasma waves. We then derive the dispersion relation of ion acoustic waves in the presence of the gas of plasmons. Coupling between these low frequency waves with plasmons is due to the existence of ponderomotive forces. We finally show that ion acoustic waves can be Landau damped by quasi-resonant plasmons. This is very similar to Landau damping of long wavelength excitations in condensates due to quasi-resonant interactions with phonons. We also show that the plasmon turbulence modifies the effective ion acoustic velocity.

17.5.1 Plasmon Kinetic Equation

Let us consider an isotropic and uniform plasma. As we have seen, the electron plasma waves can be described in terms of the fluctuations in the electron density, $\tilde{n}_e(\mathbf{r}, t)$, which obey the following equation

$$\left(\nabla^2 - \frac{1}{S_e^2}\frac{\partial^2}{\partial t^2}\right)\tilde{n}_e = \frac{\omega_p^2(\mathbf{r}, t)}{S_e^2}\tilde{n}_e, \tag{17.63}$$

where, for the sake of generality, we have considered that the electron plasma frequency is a slowly varying function of space and time, $\omega_p^2(\mathbf{r}, t) = (e^2/\epsilon_0 m_e) n_{0e}(\mathbf{r}, t)$ and $S_e = (3T_e/m_e)^{1/2}$ represents the electron thermal velocity. The space and time variations of the electron plasma density originate from the ion acoustic waves, as considered below. This definition of $\omega_p^2(\mathbf{r}, t)$ is established for generic space and time changes in the medium, but here we only focus on the linear stages of ion acoustic instabilities, for which higher order nonlinear effects such as plasmon trapping effects can be ignored. For a stationary plasma, we have $n_{0e}(\mathbf{r}, t) = n_0$, and for electron plasma perturbations of the form $\tilde{n}_e(\mathbf{r}, t) \propto \exp(i\mathbf{k}\cdot\mathbf{r} - i\omega t)$, Eq. (17.63) leads to the dispersion relation $\omega^2 = \omega_{p0}^2 + k^2 S_e^2$. However, in turbulent plasma conditions, specially if we need to describe a broad-band spectrum of electron plasma fluctuations, it is more convenient to introduce a wave kinetic equation describing the field correlations. In the same way as we did for the particle wavefunctions in Chaps. 3 and 11, we introduce here the potential auto-correlation function

$$C(\mathbf{r}, t; \mathbf{s}, \tau) = \tilde{n}_e^*(\mathbf{r} - \mathbf{s}/2, t)\, \tilde{n}_e(\mathbf{r} + \mathbf{s}/2, t), \tag{17.64}$$

and define the Wigner function for the electrostatic perturbations as the space Fourier transformation of the latter

$$F(\mathbf{k}'; \mathbf{r}, t) = \int C(\mathbf{r}, t; \mathbf{s}, \tau) \exp(-i\mathbf{k}' \cdot \mathbf{s}) d\,d\mathbf{s}. \tag{17.65}$$

Taking the inverse Fourier transformation, and setting $\mathbf{s} = 0$, we obtain

$$|\tilde{n}_e|^2 = \int F(\mathbf{k}'; \mathbf{r}, t) \frac{d\mathbf{k}'}{(2\pi)^3}. \tag{17.66}$$

For convenience, we use here $\mathbf{k}'$ for the high frequency part of the spectrum, associated with the plasmon oscillations, and ω' for the corresponding frequencies. The linear plasmon dispersion relation $\omega'(\mathbf{k}')^2 = k'^2 S_e^2 + \omega_p^2(\mathbf{r}, t)$ is assumed to apply. The symbols $\mathbf{k}$ and ω, will therefore be reserved to the low frequency oscillations of the ion acoustic type. Starting from the wave equation (17.63), we can derive an exact equation for the Wigner function, which reads

$$i\omega' \left(\frac{\partial}{\partial t} + \frac{S_e^2 \mathbf{k}'}{\omega'} \cdot \nabla \right) F = \frac{\omega_{p0}^2}{n_0} \int n_{0e}(\mathbf{k}, t)\, [F_- - F_+] \exp(i\mathbf{k} \cdot \mathbf{r}) \frac{d\mathbf{k}}{(2\pi)^3}, \tag{17.67}$$

where n_0 is equilibrium plasma density, ω_{p0} is the corresponding electron plasma frequency, and $n_{e0}(\mathbf{k})$ are the space Fourier components of the electron density fluctuations, as determined by

$$n(\mathbf{r}, t) = \int n_{e0}(\mathbf{k}) \exp(i\mathbf{k} \cdot \mathbf{r} - i\omega t) \frac{d\mathbf{k}}{(2\pi)^3}. \tag{17.68}$$

The quantities $F_\pm$ are defined by $F_\pm = F(\mathbf{k}' \pm \mathbf{k}/2)$. It is now useful to consider the case in which a single ion acoustic mode is perturbing the background plasma and modulates the plasma density. In such a case, the wave kinetic equation (17.67) is reduced to

$$\left(\frac{\partial}{\partial t} + \mathbf{v}' \cdot \nabla \right) F = -\frac{i}{\omega'} \frac{\omega_{p0}^2}{n_0} \tilde{n}_i(\mathbf{k})\, [F_- - F_+] \exp(i\mathbf{k} \cdot \mathbf{r} - i\omega t), \tag{17.69}$$

where $\mathbf{v}' = S_e^2 \mathbf{k}'/\omega'$ is the plasmon group velocity. It should be noticed that, when the length scale of the ion acoustic wave is much larger than that of the electron plasma waves, and $|\mathbf{k}| \ll |\mathbf{k}'|$, the quantities $F_\pm$ can be approximated by

$$F_\pm \simeq F(\mathbf{k}') \pm \frac{\mathbf{k}}{2} \cdot \frac{\partial F}{\partial \mathbf{k}'}. \tag{17.70}$$

In this limit, Eq. (17.69) is reduced to a kinetic equation of the Vlasov type [8].

17.5.2 Ion Oscillations

We can now study the plasma stability with respect to ion acoustic perturbations, in the presence of a broad plasmon turbulent spectrum. We start from the ion fluid equations and assume that the electrons are in Boltzmann equilibrium in the slow wave potential created by the ion oscillations. By using the standard perturbation analysis, we can then derive a dispersion relation of the form [8, 9]

$$\omega^2 = k^2 v_{\text{thi}}^2 + \frac{k^2 v_{ac}^2}{(1 + k^2 \lambda_{De}^2)} \left(1 - \frac{\tilde{n}_{e2}}{Z\tilde{n}_i}\right), \tag{17.71}$$

where $v_{\text{thi}}^2 = T_i/m_i$, is the ion thermal velocity and $v_{ac}^2 = 3T_e/m_i$ is the ion acoustic velocity. Here, we have retained second order corrections for the electron density fluctuations, which contribute to the low frequency ion acoustic oscillations and are due to the ponderomotive force terms. A simple nonlinear analysis shows that these fluctuations are determined by

$$\frac{\partial^2 \tilde{n}_{e2}}{\partial t^2} = \frac{1}{2n_0} \nabla^2 \int \left(\frac{\omega'}{k'}\right)^2 F(\mathbf{k}') \frac{d\mathbf{k}'}{(2\pi)^3}. \tag{17.72}$$

Assuming slow perturbations of the form $\exp(i\mathbf{k}\cdot\mathbf{r} - i\omega t)$, and using Eqs. (17.69) and (17.72), we obtain

$$\tilde{n}_{e2} = \frac{k^2}{2n_0\omega^2} \int \left(\frac{\omega'}{k'}\right)^2 F(\mathbf{k}') \frac{d\mathbf{k}'}{(2\pi)^3}, \quad \tilde{F} = \frac{\omega_{pe}^2}{\omega'(\omega - \mathbf{k}'\cdot\mathbf{v}')} \frac{\tilde{n}_i(\mathbf{k})}{n_0} [F_{0-} - F_{0+}]. \tag{17.73}$$

We can then write the dispersion relation (17.71) in the form

$$\omega^2 = k^2 v_{\text{th,i}}^2 + \frac{k^2 v_{ac}^2}{(1 + k^2 \lambda_{De}^2)} \left[1 - \frac{\omega_{p0}^2 k^2}{2\omega^2 n_0^2} I(\mathbf{k})\right], \tag{17.74}$$

where the integral is computed over the plasmon spectrum, viz.

$$I(\mathbf{k}) = \int \frac{\omega'[F_{0-} - F_{0+}]}{k'^2(\omega - \mathbf{k}\cdot\mathbf{v}')} \frac{d\mathbf{k}'}{(2\pi)^3}. \tag{17.75}$$

We can clearly see that this integral contains new Landau resonances, of the type $\omega = \mathbf{k}\cdot\mathbf{v}'$, corresponding to plasmons with a group velocity equal to the phase velocity of the ion acoustic wave. This is a kind of wave quasi-particle resonance, similar to that found in superfluid theory. It is also sometimes useful to introduce the concept of plasmon number density $\mathcal{N}(\mathbf{k}')$, which is the energy density of electron plasma waves divided by the wave frequency ω'. It is easy to relate it with the Wigner function $F(\mathbf{k}')$, as

$$\mathcal{N}(\mathbf{k}') = \frac{e^2}{2\omega'\epsilon_0 k^2} F(\mathbf{k}'). \tag{17.76}$$

To be exact, this expression should be divided by $\hbar$, in order for $\mathcal{N}(\mathbf{k}')$ to have the correct units. Let us simply take $\hbar = 1$ in the remaining of the chapter. Noticing that $\omega' \simeq \omega_{p0}$ for the entire plasmon spectrum (otherwise plasmons are strongly damped), we can write (17.75) in terms of the quasi-particle number density as

$$I(\mathbf{k}) = \frac{2\epsilon_0}{e^2}\omega_{p0}^2 J(\mathbf{k}) , \quad J(\mathbf{k}) = 2\Omega \int \frac{\mathcal{N}_0(\mathbf{q})}{(\omega - \mathbf{k}\cdot\mathbf{v}'_q)^2 - \Omega^2} \frac{d\mathbf{q}}{(2\pi)^3}, \tag{17.77}$$

with

$$\mathbf{v}'_q = S_e^2 \frac{\mathbf{q}}{\omega_{p0}} , \quad \Omega = \frac{S_e^2}{2\omega_{p0}} k^2. \tag{17.78}$$

For the sake of illustration, let us consider the simple case of a low temperature plasmon beam (where temperature refers here to the dispersion of the plasmon distribution around a given value of the plasmon momentum $\mathbf{k}'_0$). We assume the following mono-kinetic plasmon distribution $\mathcal{N}_0(\mathbf{k}') = (2\pi)^3 \mathcal{N}_0 \delta(\mathbf{k}' - \mathbf{k}'_0)$. Remembering that $(k'/\omega')^2 = v'^2/S_e^4$, Eq. (17.78) leads to a quite simple expression for the nonlinear term in the dispersion relation (17.74), which can now be written as

$$\omega^2 = k^2 v_{\text{thi}}^2 + \frac{k^2 u_{\text{eff}}^2}{(1 + k^2 \lambda_{De}^2)}, \tag{17.79}$$

where we have defined

$$u_{\text{eff}}(\mathbf{k}) = v_{ac} \left[1 - \frac{k^4 S_e^2}{\omega^2 n_0 m_e} \frac{\omega_{p0}^2}{(\omega - \mathbf{k}\cdot\mathbf{v}'_0)^2 - \Omega^2} \right]^{1/2} . \tag{17.80}$$

In the absence of plasmons, $F_0 = 0$, this dispersion relation would reduce to the well known result for ion acoustic waves. The interesting thing about this result is that it shows that the existence of a beam of plasmons with wavenumber k_0 will lead to a modified ion acoustic velocity, which depends on the phonon wavenumber k. Usually we have $\omega^2 < \Omega^2$, or $u_{\text{eff}} > v_{ac}$. This means that the occurrence of a rot on minimum such as those found in superfluids or in ultra-cold atoms cannot occur here. This states an important difference between the ion correlations introduced by plasmons and the atom correlations previously considered for laser-cooled atoms, or in condensed quantum fluids.

References

1. T.W. Stix, *Theory of Plasma Waves* (McGraw-Hill, New York, 1962)
2. E.M. Lifshitz, L.P. Pitaevskii, *Physical Kinetics* (Butterworth-Heinemann, Oxford, 1981)
3. G. Manfredi, Fields Inst. Commun. **46**, 263 (2005)
4. G. Manfredi, F. Haas, Phys. Rev. B **64**, 075316 (2001).

5. P.K. Shukla, B. Eliasson, Nonlinear aspects of quantum plasma physics. Phys. Uspekhi **53**, 51 (2010)
6. H. Terças, J. T. Mendonça, P. K. Shukla, Phys. Plasmas **15**, 072109 (2008)
7. J.T. Mendonça, E. Ribeiro, P.K. Shukla, J. Plasma Phys. **74**, 91 (2008)
8. J.T. Mendonça, R. Bingham, Phys. Plasmas **9**, 2604 (2002)
9. J.T. Mendonça, A. Serbeto, J.R. Davies, R. Bingham, Plasma Phys. Contr. Fusion **50**, 105009 (2008)

Chapter 18
Conclusions

In this book we have covered a wide range of phenomena in ultra-cold matter, and reviewed many of the peculiar problems associated with its three distinct phases, atomic clouds, condensates and plasmas. This allowed us to reveal a number of unexpected similarities between these three phases, associated with the equilibrium configurations, the expansion processes, the elementary excitations and the collective oscillations. Such similarities are due, in particular to the existence of a mean field, which has different physical meanings in each of the phases, but accounts for similar physical phenomena.

In non-condensed atomic clouds, the mean field is provided by an exchange of photons between atoms, which is responsible for long range collective processes, and behaves very much like an electrostatic field. In Bose Einstein condensates, the mean field is produced by short range atomic collisions, but built on a long range effective potential due to the existence of quantum coherence. Finally in Rydberg plasmas, the mean field has an electromagnetic origin and is due to the long range interactions between the charged particles in the medium. The bosons mediating the long range interactions in a plasma are virtual photons, whereas for neutral laser-cooled atoms we have real photons, associated with the multiple scattering photon process. In the same spirit we could say that the long range interactions in a condensate are mediated by virtual phonons.

We have successively examined the acoustic modes in these three phases of ultra-cold matter: the hybrid sound waves in ultra-cold clouds of laser cooled atoms, the Bogoliubov oscillations in Bose Einstein condensates, and the ion acoustic waves in ultra-cold plasmas. We have noticed their similarities, in terms of quantum dispersion and cut-off frequencies, and differences mainly related with their characteristic velocities. We have also noticed that roton minima can eventually occur in the sound dispersion relation. Classical rotons can exist in laser-cooled gas, which are formally similar to (although physically different from) the quantum rotons of superfluid matter. And we have seen that dipole-dipole atom interactions can induced roton minima in condensates. We are aware that, according to the dominant views, the roton phenomenon is strictly associated with quantum fluids.

J.T. Mendonça and H. Terças, *Physics of Ultra-Cold Matter*, Springer Series on Atomic, Optical, and Plasma Physics 70, DOI 10.1007/978-1-4614-5413-7_18,

Our research on laser cooled atoms led us however to a much open view on the notion of roton, by revealing the existence of similar sound dispersion effects in non-condensed fluids.

We have also explored the analogies between the three phases of ultra-cold matter in what concerns the quasi-equilibrium profiles and expansion mechanisms. We have noticed that both condensates and Rydberg plasmas have similar Thomas-Fermi equilibria, and that the global oscillations of both non-condensed and condensed matter obey similar dispersion laws.

Finally we have insisted on the possible use of the same theoretical methods, to deal with the physical description of these three phases. First, the Wigner-Moyal procedure, which includes the definition of a quasi-probability distribution, usually called the Wigner function, and the derivation of a wave kinetic equation which describes the space time evolution of such a distribution. This procedure is particularly useful to describe resonant wave-atom interactions, and in particular atomic Landau damping. It is remarkable that the wave kinetic equations derived by such method become so similar for the three phase of ultra-cold matter. On the other hand, quantum field theory methods can equally well be applied to these different phases, describing in similar terms the quantized elementary excitations of the different kinds of mean fields.

What are the challenges of theoretical work in ultra-cold matter? First of all, the extension of the description covered in this book to the cases of an ultra-cold Fermi gas, and to ultra-cold molecules, two areas in very intense progress. Second, a deeper understanding of quantum coherence, its possible application to the observation of quantum gravity and the development of an improved atomic interferometry. Third, the search for gaseous liquid phase transitions and for crystallization in Rydberg plasmas, the understanding of photon instabilities such as bubble formation in the non-condensed gas, and the extension of the Bose Einstein condensation principle to solid matter. Finally, the exploration of enhanced collective interactions, such as those taking place in dipolar gases, as associated with electric and magnetic dipole-dipole interactions, and their impact on the elementary excitations with particular emphasis on the search for new roton instabilities. Progress is also foreseen in the quantum field description of collective and cooperative phenomena in quantum gases and quantum plasmas. However, as in any scientific branch of knowledge, the most exciting progress will certainly come from some unexpected discovery, which will lead to a significant change of paradigm.

Appendix

A.1 Atomic Structure

Manipulation of atoms and their cooling and confinement depend very much on their internal properties. Here we review the basic features of the atomic structure. For this purpose, we consider an atom with a nucleus of charge Ze, and N electrons, where $-e$ is the electron charge. The atomic structure is described by the time independent Schrödinger equation

$$H|\psi_a\rangle = E|\psi_a\rangle, \tag{A.1}$$

where E us the energy of the atomic state $|\phi_a\rangle$, and H is the Hamiltonian of the system. In the non-relativistic approximation, H can be written as

$$H = \sum_{i=1}^{N} \left[\frac{\mathbf{p}_i^2}{2m} - \frac{Zq^2}{r_i} \right] + \sum_{i<j=1}^{N} \frac{q^2}{r_{ij}}, \tag{A.2}$$

where m is the electron mass, $\mathbf{p}_i$ the linear momentum of the electron i, r_i is its distance to the nucleus, assumed at the centre of the coordinates r_{ij} the distance between the electrons i and j, and $q^2 = e^2/4\pi\epsilon_0$, where ϵ_0 is the electric permittivity of the vacuum. In the $\mathbf{r}$-representation, the wavevector $|\psi_a\rangle$ is represented by a wave function $\psi_a(\mathbf{r}_1, \mathbf{r}_2, , , , \mathbf{r}_N)$, and the momentum operators are $\mathbf{p}_i = -i\hbar\partial/\partial\mathbf{r}_i$. Let us now define the total angular momentum of the atom by

$$\mathbf{L} = \sum_{i=1}^{N} \mathbf{l}_i = \sum_{i=1}^{N} \mathbf{r}_i \times \mathbf{p}_i . \tag{A.3}$$

It can easily be shown that, for any i and j, the following commutation relations are satisfied

$$\left[\mathbf{l}_i, \mathbf{p}_j^2\right] = 0 , \quad \left[\mathbf{l}_i, 1/r_j\right] = 0 , \quad \left[\mathbf{l}_i, 1/r_{ij}\right] = -\left[\mathbf{l}_j, 1/r_{ij}\right] . \tag{A.4}$$

J.T. Mendonça and H. Terças, *Physics of Ultra-Cold Matter*, Springer Series on Atomic, Optical, and Plasma Physics 70, DOI 10.1007/978-1-4614-5413-7,

From this we conclude that the operator **L** commutes with the Hamiltonian (A.2), or $[H, \mathbf{L}] = 0$. As a result, the eigenstates of H can also be eigenstates of the operators L^2 and L_z. The corresponding eigenvalues $\hbar^2 L(L+1)$ and $\hbar M_L$ can then be use to identify the eigenfunction or eigenvectors of H.

We also note that H is independent of the spin variables, allowing to write $[H, \mathbf{S}] = 0$, where $\mathbf{S} = \sum_i \mathbf{s}_i$ is the total electron spin. Therefore, the eigenvalues $\hbar S(S+1)$ and $\hbar M_S$ cam also be use to identify the eigenstates of the Hamiltonian of the atom. As a result, the eigenstates of H can be written as $|S, M_S, L, M_L\rangle$.

At this point, relativistic corrections to Eq. (A.2) should be introduced. In particular, they are associated with spin-orbit interactions. The corresponding Hamiltonian term, to be added to the above expression, is

$$H_{so} = \sum_{i=1}^{N} a_i (\mathbf{s}_i \cdot \mathbf{l}_i), \tag{A.5}$$

where a_i are appropriate constants, depending on $\mathbf{l}_i$ and $\mathbf{s}_i$. Now, the operators **L** and **S** do not commute with H_{so}, because the three components of $\mathbf{l}_i$ and $\mathbf{s}_i$ do not commute with each other. Therefore, the eigenstates of the new Hamiltonian $H_a = H + H_{so}$ cannot be characterized by the quantum numbers indicated above. We need to define the total angular momentum $\mathbf{J} = \mathbf{L} + \mathbf{S}$, which now commutes with H_a, and the energy eigenstates can be labeled by the eigenvalues of J^2 and M_J, as $|J, M_J\rangle$. The quantum number J is either an integer or a half-integer, in the range $|L - S| \leq J \leq L + S$, and M_J is an integer or half-integer such that $-J < M_J < J$.

The state representation in terms of J and M_J is valid as long as the total nuclear momentum **I** is zero. When this is not true, we should define the total angular momentum of the atom as $\mathbf{F} = \mathbf{I} + \mathbf{J}$. The good quantum numbers now become F and M_F, following the same hierarchy of J and M_J, namely $|J - I| \leq F \leq J + I$, and $-F < M_F < F$.

Finally, it should be noticed that the nuclear perturbation is generally very small, which means that J and M_J remain as good, although approximate, quantum numbers. Similarly, the spin-orbit energy is also in general very small, so that L and S also remain as good approximate quantum numbers. This allows us to use the eigenstate notation $|L, S, J, M_J\rangle$.

Exact solutions to the eigenvalue equations (A.1) and (A.2) are in general not possible, due to the existence of the electron-electron interaction terms in r_{ij}, which prevent the use of variable separation methods. An important approximation is provided by the *central field method*, where it is assumed that each electron moves inside the atom in the presence of an effective central potential, $U(r)$, which results from the combined effect of the attraction from the nucleus and repulsion from the other electrons. The Hamiltonian H is then replaced by

$$H = H_0 + H_1 \,, \quad H_0 = \sum_{i=1}^{N} \left[\frac{\mathbf{p}_i^2}{2m} - eU(r_i) \right], \tag{A.6}$$

where the dominant term, H_0, describes the electron motion in the central field describes by $U(r)$, and the perturbed part H_1 is defined as

$$H_1 = \sum_{i<j=1}^{N} \frac{q^2}{r_{ij}} - \sum_{i=1}^{N} \left[\frac{Zq^2}{r_i} - eU(r_i) \right] \quad \text{(A.7)}$$

representing the deviations (assumed small) from the central field model. It is clear that, for an electron moving near the nucleus, the central potential will approach the bare Coulomb potential of the nucleus,

$$\lim_{r \to 0} U(r) = Z\frac{q^2}{er}. \quad \text{(A.8)}$$

On the other hand, at large distances, it will tend to the Coulomb potential of the nucleus but screened or dressed by the other $(N-1)$ electrons, or, equivalently,

$$\lim_{r \to \infty} U(r) = [Z - (N-1)]\frac{q^2}{er}. \quad \text{(A.9)}$$

For intermediate distances, $U(r)$ will depend on the details of the electron distribution inside the atom, and it will be more difficult to model. But, for the ground state and the first excited state, as well as for highly excited electron states, this central simple central field model can be very useful. Approximate methods have been developed to determine the central field potential $U(r)$ at intermediate distances, such as the well known Thomas-Fermi and the Hartree-Fock models.

The advantage of the central field model is that the eigenstates of H_0 can easily be found, using separation of the different electron coordinates. As we can see from Eq. (A.6), H_0 takes the form of a sum of N identical Hamiltonians, $H_0 = \sum_i h_i$, each of them satisfying a reduced eigenvalue equation

$$h_i |\psi_i\rangle = \epsilon_i |\psi_i\rangle \ , \quad h_i = \frac{\mathbf{p}_i^2}{2m} - eU(r_i), \quad \text{(A.10)}$$

where $|\psi_i\rangle$ is the eigenvector of the electron i, with energy ϵ_i, and the total eigenvector of the atom is $|\psi_a\rangle = |\psi_1\rangle|\psi_2\rangle \ldots |\psi_N\rangle$.

We know that, for lower states, the eigenvalues of the individual h are labelled by the two quantum numbers n and l, such that $n = 1, 2, 3, \ldots$ and $|l\rangle = 0, 1, 2, \ldots (n-1)$. The associated eigenvalues depend on five quantum numbers, and we can write the individual electron eigenvalue equations (A.10) as

$$h|n, s, m_s, l, m_l\rangle = \epsilon_{nl}|n, s, m_s, l, m_l\rangle. \quad \text{(A.11)}$$

We can further decompose these eigenvalues as

$$|n, s, m_s, l, m_l\rangle = \frac{1}{r} R_{nl}(r)|s, m_s\rangle|l, m_l\rangle, \quad \text{(A.12)}$$

where $|s, m_s\rangle$ is the electron spin state vector, and where we can use the spherical harmonic representation

$$|l, m_l\rangle = Y_{lm_l}(\theta, \varphi) \tag{A.13}$$

in spherical coordinates $\mathbf{r} \equiv (r, \theta, \varphi)$. Finally, the radial function $R_{nl}(r)$ has to satisfy the equation

$$\frac{d^2}{dr^2} R_{nl} - \frac{l(l+1)}{r^2} R_{nl} + \frac{2m}{\hbar^2}[\epsilon_{nl} + eU(r)] = 0. \tag{A.14}$$

It is known that, for the simple case of a Coulomb attractive potential, as in (A.8), we get

$$\epsilon_{nl} \equiv \epsilon_n = -\frac{Z^2}{2n^2}\alpha^2 mc^2, \tag{A.15}$$

where $\alpha = e^2/\hbar c \simeq 1/137$ is the fine structure constant. As a result, the energy levels of the total central field Hamiltonian, H_0 can be written as

$$E_0 = \sum_{nl} N_{nl}\epsilon_{nl} \,, \quad \sum_{nl} N_{nl} = N, \tag{A.16}$$

where N_{nl} is the number of electrons in the sub-shell (n, l). Finally, the Pauli exclusion principle implies the existence of an additional condition for this occupation number, as $N_{nl} \leq 2(2l + 1)$. From this we conclude that the numbers N_{nl} define an electron configuration of the atom, with reads as

$$(n_1 l_1)^{N_1} \, (n_2 l_2)^{N_2} \, \ldots \,, \quad N_j \equiv N_{n_j l_j}. \tag{A.17}$$

Each individual electron energy state ϵ_{nl} has $2(2l + 1)$ degenerate states, due to the two possible values of the spin and to the $(2l + 1)$ values of $m_l = -l, \ldots 0, \ldots l$.

The ground state of the atom is determined by distributing the Z electrons over the different shells $n = 1, 2, 3, \ldots$, usually labelled by the capital letters $(K, L, M, N, \ldots)$, whereas the sub-shells inside each of these shells, corresponding to different values of $l = 0, 1, 2, 3, \ldots$ are usually designed by the symbols $(s, p, d, f, \ldots)$.

The first element, Hydrogen ($Z = 1$), is in the ground state $1s$, whereas Helium ($Z = 2$) is in the ground state $1s^2$, with the two electrons having opposite spin states and fully occupying the first shell K. Some families of atoms are particularly relevant to the cold atom physics. They are the *noble gases*, (He, Ne, Ar, Kr, Xe). which have full shells and are chemically inert, and the *alkaline metals*, (Li, Na, K, Rb, Cs), which have one single valence electron outside the last closed shell, and therefore are chemically very reactive. Their ground state configurations and the respective ionization potentials are given in the Table A.1.

Finally, additional information on the state of the atom is provided by the total orbital and spin angular momentum. In the Russel-Saunders notation, we write the atomic state as

$$^{2S+1}L_J, \tag{A.18}$$

where $L = 0, 1, 2, 3, \ldots$ are represented by the letters (S, P, D, F, ...), S is the total spin quantum number, and J is the total angular momentum quantum number.

Table A.1 Ground state and first ionization energy of different atomic elements

Element	Atomic number (Z)	Ground state configuration	Ionization energy (eV)
H	1	$1s^1$	13.6
He	2	$1s^2$	24.6
Li	3	(He) $2s^1$	5.4
Ne	10	(He) $2s^2\,2p^6$	21.6
Na	11	(Ne) $3s^1$	5.1
Ar	18	(Ne) $3s^2\,2p^6$	15.8
K	19	(Ar) $4s^1$	4.3
Kr	36	(Ar) $4s^2\,3d^{10}\,4p^6$	14.0
Rb	37	(Kr) $5s^1$	4.2
Xe	54	(Kr) $4d^{10}\,5s^2\,5p^6$	12.1
Cs	55	(Xe) $6s^1$	3.9
Yb	70	(Xe) $6s^2\,4f^{14}$	6.3

A.2 Quantum Theory of Radiative Transitions

The interaction of the radiation field with an atom is well understood at both classical and quantum levels [1, 2]. It can be described by an Hamiltonian operator H which contains three distinct terms

$$H = H_a + H_f + H_{\text{int}}, \tag{A.19}$$

where H_a is the Hamiltonian of the atom, H_f that of the radiation field, and H_{int} describes the interaction between the atom and radiation. Let us consider the first term. The atom Hamiltonian can be derived from the eigenvalue equation $H_a|i\rangle = E_i|i\rangle$, where $|i\rangle$ is the atomic state vector associated with a given set of quantum numbers represented by the letter i. By inserting the identity operator $\sum_j |j\rangle\langle j| = I$ in this equation, we obtain for the atom Hamiltonian

$$H_a = \sum_i E_i|i\rangle\langle i| = \sum_i E_i\sigma_{ii}, \tag{A.20}$$

where the atomic transition operator $\sigma_{ij} = |i\rangle\langle j|$ was used. For the simple but very useful case of a two-level atom, this Hamiltonian can be reduced to the form

$$H_a = \frac{1}{2}\hbar\omega\sigma_z \tag{A.21}$$

using the notation $\sigma_z = (\sigma_{22} - \sigma_{11})$. If we restrict our discussion to the dipole approximation, the interaction Hamiltonian is just given by $H_{\text{int}} = e\mathbf{r}\cdot\mathbf{E}$, where $\mathbf{r}$ represents the position of the electrons inside the atom, and $\mathbf{E}$ the radiation electric field. For a plane wave decomposition of the electromagnetic field, we have for each photon mode with frequency ω_k and wavevector $\mathbf{k}$, the electric field operator is

$$\mathbf{E}_k = iC_k\mathbf{e}_k(a_k - a_k^\dagger), \tag{A.22}$$

where a_k and $a_k^\dagger$ are the time dependent destruction and creation operators, $C_k = \sqrt{\hbar\omega_k/2\epsilon_0 V}$ is a normalization factor, and $\mathbf{e}_k$ is the unit polarization vector. The interaction Hamiltonian can then be written as

$$H_{\text{int}} = i\hbar \sum_k g_k(\sigma_+ + \sigma_-)(a_k - a_k^\dagger), \tag{A.23}$$

with $\sigma_+ \equiv \sigma_{21}$ and $\sigma_- \equiv \sigma_{12}$. and

$$g_k = \frac{1}{\hbar}(\mathbf{p}_{12} \cdot \mathbf{e}_k)C_k\ , \quad \mathbf{p}_{ij} = -\langle i|e\mathbf{r}|j\rangle. \tag{A.24}$$

It can be seen that the operator $\sigma_- = |1\rangle\langle 2|$ takes an atom from its upper energy level $|2\rangle$, and brings it into the lower level $|1\rangle$, whereas σ_+ produces the opposite transition. By looking at H_{int} we can then identify four distinct processes. One, described by the term $\sigma_+ a_k$ corresponds to the excitation of the atom and the destruction of one photon. The term $\sigma_- a_k^\dagger$ represents the opposite process. The term $\sigma_- a_k$ describes the de-excitation of the atom and the destruction of one photon, resulting in an energy loss (for the total "atom + field" system) of $2\hbar\omega$. Similarly, the forth tern, $\sigma_+ a_k^\dagger$ corresponds to an energy gain of $2\hbar\omega$. Obviously, these two last terms cannot satisfy energy conservation, and should be dropped. The neglect of these two terms is called the rotating wave approximation. In this approximation, the interaction Hamiltonian reduces to

$$H_{\text{int}} = i\hbar \sum_k g_k(\sigma_+ a_k - \sigma_- a_k^\dagger). \tag{A.25}$$

Finally, field Hamiltonian appearing in Eq. (A.19) is given by $H_f = \sum_k \hbar\omega_k(a_k^\dagger a_k + 1/2)$. We can simplify even further the picture of the atom radiation interaction by assuming that the two-levels atom interacts only with a single field mode. We can then drop the sum over the field modes $\mathbf{k}$ and write the total Hamiltonian as $H = H_0 + H_{\text{int}}$, where the unperturbed term $H_0 \equiv H_a + H_f$ and the interaction term can be are given by

$$H_0 = \hbar\omega_k\left(a^\dagger a + \frac{1}{2}\right) + \frac{1}{2}\hbar\omega\sigma_z\ , \quad H_{\text{int}} = i\hbar g(\sigma_+ a - \sigma_- a^\dagger). \tag{A.26}$$

The evolution of the quantum state of the coupled atom-radiation system, can be described by a state vector $|\psi\rangle$, which evolves in time according to the Schrödinger equation

$$i\hbar\frac{\partial}{\partial t}|\psi| = H_{\text{int}}|\psi\rangle \tag{A.27}$$

and, in the spirit of the perturbation theory, can be represented as superposition of two different states

$$|\psi\rangle = \sum_n (C_{1,n}|1,n\rangle + C_{2,n}|2,n\rangle). \tag{A.28}$$

The state vectors $|j,n\rangle$ represent the system with a radiation field mode with n photons, and the atom in two the energy state $j = 1,2$. The coefficients $C_{j,n}$, which characterize the probability of finding the system in a given state, will evolve in time due to the atom-field interaction. At this point, it is useful to introduce the interaction picture, which corresponds to the use of an unitary transformation described by

$$H' = U^{\dagger}(t) H_{\text{int}} U(t) \quad , \quad U(t) = \exp\left(-\frac{i}{\hbar} H_0 t\right). \tag{A.29}$$

From this we can derive the new interaction Hamiltonian

$$H' = i\hbar g \left(\sigma_+ a e^{i\Delta\omega t} + a^{\dagger} \sigma_- e^{-i\Delta\omega t}\right), \tag{A.30}$$

where $\Delta\omega = \omega - \omega_k$ is the detuning between the photon energy and the transition energy between the two atomic states. Such a detuning will play an important role in the laser cooling process, as shown later. Going back to the Schrödinger equation and using the interaction representation, which allows us to replace H_{int} by its transformed H', we obtain the evolution equations for the

$$\frac{\partial}{\partial t} C_{2,n} = -ig\sqrt{n+1}\, e^{i\Delta\omega t} C_{1,n+1} \,, \quad \frac{\partial}{\partial t} C_{1,n+1} = ig\sqrt{n+1}\, e^{-i\Delta\omega t} C_{2,n}. \tag{A.31}$$

These equations are formally very similar to those obtained in the quasi-classical theory, even if they display important qualitative differences. As we have seen, they were obtained using the interaction picture, and are valid in the rotating wave approximation.

Assuming that the atom is initially in the excited state $|2\rangle$, we can write $C_{2,n}(0) \equiv C_n(0)$ and $C_{1,n+1}(0) = 0$. The solutions for these particular initial conditions are then given by

$$C_{2,n}(t) = C_n(0)\left[\cos(\Omega_n t/2) - \frac{i\Delta\omega}{\Omega_n}\sin(\Omega_n t/2)\right] e^{i\Delta\omega t/2} \tag{A.32}$$

and

$$C_{1,n+1}(t) = C_n(0)\frac{2ig\sqrt{n+1}}{\Omega_n}\sin(\Omega_n t/2) e^{-i\Delta\omega t/2}, \tag{A.33}$$

where we have used the oscillating frequency $\Omega_n = \sqrt{(\Delta\omega)^2 + 4g^2(n+1)}$. The inversion of population $D(t) = \sum_n [|C_{2,n}(t)|^2 - |C_{1,n+1}(t)|^2]$ can now be calculated explicitly, with the result

$$D(t) = \sum_n |C_n(0)|^2 \left[\frac{(\Delta\omega)^2}{\Omega_n^2} + \frac{4g^2(n+1)}{\Omega_n^2}\cos(\Omega_n t)\right], \tag{A.34}$$

where $|C_n(0)|^2$ is the probability to have an initial state with n photons. In the particular case of absence of radiation at $t = 0$, we have $|C_n(0)|^2 = \delta_{n0}$, and the above expression predicts the existence of Rabi oscillations between the two energy levels of the atom, at the frequency Ω_0, even in the absence of any initial radiation. This is a purely quantum vacuum effect, which results from spontaneous de-excitation of the atom, and confirms the phenomenological arguments first advance by Einstein. On the other hand, in the case of a finite amount of photons, such that $\langle n \rangle \neq 0$ at $t = 0$, Eq. (A.34) confirms the quasi-classical Rabi oscillations induced by a nearly resonant radiation field. Moreover, it also predicts the occurrence of a collapse of the Rabi oscillations, followed by a much later revival. These *collapse-revival* features, which are not discussed here, result form the quantum character of the radiation field, and are absent in the quasi-classical description where Rabi oscillations were purely sinusoidal.

Let us now consider spontaneous emission. This is important for the laser cooling process with results for a repeated excitation-fluorescence process, where the first step is photon absorption, as described below, followed by spontaneous photon emission, as described next. The spontaneous decay of an atom from the upper to the lower energy level can take place over an infinite number of radiation field modes, and not just one as considered above. As a result, the previous Rabi oscillations at the frequency Ω_0 will disappear and will be replaced by an irreversible process. The interaction Hamiltonian for one atom in an infinity of radiation field modes can be written, in the interaction picture, as

$$H' = \hbar \sum_{\mathbf{k}} \left\{ i g_{\mathbf{k}}^*(\mathbf{r}_0) a_{\mathbf{k}} e^{i\Delta\omega t} + \text{h.c.} \right\} . \tag{A.35}$$

Here we have reintroduced the spatial dependence of the field, in order to account for the possible orientations of the wavevectors $\mathbf{k}$ with respect to the atom, corresponding to the same mode frequency $\omega_k = kc$. We have $g_{\mathbf{k}}(\mathbf{r}_0) = g_{\mathbf{k}} \exp(-i\mathbf{k}\cdot\mathbf{r}_0)$ And the state vector for the system of the atom plus radiation can be written as

$$|\psi(t)\rangle = C_2(t)\,|2, n = 0\rangle + \sum_{\mathbf{k}} C_{1,\mathbf{k}}(t)\,|1, n = 1\rangle_{\mathbf{k}}, \tag{A.36}$$

where only the vacuum field, corresponding to a number of photons $n = 0$, and the single photon state $n = 1$ in any of the possible field modes $\mathbf{k}$ was retained. Replacing this in the Schrödinger equation we obtain the evolution equations for the coefficients

$$\frac{d}{dt} C_2 = -i \sum_{\mathbf{k}} g_{\mathbf{k}}^*(\mathbf{r}_0) e^{i\Delta\omega t} C_{1,\mathbf{k}}\,, \quad \frac{d}{dt} C_{1,\mathbf{k}} = i g_{\mathbf{k}}(\mathbf{r}_0) e^{-i\Delta\omega t} C_2. \tag{A.37}$$

Integrating this last equation, and replacing the result in the first one, we can derive the approximate result

$$C_2(t) = C_2(0) \exp\left(-\frac{\Gamma}{2} t \right), \quad \Gamma = \frac{1}{\pi\epsilon_0} \frac{\omega^3 |p_{12}|^2}{3\pi^2 \hbar c^3}. \tag{A.38}$$

We can therefore show that, due to the interaction with the infinity of photon modes with the same resonant frequency, but different propagation directions, the population of the upper atomic level $|2\rangle$, determined by $\rho_{22} = |C_2(t)|^2$, will decay exponentially as $\exp(-\Gamma t)$. Such a spontaneous decay occurs in the absence of any incident photon beam. The energy initially contained in the excited atom will be randomly transferred to an infinity of radiation field modes. This shows that, if the atom is located in a cavity, where the number of spatial field modes existing for frequencies close to the transition frequency ω is reduced, we can inhibit spontaneous emission and significantly increase the value of the decay time $1/\Gamma$.

With this knowledge of the spontaneous decay, we can go back to the evolution equations, and include the spontaneous decay time. It can then be shown that, in the quasi-classical limit, the resulting evolution equations for the density matrix, $\rho_{ij} = C_i^* C_j$, are [1]

$$\frac{d}{dt}\rho_{22} = -\Gamma\rho_{22} + \frac{i}{2}\Omega_R(\rho_{21} - \rho_{12}) \ , \quad \frac{d}{dt}\rho_{21} = -\left(i\Delta + \frac{\Gamma}{2}\right)\rho_{21} + i\Omega_R\rho_{22} - \frac{i}{2}\Omega_R \tag{A.39}$$

where Ω_R is the appropriate Rabi frequency, and $\Delta = \omega_{21} - \omega_k$ is the frequency detuning. The relation between these quantities and the above frequency Ω_n can easily be found. These equations we will be used to understand the laser cooling process and to determine the laser force acting on the atoms.

Index

A
AC Stark effect, 12, 58, 254
Adiabatic/adiabatic regime, 18, 19, 122, 192–195, 285–289, 298, 302, 360
Alfvén wave
  compressional, 356
  shear, 356
Alkaline metals, 384
Allen, J.F., 3
Ambipolar approximation/ambipolar diffusion coefficient, 304, 305
Ansatz, 101, 155, 182, 195
Anti-Helmholtz configuration, 21–23
Ashkin, A., 3
Aspect ratio, 181
Atom channeling, 254
Atom effective charge, 80, 85, 123
Atomic Debye length, 95, 116

B
Bates, D.R., 332
Bias field, 21
Black hole, 215–220
Blocade
  shift, 329
  sphere, 329
Bogoliubov transformation, 205–207
Bohm potential, 157, 162, 163, 216
Bohr Magneton, 18, 103
Bohr model, 323
Born-Oppenheimer approximation, 48, 49
Bose, S.N., 3
Bottleneck/kinetic bottleneck, 331, 332
Boyd, R.W., 117
Bragg diffraction, 254, 255
Breit-Wigner formula, 72
Brownian motion, 5, 261
Brush, S.G., 313

C
Central field method, 382
Charney equation, 237, 244
Charney-Hasegawa-Mima (CHM) equation, 244
Charney, J.G., 244
Chemical potential, 29, 142, 143, 146, 148, 153, 163, 200, 238, 247
Chin, C., 69
Chirikov, B.V., 289
Chu, S., 3, 16, 55
Cohen-Tannoudji, C.N., 3
Collapse-revival, 388
Condensate depletion, 208
Conductivity tensor, 352
Coupling parameter, 118, 274, 275, 312, 313, 317, 360, 371
Critical angle, 288
Cross section
  differential, 64
  total, 66–68
Cut-off, 4, 90, 95, 171, 172, 174, 263, 332, 337, 341–343, 353–356, 379
Cyclotron motion, 282–284, 286, 330, 355

D
Dalibard, J., 4, 77
Dark state, 55
Dattner, A., 97
De Broglie wavelength, 27, 43, 70, 254, 275, 370

J.T. Mendonça and H. Terças, *Physics of Ultra-Cold Matter*, Springer Series on Atomic, Optical, and Plasma Physics 70, DOI 10.1007/978-1-4614-5413-7,

Debye length, 74, 95, 97, 116, 129, 137, 278, 358, 368, 371
Debye sphere, 279, 293
Decoherence, 253, 258–265
factor, 261
Degeneracy parameter, 275
Degree of ionization, 274, 280, 311
Density matrix operator, 12
Diamagnetic medium, 282
Diffused vorticity, 240, 242, 243
Dipole approximation, 11, 45
Dirac, P.A.M., 254
Disorder induced heating, 317–320
Doppler temperature, 9, 18
Drift velocity, 243, 244, 283, 285, 288, 347, 350, 351
Dynamical Casimir effect, 203, 209, 212–215

E
Effective electric charge, 1, 4
Einstein, A., 3
Ellis, J., 262
Extraordinary mode, 356

F
Fano, U., 69
Fermi acceleration, 288
Fermi, E., 288
Feshbach, H., 69
Feshbach resonances, 63, 69–74, 208, 210, 220
Feynman, P., 238
Field
conformal, 262–265
scalar, 217, 262, 264, 265
First Born approximation, 68
Frisch, O.R., 3

G
Gould, R.W., 97
Green's function, 148
Griffin, A., 173
Guiding center, 282–285, 288, 331
Guiding-center atom, 331

H
Hänsch, T.W., 3
Hawking radiation, 170, 203, 215–216, 220
Hawking, S.W., 216, 218
Healing length, 181, 238, 239, 243
Helmholtz free energy, 316
Henkel, N., 234
Hess, H.F., 25
Horizon, 208, 215–219, 273
Hull, G.F., 3
Hybrid phonons, 4, 5, 96, 108, 115, 128, 129, 372
Hybrid resonances, 356
Hypergeometric function, 101, 165

I
Ichimaru, S., 313, 314
Inverse scattering transform (IST), 186–189
Ioffe, 20
Ioffe trap, 20–22
Ion sphere, 313–316
IST. *See* Inverse scattering transform (IST)

J
Jackson, J.D., 250
Josephson
junction, 270
oscillations, 253, 270

K
Kapitza-Dirac scattering, 254
Kapitza, P.L., 3, 254
Kasevich, M., 55
Keck, J.C., 332
Kevredikis, P.G., 200
Khalatnikov, I.M., 173, 221
Killian, T.C., 273
Klimontovitch equation, 290–292

L
Lambda configuration, 53–55
Landau criterion, 230
Landau distance, 278, 330, 331
Landau, L.D., 115, 173, 222, 231, 323
Lane-Emden equation, 83
Langmuir waves, 90
Larmor formula, 323
Larmor frequency, 18–19
Larmor radius, 282, 283, 287, 331, 332, 347
Laser filamentation, 119
Lee, T.D., 207
Legendre polynomial, 65, 98
Le Sage, G.L., 1
Lifshitz, E.M., 323
Light grating, 254
Lippmann-Schwinger equation, 67

Loss cone, 288
L-wave, 354, 356

M
Madelung transformation, 156
Magnetic mirror, 20, 21, 287–289
Magnetron rotation, 330
Magnus force, 241
Mansbach, P., 332
Mendas, I., 332
Metcalf, H.J., 280
Metric, 203, 215–218, 261, 262, 265
Mie frequency, 100, 277
Migdall, A.L., 18
Mirror ratio, 288
Misiner, A.D., 3
Mobility tensor, 352
Modes
  anomalous, 195
  breathing, 89, 101, 102, 165
  sloshing, 100, 103, 105–106
  surface, 89, 101, 102, 165
Modulational instability, 118–120, 161, 177, 200
Molasses, 16, 23
Mollow triplet, 78
Momentum recoil, 14
Moshinsky, M., 256
Moskowitz, P., 254
Moyal, J.E., 35, 43
Multipolar field, 19–20
Myatt, C.J., 30

N
Nichols, E.F., 3
Nickel, J.C., 97
Nikuni, T., 173
Noble gases, 384
Non-neutral, 1, 2
Northrop, T.G., 288

O
Olivetti, A., 101
One-component plasma, 1, 281, 313
Onsager, L., 238
Optical Bloch equations, 9, 13
Optical theorem, 66
Ordinary wave, 162

P
Painlevé-Lemaitre metric, 217
Parker, J.V., 97
Partial wave, 65, 66
Penning trap, 281, 330, 331
Phase shifts, 65–67, 71, 73, 74, 184, 188, 192
Phase state, 267, 268
Philips, W.D., 3
Phonon
  effective mass, 229
  landau damping, 230–231
  pressure, 225
Photon spin, 250
Planck time, 261
Planetary waves, 242
Plasma, plasma frequency, electron plasma frequency, 93, 94, 97, 102–104, 119, 122, 127, 129, 134, 270, 277, 317, 320, 337, 339, 343, 358, 361, 367, 374
Plemelj formula, 366
Pohl, T., 330
Polariton, 102, 125, 128, 129, 335, 339–345
Ponderomotive force, 15, 16, 335, 345–347, 376

Q
Quantum defect, 322–324, 340
Quasi-probability, 37, 43, 92, 380

R
Raab, E.L., 23
Rabi frequency/collective Rabi frequency, 11, 12, 16, 50, 106, 110, 254, 256, 327–329, 343, 344, 389
Radiative cascade, 302, 323, 340
Raman-Nath approximation, 256
Recoil temperature, 24
Recombination coefficient, 305
Relaxation-time approximation, 295
Resonance, 11, 12, 63, 69–74, 76, 78, 98, 103, 109, 113, 120, 137, 138, 172, 208, 210, 220, 231, 246, 289, 336, 341, 353, 354, 356, 370, 376
Rossby-Tkachenko dispersion relation, 248, 249
Rossby waves, Rossby number, Rossby velocity, 237, 242–247
Rotating wave approximation, 11, 13, 49, 386, 387
Roton, roton instability, roton minimum, 115, 125–131, 221, 231–236, 379, 380
Rudakov, L.I., 351

Russell, J.S., 179
R-wave, 354, 356

S
Sagdeev, R.Z., 351
Scattering length, 2, 66–71, 73, 74, 152, 208, 210, 220
Schawlow, A.L., 3
Screening potential, 314
Second sound, 226
Self-sustained, 107
Sesko, D.W., 78
Shabat, A.S., 186
Slow light, 335, 339–345
Solitons
  bright, 179, 184, 191–192
  dark, 180, 181, 185–195, 198, 200
  grey, 182–186
  ring, 200
Sound speed, 90, 94, 97, 122, 162, 169, 170, 173, 175–177, 210, 219, 220, 222, 234, 361
  bogoliubov, 162, 167, 169, 206, 221, 230, 242
Spectroscopic notation, 65
Spherical potential well, 70
Standard map, 289
States of matter, 273
Stringari, S., 163
Stuart, J.T., 246
Superatom, 329, 330
Susceptibility/atomic susceptibility, 93–95, 116, 117, 130, 135, 168, 339, 340, 342, 363, 365, 366, 369, 372
S-wave scattering, 66, 152
Sympathetic cooling, 9, 29–33

T
Tan, S.M., 259
Teller, E., 288
Thermal gas, 145, 146, 150, 161, 172–174, 176
Thomas-Fermi approximation, 153–156, 163, 243, 249
Time refraction, 203, 208–212, 214, 215, 256, 358
Tisza, L., 221
Tkachenko speed/modes, 241–242, 248–250
Tonks-Girardeau gas, 182
Tonks, L., 97
Transition matrix, 68

U
Ultra-cold gas, 1, 2, 4, 23, 74, 79, 84, 85, 100, 111, 115–138
Ultra-cold neutral plasmas, 2
Unruh, W.G., 217, 218
Upper hybrid frequency, 349
Uys, H., 261

V
van der straten, P., 280
Van der Waals potential, 68, 235
Vlasov equation, 44, 126, 131, 292–296, 298, 309, 312, 364, 369

W
Walls, D.F., 259
Water-bar solution, 84
Wave kinetic equation, 41–43, 47–52, 55, 79, 91, 108, 141, 158–159, 161, 166, 221, 231–233, 372, 374, 375, 380
Weyl transformation, 35, 39–40
Whistler mode, 355
Wiener-Khintchine theorem, 263
Wigner, E.P., 35, 44
Wigner function, 35–37, 39–41, 43, 44, 47, 48, 52, 56, 91, 92, 170, 175, 222, 223, 232, 371–373, 375, 376, 380
Wigner-Seitz radius/Wigner crystallization, 274, 313, 316

Y
Yang, C.N., 207
Yukawa potential, 278

Z
Zakharov, V.E., 186
Zaremba, E., 173
Zeeman splitting, 355

GPSR Compliance

The European Union's (EU) General Product Safety Regulation (GPSR) is a set of rules that requires consumer products to be safe and our obligations to ensure this.

If you have any concerns about our products, you can contact us on ProductSafety@springernature.com

In case Publisher is established outside the EU, the EU authorized representative is:

Springer Nature Customer Service Center GmbH
Europaplatz 3
69115 Heidelberg, Germany

Batch number: 10755329

Printed by Printforce, the Netherlands